# Homogenization

Serguei Kozlov (1954–1995)

Series on Advances in Mathematics for Applied Sciences – Vol. 50

# Homogenization

In Memory of Serguei Kozlov

Editors

**V Berdichevsky**
*Wayne State University, USA*

**V Jikov**
*Pedagogical University, Russia*

**G Papanicolaou**
*Stanford University, USA*

World Scientific
*Singapore • New Jersey • London • Hong Kong*

*Published by*

World Scientific Publishing Co. Pte. Ltd.

P O Box 128, Farrer Road, Singapore 912805

*USA office:* Suite 1B, 1060 Main Street, River Edge, NJ 07661

*UK office:* 57 Shelton Street, Covent Garden, London WC2H 9HE

**Library of Congress Cataloging-in-Publication Data**
Homogenization : Serguei Kozlov memorial volume / V. Berdichevsky,
V. Jikov, G. Papanicolaou, editors.
p. cm. -- (Series on advances in mathematics for applied
sciences ; vol. 50)
ISBN 9810230966
1. Homogenization (Differential equations). I. Kozlov, S. M.
(Serguei M.) II. Berdichevskiĭ, V. L. (Viktor L'vovich)
III. Zhikov, Vasiliĭ Vasil'evich. IV. Papanicolaou, George.
V. Series.
QA377.H577 1999
515'.353--dc21 99-24583
CIP

**British Library Cataloguing-in-Publication Data**
A catalogue record for this book is available from the British Library.



Printed in Singapore by Uto-Print

# PREFACE

The theory of homogenization is very important in understanding the physical interactions on different length and time scales. An outstanding role in the creation of homogenization theory was played by Serguei Kozlov, a Russian mathematician whose untimely death at the age of forty interrupted a brilliant research career. This book is a tribute to him. It contains a biographical sketch of Serguei, the complete list of his publications, joint papers that were completed by his collaborators after his death, papers by several colleagues dedicated to him, and some recollections.

The editors thank Tania Semenova, Jane Berdichevsky. and Nadia Vasetsky for their help in preparing this book for publication.

*V. Berdichevsky*
*V. Jikov*
*G. Papanicolaou*

# CONTENTS

Preface v

Serguei Kozlov — A Brief Biographical Sketch xiii

Some Recollections 1

Serguei Kozlov — A Review of Scientific Contributions 11
*A. Beliaev and V. Jikov*

List of Publications of S. Kozlov 16

**Critical Path Analysis of Transport in Highly Disordered Random Media** 21
*K. M. Golden and S. M. Kozlov*

**Multiscaled Homogenization** 35
*V. V. Jikov and S. M. Kozlov*
1 Introduction . . . . . . . . . . 35
2 Some properties of almost periodic functions . . . . . . . . . 40
3 Proof of the upper estimate for $(K_N)_{hom}$ . . . . . . . . . . 44
4 Proof of lower estimate for $(K_N E)_{hom}$ . . . . . . . . . . 52
5 Proof of the main result. Examples . . . . . . . . . . 59

**The Effective Thermoconductivity and Shear Modulus of a Lattice Structure: An Asymptotic Analysis** 65
*S. M. Kozlov and G. P. Panasenko*
1 Introduction . . . . . . . . . . 65
2 Effective thermoconductivity of the two-dimensional rectangular lattice structure . . . . . . . . . . 70
3 The shear modulus of the two-dimensional rectangular lattice structure . . . . . . . . . . 80

**Multiparametric Problems of Homogenization Theory** 92
*N. S. Bakhvalov and M. E. Eglit*
1 Introduction . . . . . . . . . . 92
2 Structures and composites reinforced by a periodic system of mutually orthogonal plates and bars . . . . . . . . . . 93

2.1 Effective thermoconductivity of composites with isotropic components . . . . . 96
2.2 Effective elastic moduli . . . . . 96
2.3 Effective elastic moduli for locally isotropic materials . . . . . 99
2.4 Auxiliary lemmas . . . . . 100
3 Composites reinforced by fibers . . . . . 102
4 Parameters related to different properties of each phase . . . . 104

**Non-Linear Darcy Law in a Random Porous Medium** 107
*A. Yu. Beliaev*
1 Posing the problem . . . . . 109
2 An estimate of $G(\lambda)$ from below . . . . . 113
3 An estimate for $G(\lambda)$ from above for the case of small porosity $p$ . . . . . 114
4 An estimate for $G(\lambda)$ from above for any $p < 1$ . . . . . 117
5 The proof of exact upper estimate . . . . . 123

**Distribution of Minimum Values of Weakly Stochastic Functionals** 133
*V. Berdichevsky*
1 Introduction . . . . . 133
2 An example . . . . . 138
3 Statistical mechanics of point vortices . . . . . 140
4 The Kosterlitz-Thouless phase transition . . . . . 141
5 Brittle-to-ductile transition . . . . . 146
6 Elastic body under action of random external forces . . . . . 146
7 Homogenization problems . . . . . 148
8 Some features of stationary points of entropy functional . . . . 149
9 The basic identity . . . . . 152
10 Another derivation of the basic identity . . . . . 158
11 An estimation of the prefactor . . . . . 160
12 A counterexample . . . . . 162
13 Large excitations . . . . . 164
14 Appendices . . . . . 166

**Asymmetric Strain–Stress Distribution Function for Crystal with Random Point Defects** 179
*L. Berlyand*
1 Identical defects . . . . . . . . . . . . . . . . . . . . . 179
2 Strain and stress distributions for defects of two different kinds . . . . . . . . . . . . . . . . . . . . . 185
3 Tail and first moment of the distribution . . . . . . . . . . 186
4 Why is the distribution shifted? . . . . . . . . . . . . . . 188
5 Comparison with the statistics of gravitational field arising from a random distribution of stars . . . . . . . . . 190

**Optimal Design for Uncertain Loading Condition** 193
*A. Cherkaev and E. Cherkaeva*
1 Introduction . . . . . . . . . . . . . . . . . . . . . . . 193
2 The problem formulation . . . . . . . . . . . . . . . . . 194
3 Eigenvalue problem . . . . . . . . . . . . . . . . . . . . 198
4 Multiple eigenvalues . . . . . . . . . . . . . . . . . . . 202

**Effective Properties of a Plane Two-Phase Elastic Composites: Coupled Bulk-Shear Moduli Bounds** 214
*L. V. Gibiansky*
1 Introduction . . . . . . . . . . . . . . . . . . . . . . . 214
2 Statement of the problem and the main results . . . . . . . 215
2.1 Equations of the planar elasticity . . . . . . . . . . 215
2.2 Homogenization . . . . . . . . . . . . . . . . . . . . 218
2.3 The known bounds . . . . . . . . . . . . . . . . . . . 219
3 The translation method . . . . . . . . . . . . . . . . . . 225
3.1 Functionals . . . . . . . . . . . . . . . . . . . . . . 225
3.2 Translation method . . . . . . . . . . . . . . . . . . 228
3.3 "Embedding" into the higher-order problem . . . . . . . 232
4 Coupled bulk-shear moduli bounds . . . . . . . . . . . . . 233
4.1 Generalized Hooke's law . . . . . . . . . . . . . . . . 233
4.2 Rotationally invariant eighth-order tensors . . . . . . . 236
4.3 Quasiaffine functions . . . . . . . . . . . . . . . . . 240
4.4 Coupled bulk-shear moduli bounds . . . . . . . . . . . 242
5 Discussion and generalization . . . . . . . . . . . . . . . 248
5.1 Summary of the results . . . . . . . . . . . . . . . . 248
5.2 Optimality of the bounds . . . . . . . . . . . . . . . 250
5.3 Three-dimensional problem . . . . . . . . . . . . . . . 253

**Homogenization of the Laplace Equation in a Partially Perforated Domain** 259

*W. Jäger and A. Mikelić*

1 Statement of the problem and of the results . . . . . . . . . . 260

1.1 Introduction . . . . . . . . . . . . . . . . . . . . . . . 260

1.2 Notations, assumptions and auxiliary results . . . . . . 262

1.3 Statements of the main results . . . . . . . . . . . . . . 268

2 Proofs of the Theorems . . . . . . . . . . . . . . . . . . . 272

2.1 Proof of Theorem 1.1 and Corollaries 1.1–1.3 . . . . . . 272

2.2 Proof of Theorem 1.2 . . . . . . . . . . . . . . . . . . 274

2.3 Proof of Theorem 1.3 . . . . . . . . . . . . . . . . . . 276

**Control in the Coefficients of Linear Hyperbolic Equations via Spacio-Temporal Components** 285

*K. A. Lurie*

1 Introduction . . . . . . . . . . . . . . . . . . . . . . . . 285

2 Possible applications . . . . . . . . . . . . . . . . . . . . 286

3 The structure of the paper . . . . . . . . . . . . . . . . . 287

4 Second order hyperbolic equation with chattering coefficients . . . . . . . . . . . . . . . . . . . . . . . . 288

5 Application of the Floquet theory . . . . . . . . . . . . . 294

6 The one-dimensional wave propagation: a special use of Maxwell's system . . . . . . . . . . . . . . 298

7 The Maxwell's system for a moving dielectric medium . . . . 299

8 One-dimensional wave propagation: conservation of the determinant of the material tensor . . . . . . . . . . . 305

**Multiscale Averaging for Ordinary Differential Equations. Applications to the Spectral Theory of One-Dimensional Schrödinger Operator with Sparse Potentials** 316

*S. A. Molchanov*

1 Introduction . . . . . . . . . . . . . . . . . . . . . . . . 316

2 Technical tools . . . . . . . . . . . . . . . . . . . . . . . 324

2.1 Spectral measure . . . . . . . . . . . . . . . . . . . . 324

2.2 Phase and amplitude . . . . . . . . . . . . . . . . . . 325

3 Scattering theory . . . . . . . . . . . . . . . . . . . . . . 328

4 Tests for the different types of spectra . . . . . . . . . . . 336

4.1 Tests for $\Sigma_{ac}$ . . . . . . . . . . . . . . . . . . . . 336

4.2 Tests for $\Sigma_{sc}$ and $\Sigma_{pp}$ . . . . . . . . . . . . . . . . 339

5 Cutoff lemma . . . . . . . . . . . . . . . . . . . . . 341
5.1 Martingales and homogenization lemmas . . . . . . . . . 345
6 Spectral theory of the one-dimensional Schrödinger operators with sparse potentials. Negative energies $\lambda < 0$ . . . 361
6.1 Spectral theory of the one-dimensional Schrödinger operators with sparse potentials. Positive energies $\lambda = k^2 > 0$ . . . . . . . . . . . . . . . . . . . 372
7 Perturbation theory and decaying sparse potentials . . . . . . 389

**Remarks on an Estimate of Serguei Kozlov** 398
*U. Mosco*
1 Introduction . . . . . . . . . . . . . . . . . . . . . 398
2 Kozlov's estimate . . . . . . . . . . . . . . . . . . . 399
3 The strange exponent . . . . . . . . . . . . . . . . . 400
4 The change of metric . . . . . . . . . . . . . . . . . 401
5 The physics of $V$ . . . . . . . . . . . . . . . . . . 401
6 The invariance of $d(x,y)$ . . . . . . . . . . . . . . . 403
7 The homogeneous dimension . . . . . . . . . . . . . . 403
8 Scaled Poincaré inequalities . . . . . . . . . . . . . . 404

**Central Limit Theorem and Spectral Asymptotics for Nonlinear Stochastic Partial Differential Equation with Weak Nonlinearity** 407
*A. L. Piatnitski*
1 Introduction . . . . . . . . . . . . . . . . . . . . . 407
2 Setting of the problem . . . . . . . . . . . . . . . . . 408
3 Spectral asymptotics for $\mathcal{A}_0$ . . . . . . . . . . . . . . 410
4 Compactness results . . . . . . . . . . . . . . . . . . 415
5 Central limit theorem . . . . . . . . . . . . . . . . . 416

# SERGUEI KOZLOV — A BRIEF BIOGRAPHICAL SKETCH

Serguei Kozlov was born on May 17, 1954. He was the only child in the family. His father, the late Michael L'vovich Jakub, was a professor at the Moscow State University Law School; his mother, Ekaterina Ivanovna Kozlova, is still a professor at the Moscow Academy of Law.

In high school, Serguei decided to become a mathematician. He graduated with high honor from the Moscow State University, School of Mechanics and Mathematics (Mechmat), the best Russian educational and research center in mathematics and mechanics. In his student years, Serguei specialized in different equations. He chose Professor M. I. Vishik to be his advisor. He graduated from Mechmat in 1976, and stayed there as a graduate student. His Ph.D. thesis was concerned with the homogenization theory of differential operators with random and almost periodic coefficients. He completed his Ph.D. thesis in 1979.

After graduation, Kozlov was employed by the Department of Applied Mathematics of the Moscow Institute of Civil Engineering.

In 1989, S. Kozlov defended his Doctor of Science thesis "Homogenization of Random Structures". His scientific contributions are reviewed in this volume.

In 1991–1992, S. Kozlov had visited several universities in France. In 1992 he became a professor at the Aix-Marseilles University where he worked till his death. He committed suicide on April 7, 1995.

# SOME RECOLLECTIONS

## E.I. KOZLOVA:

I am glad to have an opportunity to describe Serguei's everyday life, although often it was only the background to his creative work.

We all claim our origins in childhood. Serguei's one was happy. He grew up surrounded by the attention and care of excessively loving parents. We strove to give a good up-bringing to our son — both in the moral and educational sense. We chose books for him to read, theaters and museums to attend, sports to participate in, etc., which were appropriate to his age.

Serezha [1] very much loved being read to, and later to read himself. He especially liked the ballads about Robin Hood, which I sang to him and he could listen to them an infinite number of times.

He did not like fairy tales with an unhappy ending. Once, he was sitting in my lap and I was reading the "Little Red Riding Hood" to him. Already knowing how it ends, Serezha did not allow the wolf to do his mischief: knocking it of my hands, the book ended up in the opposite corner of the room.

Serguei was very stubborn as a child and also later in his life. If he wanted to take his favorite toy on a walk, a big metal dump truck, which rattled deafeningly when pulled along on the sidewalk, no pleading could dissuade him. I think this toy was known to all inhabitants of Sretenka (the street we lived on). If he did not want to enter a shop with me, I could not drag him in there. And any efforts to make him wear a warm coat if he thought it was not cold enough outside to wear it were wasted: he would throw the coat on the floor and run away.

However, along with being so obstinate, he was extremely law-abiding, which distinguished him from his peers. There was a theater next to our house, on the roof of which all the neighborhood kids walked and played, although it was forbidden. This prohibition was obeyed only by Serguei. In the autumn, when we gathered maple leaves, he only took the ones trampled by pedestrians from the sidewalk, because it was not permitted to walk on the grass; only rarely would he care-

[1] The name "Serezha" is a less formal version of "Serguei" often used in everyday life.

fully pull an occasional leaf off the side of the lawn. This quality was probably inherited from his father, and I made constant efforts to overcome this inordinate attention to unimportant rules of conduct and to emphasize a sensible approach to them.

We strove to make him a person capable of saying "no", not unnecessarily submitting to the will of others, able of making independent decisions. To this end I often consulted him on serious problems and asked for his opinion even when he was a child.

One year before entering grade school, he began attending kindergarten to get used to socializing with kids his age.

Serezha learned to take care of himself rather early. Even when in grade school, he did not wake me up in the morning before leaving for school: he made his bed and fixed his breakfast himself. He knew how to cook an omelet and even managed to turn it over on the pan. When I was ill, it was Serezha and not my husband who took care of me: he went grocery shopping, cooked and served the meal. But Serezha was not a handyman. On the contrary, when something requiring handiness had to be done, he was quite awkward. In kindergarten, I could always distinguish Serguei's creations from those of other kids: a wobbly stool, an oddly embroidered napkin. As an adult, he was more adroit, but to hang up a painting, like Mark Twain, he would still require a few helpers.

Adolescence and youth changed him considerably. He quickly found a common language with his classmates and acquired many friends. He started spending more time outside with the kids, and he tried to smoke. Sometimes I was summoned to his school by the teachers to discuss "his misbehavior".

At home, he disliked talking about school, friends, or teachers. Sometimes I found out more about school in one hour from one of his friends than from Serguei during the whole year. Thus he began forming this characteristic which he later termed by the maxim: "Something unsaid can not be regretted". He stuck by this principle in the years to come.

We usually had good summer and other vacations. In June, when my husband and I worked, Serguei lived with my parents (and when little, also with a nanny) in the country in Kljaz'ma, near Moscow. There he had many friends, a bicycle, a river nearby. July we usually spent by the Baltic sea, August in Crimea by the Black Sea (more often in Sevastopol' on the north side), or in Caucasus. On these vacations we would always lead an active style of life: we took long walks in the canyons and by ravine rivers and had picnics there. Serezha loved it all.

Serezha was a good but not a brilliant student. In middle school, he did not have any special preferences in academics. As for other activities, at some point he began collecting stamps, but without any system: he just liked buying them. For a short while he took up body building, but quickly lost the enthusiasm for it, although he continued to exercise with weights practically for the rest of his life. Physically, he was very strong.

Since our son did not have any particular leanings, my husband and I considered his following in our footsteps and becoming a lawyer, since he had potential. He possessed very clear logical thinking. In order to test his abilities, his father would sometimes present him with exceptional judicial cases from the textbook for university law department juniors. It was amazing that Serezha could impeccably solve these problems having at his disposal only the knowledge of the logical concept of the law and no practical experience. This choice of occupation also would have had the advantage of our being able to help him in preparation for the subject entrance exams. However, Serguei's fate ruled differently.

It came to happen that we changed residences and moved to a neighborhood of Moscow where the only good school in the area was that with an emphasis on mathematics. Serguei transferred to this school in the beginning of 8th grade.

He had to make significant mental efforts to catch up with his classmates who had already attended this school for a number of years. This endeavor was the beginning of his future choice of occupation. By the time of graduation he had decided for sure that he would attend the Moscow State University School of Mathematics and Mechanics (Mechmat).

In the new school he made new friends. Serguei was a good student, especially in science. During the last year of school, Serezha started preparing for the MSU entrance exams and was serious about his studies.

However, he was not able to enter Mechmat right away. He did well in his mathematics exams, but got a 'fail' in physics. It was a painful blow both for Serguei and for us, more so for being so unexpected and undeserved. He was prepared exceptionally well in physics because during preparation he paid special attention to this subject. So many years have passed, but I still hear Serguei's voice over the phone: 'Mom, a fail.'

The Mechmat entrance exams were in July, those for other universities in August. The question of the second choice came up. My husband, afraid of further complications, insisted on the mathematics department of the Pedagogical Institute. I thought it would be a better to try the Moscow Institute of Electronic Industry since it had a good reputation. Finally, the latter was chosen. Serguei passed all their exams with distinction and was awarded a high stipend. The last exam was in composition. I remember the trembling with which Serezha and I approached the posted lists of those who failed, afraid to find the name "Kozlov" there. But the names on the list where so overwhelmingly of a particularly different kind that the applicants standing behind us could not abstain from connotative comments.

When he began attending the institute, Serguei studied with an enthusiasm which is easily explained by the upsets of the summer. During the first months he was selected to be part of the group with an intensive study of mathematics under the supervision of professor Maslov.

However, Serguei did not abandon his desire to study at Mechmat. His father petitioned the Rector of Moscow State University, academician Petrovsky, to transfer Serguei to Mechmat. According to the rules existing at that time, transfers were not permitted for freshmen, but the Rector promised to transfer Serguei at the beginning of the second semester if he passed all his first semester exams with distinction. Serguei complied with this condition and his wish came true: he became a student of Mechmat. The transfer procedure was not easy. It was necessary to obtain the permission for the transfer from his current Institute as well, which was difficult to accomplish. I was told that professor Maslov was very displeased by the fact that the administration allowed the transfer and thus robbed his group of their most talented student.

The difference of the first semester programs between the two universities was significant enough for Serguei to have to pass some additional exams in a few areas of mathematics, which he had to learn on his own, in a very short period of time. Serezha disliked talking about his achievements, discussing his successes and repeating praises, even in the family, but he did admit that one of the instructors was very surprised by the fact that Serguei could master a specialized mathematical topic on his own. Usually, to the question of how something went, he had a ready answer: "OK."

Maybe this, second, exertion made him accustomed to mobilizing all his intellectual power, employing all his ability. During his study at Mechmat, he passed all his exams, without exception, with distinction. He was not an ordinary student: he attended theoretical seminars, eventually becoming their most active participant. He was rarely at home, spending most of the time in the libraries. After completion of his undergraduate studies, he entered graduate school and ahead of schedule received his Ph.D. under the supervision of professor M.I. Vishik. It so happened that on the day of Serguei's Ph.D. thesis defense, I received in the mail the permission to purchase a car, "Zhiguli" (in those times, it was impossible to buy a car — they were distributed according to one's place of employment). Thus, we had an opportunity to give Serguei a car as a gift for his successful defense. He quickly mastered it and became an aggressive, but very skillful driver.

His swift and successful dissertation did not safeguard Serguei from a new blow which was very painful to him, although he attempted to conceal that. He very much wanted and hoped to remain at Mechmat, but a position was not found for him. He had to independently seek employment, and eventually started teaching at the Moscow Institute of Civil Engineering.

He continued to be active in research and started attending conferences abroad. Serguei was an exceptionally hard worker. Whichever period of his life I recall to mind, I see him at his desk for hours, almost without interruption, day after day, year after year.

After the Doctor of Science defense, which took place in Leningrad, he began working in Paris for months at a time. I lived with his family there once for two happy months. Eventually, he moved to Marseilles where he worked for the rest of his life.

Serguei got married very early — at the age of 19, during his third year at the university — and since then lived separately from us. His wife, Natasha (born Il'gisonis) was an exceptionally good person. Serezha fell in love with her in 9th grade (they went to school together). Without exaggeration, she deserved to be described by the following epithets: feminine, smart, hard-working, strong, energetic, efficient, a wonderful wife, mother, and daughter and, moreover, a talented chemistry researcher. She was loved by everybody. During their life together, she completely relieved Serguei of all the household chores. They had a very harmonious life. She died shortly before Serguei's death, and her untimely end shocked everybody and greatly affected Serguei.

They had two sons — Leonid (he is now 22, about to graduate from Mechmat) and Vladimir (14 years old, in 10th grade). They live with me now.

Judging by the letters, which were written mostly by Natasha (she took it upon herself to perform this task as well), they had a very interesting life. Natasha established scientific contacts in Marseilles and had a productive scientific life there. On week-ends they went to the country, to the beach, to theaters, or museums. Serezha's letters were very unique. He could expressively describe a town where he was, eloquently relate impressions from going to museums (which he never failed to do).

The summers they usually spent in Moscow (only once in Grenoble), living mostly in Kljaz'ma. Serezha loved the country. He showed himself to be competent at organizing reconstruction and expansion of the house. He liked to mow and sow. Almost every summer, along with Natasha and her brother, Serguei rode the motor boat on the Volga. He loved fishing, water skiing, gathering mushrooms, making camp fires. He was very sociable, liked parties, singing (although he had neither a musical ear nor a voice). He established contacts easily not only in his line of work, because he also knew how to find a common language with all kinds of people. He knew how to have fun, how to laugh, and had a sense of humor. Among his non-official photographs, it is difficult to find one where he is not smiling. This quality he had since his youth. Even in school the teachers complained of Serguei's making comments and jokes which made the whole class laugh. But even during vacations, he never abandoned his work. His colleagues often came to visit him to discuss scientific problems.

I cannot describe what I felt when I found out about Serezha's death. My older grandson and I flew to Marseilles to attend the funeral. There we were overwhelmed by the great number of things to be done. I was utterly helpless since I did not know French, and my grandson did not speak it well, and we would have been lost if it weren't for the assistance of Serezha's colleague Mark Duran and his wife Annie. They helped us all along this difficult path having to do with necessary death papers, cremation, sorting Serguei's papers, documents, bills. They put us up, helped empty out the apartment, packed the things to be mailed to Moscow, did not leave us for a minute, and took us to the airport. Such consideration is never forgotten and I am infinitely grateful to this family.

Obviously, this does not say everything about Serezha. Every person is too complicated a world even to sketch it from all sides, not even taking into consideration the private internal areas, which belong to the person alone.

Despite all the pain caused by Serezha's death, I am grateful to him for having given me forty years of happy motherhood and two grandsons. He lived a short but a very rich life. There are many happy pages in his book, but there are also sad ones.

Serezha honored the graves of his ancestors. On his every visit to Moscow, he went to the cemetery where the deceased members of his family are buried. There he found his final rest. He was cremated in Marseilles and buried in Moscow at the Himkinskoe cemetery.

Requiescat in pace.

## M.I. VISHIK:

Serguei Kozlov, an excellent mathematician with a strong character and interesting personality, is no longer with us.

Serguei came under my supervision during his third year at Moscow University and subsequently became one of my best students. He participated in my seminars, and I was Serguei's advisor for his undergraduate, as well as graduate, research. After brilliantly defending his Ph.D. thesis, Serguei went on working as a mature and creative mathematician.

Moscow State University was a crucial part of Serguei's mathematical biography. The first years of his active research career coincided with the golden years of our University. At that time, Moscow mathematicians were on the leading edge of many fields of mathematics, which enabled Serguei to communicate with a great number of outstanding scholars and he managed to absorb the best traditions of the University of that time. When Serguei was a third-year student, A.V. Fursikov, A.I. Komech and I set out to study in depth probability theory and random processes because at that time our research interests turned to statistic and stochastic solutions of the Navier-Stokes system and related partial differential equations. To this end, together with B.L. Rosovsky we organized a seminar on these subjects for the students of Moscow University. Serguei attended the seminar regularly and became an active participant. The main topic of his research during these years was nonlinear parabolic equations with random forcing (his best result in the field was a form of Girsanov's equation for stochastic equations which contained partial derivatives), and after some time he became a real expert in the probability and random processes theories and, what is more, the best among those who took part in the seminar. Serguei also attended my "big" seminar on partial differential equations and their applications. In due time he became one of the most brilliant participants and often presented his results in excellent talks. Serguei was an active participant all the time he lived in Moscow, but even when he moved to France, having gotten a professorship there, each time he came back to visit his relatives he gave a wonderful lecture at the seminar. I should mention that shortly before his tragic death, he gave a talk marvelous in both style and content. Serguei was a very well educated mathematician; he profoundly mastered many fields in mathematics. Besides the above mentioned seminars, Serguei also attended, although not as regularly, those of S. Molchanov and Ja. Sinai. During the seminars, he always listened carefully, asked the speaker sharp questions and made interesting comments. When Serguei was my graduate student, he was able to set up new problems and solve them by himself; all I had to do was discuss with him the many problems he had already solved.

Serguei got interested in the problem of homogenization after J.-L. Lions' lecture at Moscow Mathematical Society. He concentrated on the topic of homogenization of equations with stationary random coefficients. Later, the main subject of his research became the averaging problem for partial differential equations with coefficients with high frequency of oscillations and related problems from pure and applied mathematics. In his research, Serguei liked new, difficult and concrete mathematical problems, disliked completing other people's work, and hated to prove theorems of existence and come up with general theories. For example, he was very proficient in the theory of pseudo-differential operators, but did not want to work with it. He was ready to cooperate with other mathematicians, but valued very highly independence in scientific research (I remember that he once said that one of his friends would have become a brilliant mathematicians if it were not for his professor's strong influence).

Serguei and his attractive wife Natasha were both 19 when they got married. Although, being a chemist, she had a career of her own, Natasha created wonderful conditions for Serguei's scientific work. She compensated Serguei's ardor by her warm personality, tact and culture. They were both very friendly and hospitable, and hosted many merry parties, which Serguei loved. They had two sons. The younger one I knew rather closely, having made his acquaintance in Paris. He is a very talented boy; he went to school in Paris, wrote poetry in French and was one of the best in his class.

Serguei was adventurous: he loved to drive at high speed and rode a motorcycle around Marseilles. In the summer, riding a motorboat and water-skiing were favorite activities; in the winter, he went cross-country skiing in areas remote from Moscow suburbs. During one of these ski trips, on which he went with Serguei Kuksin, one of his skies broke. It started snowing heavily and it was getting dark, and, not having a compass, they lost their way. In deep snow and without one ski, Serguei soon was exhausted. Fortunately, they came across a small village where they were able to spend the night. All his relatives and friends, including me, were extremely worried when he did not come back from his skiing trip. But the next day he finally returned, cheerful as always.

Serguei was very fond of his parents' dacha in Chelyuskino, near Moscow. He began its renovation just before leaving for France and was planning, upon his return, to work there in relative isolation, coming to Moscow only to teach. Unfortunately, these dreams became a reality only for a short time.

I shall keep him in my memory as an enjoyable companion, talented mathematician and a devoted friend.

# SERGUEI KOZLOV — A REVIEW OF SCIENTIFIC CONTRIBUTIONS

A. BELIAEV
*Water Problem Institute, 3 Gubkina Str., 117971 Moscow, Russia*

V. JIKOV[1]
*Pedagogical Institute, Vladimir, Russia*

Serguei Kozlov was a scientist of outstanding and versatile talent. All his achievements were distinguished by considerable professionalism and bright individuality. Research interests of Kozlov were related to the homogenization theory, disordered structures, Ito equations, as well as other problems of mathematical physics and probability theory. He had many remarkable results in averaging of difference schemes, random walks on lattices and Fokker-Planck equation. However, the major contributions are in the homogenization of quasi-periodic and random operators. At the moment when he entered the field, the asymptotic theory of averaging of periodic operators was developed. Many of the participants of this breakthrough tried unsuccessfully to understand how to do averaging for random structures and what might be an analogy of the "cell" problem for such structures. It was Serguei Kozlov who found the answer [4, 6].[2] This was a remarkable breakthrough, shedding a new light on many old results accumulated for more than a hundred years in various areas of mechanical engineering and physics, and putting many of them into the final form.

Being a lecturer he was an initiator and a leader of various scientific seminars. With his participation every seminar became more vital and interesting. With his exceptional talent to find the most perspective mathematical aspects in complicated engineering problems, Kozlov paid great attention to modern branches of science. He initiated scientific interests and activities of his colleagues to the problems in these areas, many colleagues are much obliged to him for helpful and deep discussions and support.

---

[1]In English literature another spelling of the second author's name, Zhikov, is also used.

[2]References are made here to the list of publications of Serguei Kozlov which follows this review.

Serguei shared the point of view that there are two types of scientists: donors and recipients. In this classification, Serguei was certainly a donor. He threw out remarkable ideas. Many of them had become the basis or the starting points of further investigations. A much greater number of papers could have his name as an author, but he never agreed to be a co-author if his own contribution did not seem to him greater than that of the other participants. That is why his ideas and results appeared sometimes in papers having his name in the paragraph of acknowledgments only. Nobody heard, however, a word of reproof from him in this respect.

As for himself, Serguei applied a very high criteria of honor. Usually, his part in joint papers included either the basic points or the majority of details. He always had to ensure that his contribution was sufficient to warrant his name as co-author.

During his student years, Serguei studied Ito's partial differential equations. He investigated conditions for coincidence of measures generated by Ito's equations. Also he investigated ergodic properties of solutions and proved corresponding generalization of Girsanov's formulae. Later he returned to the matter again and again.

Since 1976 he was thrilled with a new topic, homogenization, which became the major theme of his research. His above-mentioned breakthroughs [4, 6] were followed by a series of outstanding papers. We note here his paper [9] "Homogenization and G-convergence of elliptic operators" written with Jikov, Oleinik and Kha Tien Ngoan. This paper was a milestone in the homogenization theory. Its annual citation index is still one of the highest in the field. This paper was a starting point and a guide manual for many mathematical studies.

In 1979-1983 a series of papers were written by Kozlov, some of them jointly with Jikov and Oleinik, on the homogenization and G-convergence of parabolic operators. At the same time Kozlov studied some asymptotic properties of Green functions in the homogenization problems.

In 1980 Serguei with Jikov and Zlotnik considered a model integrand that corresponds to "Sobolev spaces with varying exponents". Later this approach was developed in connection with the Lavrentiev effect in variational problems.

A problem of great interest in the homogenization theory is averaging of equations which contain randomness in the terms with low order derivatives. Kozlov developed an approach to this problem that was based on his results on reducibility of operators with periodic and almost periodic coefficients. Later, in 1988-1992, he investigated jointly with Piatnitsky the asymptotic behavior of effective turbulent diffusion in the case of small viscosity coefficient.

A significant contribution by Kozlov was his approach to the averaging of difference equations and random walk on lattices. He developed some general homogenization methods and established the main results like the compactness theorem for difference schemes. With respect to random walks he obtained various sets of sufficient conditions for the central limit theorem. In part these results were established jointly with S. Molchanov. This work relates to the period of 1984-1990. It initiated a lot of further investigations both in mathematical and applied setting.

Always Serguei was interested in spectral problems for operators of the homogenization theory and in particular, with various spectral asymptotics. In this area, his results on multidimensional spectral asymptotics should be mentioned. Also he had proved jointly with M. Shubin a coincidence of spectra for random elliptic operators.

Some investigations of Kozlov in the period of 1987-1994 were concerned with the problem of effective conductivity and related topics for disordered structures. In this problem the rigorous mathematical methods and engineering approaches are mutually useful. Kozlov's contribution to this problem, especially his results on percolation, on the homogenization of highly perturbed media and hierarchical structures, were highly appreciated by mathematicians and physicists. In these activities he cooperated with K. Golden, A. Bensoussan, A. Beliaev, G. Panasenko, A. Bourgeat, L. Berliand and L. Malozemov. Each paper carrying the name S. Kozlov immediately attracted great attention from specialists and always initiated numerous investigations which resulted in developments of his original ideas.

In 1991 Kozlov had a provisional position at the Moscow Institute of Water Problems. There he dealt with applied problems of porous flows. This field was new for him, but in short period Serguei had found the way to connect his mathematical interests and abilities with problems of hydrodynamics. This subject was one of the main topics of his further researches in University of Marseille in the last 3 years of his life.

In hydrodynamics Kozlov studied a mathematical problem of homogenization for fluid flows in random porous media. Earlier this problem had been considered for periodically perforated domains by L. Tartar, G. Allaire and others. An extension of these results to more realistic random structures was an important challenging problem of the homogenization theory. Kozlov gave a rigorous justification of Darcy's law under some probabilistic assumptions about the porous structure. A principal trouble in comparison with the periodic case is caused by the percolation properties of random medium. Another one relates to porous domains containing pores of arbitrarily large size. In this case, the permeability of the porous medium could be equal to infinity. Kozlov suggested sufficient conditions for convergence in terms of a version of Poincaré's inequality. Then he proved this inequality under some assumptions on probabilistic distribution of geometric parameters like pore sizes and so-called "chemical lengths". Later these ideas were generalized by Kozlov's colleagues for non-steady flows in random media and for some other problems of homogenization.

In connection with viscous flows in porous media Kozlov's attention was attracted by the spectral properties of Stokes operator in a randomly perforated domain. For the Laplace operator with Dirichlet boundary conditions in a random domain containing a large number of "holes" the asymptotics of spectrum bottom was studied by A.-S. Sznitman. He used, in the spirit of Donsker-Varadhan approach, a method that relates to probabilistic interpretation of the Laplacian, and proved an explicit formula for the leading term of the principal eigenvalue. That way cannot be applied to vector operators. Kozlov proposed to use a re-scaling procedure which was justified by means of direct variational methods. In this way, the results on asymptotics of the principal eigenvalue were re-established for the Laplace operator. Later, Kozlov's method was extended for more general operators. His ideas are also useful for some non-linear problems corresponding to asymptotics for random structures.

For many years the homogenization of elliptic operators in perforated domains in connection with applied problems of diffusion and percolation was in the center of his interests. In 1985 he had suggested jointly with Jikov a conjecture that the connectedness of a perforated domain is the unique sufficient condition for convergence in the homogenization problem. Later this conjecture was justified for a certain class of operators, and this result was extended and generalized to homogenization on fractals. The great interest in the latter kind of structures had appeared quite recently, and Kozlov was one of those who foresaw the importance of this area. His publication on this matter is concerned with the Serpinsky gasket. He obtained some explicit formulae and general homogenization results for that structure.

In things appertaining to other contribution of Kozlov, it is necessary to note his joint results with M. Yor on asymptotics of random number of Wiener walk, the so-called Wiener football. Also he dealt with KAM theory in relation with homogenization of nonlinear equations with almost periodic coefficients and homogenization of Fokker-Planck equation, which is more realistic than the elliptic equation for some applications.

Kozlov always strove to obtain explicit results and explicit formulae every time it was possible. In particular, he made a certain progress in studying the low-concentration asymptotics and sharpening the Maxwell formula for effective conductivity. Also he considered the asymptotic behavior of the Laplace-Dirichlet integrals and had described this problem in terms of differential geometry. A lot of explicit formulae for homogenized coefficients were justified by this approach.

Kozlov was a many-sided person, and everybody enjoyed interacting with him. He was extremely open-hearted and generous in sharing his new ideas. Collaboration with him was fruitful and pleasant for people of different ages. With his superlative talent, inexhaustible energy and wonderful capacity for work, Serguei participated in a great number of investigations. For many younger mathematicians he was a teacher and an advisor. The memory of Serguei Kozlov will be kept in the hearts of those who knew him.

## LIST OF PUBLICATIONS OF S. KOZLOV


1. S.M. Kozlov, Generalized analytic operators, *Moscow Univ. Mathem. Bulletin*, **30** (5), 45–51 (1975).

2. S.M. Kozlov, Equivalence of measures in linear Ito partial differential equations, *Moscow Univ. Mathem. Bulletin*, **32** (4), 36–40 (1970).

3. S.M. Kozlov, On stochastic partial differential equations, *Transactions of Petrovskii seminar*, Moscow State Univ. Publ. House, No. 4, 147–173 (1978).

4. S.M. Kozlov, Averaging differential operators with almost periodic rapidly oscillating coefficients, *Soviet Mathematics, Doklady*, **18** (4), 1323–1326 (1977).

5. S.M. Kozlov, Averaging differential equations with rapidly oscillating almost periodic coefficients, *Matem. Sbornik*, **107** (2), 129–218 (1978).

6. S.M. Kozlov, Averaging Random Structures, *Soviet Mathematics, Doklady*, **19** (4), 950–954 (1978).

7. S.M. Kozlov, Averaging random operators, *Matem. Sbornik*, **109** (2), 188–203 (1979).

8. S.M. Kozlov, Conductivity of 2D random structures, *Russian Mathem. Surveys*, **34** (4), 193–194 (1979).

9. S.M. Kozlov, Kha T'en Ngoan, O.A. Oleinik, V.V. Zhikov, Averaging and G-convergence of elliptic operators, *Russian Mathem. Surveys*, **34** (5), 69–147 (1979).

10. S.M. Kozlov, Homogenization of differential operators, *Russian Mathem. Surveys*, **35** (2), 254 (1980).

11. S.M. Kozlov, Homogenization of differential operators with random and almost periodic coefficients, Ph.D. Thesis, 1979, Moscow State Univ., 107 p.

12. S.M. Kozlov, Homogenization of differential operators with random and almost periodic coefficients, Summary of Ph.D. Thesis, 1979, Moscow State Univ.

13. S.M. Kozlov, Asymptotic behavior at infinity of fundamental solutions of equations with almost periodic coefficients, *Moscow Univ. Mathem. Bulletin*, **35** (4), 10–16 (1980).

14. S.M. Kozlov, Asymptotics of fundamental solutions of second-order divergence differential equations, *Math. USSR Sbornik*, **41** (2), 249–267 (1982).

15. S.M. Kozlov, Reducibility of quasiperiodic operators, *Russian Mathem. Surveys*, **36** (4), (1981).

16. S.M. Kozlov, Duality of variational problems on manifolds, *Russian Mathem. Surveys*, **37** (4), ().

17. S.M. Kozlov, Multidimensional spectral asymptotics for elliptic operators, *Sov. Math. D.*, **27** (1), 153–158 (1983).

18. S.M. Kozlov, O.A. Oleinik and V.V. Zhikov, Averaging theorems for parabolic operators, *Sov. Mathematics, Doklady*, **24** (1), 285–289 (1981).

19. S.M. Kozlov, O.A. Oleinik and V.V. Zhikov, Homogenization and G-convergence of parabolic operators, *Russian Mathem. Surveys*, **35** (4), (1980).

20. S.M. Kozlov, O.A. Oleinik and V.V. Zhikov, G-convergence of parabolic operators, *Russian Mathem. Surveys*, **36** (1), 9–60 (1983).

21. S.M. Kozlov, O.A. Oleinik and V.V. Zhikov, Sur l'homogenisation d'operateurs differintiels paraboliques a coefficients presque – periodiques *C.R.Acad. Sc.Paris*, **293** ser. I, 245–248 (1981).

22. S.M. Kozlov, O.A. Oleinik and V.V. Zhikov, Homogenization of parabolic operators, *Transaction of Moscow Math. Soc.*, **45**, 1–55 (1982).

23. S.M. Kozlov, O.A. Oleinik and V.V. Zhikov, Homogenization of parabolic operators with almost periodic coefficients, *Math. USSR Sbornik*, **117** (1), 69–85 ().

24. S.M. Kozlov, Distribution of eigenvalues of differential operators, *Russian Mathem. Surveys*, **37** (5), 185–186 (1982).

25. S.M. Kozlov and M.A. Shubin, On the functions of class S of selfadjoint operators, *Russian Mathem. Surveys*, **37** (2), 201–202 (1982).

26. S.M.. Kozlov and M.A. Shubin, Theorem of the coincidences of the spectra of random elliptic operators *Funktzional. Anal. i Ego Prilozheniya* **16** (4), 74–75 (1982).

27. S.M. Kozlov and G.S. Kulikov, On the offshore mooring structures *USSR Conference on Hydrotechnics*, 1982, Leningrad, "Energoizdat", 4 p.

28. S.M. Kozlov, Ground States of Quasiperiodic Operators, *Soviet*

*Math. Doklady*, **28** (1), 114–119 (1983).

29. S.M. Kozlov, Reducibility of quasiperiodic differential operators and averaging, *Trans. Moscow Math. Soc.*, Issue 2, 101–126 (1984).

30. S.M. Kozlov, On the duality of one type variational problems, *Funktzional. Anal. i Ego Prilozheniya*, **17** (3), 9–15 (1983).

31. S.M. Kozlov, Multidimensional spectral asymptotics for elliptic operators, *Workshop on the operators in functional spaces*, , 1982, Belorussia State Univ.

32. S.M. Kozlov, Multidimensional spectral asymptotic for elliptic operators in bounded domain, *Mathematics of USSR Izvestiya*, , **24** (1), 49–71 (1985).

33. S.M. Kozlov and G.S. Kulikov Hydrodynamical wave forces acting on the ships *VNII Morgeo, Sbornik* 1984, Riga, 5 p.

34. S.M. Kozlov and S.A. Molchanov On conditions for applicability of the central limit theorem to random walks on a lattice *Soviet Math. Doklady* **30** (2), 410–413 (1984).

35. S.M. Kozlov and M.A. Shubin, On the coincidence of the spectra of random elliptic operators *Math. USSR Sbornik* **51** (2), 455–471 (1985).

36. S.M. Kozlov, The method of averaging and walks in inhomogeneous environments, *Russian Mathem. Surveys*, **40** (2), 73–145 (1985).

37. S.M. Kozlov, Averaging of difference schemes, *Math. USSR Sbornik*, , **57** (2), 351–369 (1987).

38. S.M. Kozlov, Criterium of instability of linear operators, *Funktzional. Anal. i Ego Prilozheniya*, **21** (3), 81–82 (1987).

39.S.M. Kozlov, Spectral asympotics of random operators, *Math. Notes Acad. of Sci. USSR*, **43** (3), 234–243 (1988).

40. S.M. Kozlov, Homogenization of disordered systems, Doctor Thesis, 1989, USSR Acad. Sc., USSR, 262 p.

41. A.L. Piatnitski and S.M. Kozlov, Homogenization and vanishing viscosity, *Proc. of the Fifth International Vilnius Conf. on Prob. Theory and Math. Statistics*, **2**, MOKSLAS, 330–339 (1990).

42. S.M. Kozlov, Geometric aspects of homogenization, *Russian Mathem. Surveys*, **44** (2), 91–144 (1989).

43. S.M. Kozlov, Effective diffusion in the Fokker-Planck equation, *Math. Notes Acad. of Sc. USSR*, **45** (5), 360–368 (1989).

44. S.M. Kozlov, Homogenization and percolation, *Russian Mathem. Surveys*, **44** (4), 167–168 (1988).

45. S.M. Kozlov, Geometry and asymptotics in homogenization, *Workshop on Composite Media and Homogenization Theory*, 1990, Triest, Dal Maso, Dell'Antonio eds, Birkhauser, 205–221 (1991).

46. S.M. Kozlov, Asymptotics of Laplace-Dirichlet integrals, *Preprint ICTP*, IC90/16, 1–7 (1990).

47. S.M. Kozlov, Asymptotics of Laplace-Dirichlet integrals, *Funktsional. Analiz i Ego Prilozheniya*, **24** (2), 115–125 (1990).

48. S.M. Kozlov and A.L. Piatnitski, Averaging on a background of vanishing viscosity, *Math. USSR Sbornik*, **70** (1), 241–261 (1991).

49. L. Berlyand and S.M. Kozlov, Asymptotics of homogenized moduli for the elastic chess-board composite, *Arch. Rational Mech. Anal.*, **118**, 95–112 (1992).

50. S.M. Kozlov and J. Vucan, Explicit formula for effective thermoconductivity on quadratic lattice structure, *C.R. Acad. Sci. Paris Ser.1*, **314**, 281–286 (1992).

51. S.M. Kozlov and G.P. Panasenko, Corrections to the strength material theory for the lattice structure, *Publ. du Lab. d'Anal. Num.*, Univ. Paris, VI, 1991, R91020.

52. S.M. Kozlov, KAM theory and homogenization of nonlinear partial differential equation, *Preprint Forsh. Math. ETH*, Zurich, 1991.

53. K. Golden K and S.M. Kozlov, Long range conductivity model with infinitely many thresholds, Submitted *to Phys. ReVol. Let.*

54. S.M. Kozlov and M. Yor, Winer football, *Teor. Veroiatn. i ee Primenen.*, **3**, (1992).

55. S.M. Kozlov, O.A. Oleinik and V.V. Zhikov, Homogenization of differential operators and integral functionals, Springer-Verlag, Berlin, New York, 1993.

56. S.M. Kozlov, J.W. Pitman and M. Yor, Brownian interpretation of an elliptic integral, *Seminar on Stochastic Progresses*, 1991, edt. E. Cinlar, K.L. Chung. Birkhauser, 1992.

57. S.M. Kozlov and A.L. Piatnitski, Effective diffusion for a parabolic operator with periodic potential, *SIAM J. Appl. Math.*, **53** (2), 401–418 (1993).

58. A. Bensoussan and S.M. Kozlov, Diffusion effective dans un flux périodique, *C. R. Acad. Sci. Paris*, , **315**, Série I, 765–768 (1992).

59. S.M. Kozlov, A central limit theorem for Multiscaled Permeability, *Intern. Series of Numerical Math.*, **114**, 117–127 (1993).

60. S.M. Kozlov, On the domain of variation of added masses,

polarization and effective characteristics of composites, *J. Appl. Math. Mechs.*, **56** (1), 102–107 (1992).

61. K. Golden and S.M. Kozlov, Percolation analysis of the effective permeability of porous media, 1994, Prepubl. 94 – 22, Centre de Mathématiques et d'Informatique, Université de Provence, 18 p.

62. K. Golden and S.M. Kozlov, Long range conductivity model with infinitely many thresholds, 1994, Prepubl. 94 – 21, Centre de Mathématiques et d'Informatique, Université de Provence. 11 p.

63. S.M. Kozlov, Sur la forme invariante des principes variationnels de l'homogénéisation, 1994, Prepubl. 94-20, Centre de Mathématiques et d'Informatique, Université de Provence. 11 p.

64. A.Yu. Belyaev and S.M. Kozlov, Low concentration limit for Dirichlet homogenization problem, Transactions of the II Workshop on Homogenization Theory, edt. G.Dal Maso, G.Dell'Antonio, Birkhauser, 1994.

65. A.Yu. Belyaev and S.M. Kozlov, Hierarchical structures and estimates for homogenized coefficients, *Russian J. Math. Phys.*, **1** (1), 5–18 (1993).

66. S.M. Kozlov and L.A. Malozemov, Hierarchical framework structures, *Matem. Zametki*, **51** (2), 53–58 (1992).

67. A. Bourgeat, S.M. Kozlov and A. Mikelic, Effective equations of two-phase flow in random media, *Calc. Var. and PDE's*, **3**, 385–406 (1996).

68. S.M. Kozlov and A.L. Piatnitski, Degeneration of effective diffusion in the presence of periodic potential, *Ann. Inst. Henri Poincaré, Probability and Statistics*, **32** (5), 571–587 (1996).

69. A.Yu. Belyaev and S.M. Kozlov, The homogenization of Stokes equations in a random domain, *Russian Mathem. Surveys*, **48** (4), 211–212 (1993).

70. A.Yu. Belyaev and S.M. Kozlov, Darcy equatoin for random porous media, *Comm. Pure Appl. Math.*, **49** (1), 1–34 (1996).

71. S.M. Kozlov, Harmonization and homogenization on fractals, *Comm. Math. Phys.*, **153**, 339–357 (1993).

# CRITICAL PATH ANALYSIS OF TRANSPORT IN HIGHLY DISORDERED RANDOM MEDIA

K.M. GOLDEN
*University of Utah, Department of Mathematics, 155 S 1400 E RM 233, Salt Lake City, UT 84112-0090 USA*

S.M. KOZLOV

In problems of conduction and fluid flow through complex random media, systems with a broad distribution in the local properties are often encountered. Here we introduce a continuum percolation model for such media, which is exactly solvable for the effective transport properties in the high disorder limit. The model represents such systems as fluid flowing through consolidated granular media and fractured rocks, as well as electrical conduction in some matrix-particle composites near critical volume fractions. Moreover, the results provide a rigorous basis for the widely used Ambegaokar, Halperin, and Langer critical path analysis [1]. This method rests on the proposition that transport in media with a broad range of local conductances $g$ is dominated by a critical conductance $g_c$, which is the smallest conductance such that the set $\{g \mid g > g_c\}$ percolates, and in our model this proposition is rigorously established.

A central issue in the theory of transport in disordered materials is to determine the effective properties, such as electrical conductivity and fluid permeability, from some knowledge of the microstructural features of the material. In many problems of technological importance, one meets systems which display a wide distribution in the local properties characterizing the system. For example, in porous media there is often a wide range of pore and neck sizes through which the fluid must flow [2, 3, 4, 5, 6, 7, 8]. As another example, in resistor network models of hopping conduction in amorphous semiconductors, the bonds of the network are assigned a wide range of conductivities [1, 9, 10]. A powerful idea which has been widely used [2] to estimate the effective properties of such systems is the *critical path analysis*, which was first introduced by Ambegaokar, Halperin and Langer (AHL) [1]. They proposed that transport in a medium with a broad range of local conductances $g$ is dominated by a critical value $g_c$, which is the smallest conductance such that the set $\{g \mid g > g_c\}$ percolates, or forms a connected cluster which spans the sample. This cluster is called the critical path. Then the problem of estimating transport in a highly disordered medium with a wide range of local conductances is reduced to a percolation problem with threshold value $g_c$. Critical path analysis was further developed in the context of amorphous semiconductors in [9] and [10], and its accuracy was numerically confirmed for various conductance distributions in [11]. It has also been used to analyze the permeability and electrical conductivity of porous media, such as sandstones, as obtained through mercury injection [12, 13, 14, 15], as well as fractured rocks with a broad distribution of fracture apertures [16], and porous media saturated with a non-Newtonian fluid [17].

While critical path analysis has been applied with substantial success, there has been little rigorous work on the fundamental observation that the critical conductance $g_c$ dominates the effective behavior. Here we introduce a continuum percolation model of conducting and porous random media which is exactly solvable in the high disorder limit, and the result we obtain for the effective conductivity or permeability rigorously establishes the AHL principle for this model. Furthermore, our model closely represents an important class of porous materials, including consolidated granular media and some fractured rocks, where the easiest flow paths or channels exhibit a complex random topology, similar to a Voronoi network, as in Fig. 1 [2, 18, 19, 20, 6].

We obtain our model as a "long-range" generalization of the random checkerboard in $\mathbf{R}^2$ [21, 22, 23, 24, 25, 26, 27], where the squares are assigned the conductivities 1 with probability $p$ and $\delta > 0$ with probability $1-p$. The random checkerboard has been used to model conducting materials which exhibit critical behavior too rich to be accurately handled by random resistor networks, such as graphite (conducting) particles embedded in a polymer (insulating) matrix. For example, the presence of both corner and edge connections between squares produces two percolation thresholds, with distinct asymptotic behaviors of the effective conductivity as $\delta \to 0$ (or $\delta \to \infty$) in the different regimes of $p$ separated by these thresholds [22, 24, 25, 26]. In our model we allow arbitrarily long-range connections between squares, which leads to infinitely many thresholds, and rather complex asymptotic behavior, which we can nevertheless obtain exactly. Our analysis is based on the variational formulation of effective properties, which allows us to obtain bounds required for the asymptotics, by constructing trial fields which exploit relevant percolation structures [25, 26, 27]. A key com-

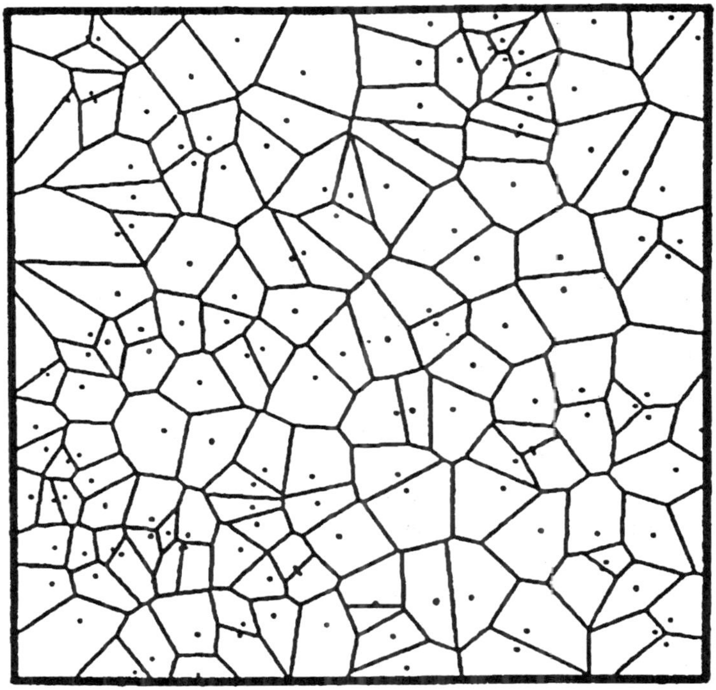

Figure 1: Two dimensional Voronoi tessellation (from [2]). The boundaries of the polygonal grains are formed from points which are equidistant to the dots in two neighboring grains. These boundaries form the channels of easiest flow in consolidated granular media and fractured rocks

ponent of our percolation analysis is to connect our model to a Poisson distribution of discs in the plane.

Whereas the standard checkerboard model allows for only two types of connections between conducting particles (squares), our generalization allows for arbitrarily many. This leads us to expect that our model may serve as a good representation for high contrast matrix-particle composites, particularly in the regime where the conducting particles "percolate," yet are of low enough volume fraction so that the conducting phase is only "partially connected" [27], where there is a broad distribution of connection types between the conducting particles. In such regimes the effective conductivity has been observed to vary over many orders of magnitude [28] over small volume fraction ranges, as does our model. More precisely, in the high disorder limit, the paths of easiest current flow in our model form a (continuum) Voronoi network with a broad distribution of "bond" conductivities. This network can be identified with the one obtained from joining the centers of "touching" conducting particles. The quality of the connection between the particles determines the conductivity of the path joining them. Our model can be modified to handle such problems, and our variational analysis, which is quite general, can be suitably applied to this and other related problems.

We now formulate the mathematical model. For simplicity, we give the formulation for electrical conductivity in $d = 2$, but the model and result can be carried over to $d = 3$, as well as fluid flow in a porous medium obeying Darcy's law, which will be discussed later. Consider the checkerboard of unit white squares in $\mathbf{R}^2$, with centers of the squares being the points of the lattice $\mathbf{Z}^2$. Randomly color the squares red with probability $p$, where the probability of coloring one square is independent from any other. Then for $x \in \mathbf{R}^2$, let $S(x)$ be the distance from $x$ to the boundary of the nearest red square, with $S(x) = 0$ if $x$ is inside a red square. Then define the local conductivity $\sigma(x)$ as

$$\sigma(x) = e^{\lambda S(x)}. \tag{1}$$

It is useful to think of the red squares as insulating particles, as we will be considering asymptotics as $\lambda \to +\infty$ (although we could just as easily consider $\lambda \to -\infty$). The medium defined by (1) can be thought of as being divided into grains associated with each red square, where each grain is the set of all points for which the distance to its red square is smaller than the distance to any other red square. The boundaries between these grains, where the distance to a red square is maximal, form the "channels" through which current (or fluid) passing through the medium will tend to flow. For small $p$, the set of boundaries forms a Voronoi network, as shown in Fig. 1. The points in the figure represent the red squares.

Our goal is to find the $\lambda \to +\infty$ asymptotics of the effective conductivity $\sigma^*(p)$ of the medium in (1), which is defined as follows (e.g. [29, 30]). Let $E(x)$ and $J(x)$ be the stationary random electric and current fields in the medium satisfying $J(x) = \sigma(x)E(x), \nabla \cdot J = 0, \nabla \times E = 0$, and $< E(x) >= e_k$, where $e_k$ is a unit vector in the $k^{th}$ direction, and $< \cdot >$ denotes ensemble or infinite volume average. Then the effective conductivity $\sigma^*$ is defined via

$$< J >= \sigma^* < E > . \tag{2}$$

In order to state the results for the asymptotics and to more fully describe our model, we first consider standard nearest neighbor site percolation on $\mathbf{Z}^2$ [31, 32, 33], which is equivalent to the percolation of nearest neighbor red squares (connected along an edge), with percolation threshold $p_c^s \approx 0.59$. Now we relax the nearest neighbor restriction for connectedness, and consider a generalized definition of percolation of the red squares. We say that two red squares $\hat{x}$ and $\hat{y}$ with centers $x$ and $y$ in $\mathbf{Z}^2$, are $r$-connected if there is a sequence $\hat{x}_0, \ \hat{x}_1, \cdots, \hat{x}_n, \ \hat{x}_0 = \hat{x}, \ \hat{x}_n = \hat{y}$ of red squares connecting them such that $dist\{\hat{x}_i, \ \hat{x}_{i+1}\} \leq r$, where $dist\{\hat{x}_i, \ \hat{x}_{i+1}\}$ means the shortest distance between the boundaries of $\hat{x}_i$ and $\hat{x}_{i+1}$. With $\theta_r(p)$ the infinite cluster density of $r$-connected red squares, we define $p_c(r)$ as the percolation threshold for $\theta_r(p)$. We shall be concerned with a particular sequence $r_j, \ j \geq 1$, defined by squares which are increasingly distant from $\hat{0}$, the square centered at the origin, with $r_1 = 0, \ r_2 = 1, \ r_3 = \sqrt{2}, \ r_4 = 2, \ r_5 = \sqrt{5}, \ r_6 = \sqrt{8}, \cdots$. Note that it suffices to consider squares with centers $(m, n) \in \mathbf{Z}^2$ with $n \geq m \geq 0, \ n > 0$. For simplicity we denote $p_c(r_j)$ as $p_c(j)$, and we also replace the term $r_j$-connected by $j$- connected. Note that the $j = 1$ $(r_1 = 0)$ case includes both the nearest neighbor case above and the next nearest neighbor (diagonal) case (since the distance between connected squares is 0 in both cases), so that $p_c(1) = 1 - p_c^s \simeq 1 - 0.59 = 0.41$ [25, 26, 27]. Furthermore $p_c(j+1) < p_c(j)$, which can be obtained from [34]. From analysis of a

Poisson distribution of discs in the plane (see below), one can also find that

$$p_c(j) \sim \frac{\pi \mu_c}{j} , \qquad j \to \infty , \tag{3}$$

where $\mu_c$ is the critical percolation intensity for unit discs [33].

The last ingredient we need to state the results is the notion of critical values of $S(x)$, which are associated with the $p_c(j)$. These values $S_c(j)$ are defined by the observation that for $p > p_c(j)$, the set $R_j = \{x \in \mathbf{R}^2 : S(x) \leq S_c(j)\}$ percolates in $\mathbf{R}^2$, where for $j$ associated with $(m,n) \in \mathbf{Z}^2$ as above,

$$S_c(j) = r_j/2 = \sqrt{(1-\delta_{m0})(m-1)^2 + (n-1)^2}\Big/2 , \tag{4}$$

where $\delta_{m0}$ is the Kroneker delta. By percolating in $\mathbf{R}^2$ we mean that $R_j$ contains an infinite polygonal line joining vertices of the red squares in $R_j$. Note that $R_1$ is just the set of red squares, which percolates in the above sense when $p > p_c(1)$. Now in terms of the $S_c(j)$, define a step function $S_c(p)$ via

$$S_c(p) = S_c(j) , \quad p_c(j) < p < p_c(j-1) , \ j \geq 1, \tag{5}$$

where $p_c(0) = 1$. These critical distances $S_c(j)$ in our model correspond via (1) to the critical conductances $g_c$ in the AHL theory.

The principal results of our investigation are now stated as follows. For the effective conductivity $\sigma^*(p)$ of the medium $\sigma(x) = e^{\lambda S(x)}$, we have for $p \neq p_c(j)$,

$$\frac{1}{\lambda} \, log \, \sigma^*(p) \sim S_c(p), \qquad \lambda \to \infty, \tag{6}$$

which establishes the validity of the AHL critical path analysis in the high disorder limit for our model. Furthermore, we have the following $p \to 0$ asymptotics for the exponents,

$$S_c(p) \sim \sqrt{\frac{\mu_c}{p}} , \qquad p \to 0 , \tag{7}$$

where $p$ is the density of red squares (i.e., with units of inverse square length). Via (6) and (7), we see that $\sigma^*(p)$ for our long range checkerboard model exhibits infinitely many thresholds $p_c(j) \to 0$ as $j \to \infty$

with an infinite set of asymptotics as $\lambda \to \infty$. If instead of $\lambda \to \infty$ we wish to put $\lambda = 1$ so that $\sigma(x) = e^{S(x)}$, and consider the asymptotics of $\sigma^*(p)$ as $p \to 0$, we also find that

$$\sqrt{p}\; log\; \sigma^*(p) \sim \sqrt{\mu_c}\ , \qquad p \to 0 \tag{8}$$

We now give the analysis which leads to these results. The idea is to exploit the variational definition of $\sigma^*$ equivalent to (2) and its dual, to obtain upper and lower bounds on $\sigma^*$. Let $\Lambda_N = [0, N] \times [0, N] \subset \mathbf{R}^2$. Then the variational form of (2) is

$$\sigma^* = \lim_{N\to\infty} \frac{1}{N^2} \inf_{u\in\mathcal{P}} \int_{\Lambda_N} \sigma(x)\ |\nabla u|^2 dx\ , \tag{9}$$

where $\mathcal{P} = \{$continuous potentials $u$ on $\Lambda_N : u(0, x_2) = 0\ ,\ u(N, x_2) = N, \forall x_2 \in [0, N]\}$. We obtain bounds by inserting trial $u$ into (9). To describe the construction, we recall certain properties of standard site percolation. It has been shown [35, 36] that for $p > p_c^s$ the number $\alpha_N$ of disjoint paths through the occupied phase, (formed from sequences of nearest neighbor, occupied sites) which cross $\Lambda_N$ horizontally or vertically, or disjoint crossings, satisfies (roughly speaking) $\alpha_N = O(N)$ as $N \to \infty$. In our generalized model, recall that for $p > p_c(j), R_j = \{x \in \mathbf{R}^2 : S(x) \leq S_c(j)\}$ percolates in $\mathbf{R}^2$. In this case, the number $\alpha_N(j)$ of disjoint crossings of $\Lambda_N$ by j-connected red squares also satisfies $\alpha_N(j) = O(N)$ as $N \to \infty$. We call the associated disjoint subsets of $R_j$ that cross $\Lambda_N$ pink "j-chains" (we say "pink" because such sets contain both red squares and parts of white ones). It will be necessary for our purposes to consider only those chains which cross vertically. Note that these j-chains can be viewed as ribbons which at some points may have zero width, for example at corner connections between red squares with $j = 1$ and $r_1 = 0$, with $p_c(1) = 1 - p_c^s < p < p_c^s$. In this case, note that both $R_1$ and its complement in $\mathbf{R}^2$ percolate, with similar behavior for $j > 1$. (In standard nearest neighbor lattice percolation models, double connectivity, or simultaneous percolation of the two phases, does not occur in two dimensions, but only in higher dimensions.) Now, the trial $u$ is constructed as follows, using a similar technique to that which has been used for the random checkerboard [25]. On the j-chains $u$ increases linearly across the chain, in such a way that the total contribution to (9) of $|\nabla u|^2$ on the pink j-chains is $O(N^2)$. In each region between the j-chains, $u$ is a constant (so that

$\nabla u = 0$), which, along with additive constants for $u$ on the j-chains, are chosen so that $u$ is continuous, and it "steps up" as one moves across $\Lambda_N$. Constructing $u$ in the neighborhood of points where the j-chains have zero thickness is handled with asymptotic expansions, which are patched continuously to the rest of $u$, as in [25] for the so-called Laplace-Dirichlet integral. For such $u$ then, the integrand in (9) is zero off the j-chains, and for $x$ in the pink j-chains,

$$\sigma(x) \leq e^{\lambda S_c(j)}, \tag{10}$$

which leads to the inequality

$$\sigma^*(p) \leq C_1 e^{\lambda S_c(j)}\ , \qquad p > p_c(j)\ , \tag{11}$$

for some $C_1 > 0$ (depending on $p$), with $j \geq 1$. Another trial potential which gives the same type of bound (pointed out to us by a referee) is the actual potential for a medium which has unit conductivity in $R_j$ and is superconducting off $R_j$.

To get the lower bound, we first note that a dual formulation of (9) can be obtained by replacing $\sigma^*$ and $\sigma(x)$ by $(\sigma^*)^{-1}$ and $\sigma^{-1}(x)$, respectively. The key observation in the analysis now is that for $p < p_c(j-1), j \geq 1$ (with $p_c(0) = 1), W_j = \{x \in \mathbf{R}^2 : S(x) \geq S_c(j)\}$ percolates in $\mathbf{R}^2$, which can be seen as follows. When $p < p_c(j-1)$, $R_{j-1}$ cannot percolate, $j \geq 2$. In this case, easy geometric reasoning shows that there must exist infinite chains of white cells, such that the minimal thickness $r_j$ of these white chains (meaning that discs of radius $r_j/2$ percolate in these chains) is $2S_c(j)$. Then $Wj = \{x \in \mathbf{R}^2 : S(x) \geq r_j/2 = S_c(j)\}$, which contains this set of white chains, percolates in $\mathbf{R}^2$. Now constructing $u$ similar to that above, one obtains

$$[\sigma^*(p)]^{-1} \ \leq\ C_2 e^{-\lambda S_c(j)}\ , \qquad p < p_c(j-1)\ , \tag{12}$$

for some $C_2 > 0$ (depending on $p$) , with $j \geq 1$. Combining (11) and (12) yields (6).

In order to obtain the asymptotic behaviors of the thresholds in (3) and the exponents in (7), we connect our work to the analogous problem for a Poisson distribution of discs in $\mathbf{R}^2$. Let $\{x_k\}_{k=1}^{\infty}$ be a set of Poisson distributed red points in the plane, with intensity $\mu$. First we define, analogously to (1), $S_\mu(x) = dist\{x,\ \text{nearest } x_k\}$ and $\sigma_\mu(x) = e^{\lambda S_\mu(x)}$. Let $S_\mu^c$ be the smallest $h$ for which $\{x \in \mathbf{R}^2 : S_\mu(x) \leq h\}$ percolates.

Then $S^c_\mu$ coincides with $r^c_\mu$, the minimum radius such that the discs of radius $r^c_\mu$ centered at the $x_k$ percolate. The above arguments used for the long range checkerboard yield for the effective conductivity $\sigma^*_\mu$ in this case

$$\frac{1}{\lambda}\, log\ \sigma^*_\mu \sim S^c_\mu = r^c_\mu\ , \qquad \lambda \to \infty. \tag{13}$$

We remark that, via the scaling properties of the Poisson model, we can replace $\lambda S_\mu(x)$ by $\frac{\lambda}{\sqrt{\mu}}\ S_1(x)$, so that we may set $\lambda = 1$ and consider asymptotics as $\mu \to 0$, with a result analogous to (13).

It is now useful to note that the above Poisson model can be obtained by rescaling our checkerboard model, where the red squares of our model correspond to the $x_k$ of the Poisson model as follows. On the scaled lattice $h\mathbf{Z}^2, h > 0$, let the density of red squares be $p/h^2 = \mu$ . As $p \to 0$, with $h = \sqrt{p/\mu} \to 0$ as well, $\hat{S}(x,p) = h\ S(x/h,p) \to S_\mu(x)$. Then the critical values also converge, $\hat{S}_c(p) \to S^c_\mu$ as $p \to 0$. Then with $\hat{S}_c = hS_c$ and $h = \sqrt{p/m}$, setting $\mu = 1$ yields $\sqrt{p}S_c(p) \to r^c_1$ as $p \to 0$, which is equivalent to (7), since $\mu_c = (r^c_1)^2$. Furthermore, the critical density of red points on the rescaled lattice is $p_c(j)/h^2$, so that $\mu_c = \lim_{j\to\infty} \frac{p_c(j)}{h^2}$. Note that $p_c(j)$ is the critical $p$ for which disks of radius $r_j$ percolate. So if we rescale the lattice with $h \sim 1/r_j$ as $j \to \infty$, then unit discs percolate, so that

$$\mu_c = \lim_{j\to\infty}\ p_c(j) r_j^2\ . \tag{14}$$

To relate $r_j$ to $j$, we note that there are $O(j)$ integer points inside the disc of radius $r_j$ as $j \to \infty$, so that

$$j \sim \pi r_j^2\ , \quad j \to \infty\ , \tag{15}$$

which combined with (14) yields (3).

In closing we wish to make a few remarks. Presumably an effective medium approach as in [22] could give an accounting of the behavior of our model for small $j$. However, as $j$ grows, the number of configurations of squares that must be considered grows extremely rapidly, and numerical calculations become intractable.

An interesting question is the transition between different exponents for large $\lambda$ as $p$ crosses the threshold $p_c(j)$ . We remark that, for example, the constant $C_1$ in estimate (11) diverges like $\xi(p)$ as $p \to p_c(j)^+$, where $\xi(p)$ is the correlation length for $j-$percolation.

Finally, for fluid flow (with unit viscosity) in porous media [2] obeying Darcy's law $v = -K(x)\nabla P$, where $K(x)$ is the local permeability, $v$ is the fluid velocity satisfying $\nabla \cdot v = 0$, and $P$ is the pressure (including gravity), one is interested in the effective permeability $K^*$, defined analogously to (2). As briefly mentioned earlier, if $K(x)$ has the form (1), then for large $\lambda$ it is a close model for flow through consolidated granular media, where the grains themselves are permeable, with decreasing permeability as one approaches a hard core. Fig. 2 shows a computer simulation of a grain consolidation process [19]. The sequence (a) - (d) shows increasing consolidation, and correspondingly decreasing porosity. As $\lambda \to \infty$ the network of easiest flow paths in our model closely resembles the configuration in (d), which itself is similar to many types of sedimentary rocks, including Devonian sandstone [19].

**Reminiscence: (from K.G.)**

The work represented in this paper evolved over a period of a few years beginning in August 1991, when I visited Serguei Kozlov in Moscow, which turned into a very eventful time. During our initial investigations of the model considered here, our work was interrupted by the coup against Gorbachev, which led to the rise of Yeltsin, and the fall of the Soviet Union. During the first day, Serguei and I ventured into Red Square and saw the tanks assembled, on the last day we attended Yeltsin's speech from the Russian "White House," and there were some very tense times in between. It was a most memorable sequence of events, and I am grateful that I was able to experience it with Serguei and his family. We continued this work after Serguei moved to the south of France, and had some very enjoyable and productive periods there. During the times that we spent together, I came to greatly appreciate the depth, clarity, and power of Serguei's ideas, as well as his friendship. His untimely death is truly a loss, and he is deeply missed.

**Acknowledgments:**

K. M. Golden would like to thank M. Aizenman and L. Berlyand for helpful conversations on percolation theory and checkerboard models, and also acknowledges the warm hospitality at the Moscow Civil Engineering Institute, the Université de Toulon, and the Université de

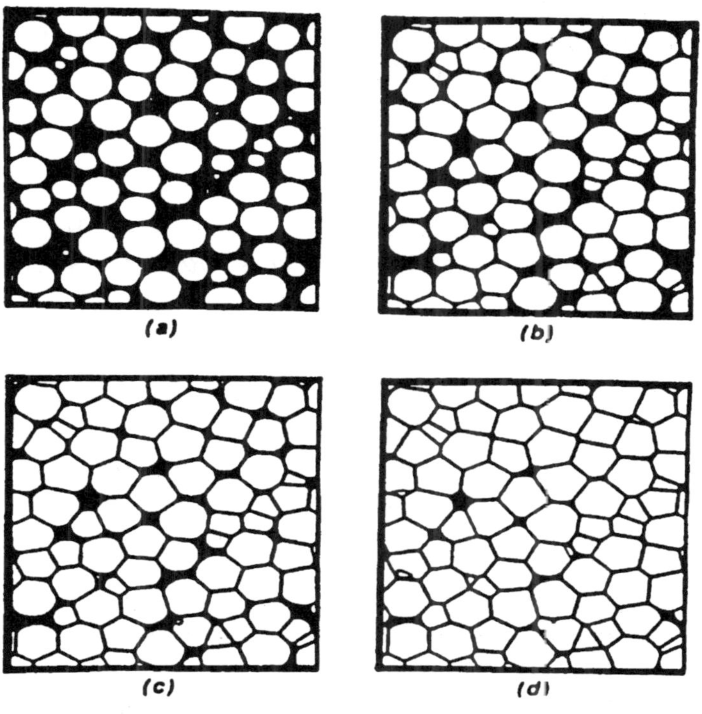


Figure 2: Computer simulation of the consolidation of spherical grains (from [19]), showing decreasing porosity $\varphi$, with (a) $\varphi = 0.364$, (b) $\varphi = 0.200$, (c) $\varphi = 0.100$ and (d) $\varphi = 0.030$

Provence, where most of this work was completed. S. M. Kozlov would have liked to thank G. Dagan for helpful discussions they had on porous media problems at Oberwolfach. L. Schwartz, M. Sahimi, VCH Publishers, and the Amercian Physical Society are kindly acknowledged for use of the figures that appear here. Finally, K. M. Golden gratefully acknowledges support from NSF Grants DMS-9307324, DMS-9622367, and OPP-9725038, and ONR Grant N000149310141.

## References

1. V. Ambegaokar, B.I. Halperin, and J.S. Langer, Hopping conductivity in disordered systems, *Phys. Rev. B*, **4**, 2612–2620 (1971).
2. M. Sahimi. Flow and Transport in Porous Media and Fractured Rock, VCH, Weinheim, 1995.
3. F.A.L. Dullien. Porous Media, Fluid Transport and Pore Structure, Academic Press, New York, 1992.
4. I. Fatt, The network model of porous media, *Trans. Am. Inst. Min. Metall. Pet. Eng.*, **207**, 144–177 (1956).
5. J. Koplik, Creepeng flow in two-dimentional networks, *J. Fluid Mech.*, **119**, 219–247 (1982).
6. S. Feng, B.I. Halperin, and P.N. Sen, Transport properties of continuum systems near the percolation threshold, *Phys. Rev. B*, **35**, 197–214 (1987).
7. G. Dagan, Flow and Transport in Porous Formations, Springer-Verlag, New York, 1989.
8. G. Matheron, Éléments pour une Théorie des Milieux Poreux, Masson et $C^{ie}$, Paris, 1967.
9. V.K. Shante, Hopping conduction in quasi-one-dimensional disordered compounds, *Phys. Rev. B*, **16**, 2597–2612 (1977).
10. S. Kirkpatrick, Models of disordered materials, In R. Balian, R. Maynard, and G. Toulouse, editors, Ill-Condensed Matter. North-Holland, New York, 1983.
11. D. Berman, B.G. Orr, H.M. Jaeger, and A.M. Goldman, Conductances of filled two-dimensional networks, *Phys. Rev. B*, **33**, 4301 (1986).
12. A.J. Katz and A.H. Thompson, Quantitative prediction of permeability in porous rock, *Phys. Rev. B*, **34**, 8179 (1986).

13. A.J. Katz and A.H. Thompson, Prediction of rock electrical conductivity from mercury injection experiments, *J. Geophys. Res. B*, **92**, 599 (1987).
14. J.R. Banavar and D.L. Johnson, Characteristic pore sizes and transport in porous media, *Phys. Rev. B*, **35**, 7283 (1987).
15. P. Le Doussal, Permeability versus conductivity for porous media with wide distribution of pore sizes, *Phys. Rev. B*, **39**, 4816 (1989).
16. E. Charlaix, E. Guyon, and S. Roux, Permeability of a random array of fractures of widely varying apertures, *Trans. Porous Media*, **2**, 31 (1987).
17. M. Sahimi, Nonlinear transport processes in disordered media, *AIChE J.*, **39**, 369 (1993).
18. A.R. Kerstein, Equivalence of the void percolation problem for overlapping spheres and a network problem, *J. Phys. A*, **16**, 3071–3075 (1983).
19. J.N. Roberts and L.M. Schwartz, Grain consolidation and electrical conductivity in porous media, *Phys. Rev. B*, **31**, 5990–5997 (1985).
20. G.R. Jerauld, J.C. Hatfield, L.E. Scriven, and H. T. Davis, Percolation and conduction on Voronoi and triangular networks: a case study in topological disorder, *J. Phys. C*, **17**, 1519–1529 (1984).
21. A.M. Dykhne, Conductivity of a two-dimensional two-phase system, *Sov. Phys. JETP*, **32**, 63 (1971).
22. P. Sheng and R.V. Kohn, Geometric effects in continuous-media percolation, *Phys. Rev. B*, **26**, 1331 (1982).
23. J.B. Keller, Effective conductivity of periodic composites composed of two very unequal conductors, *J. Math. Phys.*, **28**, 2516 (1987).
24. S.A. Molchanov, Ideas in the theory of random media, *Acta Appl. Math.*, **22**, 139 (1991).
25. S.M. Kozlov, Geometric aspects of homogenization, *Russ. Math. Surv.*, **44**, 91 (1989).
26. L. Berlyand and K. Golden, Exact result for the effective conductivity of a continuum percolation model, *Phys. Rev. B*, **50**, 2114–2117 (1994).

27. K. Golden, Scaling law for conduction in partially connected systems, *Physica A*, **207**, 213–218 (1994).
28. D.S. Mclachlan, M.Blaszkiewicz, and R.E. Newnham, Electrical resistivity of composites, *J. Am. Ceram. Soc.*, **73**, 2187 (1990).
29. S.M. Kozlov, Averaging random structures (in Russian), *Dokl. Akad. Nauk SSSR*, , **241**, 1016 (1978).
30. K. Golden and G. Papanicolaou, Bounds for effective parameters of heterogeneous media by analytic continuation, *Comm. Math. Phys.*, **90**, 473–491 (1983).
31. D. Stauffer and A. Aharony. Introduction to Percolation Theory, Second Edition. Taylor and Francis Ltd., London, 1992.
32. H. Kesten. Percolation Theory for Mathematicians. Birkhäuser, Boston, 1982.
33. G. Grimmett. Percolation. Springer-Verlag, New York, 1989.
34. M. Aizenman and G. Grimmett, Strict monotonicity for critical points in percolation and ferromagnetic models, *J. Stat. Phys.*, **63**, 817–835 (1991).
35. G. Grimmett and H. Kesten, First-passage percolation, network flows and electrical resistances, *Z. Wahr.*, **66**, 335–366 (1984).
36. J.T. Chayes and L. Chayes, Bulk transport properties and exponent inequalities for random resistor and flow networks, *Comm. Math. Phys.*, **105**, 133–152 (1986).

# MULTISCALED HOMOGENIZATION

V.V. JIKOV
*Pedagogical Institute, Vladimir, Russia*

S.M. KOZLOV

This paper develops the ideas which were first appeared in S. Kozlov's paper [5].

## 1 Introduction

Consider the divergent operator

$$\nabla \cdot (K_N(x)\nabla),$$

where the *scalar* coefficient $K_N$ has the structure

$$K(x) = K_N(x) = a^N\left(\frac{x}{\xi_1}\right) a^N\left(\frac{x}{\xi_2}\right) \dots a^N\left(\frac{x}{\xi_N}\right), \quad x \in \mathbb{R}^d. \tag{1.1}$$

Here $a(x) = a^N(x) > 0$ is a periodic function with respect to the cell of periodicity

$$\square = \{x \in \mathbb{R}^d, |x_i| < \frac{1}{2}, i = 1, 2, \dots, d\},$$

$\xi_1, \xi_2, \dots, \xi_N > 0$ are *scale factors*, $\xi_N \to 0$ as $N \to \infty$.

For each fixed $N$ the coefficient $K_N(x)$ is an almost-periodic function. According to homogenization theory [1], we have some homogenized matrix $(K_N E)_{hom}$ ($E$ is an identical matrix) and we are interested in the calculation of the limit

$$\lim_{N\to\infty} (K_N E)_{hom}. \tag{1.2}$$

Each factor $a^N\left(\frac{x}{\xi_i}\right)$ in (1.1) is associated with some *fluctuation* or *imhomogeneity* of size $\xi_i$. So, the limiting ideal medium contains imhomogeneities of any small size. These ideal mediums are similar to *differential schemes* introduced on physical level, by Bruggemann [2] in 1935.

We assume that

$$a^N = 1 + O(\frac{1}{\sqrt{N}}) \text{ in } W^{1,\infty(\square)}, \tag{1.3}$$

$$< a^N >, \quad \left\langle \frac{1}{a^N} \right\rangle \leq 1 + \frac{\beta}{N} = h, \tag{1.4}$$

where

$$< \cdot >= \int_{\square} \cdot \, dx,$$

$$\sum_{k=1}^{\infty} k \left( \frac{\xi_k}{\xi_{k-1}} \right)^2 < \infty, \tag{1.5}$$

$$\left.\begin{array}{l} \text{the scale factors } \xi_1, \xi_2, \ldots \text{ are rationally independent, i.e.} \\ \text{the equality } \xi_1 n_1 + \ldots \xi_k n_k, \text{ where } n_i \in \mathbf{Z} \\ \text{holds only if } n_1 = n_2 \ldots n_k. \end{array}\right\} \tag{1.6}$$

Condition (1.5), (1.6) are satisfied if we take $\xi_k = e^{-k^\alpha}$ and, in general, $\xi_k = q^{-k^\alpha}$, $q > 1$ is a transcendental number, $\alpha > 1$.

Under conditions (1.3)-(1.6) we find the connection of the limit (1.2) with the limit of the so-called *reiterated homogenized matrices.* Let us formulate this notion, see [3].

For each periodic matrix $a(x)E = a^N(x)E$ we introduce the constant matrix

$$A_1 = (aE)_{hom}$$

via the usual relation

$$\eta \cdot A_1 \eta = \inf_{u \in H^1_{per}(\square)} < a|\eta + \nabla u|^2 >, \quad \forall \eta \in \mathbb{R}^d, \tag{1.7}$$

where $H^1_{per}(\square)$ is a Sobolev space of periodic functions with norm $< |u|^2 + |\nabla u|^2 \rangle^{1/2}$.

The Euler equation of variational problem (1.7) has the form

$$\nabla \cdot (a(\eta + \nabla u)) = 0, \quad u \in H^1_{per}(\square).$$

If we set $u = \mathbf{u}\cdot\xi$, then for the vector $\mathbf{u}$ we get the following periodic problem

$$\nabla \cdot (a(E + \nabla \mathbf{u})) = 0,$$

$$\nabla \mathbf{u} = \begin{pmatrix} \frac{\partial u_1}{\partial x_1} & \cdots & \frac{\partial u_d}{\partial x_1} \\ \vdots & & \\ \frac{\partial u_1}{\partial x_d} & \cdots & \frac{\partial u_d}{\partial x_d} \end{pmatrix},$$

moreover,

$$A_1 = \left\langle a(E + \nabla \mathbf{u})^T (E + \nabla \mathbf{u}) \right\rangle = \langle a(E + \nabla \mathbf{u}) \rangle ,$$

where $B^T$ denotes transposition to the matrix $B$. As well known, $A_1$ is an elliptic matrix, and by induction we can define the following reiterated homogenized matrices:

$$A_0 = E, \quad A_1 = (aE)_{hom}, \quad A_2 = (aA_1)_{hom}, \ldots, A_k = (aA_{k-1})_{hom},$$
$$\eta A_k \eta = \inf_{u \in H^1_{per}(\square)} < a(\eta + \nabla u) A_{k-1} (\eta + \nabla u) > . \tag{1.8}$$

As above this means that we have to find solution $\mathbf{u}_k$ of the problem

$$\nabla \cdot (aA_{k-1}(E + \nabla \mathbf{u}_k)) = 0 \tag{1.9}$$

and after that we can pose

$$A_k =< a(E + \nabla \mathbf{u}_k)^T A_{k-1} (E + \nabla \mathbf{u}_k) > . \tag{1.10}$$

One can realize the special simplification in the isotropic case, when $A_k = A_1^k$ ($A_1$ in this case is a number).

One of our main results here is the following estimate.

**Lemma 1.1** *Let conditions (1.3)-(1.6) be satisfied. Then the estimate holds*

$$A_N - O(\frac{1}{N}) \le (K_N E)_{hom} \le A_N + O(\frac{1}{N}).$$

Basically this estimate enables us to calculate the $\lim_{N\to\infty} (K_N)_{hom}$ as far as information on $A_N$ is provided. In order to get such information we accept the additional condition

$$(a^N C)_{hom} = C + \frac{M(C)}{N} + o\left(\frac{1}{N}\right) \tag{1.11}$$

for any positively defined matrix $C$ where the remainder is uniformly small with respect to $C$, and $C$ obeys the inequalities

$$\nu^{-1}E < C < \nu E, \quad \nu > 0.$$

In relation (1.11) $M(C)$ is some symmetric matrix which depends on $C$. Matrix function M(C) is assumed to be of class $C^1$. Note that in the isotropic case

$$M(E) = m\,E, \quad m \in \mathbb{R}.$$

Note also the "diagonal case" i.e. the case when $M(C)$ is diagonal for any diagonal $C$.

To solve the problem of the limiting behavior of $A_N$, let us consider the matrix equation

$$\frac{dA(t)}{dt} = M(A), \quad A(0) = E, \tag{1.12}$$

which was introduced by Avellaneda [4] in the similar situation.

**Lemma 1.2** *Suppose conditions on $M$ enumerated above are satisfied. Then*

$$\lim_{N\to\infty} A_N = A(1).$$

Combining results of our two lemmas we get

**Theorem 1.1** *Under condition (1.3)-(1.6), (1.11) and assuming existence and uniqueness of the solution to the problem (1.12), we have the limiting relation*

$$\lim_{N\to\infty} (K_N E)_{hom} = A(1).$$

In the isotropic case (1.12) has an explicit solution, and we resume this case as

**Corollary 1.2** *If $M(E) = mE$ and $m \in \mathbb{R}$, then*

$$\lim_{N\to\infty} (K_N E)_{hom} = e^m E.$$

Let us discuss a particular situation when condition (1.11) is satisfied.

Suppose that

$$a^N = 1 + \frac{f(x)}{\sqrt{N}} + \frac{g(x)}{N} + O\left(\frac{1}{N^{3/2}}\right) \tag{1.13}$$

$f(x), g(x) \in W^{1,\infty}_{per}(\square)$ and the remainder is also controlled in $W^{1,\infty}_{per}(\square)$. Then one can show that (1.11) is true, but explicit representation for $M(C)$ is rather complicated.

However, in one important case the matrix $M(E)$ is isotropic and the answer can be given in an explicit form.

We say that $f$ is cubic symmetric if $f$ is symmetric with respect to the arguments $x_1, \ldots, x_d$ and even relatively each of its.

**Theorem 1.3** *If $f$ is cubic symmetric, then*

$$M(E) = \left(< g > - \frac{< f^2 >}{d}\right) E$$

*and therefore*

$$\lim_{N\to\infty} (K_N E)_{hom} = e^{<g> - \frac{\langle f^2 \rangle}{d}} E.$$

The following example is of special interest. Consider the coefficient

$$a(x) = e^{\frac{f(x)}{\sqrt{N}}}.$$

In this case

$$K_N(x) = e^{\frac{f(\frac{x}{\xi_1}) + \cdots f(\frac{x}{\xi_N})}{\sqrt{N}}}$$

and from Theorem 1.3 we get:

$$\lim_{N\to\infty} (K_N E)_{hom} = e^{<f^2>(\frac{1}{2} - \frac{1}{d})} = \lim_{N\to\infty} \left\langle K_N^{\frac{d-2}{d}} \right\rangle^{\frac{d}{d-2}}$$

It is called the Landau formula.

Theorem 1.1 is true also for other coefficients $K_N$ constructed by the same principle (1.1). For instance, it is interesting to consider a situation when the initial coefficient $a^N(x)$ takes only two values: 0 and 1, in other words,

$$a^N(x) = \begin{cases} 0 \text{ if } x \in F_N, \\ 1 \text{ if } x \in \square \setminus F_N, \end{cases} \tag{1.14}$$

where $F_N$ is a closed periodic domain, $\mathbb{R}^d \setminus F_N$ is connected. Problems of this kind we discuss at the end of section 5. Note here only that instead of Landau's formula the law of Bruggemann arises.

## 2 Some properties of almost periodic functions

1. If we take a periodic function $f \in L^2(\square)$ and rationally independent number $\xi_1, \xi_2$, then the product

$$f\left(\frac{x}{\xi_1}\right) f\left(\frac{x}{\xi_2}\right)$$

will be already an almost-periodic function. Thus, we deal with homogenization of differential operators with almost-periodic coefficients.

Let us recall some notion from theory of almost periodic functions.

Denote by $Trig(\mathbb{R}^d)$ the set of all trigonometric polynomials, i.e. functions $u(x)$ that can be represented as a finite sum

$$u(x) = \sum u_\lambda \exp(ix \cdot \lambda), \quad u_\lambda \in \mathrm{C}, \quad \lambda \in \mathbb{R}^\alpha$$

**Definition 2.1** *A function $f(x) \in L^1_{loc}(\mathbb{R}^\alpha)$ possesses a mean value $< f >$ if the limit*

$$\lim_{t\to\infty} \frac{1}{t^d} \int\limits_{t\square} f(x)dx = < f >$$

*exists.*

It is easy to see that any trigonometric polynomial has a mean value equal to $u_0$.

Consider the set of all function $f(x) \in L^2_{loc}(\mathbb{R}^d)$ with a finite norm

$$\|f\|_2 = \overline{\lim_{t\to\infty}} \left( t^{-d} \int\limits_{t\square} |f|^2 dx \right)^{\frac{1}{2}} \quad \text{(the so-called Bezikovich norm).}$$

**Definition 2.2** *A function $f(x) \in L^2_{loc}(\mathbb{R}^d)$ is called almost-periodic (in the sense of Bezikovich) if there is a sequence of trigonometric polynomials converging to $f$ in the Bezikovich norm. The space of such functions is denoted by $B^2(\mathbb{R}^d)$.*

The functional $\|\cdot\|$ possesses all the properties of a norm, except that the equality $\|f\|_2 = 0$ does not imply $f(x) = 0$ almost everywhere in $\mathbb{R}^d$. Function $f_1, f_2$ will be call equivalent if the Bezikovich norm of their difference $f_1 - f_2$ is equal to zero.

We introduce also the norm

$$\|f\|_1 = \overline{\lim_{t\to\infty}} t^{-d} \int_{t\square} |f(x)|dx, \quad f \in L^1_{loc}(\mathbb{R}^d)$$

By an elementary argument one can show that the functional of the mean value is continuous with respect to the norm $\|\cdot\|_1$. Therefore, if $\|f - f_k\|_1 \to 0$ and $f_k$ has a mean value then $f$ also has a mean value, and

$$< f >= \lim_{k\to\infty} < f_k > .$$

Let us study the space $B^2(\mathbb{R}^d)$.

By definition, for any given $f, g \in B^2(\mathbb{R}^d)$ there exist elements $f^\delta, g^\delta \in Trig(\mathbb{R}^d)$ such that

$$\|f - f^\delta\|_2 \to 0, \quad \|g - g^\delta\|_2 \to 0 \text{ as } \delta \to 0.$$

Therefore the inequality

$$\|uv\|_1 \le \|u\|_2^{1/2}\|v\|_2^{1/2}$$

implies that $\|fg - f^\delta g^\delta\|_1 \to 0$. It follows that the product of any two function from $B^2(\mathbb{R}^d)$ has a mean value. This fact allows us to define the bilinear form $< fg >$ on the space $B^2(\mathbb{R}^d)$.

This bilinear form possesses all the properties of a scalar product, if we consider it on the classes of equivalence; and therefore $B^2(\mathbb{R}^d)$ becomes a Hilbert space.

The exponential functions $\exp(i\lambda \cdot x)$ form an orthogonal basis (although an uncountable one). Consequently, each function $f \in B^2$ can be represented as the corresponding Fourier series:

$$f = \sum f_\lambda \exp(i\lambda \cdot x), \tag{2.1}$$

where the sum is taken over $\lambda$ belonging to a countable subset of $\mathbb{R}^d$; and the series converges in the Bezikovich norm. The Parseval equality

$$\left\langle |f|^2 \right\rangle = \sum |f_\lambda|^2$$

is also valid in this case. Conversely, if $\sum |f_\lambda|^2 < \infty$, then there exists a function $f \in B^2$, which Fourier series has the form (2.1).

2. Let us return to the product

$$F(x) = f\left(\frac{x}{\xi_1}\right) f\left(\frac{x}{\xi_2}\right),$$

where $f$ is periodic, $f \in L^2(\square), \xi_1, \xi_2$ are rationally independent. The function $F$ belongs to $L^1_{loc}(\mathbb{R}^d)$ only, but one can claim that in certain sense it is an element of the space $B^2(\mathbb{R}^d)$. Really, let

$$f\left(\frac{x}{\xi_j}\right) = \sum_{k \in \mathbf{Z}^d} f_k \exp(2\pi i \frac{k}{\xi_j} \cdot x), \quad j = 1, 2,$$

be Fourier expansions. After formally multiplication of these series we get the series

$$\sum_{k,l} f_k f_l \exp\left(2\pi i (\frac{k}{\xi_1} + \frac{k}{\xi_2}) \cdot x\right) = \sum_\lambda F_\lambda \exp(i\lambda \cdot x)$$

Due to the inequality

$$\frac{k}{\xi_1} + \frac{k}{\xi_2} \neq \frac{\tilde{k}}{\xi_1} + \frac{\tilde{l}}{\xi_2} \text{ for } k \neq \tilde{k} \text{ or } l \neq \tilde{l},$$

we have

$$\sum_\lambda |F_\lambda|^2 = \sum_{k,l} |f_k f_l|^2 = \left(\sum_k |f_k|^2\right)^2.$$

Consequently, the series $\sum F_\lambda \exp(i\lambda \cdot x)$ converges in $B^2(\mathbb{R}^d)$; the corresponding sum can be considered as a product

$$F(x) = f\left(\frac{x}{\xi_1}\right) f\left(\frac{x}{\xi_2}\right),$$

moreover,

$$\left\langle (F(x))^2 \right\rangle = \left\langle f\left(\frac{x}{\xi_1}\right)\right\rangle \left\langle f\left(\frac{x}{\xi_2}\right)\right\rangle = \langle f(x)\rangle^2 .$$

Analogously, if the $f_i(x) \in L^2(\square), \xi_i$ are rationally independent, $i = 1, 2, \dots k$, then

$$\left\langle f_1\left(\frac{x}{\xi_1}\right) f_2\left(\frac{x}{\xi_2}\right) \cdots f_k\left(\frac{x}{\xi_k}\right)\right\rangle =$$
$$= \left\langle f_1\left(\frac{x}{\xi_1}\right)\right\rangle \left\langle f_2\left(\frac{x}{\xi_2}\right)\right\rangle \cdots \left\langle f_k\left(\frac{x}{\xi_k}\right)\right\rangle = \langle f_1(x)\rangle \langle f_2(x)\rangle \cdots \langle f_k(x)\rangle . \tag{2.2}$$

3. Since the initial coefficient $K_N$ is almost periodic, we must recall the definition of homogenized matrix in the almost periodic case.

Consider a vector $f \in \left(B^2(\mathbb{R}^d)\right)^d$. Then the Fourier expansion of $f$ has the form (2.2), where $f_\lambda$ is now an element of $C^d$. For $\lambda \neq 0$ the vector $f_\lambda$ can be represented as a sum of two mutually orthogonal components:

$$f_\lambda = a_\lambda + b_\lambda$$

where $a_\lambda = t\lambda, t \in C, b_\lambda \perp \lambda$. Obviously, $\exp(ix \cdot \lambda)a_\lambda$ is a potential vector and $\exp(ix \cdot \lambda)b_\lambda$ is a solenoidal one. Therefore, we have the orthogonal decomposition of $f$:

$$f =< f > + \sum_{\lambda\neq 0} a_\lambda \exp(ix \cdot \lambda) + \sum_{\lambda \neq 0} b_\lambda \exp(ix \cdot \lambda). \tag{2.3}$$

Both of the series in the right-hand side represent a vector valued function from $(B^2)^d$, since $\sum |a_\lambda|^2 + |b_\lambda|^2 = \sum |f_\lambda|^2 < \infty$.

Denote by $V^2_{pot}$ (respectively, $V^2_{sol}$) the closure in $(B^2)^d$ of the set formed by all potential (respectively, solenoidal) polynomials with zero mean value; then (2.3) implies the following orthogonal decomposition

$$(B^2)^d = V^2_{pot} \otimes V^2_{sol} \otimes \mathbb{R}^d.$$

Now we can define the homogenized matrix, namely,

$$\eta(K_N E)_{hom}\eta = \inf_{v \in V^2_{pot}} \left\langle K_N(\eta + v)^2 \right\rangle , \tag{2.4}$$

$$\eta\left((K_N E)_{hom}\right)^{-1}\eta = \inf_{v\in V^2_{sol}} \left\langle K_N^{-1}(\eta+v)^2\right\rangle, \tag{2.5}$$

From these formulas we get the following variational principles to estimate the homogenized matrix:

*If $v$ is a potential matrix (i.e. the columns belong to $V^2_{pot}$), then*

$$(K_N E)_{hom} \le \left\langle K_N(E+v)^T(E+v)\right\rangle. \tag{2.6}$$

*If $w$ is a solenoidal matrix (i.e. the columns belong to $V^2_{sol}$), then*

$$(K_N E)_{hom} \ge \left\langle K_N^{-1}(E+w)^T(E+w)\right\rangle^{-1}. \tag{2.7}$$

## 3 Proof of the upper estimate for $(K_N)_{hom}$

1. We start with some simple properties of the homogenized matrices $A_k$ and solutions of the corresponding periodic problems (1.9).

**Proposition 3.1** *For any $k = 1, 2, \dots N$ we have*

$$\nu^{-1} \le A_k \le \nu E, \tag{3.1}$$

*where $\nu > 0$ is independent on $k, N$.*

**Proof.** From condition (1.4) and variational representations (1.7), (1.8) we get

$$\begin{aligned}
&A_1 \le < a > E \le \left(1+\frac{\beta}{N}\right) E = hE \\
&A_2 \le hA_1 \le h^2 E \\
&\vdots \\
&A_k \le h^{k-1} A_{k-1} \le h^k E.
\end{aligned}$$

Since $h^k \le e^\beta$, the right estimate (3.1) is proved. The left one can be obtained from the dual representation for $A_k$, see section 4.

Now we establish some estimates for solution of periodic problem (1.9). It is convenient to consider this question in more general form, namely, for the following vector equation:

$$\nabla \cdot (aC(E + \nabla \mathbf{u}))) = 0, \quad \nu^{-1}E \leq C \leq \nu E, \quad \nu > 0. \tag{3.2}$$

As a preliminary, note the simple fact.

**Proposition 3.2** *Let $v$ be a solution of the scalar periodic problem*

$$v \in H^1_{per}(\square), \quad \nabla \cdot (aC\nabla v) = f, \quad f \in L^2(\square), \quad < f >= 0.$$

*Then*

$$\|v\|_{H^2_{per}(\square)} \leq l\|f\|_{L^2(\square)}, \tag{3.3}$$

*where $l$ is independent of $C$ and $N$.*

**Proof.** Set $X = \{f \in L^2(\square), < f >= 0\}$; then the operator

$$\nabla \cdot (C\nabla) : H^2_{per}(\square) \to X$$

has an inverse one. The operator

$$\nabla \cdot ((a-1)C\nabla) : H^2_{per} \to X$$

has a small norm because of condition (1.3). Therefore the operator $\nabla \cdot (aC\nabla) : H^2_{per} \to X$ has an inverse one and estimate (3.3) holds.

Now we study vector periodic problem (3.2).

As the first approximation to the solution $\mathbf{u}$, we take the solution of the following more simple problem:

$$\nabla \cdot (C(\nabla z + aE)) = 0. \tag{3.4}$$

We conclude from condition (1.3) that

$$\|\nabla z\|_{H^1(\square)} \leq \frac{c_1}{\sqrt{N}}.$$

Next, we have

$$aC(E + \nabla z) = C(\nabla z + aE) + (a-1)\nabla z,$$

$$\nabla \cdot (aC(\nabla z - \nabla u)) = \nabla \cdot (a-1)\nabla z = O\left(\frac{1}{N}\right) \text{ in } L^2(\square).$$

Proposition 3.2 implies that

$$\|\nabla z - \nabla u\|_{H^1(\square)} = O\left(\frac{1}{N}\right). \tag{3.5}$$

Thus, we proved the following result

**Proposition 3.3** *Let* $\mathbf{u}_k (k = 1, 2, \ldots N)$ *be solutions of periodic problems (1.9). Then*

$$\|\mathbf{u}_k\|_{H^2_{per}(\square)} \leq \frac{c_1}{\sqrt{N}}, \tag{3.6}$$

*where* $c_1$ *is independent of* $k, N$.

2. Now we proceed to a proof the upper bound for $(K_N)_{hom}$. The problem is to find a suitable almost periodic potential matrix $v(x)$ and apply the variational principle (2.6).

Let us denote by $\mathbf{u}_k(x)(k = 1, 2, \ldots N)$ the solution of (1.9),

$$\varphi_j(x) = \mathbf{u}_{N-j+1}\left(\frac{x}{\xi_j}\right), \quad \psi_j(x) = \nabla \mathbf{u}_{N-j+1}\left(\frac{x}{\xi_j}\right), \quad j = 1, 2, \ldots, N. \tag{3.7}$$

We are going to estimate $(K_N)_{hom}$ from above using a test function of the form

$$v(x) = \nabla w,$$

$$w = \sum_{k=1}^{N} \xi_k \varphi_k(x) \left[(E + \psi_{k-1}(x)) \ldots (E + \psi_1(x))\right] =$$

$$= \sum_{k=1}^{N} \xi_k \varphi_k(x) \prod_{j=k-1}^{1} (E + \psi_j).$$

As far as only the gradient enters to the formula for the homogenized matrix we can add a contact vector, assuming the normalization condition:

$$< a u_k >= 0. \tag{3.8}$$

Let us find the gradient of $w$. We have

$$E + v = E + \nabla w = Q + R,$$

where

$$Q = E + \sum_{k=1}^{N} \psi_k \prod_{j=k-1}^{1} (E + \psi_j)$$

and $R$ is the matrix with the rows

$$R = \sum_{k=1}^{N} \xi_k \varphi_k \frac{\partial}{\partial x_s} \left( \prod_{j=k-1}^{1} (E + \psi_j) \right)$$

Note that $Q$ could be presented in the form

$$Q = \prod_{j=N}^{1} (E + \psi_j) = (E + \psi_N) \dots (E + \psi_1).$$

Due to the variational principle (2.6), we have the matrix inequality:

$$\begin{aligned} (K_N E)_{hom} &\leq \left\langle K_N (E + v)^T (E + v) \right\rangle = \left\langle K_N (Q + R)^T (Q + R) \right\rangle = \\ &= \left\langle K_N Q^T Q \right\rangle + \left\langle K_N R^T R \right\rangle + \left\langle K_N (Q^T R + R^T Q) \right\rangle \end{aligned} \tag{3.9}$$

We will see later that the first term in (3.9) gives the principal contribution to the sum. Now let us calculate it explicitly. We have

$$a_i = a \left( \frac{x}{\xi_i} \right),$$

$$K_N Q^T Q = a_1 a_2 \dots a_N (E + \psi_1)^T \dots (E + \psi_N)^T (E + \psi_N) \dots (E + \psi_1).$$

By (3.7), (1.10), (2.2), we get

$$\begin{aligned} &< a_N (E + \psi_N)^T (E + \psi_N) >=< a (E + \nabla \mathbf{u}_1)^T (E + \nabla \mathbf{u}_1) >= A_1, \\ &< a_{N-1} (E + \psi_{N-1})^T A_1 (E + \psi_{N-1}) >= \\ &< a (E + \nabla \mathbf{u}_2)^T A_1 (E + \nabla \mathbf{u}_2) >= A_2, \\ &\quad \vdots \\ &\left\langle K_N Q^T Q \right\rangle =< a_1 (E + \psi_1) A_{N-1} (E + \psi_1) >= \\ &=< a (E + \nabla \mathbf{u}_{N-1})^T (E + \nabla \mathbf{u}_{N-1}) >= A_N. \end{aligned}$$

In order to estimate the other terms in (3.9) recall some properties of the matrix norms. Let, by definition, for any matrix $A = \{a_{ij}\}$

$$|A| = \sqrt{\frac{\sum\limits_{i,j=1}^{d} a_{ij}^2}{d}} = \sqrt{\frac{1}{d}\mathrm{tr}(AA^T)} = \sqrt{\frac{1}{d}\mathrm{tr}(A^TA)}.$$

By this definition we have

$$\begin{gathered} |A\xi| \leq \sqrt{d}|A||\xi|, \quad |\xi A| \leq |A||\xi| \\ \left(|\xi| = (\xi_1^2 + \dots \xi_d^2)^{1/2}\right), \end{gathered} \tag{3.10}$$

$$|AB| \leq \sqrt{d}|A||B|, \tag{3.11}$$

$$A^TA \leq d|A|^2E. \tag{3.12}$$

The chosen normalization is justified by the following property

$$\left.\begin{array}{l} \text{if matrix } A \text{ is symmetric and} \\ A^2 \geq E, \text{ then for any } B \\ |B| \leq |AB|. \end{array}\right\} \tag{3.13}$$

Now let us estimate the second and the third terms in the right hand side of (3.9). For the second term from (3.11) we have

$$\left|\left\langle K_N R^T R\right\rangle\right| \leq \left\langle K_N |R^T R|\right\rangle \leq \sqrt{d}\left\langle K_N |R|^2\right\rangle = \frac{1}{d}\sum_{s=1}^{d}\left\langle K_N |R_s|^2\right\rangle,$$

where $R_s$ is a row of the matrix $R$.

Next, we have

$$R_s = \sum_{k=1}^{N} \xi_k \varphi_k \frac{\partial}{\partial x_s}\left(\prod_{j=k-1}^{1}(E + \psi_j)\right) = \sum_{k=1}^{N} \xi_k B_k, \tag{3.14}$$

$$B_k = \varphi_k \frac{\partial}{\partial x_s}\left(\prod_{j=k-1}^{1}(E + \psi_j)\right) \tag{3.15}$$

It is easy to see that

$$\langle K_N B_k \cdot B_l\rangle = 0 \quad \text{if } k \neq l. \tag{3.16}$$

Really, if $l > k$, then the integrand in (3.16)contains only one co-factor depending on $x/\xi_e$, namely, $a_l\varphi_l$. According to (3.8), we have $< a_l\varphi_l >= 0$, and therefore orthogonality (3.16) takes place. Thus, we

$$\left\langle K_N|R_s|^2\right\rangle = \sum_{k=1}^{N} \xi_k^2 < K_N B_k^2 > . \tag{3.17}$$

Let us calculate each term in this sum. From (3.15) we have

$$B_k = \sum_{i=1}^{k-1} B_{ki}, \quad B_{ki} = \varphi_k \left[ \prod_{j=k-1}^{i+1} (E+\psi_j) \frac{\partial \psi_i}{\partial x_s} \prod_{j=i-1}^{1} (E+\psi_j) \right].$$

It is necessary to estimate the following scalar products:

$$\langle K_N B_{ki} B_{kj} \rangle, \quad i,j = 1,\dots,k-1. \tag{3.18}$$

First consider the case of $i = j$. We have

$$|B_{ki}| \le$$

$$d^{3/2}|\varphi_k| \left| \frac{\partial \psi_i}{\partial x_s} \right| |(E+\psi_{k-1})\cdots(E+\psi_{i-1})|\,|(E+\psi_{i-1})\cdots(E+\psi_1)|.$$

According to Proposition 3.1,

$$E \le \nu A_k \quad (k = 1,2,\dots,N).$$

Therefore, from property (3.3), we get

$$|(E+\psi_{k-1})\cdots(E+\psi_{i+1})| \le \nu^{1/2} \left| A_{N-k+1}^{1/2}(E+\psi_{k-1})\cdots(E+\psi_{i+1}) \right|,$$

$$|(E+\psi_{i+1})\cdots(E+\psi_1)| \le \nu^{1/2}|A_{N-i+1}^{1/2}(E+\psi_{i-1})\cdots(E+\psi_1)|.$$

Consequently,

$$\left\langle K_N|B_{ki}|^2\right\rangle \le d^3\nu^2 < a_{k+1}\dots a_N > \lambda\mu_i\, J_i||G_i| \tag{3.19}$$

where

$$
\begin{aligned}
&\lambda =< a_k\varphi_k^2 >=< au_{N-k+1}^2 >,\\
&\mu_i = \left\langle a_i \left|\frac{\partial \psi_i}{\partial x_s}\right|^2 \right\rangle = \frac{1}{\xi_i^2}\left\langle a \left|\frac{\partial}{\partial x_s}\nabla u_{N-i+1}\right|^2 \right\rangle,\\
&J_i = \left\langle a_1 a_2 \dots a_{i-1} A_{N-i+1}^{1/2}(E+\psi_{i-1})\dots(E+\psi_1)\right\rangle,\\
&G_i = \left\langle a_{i+1} \dots a_{k-1} A_{N-k+1}^{1/2}(E+\psi_{k-1})\dots(E+\psi_{i+1})\right\rangle.
\end{aligned}
$$

The quantities $J_i, G_i$ can be calculated in an explicit form. Really, we have

$$
J_i = \frac{1}{d} tr \left\langle a_1 a_2 \dots a_{i-1}(E+\psi_1)^T \dots \right.
$$
$$
\left. (E+\psi_{i-1})^T A_{N-i+1}(E+\psi_{i-1})\dots(E+\psi_1)\right\rangle.
$$

Using definition (3.7) of function $\psi_i$, by the sequential integration we get

$$
\begin{aligned}
&\left\langle (E+\psi_{i-1})^T a_{i-1} A_{N-i+1}(E+\psi_{i-1})\right\rangle =\\
&\qquad = \left\langle (E+\nabla \mathbf{u}_{N-i+2})^T a A_{N-i+1}(E+\nabla \mathbf{u}_{N-i+2})\right\rangle = A_{N-i+2}\\
&\left\langle (E+\psi_{i-2})^T a_{i-2} A_{N-i+2}(E+\psi_{i-2})\right\rangle =\\
&\qquad = \left\langle (E+\nabla \mathbf{u}_{N-i+3})^T a A_{N-i+2}(E+\nabla \mathbf{u}_{N-i+3})\right\rangle = A_{N-i+3}\\
&\vdots\\
&\left\langle (E+\psi_1)^T a_1 A_{N-1}(E+\psi_1)\right\rangle =\\
&\qquad = \left\langle (E+\nabla \mathbf{u}_N)^T a A_{N-1}(E+\nabla \mathbf{u}_N)\right\rangle = A_N.
\end{aligned}
$$

Hence

$$
J_i = \frac{1}{d} tr A_N \le \nu.
$$

In the similar way,

$$
G_i = \frac{1}{d} tr A_{N-i} \le \nu.
$$

Then from (3.19), taking into account the inequalities,

$$
\lambda \le \frac{c_1}{N}, \quad \mu_i \le \frac{c_1}{\xi_i^2 N},
$$

we get the desired estimate

$$\left\langle K_N |B_{ki}|^2 \right\rangle \le \frac{c_1}{\xi_i^2 N^2}.$$

Scalar product (3.18) for $i \neq j$ we estimate by the Cauchy inequality:

$$|\langle K_N B_{ki} \cdot B_{kj} \rangle| \le \frac{c_1}{N^2 \xi_i \xi_j}.$$

Therefore we get

$$\begin{aligned} \langle K_N B_k B_k \rangle = \left\langle K_N \sum_{i=1}^{k-1} B_{ki} \cdot \sum_{j=1}^{k-1} B_k \right\rangle = \sum_{i,j=1}^{k-1} \langle K_N B_{ki} \cdot B_{kj} \rangle \le \\ \le \frac{c_1}{N^2} \sum_{i,j=1}^{k-1} \frac{1}{\xi_i \cdot \xi_j} = \frac{c_1}{N^2} \left( \sum_{j=1}^{k-1} \frac{1}{\xi_j} \right)^2. \end{aligned}$$

Taking into account relation (3.17) and condition (1.5) we have

$$\langle K_N |R_s|^2 \rangle = \sum_{k=1}^{N} \xi_k^2 \langle K_N |B_k|^2 \rangle \le$$

$$\le \frac{c_1}{N^2} \sum_{k=1}^{N} \xi_k^2 \left( \sum_{j=1}^{k-1} \frac{1}{\xi_j} \right)^2 \le \frac{c_1}{N^2} \sum_{k=1}^{\infty} k \left( \frac{\xi_k}{\xi_{k-1}} \right)^2 \le \frac{\hat{c}_1}{N^2}.$$

At last, the term $\left\langle K_N (Q^T R + R^T Q) \right\rangle$ in sum (3.11) can be easily estimated. By the Cauchy inequality

$$Q^T R + R^T Q \le \varepsilon Q^T Q + \varepsilon^{-1} R^T R \le \varepsilon Q^T Q + \varepsilon^{-1} d^{3/2} |R|^2 E.$$

Setting $\varepsilon = \frac{1}{N}$, we obtain

$$\left| \left\langle K_N (Q^T R + R^T Q) \right\rangle \right| \le \frac{c_1}{N}.$$

Thus, we proved the inequality

$$\langle K_N \rangle_{hom} \le \left\langle K_N Q^T Q \right\rangle + O\left(\frac{1}{N}\right) = A_N + O\left(\frac{1}{N}\right).$$

## 4 Proof of lower estimate for $(K_N E)_{hom}$

1. Now we proceed to the proof of the lower bound for $(K_N E)_{hom}$. Again the problem is to construct a suitable almost periodic solenoidal matrix $w$ and apply the variational principle (2.7).

Let us consider the dual representation for the homogenized matrix $A_1 = (aE)_{hom}$,

$$\xi A_1^{-1}\xi = \inf_{p \in L^2_{sol}(\square), <p>=0} \left\langle \frac{1}{a}(\xi + p)^2 \right\rangle ,$$

where by definition

$$L^2_{sol}(\square) = \left\{ p \in \left(L^2_{per}(\square)\right)^d, < p \cdot \nabla\varphi >= 0 \quad \forall\varphi \in H^1_{per}(\square) \right\}.$$

The corresponding Euler equation has the form

$$\left\langle \frac{1}{a}(\xi + p) \cdot \varphi \right\rangle = 0 \quad \forall\varphi \in L^2_{sol}, \quad < \varphi >= 0,$$

or

$$\operatorname{rot}\left(\frac{1}{a}(\xi + p)\right) = 0, \quad p \in L^2_{sol}, \quad < p >= 0.$$

Recall that for any vector $v \in \left(L^2_{loc}(R^d)\right)^d$ the relation

$$(\operatorname{rot} v, \varphi)_{ij} = -\int_{\mathbb{R}^d} \left( v_i \frac{\partial\varphi}{\partial x_j} - v_j \frac{\partial\varphi}{\partial x_i} \right) dx, \quad \varphi \in C_0^\infty(\mathbb{R}^d),$$

define a skew-symmetric matrix $\operatorname{rot} v$, whose elements belong to $H^{-1}_{loc}(R^d)$.

Setting $p = P\xi$, for the matrix $P$ we get the following periodic problem

$$\operatorname{rot}\left(\frac{1}{a}(E + P)\right) = 0, \quad \nabla \cdot P = 0, \quad < P >= 0,$$

moreover,

$$A_1^{-1} = \left\langle \frac{1}{a}(E + P)^T(E + P) \right\rangle .$$

We consider also the dual representation for matrices $A_k$:

$$\xi A_k^{-1}\xi = \inf_{p\in L^2_{sol}(\Box),<p>=0} \left\langle (\xi+p)\frac{1}{a}A_{k-1}^{-1}(\xi+p)\right\rangle, \quad k=2,\dots,N.$$

The corresponding matrix Euler equation has the form

$$\begin{cases} \operatorname{rot}\left(\frac{1}{a}A_{k-1}^{-1}(E+P_k)\right)=0, \\ \nabla\cdot P_k=0, \langle P_k\rangle=0, \end{cases} \tag{4.1}$$

moreover,

$$A_k^{-1}=\left\langle \frac{1}{a}(E+P_k)^T A_{k-1}^{-1}(E+P_k)\right\rangle.$$

It is easy to check that

$$\|P_k\|_{H^1(\Box)} \le \frac{c_1}{\sqrt{N}}.$$

It fact, the matrix $P_k$ can be expressed by the solution $\mathbf{u}_k$ of problem (1.9), namely,

$$E+P_k=a(E+\nabla u_k)A_k^{-1}$$

and therefore (see Proposition 3.3)

$$P_k=a(E+\nabla u_k)A_k^{-1}-E=O\left(\frac{1}{\sqrt{N}}\right) \text{ in } H^1(\Box).$$

Let us recall the classical result about the representation of a solenoidal periodic vector as the divergence of a skew-symmetric matrix (in the three-dimensional case – as an image of the curl operator).

**Proposition 4.1** *Every vector* $v\in L^2_{sol}(\Box),<v>=0$, *admits the representation*

$$v_j=\frac{\partial}{\partial x_i}\alpha_{ij}, \quad \alpha_{ij}=\alpha_{ji}, \quad \alpha_{ij}\in H^1_{per}(\Box), \quad <\alpha_{ij}>=0,$$

$$\|\alpha_{ij}\|_{H^1_{per}(\Box)} \le c_0\|v\|_{L^2(\Box)}.$$

**Proof.** Let

$$v = \sum_{k \neq 0} v^k \exp\left(i \frac{k \cdot x}{2\pi}\right)$$

be a Fourier series expansion. Since $v^k \exp\left(i\frac{k \cdot x}{2\pi}\right)$ is a solenoidal vector, $v^k \perp k$, and therefore the vector $v^k$ can be written in the form

$$v_j^k = \alpha_{ij}^k, \qquad \text{where } \alpha_{ij}^k = \frac{v_j^k k_i - v_i^k k_j}{|k|^2}.$$

Now the required matrix $\{\alpha_{ij}(x)\}$ can be defined by the formula

$$\alpha_{ij}(x) = -2\pi i \sum_{k \neq 0} \alpha_{ij}^k \exp\left(i \frac{k \cdot x}{2\pi}\right)$$

and proposition 4.1 is proved.

The inverse statement is quite trivial.

**Proposition 4.2** *If* $\alpha = \{\alpha_{ij}\}$ *in* $\left(H_{per}^1(\Box)\right)^{d^2}$, $\alpha_{ij} = -\alpha_{ji}$, *then* $\nabla \cdot (\nabla \cdot \alpha) = 0$.

**Proof.** We have by definition $(\nabla \cdot \alpha)_j = \frac{\partial}{\partial x_i}\alpha_{ij}$

$$\nabla \cdot (\nabla \cdot \alpha) = \frac{\partial}{\partial x_j}\left(\frac{\partial}{\partial x_i}\alpha_{ij}\right) = \frac{\partial^2 \alpha_{ij}}{\partial x_j \partial x_i} - \frac{\partial^2 \alpha_{ij}}{\partial x_j \partial x_i} = 0.$$

and Proposition 4.2 is proved.

Now, instead of the solenoidal vector $v$, we consider the solenoidal matrix

$$v = \{v_{ij}\}, \quad \frac{\partial}{\partial x_i} v_{ij} = 0, \quad < v >= 0.$$

This matrix admits the representation

$$v_{ij} = \frac{\partial}{\partial x_k} b_{kij}, \quad b_{kij} = -b_{ikj}, \quad b_{kij} = H_{per}^1(\Box),$$

or, in short, $v = \nabla \cdot B$.

Let us define action of such tensor $B = \{b_{kij}\}$ on an arbitrary matrix $A = \{a_{ij}\}$ by the formula

$$BA = D = \{D_{kis}\}, \quad D_{kis} = b_{kij} a_{js}.$$

It is clear that this action preserve the property of skew-symmetry, i.e.

$$D_{kis} = -D_{iks}. \tag{4.2}$$

Next, we have

$$\nabla \cdot (BA) = \tfrac{\partial}{\partial x_k} D_{kis} = \tfrac{\partial}{\partial x_k}(b_{kij}a_{js}) =$$
$$\tfrac{\partial}{\partial x_k}(b_{kij})a_{js} + b_{kij}\tfrac{\partial}{\partial x_k}a_{js} = \nabla \cdot (B)A + B \times \nabla A, \tag{4.3}$$

where by definition

$$(B \times \nabla A)_{is} = b_{kij}\frac{\partial}{\partial x_k}a_{is}.$$

In what follows only it is important the estimate

$$|B \times \nabla A|^2 \leq l|B|^2 \sum_{s=1}^{d} \left|\frac{\partial}{\partial x_s}A\right|^2, \quad |B| = \frac{1}{d}b_{kij}b_{kji}. \tag{4.4}$$

At last the key equality

$$\nabla \cdot (BA) = 0$$

follows from (4.2).

2. Now we are in a position to construct the suitable solenoidal matrix $w$.

Consider the solution $P_k$ of periodic problem (4.1). The matrix $P_k$ admits the representation as the divergence of a tensor $B_k$ satisfies above mentioned property of skew-symmetry:

$$P_k = \nabla \cdot B_k, \quad \left\langle \frac{B_k}{a} \right\rangle = 0.$$

Now we set

$$\varphi_k(x) = B_{N-k+1}\left\langle \frac{x}{\xi_k} \right\rangle$$
$$\psi_k(x) = P_{N-k+1}\left\langle \frac{x}{\xi_k} \right\rangle, \quad k = 1, 2, . \ ., N,$$
$$Z(x) = \sum_{k=1}^{N} \xi_k \varphi_k \prod_{j=k-1}^{1} (E + \varphi_j).$$

Define the required solenoidal test-matrix $w$ by

$$w = \nabla \cdot (Z).$$

We get

$$E + w = \prod_{j=N}^{1} (E + \psi_j) + R = Q + R,$$

$$R = \sum_{k=1}^{N} \xi_k \varphi_k \times \nabla \left( \prod_{j=k-1}^{1} (E + \psi_j) \right) = \sum_{k=1}^{N} \xi_k R_k, \qquad (4.5)$$

where

$$R_k = \varphi_k \times \nabla \left[ \prod_{j=k-1}^{1} (E + \psi_j) \right]. \qquad (4.6)$$

As above, we have

$$\left\langle K_N^{-1} (E + w)^T (E + w) \right\rangle =$$
$$=< K_N^{-1} Q^T Q > + < K_N^{-1} R^T R > + \left\langle K_N^{-1} (Q^T R + R^T Q) \right\rangle \quad (4.7)$$

and it is necessary to calculate the main term $< K_N^{-1} Q^T Q >$ and estimate the remainders.

We calculate the main term

$$< K_N^{-1} Q^T Q^T >=$$
$$\left\langle \frac{1}{a_1 a_2 \ldots a_N} N (E + \psi_1)^T \ldots (E + \psi_N)^T (E + \psi_N)(E + \psi_1) \right\rangle$$

by sequential integration over $\frac{x}{\xi_N}, \frac{x}{\xi_{N-1}}, \ldots, \frac{x}{\xi_1}$. As a result, we obtain the exact equality:

$$< K_{-1} Q^T Q >= A_N^{-1}.$$

Let us show how the term $< K_N^{-1} R^T R >$ can be estimated. We have

$$< K_N^{-1} R^T R >= \sum_{k,l=1}^{N} \xi_k \xi_l < K_N^{-1} R_k^T R_l >= \sum_{k=1}^{N} \xi_k^2 < K_N^{-1} R_k^T R_k >, \tag{4.8}$$

because

$$< K_N^{-1} R_k^T R_l >= 0 \text{ for } k \neq l.$$

Let us estimate each term in sum (4.8). From inequalities (3.11), (4.4) we get

$$\left| < K_N^{-1} R_k^T R_k > \right| \leq \sqrt{d} \left\langle K_N^{-1} |R_k|^2 \right\rangle , \tag{4.9}$$

$$|R_k|^2 = \left| \varphi_k \times \nabla( \prod_{j=k-1}^{1} (E + \psi_j)) \right|^2 \leq l |\varphi_k|^2 \sum_{s=1}^{d} \left| \frac{\partial}{\partial x_s} \prod_{j=k-1}^{1} (E + \psi_j) \right|^2 \leq$$
$$\leq l d^3 \varphi_k^2 \sum_{s=1}^{d} \left[ \sum_{i=1}^{k-1} \left| \frac{\partial \psi_i}{\partial x_s} \right| \left| \prod_{j=k-1}^{i+1} (E + \psi_j) \right| \left| \prod_{j=k-1}^{1} (E + \psi_j) \right| \right]^2 . \tag{4.10}$$

Since (see Proposition 3.1)

$$E \leq \nu A_k^{-1},$$

by property (3.13) we get

$$|(E + \psi_{k-1}) \ldots (E_i + \psi_{j+1})| \leq \nu^{\frac{1}{2}} \left| A_{N-k+1}^{-\frac{1}{2}} (E + \psi_{k-1}) \ldots (E + \psi_{i+1}) \right| ,$$

$$|(E + \psi_{i-1}) \ldots (E + \psi_{+}1)| \leq \nu^{\frac{1}{2}} \left| A_{N-i+1}^{-\frac{1}{2}} (E + \psi_{i-1}) \ldots (E + \psi_{i+1}) \right| .$$

From here and (4.9), (4.10) by inequality Cauchy we obtain

$$\left\langle K_N^{-1} |R_k|^2 \right\rangle \leq c_1 \lambda \left\langle \frac{1}{a} \right\rangle^{N-k} \left( \sum_{i=1}^{k-1} \mu_i^{\frac{1}{2}} J_i^{\frac{1}{2}} G_i^{\frac{1}{2}} \right)^2 , \tag{4.11}$$

where

$$\lambda = \left\langle a_k^{-1}|\varphi_k|^2\right\rangle = \left\langle a^{-1}|B_{N-k+1}|^2\right\rangle,$$

$$\mu_i = \sum_{s=1}^{d}\left\langle \frac{1}{a_i}\left|\frac{\partial\psi_i}{\partial x_s}\right|^2\right\rangle = \frac{1}{\xi_i^2}\sum_{s=1}^{d}\left\langle \frac{1}{a}\left|\frac{\partial P_{N-i+1}}{\partial x_s}\right|^2\right\rangle,$$

$$J_i = \left\langle \frac{1}{a_1 a_2 \dots a_{i-1}} A_{N-i+1}^{-\frac{1}{2}}(E+\psi_{i-1})\dots(E+\psi_1)\right\rangle,$$

$$G_i = \left\langle \frac{1}{a_{i+1}\dots a_{k-1}} A_{N-k+1}^{-\frac{1}{2}}(E+\psi_{k-1})\dots(E+\psi_1)\right\rangle.$$

The quantities $J_i, G_i$ can be calculated in explicit form. For instance,

$$J_i = \frac{1}{d}tr\left\langle \frac{1}{a_1\dots a_{i-1}}(E+\psi_1)^T\dots(E+\psi_{i-1})^T A_{N-i+1}^{-1}(E+\psi_{i-1})\dots(E+\psi_1)\right\rangle,$$

and the sequential integration gives:

$$\left\langle (E+\psi_{i-1})\frac{1}{a_{i-1}}A_{N-i+1}^{-1}(E+\psi_{i-1})\right\rangle =$$
$$= \left\langle (E+P_{N-i+2})^T\frac{1}{a}A_{N-i+1}^{-1}(E+P_{N-i+2})\right\rangle = A_{N-i+2}^{-1},$$
$$\left\langle (E+\psi_{i-2})\frac{1}{a_{i-2}}A_{N-i+2}^{-1}(E+\psi_{i-2})\right\rangle =$$
$$= \left\langle (E+P_{N-i+3})^T\frac{1}{a}A_{N-i+2}^{-1}(E+P_{N-i+2})\right\rangle = A_{N-i+2}^{-1},$$

$$\vdots$$

$$\left\langle (E+\psi_1)\frac{1}{a_1}A_{N-1}^{-1}(E+\psi_1)\right\rangle = \left\langle (E+P_N)^T\frac{1}{a}A_{N-1}^{-1}(E+P_N)\right\rangle = A_N^{-1}.$$

Hence,

$$J_i = \frac{1}{d}trA_N^{-1} \le \nu.$$

In the similar way,

$$G_i = \frac{1}{d}trA_{N-i}^{-1} \le \nu.$$

Then, taking into account the inequalities

$$\lambda \leq \frac{c_1}{N} \quad \mu_i \leq \frac{c_1}{\xi_i^2 N}$$

we get from (4.11) and (4.8):

$$\left\langle K_N^{-1}|R_k|^2\right\rangle \leq \frac{c_1}{N^2}\left(\sum_{i=1}^{k-1}\frac{1}{\xi_i}\right)^2 \leq \frac{c_2 k}{N^2\xi_{k-1}^2}$$
$$\left\langle K_N^{-1}R^T R\right\rangle \leq \frac{c_3}{N^2}\left(\sum_{k=1}^{N} k\frac{\xi_k}{\xi_{k-1}}\right)^2 \leq \frac{\hat{c}_4}{N^2}$$

At last, the term $\left\langle K_N(Q^T R + R^T Q)\right\rangle$ in sum (4.7) can be easily estimated and we obtain the desired inequality

$$\left\langle K_N^{-1}(E+w)^T(E+w)\right\rangle \leq A_N^{-1} + O\left(\frac{1}{N}\right)$$

and Lemma 1.1 is proved.

## 5 Proof of the main result. Examples

1. It remains to prove Lemma 1.2 in order to get the main result (Theorem 1.1).

From additional condition (1.11) we get

$$\begin{cases} A_0 = E \\ A_1 = E + \dfrac{M(E)}{N} + o\left(\frac{1}{N}\right) \\ A_2 = A_1 + \dfrac{M(A_1)}{N} + o\left(\frac{1}{N}\right) \\ \vdots \\ A_N = A_{N-1} + \dfrac{M(A_{N-1})}{N} + o\left(\frac{1}{N}\right). \end{cases}$$

These relations can be written as follows:

$$\frac{A_{k+1} - A_k}{\frac{1}{N}} = M(A_k) + o\left(\frac{1}{N}\right), \quad k = 0, 1, \ldots, N-1. \tag{5.1}$$

Let us divide the segment $[0,\ 1]$ on $N$ subsegment by point $t = \frac{k}{N}$, $k = 0, 1, \ldots, N$. Then relation (5.1) gives the Euler procedure for solving of equation (1.12). In fact, if we define a continuous function $A_N(t)$ $(t \in [0, 1])$ which is equal to $A_k$ for $t = \frac{k}{N}$ and linear on each segments $\left[\frac{k}{N}, \frac{k+1}{N}\right]$, $k = 0, 1, \ldots, N-1$, then we can easily check that

$$\lim_{N\to\infty} \sup_{0\le t\le 1} |A_N(t) - A(t)| = 0,$$

where $A(t)$ is the solution of matrix Cauchy problem (1.12). Lemma 1.2 is proved and thereby Main Theorem 1.1 is proved, too.

2. Now we study in detail the case when the coefficient $a = a^N$ has the form (1.13). Our goal is show that additional condition (1.11) holds here. We consider periodic problem (3.2) because it is a matrix Euler equation for the variational problem

$$(aC)_{hom} = \inf_{u\in H^1_{per}(\Box)} \langle(\xi + \nabla u)aC(\xi + \nabla u)\rangle .$$

We have already got some results about problem (3.2) in section 3. Now we make more precise it. In the present case equation (3.4) for $z$ takes the form

$$\nabla \cdot C(\nabla z + \frac{f}{\sqrt{N}}E) = 0$$

and therefore

$$z = \frac{\chi}{\sqrt{N}},$$

where $\chi$ is the solution of periodic problem

$$\nabla \cdot C(\nabla \chi + fE) = 0. \tag{5.2}$$

From (1.13) for $a^N$ and estimate (3.5) we have

$$
\begin{aligned}
(aC)_{hom} &= \langle aC(E+\nabla u)\rangle = \langle aC(E+\nabla z)\rangle + O\left(\frac{1}{N^{3/2}}\right) = \\
&= \left\langle \left(1+\frac{f}{N^{1/2}}+\frac{g}{N}+O\left(\frac{1}{N^{3/2}}\right)\right) C\left(E+\frac{\nabla\chi}{N^{1/2}}\right)\right\rangle + O\left(\frac{1}{N^{3/2}}\right) = \\
&= C + \frac{C}{N} < g > + \frac{C < f\nabla\chi >}{N} + O\left(\frac{1}{N^{3/2}}\right)
\end{aligned} \tag{5.3}
$$

and therefore

$$
M(C) = < g > C + C < f\nabla\chi > \tag{5.4}
$$

Let us analyze equation (5.2) for $\chi$. From the Fourier series expansion

$$
f = \sum_{k\in Z^d} f_k \exp(2\pi i k\cdot x)
$$

we easily find

$$
\frac{\partial\chi_i}{\partial x_j} = -\sum_k \frac{1}{2\pi}\frac{(e^i\cdot Ck)k_j}{k\cdot Ck} f_k \exp(2\pi i k\cdot x),
$$

$$
(C < f\nabla x >)_{ij} = -\sum_k \frac{(e^i\cdot Ck)(e^j\cdot Ck)}{k\cdot Ck}|f_k|^2. \tag{5.5}
$$

From here we see that, if $f$ is cubic symmetric and $C = E$, then

$$
< f\nabla\chi > = -\frac{1}{d} < f^2 >,
$$

$$
M(E) = \left[< g > -\frac{1}{d} < f^2 >\right] E.
$$

One can check that the operator $M(C)$ satisfies the Lipschitz condition.

Thereby theorem 1.5 is proved.

It is clear from (5.5) that, if $f$ is even, then $M(C)$ is diagonal for any diagonal $C$. For the diagonal matrix

$$
C = \operatorname{diag}(\lambda_1, \lambda_2, \dots, \lambda_d)
$$

we have

$$M(C) = \mathrm{diag}(\mu_1, \mu_2, \ldots, \mu_d),$$

$$\mu_i = \lambda_i < g > - \sum_{k \in Z^d} \frac{1}{2\pi} \frac{\lambda_i 2k_i^2}{\lambda_1 k_1^2 + \lambda_2 k_2^2 + \ldots + \lambda_d k_d^2} |f_k|^2.$$

Let us consider some simple examples in which $f$ is not cubic symmetrical.

**Example 1.** Let $d = 2$, $f(x) = \cos \frac{x_1}{2\pi}$, $g = \frac{f^2}{2}$. Then

$$M(C) = \frac{1}{4} C - \frac{1}{4\pi} \begin{pmatrix} c_{11} & 0 \\ 0 & 0 \end{pmatrix}$$

Matrix equation (1.12) is linear in this case.

**Example 2.** Let $d = 2$, $f(x) = \cos(x_1 + x_2)$, $g = \frac{f^2}{2}$. Then

$$M(C) = \frac{1}{4} C - \frac{1}{4\pi(c_{11} + 2c_{12} + c_{22})}$$

$$\begin{pmatrix} (c_{11} + c_{12})^2 & (c_{11} + c_{12})(c_{11} + c_{22}) \\ (c_{11} + c_{12})(c_{12} + c_{22}) & (c_{12} + c_{22})^2 \end{pmatrix}$$

Matrix equation (1.12) has no diagonal solutions.

**Example 3.** Let $d = 2$, $f = \cos(2x_1 + x_2) + \cos(2x_1 - x_2)$, $g = \frac{f^2}{2}$.

The function $f$ is even and we can consider diagonal matrices $C = \mathrm{diag}(\lambda_1, \lambda_2)$ only. In this case we have

$$\begin{cases} \lambda_1 = \frac{1}{2}\lambda_1 - \frac{2\lambda_1^2}{\pi^2(4\lambda_1 + \lambda_2)} \\ \lambda_2 = \frac{1}{2}\lambda_2 - \frac{\lambda_2^2}{2\pi(4\lambda_1 + \lambda_2)} \end{cases}$$

This matrix equation has no isotropic solutions.

3. Consider a “soft” problem when the coefficient $a_N$ has the form (1.14). We restrict it to the case of

$$a(x) = a^N(x) = \begin{cases} 0 & \text{if } x \in \frac{B}{\sqrt{N}} \\ 1 & \text{if } x \in \square \setminus \frac{B}{\sqrt{N}}, \end{cases}$$

where $B$ is the closure of a piecewise smooth domain in $\mathbb{R}^d$.

The homogenized matrix $(aC)_{hom}$ is defined by the same formula

$$\xi(aC)_{hom}\xi = \inf_{u\in H^1_{per}(\square)} \langle a(\xi+\nabla u)C(\xi+\nabla u)\rangle$$

and the representation of Maxwell-Rayleigh

$$(aC)_{hom} = C + \frac{M(C)}{N} + o\left(\frac{1}{N}\right)$$

holds. One can check that main Theorem 1.1 remains true.
**Example 4.** Let $B$ be a ball. Then

$$M(E) = -\frac{|B|d}{d-1}.$$

Consequently, in this case

$$\lim_{N\to\infty}(K_N E)_{hom} = \exp(-|B|\frac{d}{d-1})E.$$

Let us calculate a concentration of pores in the media:

$$c = 1 - \lim_{N\to\infty}\left\langle a^N\right\rangle^N = 1 - \lim_{N\to\infty}\left(1-\frac{|B|}{N}\right)^N = 1 - e^{-|B|}$$

and thus

$$\lim_{N\to\infty}(K_N E)_{hom} = (1-c)^{\frac{d}{d-1}} \quad \text{(the law Bruggemann [2])}.$$

**Example 5.** Let $B$ be an ellipsoid with center at the origin and principal axes $2b_1, 2b_2, \ldots, 2b_d$ which are parallel to the coordinate ones. In this case $M(C)$ is diagonal for any diagonal $C = \operatorname{diag}(\lambda_1, \ldots, \lambda_d)$. We have

$$M(C) = C\,\operatorname{diag}(\mu_1, \ldots, \mu_d), \quad \mu_i = \frac{|B|}{1-\delta_i},$$

$$\delta_i = \frac{1}{2\omega_d|detC|^{\frac{1}{2}}}\int_0^\infty \frac{ds}{\left(s+\frac{b_i^2}{\lambda_i}\right)\sqrt{\left(s+\frac{b_1^2}{\lambda_1}\right)\left(s+\frac{b_2^2}{\lambda_2}\right)\cdots\left(s+\frac{b_d^2}{\lambda_d}\right)}},$$

$\omega_d$ is the volume of a unit ball in $\mathbb{R}^d$.

## References


1. V. Jikov, S. Kozlov and O. Oleinik, Homogenization of Differential Equations and Variational Functionals, Springer-Verlag, Berlin, 1994.
2. D.A.G. Bruggemann, Berechung verschiedener physikalischer Konstanten von heterogenen Substanten, *Ann. Physik*, **24**, 633–652 (1935).
3. A. Bensoussan, J.L. Lions and G. Papanicolaou, Asymptotic analysis for periodic structure, North Holland, Amsterdam, 1978.
4. M. Avellaneda, Iterated homogenization, differential effective medium theory and applications, *Comm. Pure Appl. Math.*, **XL**, 527–554 (1987).
5. S. Kozlov, Multiscaled approach in homogenization. *Proceeding of the Second Workshop on Composite Media and Homogenization Theory*, ICTP, Trieste, Italy, September 20 - October 1, 1993. World Scientific, 217–229 (1995).

# THE EFFECTIVE THERMOCONDUCTIVITY AND SHEAR MODULUS OF A LATTICE STRUCTURE: AN ASYMPTOTIC ANALYSIS[1]

S.M. KOZLOV
G.P. PANASENKO
*Equipe d'Analyse Numerique Lyon - Saint Etienne, CNRS UMR 5585, Universite Jean Monnet, 23, rue P. Michelon, 42023, Saint Etienne, France*

Herein we discuss the effective thermoconductivity and effective shear modulus of the rectangular (quadratic) lattice structure with the width of the bond equal to a small parameter. We give the asymptotic expansion of the effective thermoconductivity and effective shear modulus which are determined in terms of homogenization theory. First order approximation for the effective thermoconductivity is well known [1,2]. It follows from the principle of splitting of the homogenized operator. The main goal of the first part of this work is to find the correctors to this approximation. We give an explicit form of the second order approximation and of the complete asymptotic expansion. It corresponds to the influence of the orthogonal strips on the effective coefficient. In the second section we calculate the effective shear modulus of the same structure. Even the first order approximation was unknown. We obtain here this approximation as well as the complete asymptotic expansion.

## 1 Introduction

Lattice structures have been widely adopted in engineering today (trusses of bridges and industrial installations, frameworks of housing, supports of electric power lines, etc.). These structures consist of a system of connected rods (strips in two–dimensional case). The mathematically rigorous modeling of frameworks depending on two small parameters started in [11] and was then developed in [1,3-7,12-14]. These papers consider a periodic framework as a domain with two (or more) small parameters: $\varepsilon$ (the ratio of a periodicity size to the characteristic

[1]The first part of the work was done during the stay of S.M. Kozlov at the Laboratory of the Numerical Analysis of the University Paris–6. At that time S.M. Kozlov and G.P. Panasenko were affiliated with Moscow Institute of Civil Engineering and the School of Mechanics and Mathematics of Moscow State University, correspondingly. The second part of this work has been done during the stay of G.P. Panasenko at the University of Provence.

size of a problem) and $\delta$ (the ratio of the diameter of a rod to the period of structure). Such framework is called a lattice structure.

For example, the simplest lattice structure (the so – called two – dimensional rectangular lattice structure)

$$B_{\varepsilon,\delta} = \bigcup_{k=-\infty}^{+\infty} \left(\{|x_1 - k\varepsilon| < \varepsilon\delta/2\} \cup \{|x_2 - k\varepsilon| < \varepsilon\delta/2\}\right)$$

consists of two periodic systems of thin strips oriented in the coordinate directions. As small parameters, we take the constant step $\varepsilon$ of the strips (i.e. the distance between the axes of the neighboring strips), and the ratio $\delta$ of the width of the strip to the periodic step $\varepsilon$.

Partial differential equations posed in these domains have been homogenized. The principle of splitting of homogenized operator [11] gives the opportunity to write an explicit formula for the first term of the asymptotics of this operator.

The typical result on asymptotic analysis of lattice structures is as follows.

**Theorem 1.1** *([1]) Let $G$ be a bounded domain in $R^2$ with a sufficiently regular boundary. Consider the problem*

$$\Delta v_{\varepsilon,\delta} = f(x), \quad x \in G \cap B_{\varepsilon,\delta}; \quad \frac{\partial v_{\varepsilon,\delta}}{\partial n} = 0, \quad x \in G \cap \partial B_{\varepsilon,\delta};$$

$$v_{\varepsilon,\delta} = 0, \quad x \in G \cap \partial \bar{B}_{\varepsilon,\delta}, \ \text{where } f \in C^1\left(\bar{G}\right). \tag{1.1}$$

*Let $v_0$ be the solution of the so–called homogenized problem*

$$\Delta v_0 = 2f(x), \quad x \in G; \quad v_0 = 0, \quad x \in \partial G. \tag{1.2}$$

*Then the estimate holds:*

$$\frac{\|v_{\varepsilon,\delta} - v_0\|_{L_2\left(G \cap B_{\varepsilon,\delta}\right)}}{\sqrt{\operatorname{meas}\left(G \cap B_{\varepsilon,\delta}\right)}} \le C\left(\sqrt{\delta} + \sqrt{\varepsilon}\right), \tag{1.3}$$

*where the constant $C$ does not depend on the small parameters.*

This type of convergence of the solution $v_{\varepsilon,\delta}$ of the initial problem to the solution of homogenized problem $v_0$ is called $L$-convergence([15]). The normalization factor $\frac{1}{\sqrt{\operatorname{meas}(G \cap B_{\varepsilon,\delta})}}$ here is important because the measure of the intersection $G \cap B_{\varepsilon,\delta}$ is small; it is equal to $2\delta - \delta^2$.

Note that the homogenized equation can be multiplied by $\delta$ and it takes a form

$$\delta \Delta v_0 = 2\delta f(x), \quad x \in G, \quad v_0 = 0, \quad x \in \partial G. \tag{1.4}$$

On the other hand we can consider the problem (1.1) with the finite $\delta$ (i.e. $\delta$ has a magnitude 1). Such problems can be studied by a classical homogenization method (as the unique small parameter $\varepsilon$ tends to zero). The typical result in this case is as follows ([1]).

**Theorem 1.2** *([1]) Let $G$ be a bounded domain in $R^2$ with a sufficiently regular boundary. Consider the problem (1.1) with $f \in C^1\left(\bar{G}\right)$. Let $B$ be the dilated lattice structure $B_{\varepsilon,\delta}$*

$$B = \left\{\xi \in R^2, \xi \in \bigcup_{k=-\infty}^{+\infty} \left(\{|\xi_1 - k| < t/2\} \cup \{|\xi_2 - k| < t/2\}\right)\right\}.$$

*Let $U$ be a solution of the so-called cell problem*

$$\Delta U = 0, \quad \xi = B; \quad \frac{\partial U}{\partial n} = 0, \quad \xi \in \partial B \tag{1.5}$$

*with $U - \xi_1$ being periodic in both variables $\xi_1, \xi_2$ and let $\hat{K}$ be the so-called effective conductivity:*

$$\hat{K} = \left\langle \frac{\partial}{\partial \xi_1} U \right\rangle = \int_Q \frac{\partial U}{\partial \xi_1} d\xi_1 d\xi_2 \tag{1.6}$$

*where*

$$Q = B \cap (-1/2, 1/2)^2.$$

*Let $w_0$ be the solution of the so-called homogenized problem*

$$\hat{K} \Delta w_0 = meas\,(B \cap Q)\, f(x), \quad x \in G; \quad w_0 = 0, \quad x \in \partial G. \tag{1.7}$$

*Then the estimate holds:*

$$\|v_{\varepsilon,\delta} - w_0\|_{L_2\left(G \cap B_{\varepsilon,\delta}\right)} \le C\sqrt{\varepsilon}, \tag{1.8}$$

*where the constant $C$ does not depend on the small parameter.*

The asymptotic analysis of the problem (1.5) as $\delta$ tends to zero gives the formula (see [1, 12]):

$$\hat{K} = \delta + O\left(\delta^2\right),$$

i.e. the first term of the asymptotic expansion of the effective coefficient is described by the coefficient of the equation (1.4). This equation is an asymptotic approximation of the first order for the equation (1.7).

Below we seek the following terms of the asymptotic expansion of the coefficients of the homogenized operator for the two–dimensional rectangular lattice structure. It corresponds to the contribution of the common parts of the orthogonal strips of the structure.

Now we formulate the main results. We prove the following theorem:

**Theorem 2.1.** *[1] For $\hat{K}$ the following asymptotic expansion holds*

$$\hat{K} = \delta + k_g\delta^2 + O\left(\delta^3\right)$$

*where*

$$k_g = 4 \int\limits_0^{\frac{1}{\pi}a\sinh 1} x\sqrt{cosech^2\pi x - 1}dx.$$

Moreover, the complete asymptotic expansion is given by the formula (Theorem 2.2): for any natural $n$

$$\hat{K} = \delta + \frac{k_g\delta^2}{1 + 2c_1\delta} + O\left(\delta^n\right),$$

where

$$c_1 = \frac{1}{4} - \frac{1}{2\pi}\lg 2.$$

Analogous problem is related to the system of equations of elasticity defined on the lattice structure $B_{\varepsilon,\delta}$. The effective tensor of elasticity moduli is defined in [1]. The principle of splitting gives the opportunity to write an explicit formula for the terms of order $\delta$ of the effective tensor of elasticity moduli. Nevertheless some of these moduli have a magnitude $o\left(\delta\right)$ and they are not described by the principle of splitting. For example this situation takes place for shear modulus.

In the second section we calculate the effective shear modulus of the same structure. Even the first order approximation was unknown.

We obtain here this approximation as well as the complete asymptotic expansion.

Let $L$ be the two–dimensional operator of isotropic elasticity with constant coefficients:

$$L = \frac{\partial}{\partial x_i}\left(A_{ij}\frac{\partial}{\partial x_j}\right),$$

where

$$A_{11} = \begin{pmatrix} \lambda + 2\mu & 0 \\ 0 & \mu \end{pmatrix}, \quad A_{12} = \begin{pmatrix} 0 & \lambda \\ \mu & 0 \end{pmatrix},$$

$$A_{21} = \begin{pmatrix} 0 & \mu \\ \lambda & 0 \end{pmatrix}, \quad A_{22} = \begin{pmatrix} \mu & 0 \\ 0 & \lambda + 2\mu \end{pmatrix},$$

$\lambda, \mu$ are positive Lame's constants.

Consider the homogenized operator

$$\bar{L} = \frac{\partial}{\partial x_i}\left(\bar{A}_{ij}\frac{\partial}{\partial x_j}\right),$$

corresponding to the homogeneous Neumann boundary condition at the boundary $\partial B$ of a lattice domain $B$. The coefficients $\bar{A}_{ij}$ of the homogenized operator $\bar{L}$ are defined on the base of solutions of the corresponding cell problems (c.f. [1]). In particular, the effective shear modulus of a lattice structure could be determined as an element $\bar{a}_{11}^{22}$ of the matrix $\bar{A}_{11}$, equal to

$$\bar{\mu} = 2\mu \left\langle e\left(U\right)\right\rangle, \tag{1.9}$$

where $U\left(\xi\right)$ is a $2D$ vector-valued function $\left(u_1\left(\xi\right), u_2\left(\xi\right)\right)^T$ ($T$ is the transposition symbol), which satisfies a system of equalities

$$LU = 0, \quad \xi \in B; \quad \frac{\partial U}{\partial \nu} = n_i A_{ij}\frac{\partial U}{\partial \xi_i} = 0, \quad \xi \in \partial B, \tag{1.10}$$

and

$$U - \begin{pmatrix} 0 \\ \xi_1 \end{pmatrix}$$

is a 1-periodic vector-function of the variables $\xi_1$ and $\xi_2$; $(n_1, n_2)$ is the outside normal to $\partial B$, i.e. $n_i \in \{-1, 0, 1\}$;

$$e\left(U\right) = \frac{1}{2}\left(\frac{\partial U_1}{\partial x_2} + \frac{\partial U_2}{\partial x_1}\right)$$

is a shear strain.

**Theorem. 3.1** *Let* $\Pi$ *be an infinite cross*

$$\Pi = \left\{ z \in R^2,\ |z_1| < 1/2 \quad or \quad |z_2| < 1/2 \right\},$$

$$L_z = \frac{\partial}{\partial z_i}\left(A_{ij}\frac{\partial}{\partial z_j}\right), \quad \frac{\partial}{\partial \nu_z} = n_i\left(A_{ij}\frac{\partial}{\partial z_j}\right);$$

*let* $u(z)$ *be a 2D vector-valued function, which is the solution of the homogeneous elasticity problem*

$$L_z u = 0, \quad z \in \Pi; \quad \frac{\partial u}{\partial \nu_z} = 0, \quad z \in \partial\Pi$$

*with the following asymptotic behavior at the infinity*

$$\pm 3 \begin{pmatrix} -z_1 z_2 \\ \frac{z_1^2}{2} + \frac{\lambda z_2^2}{2(\lambda+2\mu)} \end{pmatrix} + r_1^{\pm}(z) + o\left(z_1^{-1}\right), \quad z_1 \to \pm\infty,$$

$$\pm 3 \begin{pmatrix} \frac{z_2^2}{2} + \frac{\lambda z_1^2}{2(\lambda+2\mu)} \\ -z_1 z_2 \end{pmatrix} + r_2^{\pm}(z) + o\left(z_2^{-1}\right), \quad z_2 \to \pm\infty,$$

*where* $r_q^{\pm}$ *are rigid displacements, i.e.*

$$r_q^{\pm}(z) = \begin{pmatrix} h_q^{\pm} \\ p_q^{\pm} \end{pmatrix} + s_q^{\pm}\begin{pmatrix} -z_2 \\ z_1 \end{pmatrix},$$

$h_q^{\pm}, p_q^{\pm}, s_q^{\pm}$ *are constants.*

*Then the effective shear modulus can be represented in a form*

$$\bar{\mu} = \left( \frac{4\mu(\lambda+\mu)}{\lambda+2\mu} + 2\mu \int_{\Pi} e_z(u)dz \right)\delta^3 + O\left(\delta^4\right), \quad \delta \to +0$$

*where*

$$e_z(u) = \frac{1}{2}\left(\frac{\partial u_1}{\partial z_2} + \frac{\partial u_2}{\partial z_1}\right).$$

## 2 Effective thermoconductivity of the two-dimensional rectangular lattice structure

Consider the dilated two-dimensional rectangular lattice structure with the width of a bond equal to $\delta$ [11]:

$$B = \bigcup_{k=-\infty}^{+\infty} \left(\{|x_1 - k| < \delta/2\} \cup \{|x_2 - k| < \delta/2\}\right), \qquad (2.1)$$

where $\delta$ is a small parameter.

We discuss homogenization theory for the Laplacian on $B$ with the homogeneous Neumann boundary conditions. Then in order to find homogenized matrix which is due to the symmetry equal to $\bar{K}\delta_{ij}$ we should solve a sell problem

$$\Delta U = 0, \quad x \in B; \quad \frac{\partial U}{\partial n} = 0, \quad x \in \partial B \tag{2.2}$$

with $R - x_1$ being periodic in both of the variables $x_1, x_2$. Then the homogenized coefficient which is in that case effective thermoconductivity is equal to

$$\hat{K} = \left\langle \frac{\partial}{\partial x_1} U \right\rangle = \int_Q \frac{\partial U}{\partial x_1} dx_1 dx_2, \tag{2.3}$$

where

$$Q = B \cap (-1/2, 1/2)^2 .$$

Now we can formulate and prove the first result.

**Theorem 2.1** *For $\hat{K}$ the following asymptotic expansion holds*

$$\hat{K} = \delta + k_g \delta^2 + O\left(\delta^3\right), \tag{2.4}$$

*where*

$$k_g = 4 \int_0^{\frac{1}{\pi} a \sinh 1} x \sqrt{\operatorname{cosech}^2 \pi x - 1} dx. \tag{2.5}$$

First term in (2.4) is exactly what the principle of splitting of homogenized operator predicts [11], and the second term is a consequence of more precise homogenization theory.

**Proof**

In order to prove the theorem we construct an asymptotic expansion of the solution (2.2). As in [2], we seek the asymptotic solution of (2.2) in a form of series

$$U \approx v(x_1) \rho\left(\frac{x_1}{\delta}\right) + \sum_{l=0}^{\infty} t^l N_l \left(\frac{x}{\delta}\right) \tag{2.6}$$

where

$$v(x_1) \approx \sum_{l=0}^{\infty} t^l v_l(x_1) \tag{2.7}$$

$$v_0(x_1) = x_1, \quad x_1 \in (-\infty, +\infty) \tag{2.8}$$

$$v_l(x_1) = a_l x_1 + b_l, \quad x_1 \in (0,1). \tag{2.9}$$

Functions $N_l(x)$ are assumed to be functions of boundary layer type. This means that we require an estimate

$$\left| N_l\left(\frac{x}{t}\right) \right| \le \hat{c}_1 e^{-\hat{c}_2 |x|/t}, \quad \hat{c}_1, \hat{c}_2 > 0. \tag{2.10}$$

We expand linear on $(0,1)$ functions $v_l(x_1)$ onto $R^1$ in such a way that $v_l(x_1)$ is 1-periodic, $l = 1, 2, \ldots$. This means that now defined on $R^1$ function $v_l$ has different limiting values at zero from the left and from the right.

let us denote those values as

$$v_l^{\pm} = \lim_{x_1 \to \pm 0} v_l(x_1) = \pm b_l. \tag{2.11}$$

Function $\rho(z)$ is an auxiliary function and has the following properties:

$$\rho(z) \in C^\infty\left(R^1\right), \quad \rho \text{ is even}, \quad \rho(z) = \begin{cases} 0, & |z| \le 0.75 \\ 1, & z_1 \ge 1 \end{cases}$$

Posing $z = x/\delta$ we assume that $N_l(z)$ in (2.6) is defined on an infinite cross

$$\Pi = \left\{ z \in R^2 : |z_1| < 1/2 \text{ or } |z_2| < 1/2 \right\}.$$

Substituting (2.6) into (2.1) we obtain for $N_l$ the following problems

$$\Delta N_l = -\frac{\partial^2 \left(\rho(z_1) V_l(z_1)\right)}{\partial z_1^2}, \quad z \in \Pi; \quad \frac{\partial N_l}{\partial n_z} = 0, \quad z \in \partial\Pi \tag{2.12}$$

where the posed

$$V_l(z_1) = z_1 a_{l-1} + v_l^+ \theta_+(z_1) + v_l^- \theta_-(z_1)$$

$$\theta_+(z_1) = \begin{cases} 1, & z_1 > 0 \\ 0, & z_1 \le 0 \end{cases}, \quad \theta_-(z_1) = \begin{cases} 1, & z_1 < 0 \\ 0, & z_1 \ge 0 \end{cases} \tag{2.13}$$

and $v_l^{\pm}, a_l$ are defined as (2.9) and (2.11).

Now let us discuss a solution of the problem

$$\Delta \bar{N}_l = -\frac{\partial^2 (\rho(z_1) z_1)}{\partial z_1^2}, \quad a_{l-1} z \in \Pi; \quad \frac{\partial \bar{N}_l}{\partial n_z} = 0, \quad z \in \partial\Pi. \tag{2.14}$$

As any harmonic function in a strip with the Neumann boundary conditions the functions $\bar{N}_l$ tend exponentially to some constants $c_l^{\pm}$ as $z_1 \to \pm\infty$ (see [1]). Since $\bar{N}_l$ are odd functions with respect to $z_1$ we obtain the following relations for these limit constants: $c_l^+ = -c_l^- = c_l$ and $\bar{N}_l$ approach zero exponentially when $z_2 \to \pm\infty$. Let us choose

$$v_l^{\pm} = c_l^{\pm} \tag{2.15}$$

then by definition the function

$$\Delta N_l = \Delta \bar{N}_l(z) - \rho(z_1)\left(v_l^+ \theta_+(z_1) + v_l^- \theta_-(z_1)\right)$$

satisfies the equation (2.12) and the estimate (2.10). Now piecewise linear functions $v_l(x_1)$ are completely defined by the condition of periodicity which leads to the relation

$$a_l = -\left(v_l^+ - v_l^-\right) = -2c_l. \tag{2.16}$$

This procedure means that having by induction the constant $a_{l-1}$ we can find $\bar{N}_l$ from the equation (2.14) and then the constants $c_l^{\pm}$. Hence we get $a_l$ defined by (2.16) and we can define $N_l(z)$ and make the next step of the induction. Starting from $v_0(x_1) = x_1$ and $N_0 = 0$ we obtain the full asymptotic expansion. Rather standard estimates in $H^1(Q)$ show that this is actually the expansion of solution $U$ (see [1], [2] where more general case is discussed).

Now we pose

$$u = \bar{N}_1(z) + \rho(z_1) z_1. \tag{2.17}$$

Since, according to 2.16, $a_0 = 1$, we can pose the following homogeneous problem for $u$:

$$\Delta u = 0, \quad z \in \Pi; \quad \frac{\partial u}{\partial n_z} = 0, \quad z \in \partial\Pi. \tag{2.18}$$

Then as it follows from (2.17) and from properties of $\bar{N}_1$ we obtain for $u$ the following exponential relations (i.e. $u$ converges exponentially to the values prescribed below)

$$u \to 0 \ (z_2 \to \pm\infty), \quad u - z_1 \to \pm c_1 (z_2 \to \pm\infty). \tag{2.19}$$

Above that we have the relation

$$u = \bar{N}_1(z) + \rho(z_1) z_1 = N_1 + V_1(z_1)\rho(z_1), \qquad (2.20)$$

which follows from (2.13), (2.17). Substitute now our asymptotic expansion to the formula for the homogenized coefficient, where $z = x/\delta$. We obtain

$$\hat{K} = \left\langle \frac{\partial U}{\partial x_1} \right\rangle = \left\langle \frac{\partial}{\partial z_1} (z_1 \rho(z_1) + v_1(z_1)\rho(z_1) + N_1(z)) \right\rangle + +O\left(t^3\right) =$$

$$= \left\langle \frac{\partial}{\partial z_1} \left( z_1 \rho(z_1) + \left( a_1 x_1 + v_l^+ \theta_+(z_1) + v_l^- \theta_-(z_1) \right) \rho(z_1) + N_1(z) \right) \right\rangle +$$

$$O\left(\delta^3\right) = \left\langle \frac{\partial}{\partial z_1} (V_1(z)\rho(z_1) + N_1(z) + \delta \rho(z_1) a_1 z_1) \right\rangle + O\left(\delta^3\right) =$$

$$= \left\langle \frac{\partial u}{\partial z_1} \right\rangle + \delta a_1 \left\langle \frac{\partial}{\partial z_1} (\rho(z_1) z_1) \right\rangle + O\left(\delta^3\right) =$$

$$= \left\langle \frac{\partial u}{\partial z_1} \right\rangle + \delta^2 a_1 + O\left(\delta^3\right).$$

Finally, from (2.16) we get the following formula

$$\hat{K} = \left\langle \frac{\partial u}{\partial z_1} \right\rangle - 2\delta^2 c_1 + O\left(\delta^3\right). \qquad (2.21)$$

Calculating the average in the right hand side, we obtain

$$\left\langle \frac{\partial u}{\partial z_1} \right\rangle = \int_Q \frac{\partial u}{\partial z_1} dx_1 dx_2 = \int_{-\delta/2}^{\delta/2} dx_2 \int_{-1/2}^{1/2} dx_1 \frac{\partial u}{\partial z_1} + 2 \int_{\delta/2}^{1/2} dx_2 \int_{-\delta/2}^{\delta/2} dx_1 \frac{\partial u}{\partial z_1} =$$

$$= \delta^2 \int_{-1/2}^{1/2} dz_2 \int_{-1/2\delta}^{1/2\delta} dz_1 \frac{\partial u}{\partial z_1} + 2t^2 \int_{1/2}^{1/2\delta} dz_2 \int_{-1/2}^{1/2} dz_1 \frac{\partial u}{\partial z_1} =$$

$$= 2\delta^2 \int_{-1/2}^{1/2} dz_2 u\left(\frac{1}{2\delta}, z_2\right) + 4\delta^2 \int_{1/2}^{1/2\delta} dz_2 u\left(\frac{1}{2\delta}, z_2\right) =$$

$$= 2\delta^2\left(\frac{1}{2\delta}+c_1\right)+4\delta^2\int\limits_{1/2}^{1/2\delta} dz_2 u\left(\frac{1}{2\delta},z_2\right)+O\left(e^{-\hat{c}/\delta}\right), \quad \hat{c}>0;$$

here we have used (2.19). Tending $\delta$ to 0 and taking into account the exponential estimate for $u$ we get

$$\hat{K}=\delta+4\delta^2\int\limits_{1/2}^{\infty} dz_2 u\left(\frac{1}{2},z_2\right)+O\left(\delta^3\right). \tag{2.22}$$

Now we are going to find the solution $u$ and consequently the constant $c_1$ explicitly. First we note that a harmonic function defined in a strip which has a linear growth and satisfies the Neumann boundary conditions is exactly $x_1$. This gives an idea to find a conformal map of $\Pi$ onto a strip:

$$\Pi_h=\{z\in \mathrm{C},\ z=x_1+ix_2,\ |x_2|<h\}.$$

We construct this mapping $w$: $\Pi_h\cong\Pi$ in two steps and define the parameter $h$ at the end of the construction. let us denote by $\Pi_+=\{z\in\Pi,\ x_2>0\}$. Then the transformation

$$w(\varsigma)=\frac{1}{\pi}\left(a\tan\frac{\varsigma}{\sqrt{\varsigma^2-2}}+a\tanh\frac{\varsigma}{\sqrt{\varsigma^2-2}}\right) \tag{2.23}$$

maps upper half plane $\{\mathrm{Im}\varsigma>0\}$ onto $\Pi_+$ (see [6], p.190). Moreover, the image of the segment $[-1,1]$ is the real axis $R$. Hence, due to the symmetry principle, $w(\varsigma)$ maps also $\mathrm{C}\setminus\{(-\infty,-1)\cup(1,+\infty)\}$ onto $\Pi$. The second step is the mapping

$$\varsigma(z)=i\,\mathrm{sh}\frac{\pi z}{2h} \tag{2.24}$$

which transforms $\Pi_h$ onto $\mathrm{C}\setminus\{(-\infty,-1)\cup(1,+\infty)\}$ (see [6], p.188). Comparing (2.24) and (2.23) we obtain the desired transformation

$$w(z)=\frac{1}{\pi}\left(a\tan\left(\frac{i\,\mathrm{sh}\frac{\pi z}{2h}}{\sqrt{-\mathrm{sh}^2\frac{\pi z}{2h}-2}}\right)+a\tanh\left(\frac{i\,\mathrm{sh}\frac{\pi z}{2h}}{\sqrt{-\mathrm{sh}^2\frac{\pi z}{2h}-2}}\right)\right) \tag{2.25}$$

which maps $\Pi_h$ onto $\Pi$. Now we pose

$$\alpha(z) = \frac{1}{\pi} \frac{i \operatorname{sh}\frac{\pi z}{2h}}{\sqrt{-\operatorname{sh}^2\frac{\pi z}{2h} - 2}}. \tag{2.26}$$

then

$$w(z) = \frac{1}{\pi}\left(a\tan\alpha(z) + a\tanh\alpha(z)\right). \tag{2.27}$$

As far as the angles are preserved in the conformal mapping we obtain the harmonic function

$$u_* = \operatorname{Re} w^{-1}(w(z)) \tag{2.28}$$

satisfying the Neumann boundary condition on the boundary of $\Pi$. Now we are going to investigate the behavior of $u_*$ at the infinity. From the formulae

$$a\tan = \frac{1}{2i}\lg\frac{1+i\alpha}{1-i\alpha}, \quad a\tanh\alpha = \frac{1}{2i}\lg\frac{1+\alpha}{1-\alpha} \tag{2.29}$$

we have $w \to \infty$ as

$$\alpha \to \pm i \quad \text{or} \quad \alpha \to \pm 1. \tag{2.30}$$

From (2.26) we obtain that

$$\alpha \to \pm i \quad \text{as} \quad z \to \pm ih \tag{2.31}$$

and

$$\alpha \to \pm 1 \quad \text{as} \quad z \to \pm\infty \tag{2.32}$$

Assume that $z \to +\infty$, then $\alpha \to +1$ and we have

$$w(z) = \frac{1}{\pi}\left(\frac{1}{2i}\lg i + \frac{1}{2}\lg 2 - \frac{1}{2}\lg(1-\alpha)\right). \tag{2.33}$$

From (2.26), (2.29) we obtain

$$\alpha(z) = 1 - \operatorname{sh}^2\frac{\pi z}{2h}(1 - O(1)) \quad \text{as } z \to +\infty, \tag{2.34}$$

so (2.33) gives us

$$w(z) = \frac{1}{\pi}\left(\frac{1}{2i}i\frac{\pi}{2} + \frac{1}{2}\lg 2 + \lg\operatorname{sh}^2\frac{\pi z}{2h}\right)(1 + O(1)) =$$

$$= \left(\frac{1}{4} - \frac{1}{2\pi}\lg 2 + \frac{z}{2h}\right)(1 + O(1)). \tag{2.35}$$

This means that we have the following values

$$h = \frac{1}{2}, \quad c_1 = \frac{1}{4} - \frac{1}{2\pi}\lg 2. \tag{2.36}$$

Now in order to calculate the integral (2.22) we should find on the $z$-plane the image of the point $(1/2, 1/2)$ of the $w$-plane. Clearly it is a singular point of the derivative of $w$. On the line $x_2 = 1/2$ we have

$$i \operatorname{sh}\pi z = -\operatorname{ch}\pi x_1$$

and

$$\alpha(z) = \frac{-\operatorname{ch}\pi x_1}{\sqrt{\operatorname{ch}^2 x_1 - 2}}. \tag{2.37}$$

It means that the singular point of the derivative of $w$ is the point

$$z_* = \left(\frac{1}{\pi} a\cosh\sqrt{2}, \frac{i}{2}\right).$$

Now according to (2.31) we obtain that the image of the segment

$$\left\{x_2 = \frac{1}{2}, \quad 0 < x_1 < \frac{1}{\pi} a\cosh\sqrt{2}\right\}$$

is the ray

$$\left\{w_1 = \frac{1}{2}, \quad \frac{1}{2} < w_2 < +\infty\right\}.$$

Changing the variable according to (2.28) we obtain

$$I = \int_{1/2}^{\infty} u\left(\frac{1}{2}, x_2\right) dx_2 = \operatorname{Im}\int_{1/2}^{\frac{1}{2}+i\infty} z(w) dw = -\operatorname{Im}\int_{i/2}^{z_*} z w'(z) dz. \tag{2.38}$$

Calculate now $w'(z)$. From (2.25), (2.26), (2.37) we get

$$w'(z) = \frac{1}{\pi}\left(\frac{1}{1+\alpha^2} + \frac{1}{1-\alpha^2}\right)\alpha' = \frac{2}{\pi}\frac{\alpha'}{1-\alpha^4}. \tag{2.39}$$

The relation (2.37) yields

$$\alpha'(z) = \frac{2\mathrm{sh}\pi x_1}{\sqrt{\left(\mathrm{ch}^2\pi x_1 - 2\right)^3}}. \tag{2.40}$$

Finally applying the relations (2.39), (2.40) we get for $I$ the following expression

$$I = \int_0^{a_*} x\sqrt{\mathrm{cosech}^2 x\pi - 1}dx, \quad \text{where } a_* = \frac{1}{\pi}a\sinh 1,$$

and the Theorem 2.1 is proved.

**Remark.** According to the Theorem 2.1

$$\hat{K} = \delta + k_g\delta^2 + O\left(\delta^3\right)$$

where $k_g > 0$. This formula could be compared with the simple approximate formula of the strength material theory

$$\hat{K}_{sm} = \delta.$$

We see that the precise value is greater and this is due to the boundary layer contribution characterized by the coefficient $k_g$.

Now we can construct the complete asymptotic expansion of $\hat{K}$. The following theorem can be proved.

**Theorem 2.2** *Assume that $c_1$ is defined by (2.36), and $k_g$ is defined by (2.5). Then for all natural $n$ the following asymptotic formula holds true:*

$$\hat{K} = \delta + \frac{k_g\delta^2}{1 + 2c_1\delta} + O\left(\delta^n\right). \tag{2.41}$$

Thus the complete asymptotic expansion differs from (2.4) by the term $2c_1\delta$ in the denominator only.

**Proof**

From (2.14) we obtain:

$$\bar{N}_l = \bar{N}_1 a_{l-1}, \quad c_l = a_{l-1}c_1$$

and therefore (2.15),(2.16)

$$a_l = -2c_l = -2a_{l-1}c_1,$$

$$v_l^{\pm} = \pm c_l = \pm a_{l-1} c_1,$$

then (2.14) yields

$$V_l(s_1) = z_1 a_{l-1} + a_{l-1} c_1 \theta_+(z_1) - a_{l-1} c_1 \theta_-(z_1) = a_{l-1} V_l(s_1),$$

$$N_l(z) + V_l(z_1)\rho(z_1) =$$

$$= \bar{N}_l(z) + \rho(z_1)\left(v_l^+ \theta_+(z_1) + v_l^- c_1 \theta_-(z_1)\right) + V_l(z_1)\rho(z_1) =$$

$$= a_{l-1}\bar{N}_1 + a_{l-1}\rho(z_1)(c_1\theta_+(z_1) - c_1\theta_-(z_1)) + a_{l-1}V_1(z_1)\rho(z_1) =$$

$$= a_{l-1}\left(\bar{N}_1 + z_1\rho(z_1)\right) = a_{l-1}u.$$

Hence,

$$\hat{K} = \left\langle \frac{\partial U}{\partial x_1} \right\rangle \approx \delta^{-1} \left\langle \frac{\partial}{\partial z_1} \sum_{l=0}^{\infty} \delta^l \left(V_1(z_1)\rho(z_1) + N_l(z)\right) \right\rangle \approx$$

$$\approx \delta^{-1} \left\langle \frac{\partial}{\partial z_1} \sum_{l=0}^{\infty} \delta^l a_{l-1} u \right\rangle \approx \left\langle \frac{\partial u}{\partial z_1} \right\rangle \sum_{l=1}^{\infty} \delta^{l-1} a_{l-1} \approx \left\langle \frac{\partial u}{\partial z_1} \right\rangle \sum_{l=0}^{\infty} \delta^l a_l,$$

$$a_0 = 1, \quad a_l = -2c_1 a_{l-1} = (-2c_1)^l.$$

Hence,

$$\hat{K} \approx \left\langle \frac{\partial u}{\partial z_1} \right\rangle \sum_{l=0}^{\infty} (-2c_1\delta)^l = \left\langle \frac{\partial u}{\partial z_1} \right\rangle \frac{1}{1 + 2c_1\delta}$$

where

$$\left\langle \frac{\partial u}{\partial z_1} \right\rangle = 2\delta^2 \left(\frac{1}{2\delta} + c_1\right) + 4\delta^2 \int_{1/2}^{1/2\delta} dz_2 u\left(\frac{1}{2}, z_2\right) + O(\delta^n) =$$

$$= \delta + 2\delta^2 c_1 + 4\delta^2 I + O(\delta^n),$$

$$c_1 = \frac{1}{4} - \frac{1}{2\pi}\lg 2, \quad I = \int_0^{\frac{1}{\pi} a \sinh 1} x\sqrt{\operatorname{cosech}^2 \tau x - 1}\, dx.$$

Theorem 2.2 is proved.

# 3 The shear modulus of the two–dimensional rectangular lattice structure

Consider now an analogous problem for the two–dimensional operator of isotropic elasticity with constant coefficients:

$$L = \frac{\partial}{\partial x_i}\left(A_{ij}\frac{\partial}{\partial x_j}\right),$$

where

$$A_{11} = \begin{pmatrix} \lambda + 2\mu & 0 \\ 0 & \mu \end{pmatrix}, \quad A_{12} = \begin{pmatrix} 0 & \lambda \\ \mu & 0 \end{pmatrix}$$

$$A_{21} = \begin{pmatrix} 0 & \mu \\ \lambda & 0 \end{pmatrix}, \quad A_{22} = \begin{pmatrix} \mu & 0 \\ 0 & \lambda + 2\mu \end{pmatrix}$$

$\lambda, \mu$ are Lame's constants. Let (as above)

$$B_{\varepsilon,\delta} = \bigcup\nolimits_{k=-\infty}^{+\infty} \left(\{|x_1 - k| < \delta/2\} \cup \{|x_2 - k| < \delta/2\}\right) \tag{3.1}$$

be a lattice structure with the thickness of strips $t \ll 1$, and the periodic cell

$$Q = \{x = (x_1,x_2)\,,\; x \in B,\; x_i \in [1/2, 1/2]\,,\; i = 1, 2\}\,. \tag{3.2}$$

Consider the homogenized operator

$$\bar{L} = \frac{\partial}{\partial x_i}\left(\bar{A}_{ij}\frac{\partial}{\partial x_j}\right),$$

corresponding to the homogeneous Neumann boundary condition at the boundary $\partial B$ of a lattice domain $B$. The coefficients $A_{ij}$ of the homogenized operator $\bar{L}$ are defined on the base of solution of the corresponding cell problems (c.f. [1]). In particular the effective shear modulus of a lattice structure could be determined as an element $\bar{a}_{11}^{22}$ of the matrix $\bar{A}_{11}$, equal to

$$\bar{\mu} = 2\mu \left\langle e\left(U\right)\right\rangle, \tag{3.3}$$

where $U\left(x\right)$ is a 2D vector-valued function $\left(u_1\left(x\right), u_2\left(x\right)\right)^T$ ($T$ is the transposition symbol), that satisfies a system of equalities

$$LU = 0, \quad x \in B; \quad \frac{\partial U}{\partial \nu} = n_i A_{ij}\frac{\partial U}{\partial x_j} = 0, \quad x \in \partial B, \tag{3.4}$$

and

$$U-\begin{pmatrix}0\\x_1\end{pmatrix}$$

is a 1-periodic vector-function of the variables $x_1$ and $x_2$; $(n_1, n_2)$ is the outside normal to $\partial B$, i.e. $n_i \in \{-1, 0, 1\}$;

$$e(U)=\frac{1}{2}\left(\frac{\partial U_1}{\partial x_2}+\frac{\partial U_2}{\partial x_1}\right) \tag{3.5}$$

is a shear strain.

**Theorem 3.1** *Let $\Pi$ be an infinite cross*

$$\Pi=\left\{z\in R^2, \quad |z_1|<1/2 \quad or \quad |z_2|<1/2\right\},$$

$$L_z=\frac{\partial}{\partial z_i}\left(A_{ij}\frac{\partial}{\partial z_j}\right), \quad \frac{\partial}{\partial \nu_z}=n_i\left(A_{ij}\frac{\partial}{\partial z_j}\right);$$

*let $u(z)$ be a 2D vector-valued function, which is the solution of the homogeneous elasticity problem*

$$L_z=0, \quad z\in\Pi; \quad \frac{\partial u}{\partial \nu_z}=0, \quad z\in\partial\Pi,$$

*with the following asymptotic behavior at the infinity*

$$\pm 3\begin{pmatrix}-z_1z_2\\ \frac{z_1^2}{2}+\frac{\lambda z_2^2}{2(\lambda+2\mu)}\end{pmatrix}+r_1^{\pm}(z)+o\left(z_1^{-1}\right), \quad z_1\to\pm\infty,$$

$$\pm 3\begin{pmatrix}\frac{z_2^2}{2}+\frac{\lambda z_1^2}{2(\lambda+2\mu)}\\ -z_1z_2\end{pmatrix}+r_2^{\pm}(z)+o\left(z_2^{-1}\right), \quad z_2\to\pm\infty,$$

*where $r_q^{\pm}$ are rigid displacements, i.e.*

$$r_q^{\pm}(z)=\begin{pmatrix}h_q^{\pm}\\p_q^{\pm}\end{pmatrix}+s_q^{\pm}\begin{pmatrix}-z_2\\z_1\end{pmatrix},$$

*$h_q^{\pm}, p_q^{\pm}, s_q^{\pm}$ are constants.*

*Then the effective shear modulus can be represented in a form*

$$\bar{\mu}=\left(\frac{4\mu(\lambda+\mu)}{\lambda+2\mu}+2\mu\int_{\Pi} e_z(u)dz\right)\delta^3+O\left(\delta^4\right), \quad \delta\to+0,$$

*where*

$$e_z(u)=\frac{1}{2}\left(\frac{\partial u_1}{\partial z_1}+\frac{\partial u_2}{\partial z_2}\right).$$

**Proof**

To prove the theorem we shall find an asymptotic expansion of the solution of the problem (3.4).

Note that there is a symmetry of the cell problem, which yields ([1], p.217) the following properties of the solution: the first component $u_1$ of the vector $U$ is even in $x_1$ and odd in $x_2$; the second component $u_2$ is even in $x_2$ and odd in $x_1$.

Analogously to [12,13], we seek the asymptotic solution of the problem (3.4) in a form

$$U = \sum_{m=1}^{2} \rho\left(\frac{x_m}{\delta}\right) v_m^\delta(x) + \sum_{l=0}^{\infty} \delta^l N_l\left(\frac{x}{\delta}\right), \quad x \in B \quad and \quad x \in Q,$$

where $v_m^\delta(x)$ is an interior ansatz, which satisfies the equation and boundary conditions (3.4) at the strips $B$ parallel to the axis $Ox_m$, $N_l(x/\delta)$ are the boundary layer correctors, they decay exponentially to zero out of the point $(0,0)$, $\rho(z) \in C^\infty(R_1)$,

$$\rho(z) = \begin{cases} 0 & \text{for } |z| \le \frac{3}{4} \\ 1 & \text{for } |z| \le 1 \end{cases}$$

$v_m^\delta$, $N_l$ are 2D vectors. The functions $v_m^\delta(x)$ have the following form (c.f. [12])

$$v_m^\delta(x) = V_m(x_m) + \sum_{l=0}^{\infty} \delta^l N_{lm}\left(\frac{x_{\bar m}}{\delta}\right)\frac{d^l V_m}{dx_m^l},$$

$$V_m(x_m) = \sum_{l=0}^{\infty} \delta^l V_{lm}(x_m),$$

where $N_{lm}, m = 1,2$ are $2\times 2$ matrices, $V_m, V_{lm}$ are the 2D vectors, $\bar m = 3-m$. It follows from [12] that the first components of the vectors $V_1, V_{l1}$ are linear functions, and the second components are the third order polynomials; on the contrary, the first component of the vectors $V_2, V_{l2}$ are the third order polynomials and the second components are linear. Then the matrices $N_{lm}$ are calculated ([12], p.92) for all values $l$ that we need. Taking into consideration the symmetry of the problem we can write the precise formula of $v_m^\delta(x)$: Let $I_+ = (0,1/2), I_- = (0,-1/2)$ and $d_i^q, g_i^q, f_i^q$, $i=1,2$, $q=\pm$ are constants that should be defined; then

$$v_1^\delta = \begin{pmatrix} -x_2\left(d_1^q + 2_1^q x_1 + 3f_1^q x_1^2\right) + \frac{f_1^q}{\lambda+2\mu}\left((3\lambda+4\mu)\,x_2^3 - \frac{11\lambda+12\mu}{4}\delta^2 x_2\right) \\ c_1^q + d_1^q x_1 + g_1^q x_1^2 + f_1^q x_1^3 + \left(g_1^q + 3f_1^q x_1\right)\frac{\lambda}{\lambda+2\mu}\left(x_2^2 - \frac{\delta^2}{12}\right) \end{pmatrix},$$

$$x_1 \in I_q, \quad q = \pm, \tag{3.6}$$

$$v_2^\delta = \begin{pmatrix} c_2^q + d_2^q x_2 + g_2^q x_2^2 + f_2^q x_2^3 + (g_1^q + 3 f_2^q x_2) \frac{\lambda}{\lambda+2\mu} \left(x_1^2 - \frac{\delta^2}{12}\right) \\ -x_1 \left(d_2^q + 2_2^q x_2 + 3 f_2^q x_2^2\right) + \frac{f_2^q}{\lambda+2\mu} \left((3\lambda+4\mu) x_1^3 - \frac{11\lambda+12\mu}{4} \delta^2 x_1\right) \end{pmatrix},$$

$$x_2 \in I_q, \quad q = \pm, \tag{3.7}$$

and the symmetry implies:

$$c_i^+ = -c_i^-, \quad d_i^+ = d_i^-, \quad g_i^+ = -g_i^-, \quad f_i^+ = f_i^-;$$

the 1-periodicity conditions for the vector-valued functions $v_1^\delta + (0, x_1)^T$ and $v_2^\delta$ yields the equalities

$$8c_i^+ + 4d_i^+ + 2_i^+ + f_i^+ = 4\delta_{i1}, \quad 2g_i^+ + 3f_i^+ = 0. \tag{3.8}$$

It could be checked easily that $v_i^\delta$ are the solutions of the homogeneous elasticity Neumann problems in the strips $\{x_i \in R, \ |x_{3-i}| < \delta/2\}$

Introduce the rapid variables $z = x/\delta$ in (3.6 ,3.7 ). We get $v_i^\delta(x) = V_i^\delta(z)$, i.e.

$$V_1^\delta(z) = \begin{pmatrix} -\delta d_1^q z_2 - 2\delta^2 g_1^q z_1 z_2 - \delta^3 f_1^q \left(3z_1^2 z_2 - \frac{3\lambda+4\mu}{\lambda+2\mu} z_2^3 + \frac{11\lambda+12\mu}{4(\lambda+2\mu)} z_2\right) \\ c_1^q + \delta d_1^q z_1 + \delta 2 g_1^q \left(z_1^2 + \frac{\lambda}{\lambda+2\mu}\zeta_2\right) + \delta^3 f_1^q \left(z_1^3 + \frac{3\lambda_{z_1}}{\lambda+2\mu}\zeta_2\right) \end{pmatrix},$$

$$V_2^\delta(z) = \begin{pmatrix} c_2^q + \delta d_2^q z_1 + \delta 2 g_2^q \left(z_2^2 + \frac{\lambda}{\lambda+2\mu}\zeta_1\right) + \delta^3 f_2^q \left(z_2^3 + \frac{3\lambda_{z_2}}{\lambda+2\mu}\zeta_1\right) \\ -\delta d_2^q z_2 - 2\delta^2 g_2^q z_1 z_2 - \delta^3 f_2^q \left(3z_2^2 z_1 - \frac{3\lambda+4\mu}{\lambda+2\mu} z_1^3 + \frac{11\lambda+12\mu}{4(\lambda+2\mu)} z_1\right) \end{pmatrix},$$

here $\zeta_2 = z_2^2 - 1/12; \quad \zeta_1 = z_1^2 - 1/12$

We seek constants $c_i^q, d_i^q, g_i^q, f_i^q$ in a form of the asymptotic series

$$\sum_{l=0}^\infty \delta^l c_{il}^q, \quad \sum_{l=0}^\infty \delta^l d_{il}^q, \quad \sum_{l=0}^\infty \delta^l g_{il}^q, \quad \sum_{l=0}^\infty \delta^l f_{il}^q.$$

Introduce the notation

$$V_{1l}^t(z) =$$

$$= \begin{pmatrix} -d_{1,l-1}^q z_2 - 2g_{1,l-2}^q z_1 z_2 - f_{1,l-3}^q \left(3z_1^2 z_2 - \frac{3\lambda+4\mu}{\lambda+2\mu} z_2^3 + \frac{11\lambda+12\mu}{4(\lambda+2\mu)} z_2\right) \\ c_{1l}^q + d_{1,l-1}^q z_1 + g_{1,l-2}^q \left(z_1^2 + \frac{\lambda}{\lambda+2\mu}\zeta_2\right) + \delta^3 f_{1,l-3}^q \left(z_1^3 + \frac{3\lambda_{z_1}}{\lambda+2\mu}\zeta_2\right) \end{pmatrix},$$

$$V_{21}^t(z) =$$

$$= \begin{pmatrix} c_{2l}^q + d_{2,l-1}^q z_1 + g_{2,l-2}^q \left(z_2^2 + \frac{\lambda}{\lambda+2\mu}\zeta_1\right) + \delta^3 f_{2,l-3}^q \left(z_2^3 + \frac{3\lambda_{z_2}}{\lambda+2\mu}\zeta_1\right) \\ -d_{2,l-1}^q z_2 - 2g_{2,l-2}^q z_1 z_2 - f_{2,l-3}^q \left(3z_2^2 z_1 - \frac{3\lambda+4\mu}{\lambda+2\mu} z_1^3 + \frac{11\lambda+12\mu}{4(\lambda+2\mu)} z_1\right) \end{pmatrix},$$

The constants with negative indices are equal to zero.

The vectors $N_l(z)$ are the solution to the problems

$$L_z N_l = -L_z\left(\sum_{m=1}^{2} V_{ml}(z)\,\rho(z_m)\right), \quad z \in \Pi,$$

$$\frac{\partial N_l}{\partial \nu_z} = -\frac{\partial}{\partial \nu_z}\left(\sum_{m=1}^{2} V_{ml}(z)\,\rho(z_m)\right), \quad z \in \partial\Pi. \tag{3.9}$$

The orthogonality of the right hand side function of (3.9) to the kernel of the problem (3.9) is necessary and sufficient for the existence of the bounded energy integral (c.f. [10], [12, 13]), i.e. the right hand side function should be orthogonal to the constant vectors and to the vector $(-z_2, z_1)^T$. Let $\Phi$ be one of these vectors. Since the right hand side functions of (3.9) are equal to zero outside the square $(-1,1)^2$ and $\rho(\pm 1) = 1$ we get

$$-\int_{\Pi} (V_{ml}(z)\,\rho(z_m), \varphi) dz + \int_{\partial\Pi} \left(\frac{\partial}{\partial \nu_z}\left(\sum_{m=1}^{2} V_{ml}(z)\,\rho(z_m)\right), \varphi\right) dz =$$

$$\sum_{m=1}^{2} \int_{\{z_m=\pm1, |z_{3-m}|<\frac{1}{2}\}} \left(\frac{\partial V_{ml}}{\partial \nu_z}, \varphi\right) dz.$$

We obtain the solvability condition

$$\sum_{m=1}^{2} \int_{\{z_m=\pm1, |z_{3-m}|<\frac{1}{2}\}} \left(\frac{\partial V_{ml}}{\partial \nu_z}, \varphi\right) dz = 0. \tag{3.10}$$

Calculating $\partial V_{ml}/\partial \nu_z|_{z_m=\pm1}$ we get

$$\frac{\partial V_{1l}}{\partial \nu_z}|_{z_1=\pm1} = \pm(A_{11}\frac{\partial V_{1l}}{\partial z_1} + A_{12}\frac{\partial V_{1l}}{\partial z_2})|_{z_1} = \pm1 =$$

$$\pm(g^{\pm}_{1,l-2} - \frac{8\mu(\lambda+\mu)}{\lambda+2\mu} z_2, 0)^T$$

$$f^{\pm}_{1,l-3}(-\frac{24\mu(\lambda+\mu)}{\lambda+2\mu} z_1 z_2, -\frac{\mu(\lambda+\mu)}{\lambda+2\mu}(12z_2^2 - 3))^T)|z_1 =$$

$$= \pm1 = \pm\frac{4\mu(\lambda+\mu)}{\lambda+2\mu}\left(-2g^{\pm}_{1,l-2} z_2 - 6f^{\pm}_{1,l-3} z_1 z_2, 3\left(z_2^2 - \frac{1}{4}\right) f^{\pm}_{1,l-3}\right)^T,$$

$$\left.\frac{\partial V_{2l}}{\partial \nu_z}\right|_{z_2=\pm 1} = \frac{4\mu(\lambda+\mu)}{\lambda+2\mu}\left(3\left(z_1^2-\frac{1}{4}\right)f^{\pm}_{1,l-3}, -2g^{\pm}_{2,l-2}z_1 - 6f^{\pm}_{2,l-3}z_1z_2\right)^T.$$

Taking into consideration that $f^{+}_{i,l-3} = f^{1}_{i,l-3}$, $g^{+}_{i,l-2} = -g^{-}_{i,l-2}$, we observe that the solvability condition (3.10) is satisfied for $\varphi = const$ naturally, and for $\varphi = (-z_2, z_1)^T$ we get

$$g^{+}_{1,l-2} = g^{+}_{2,l-2}. \tag{3.11}$$

The solutions $N_l$ of the problems (3.9) stabilize some rigid displacements as $|z| \to +\infty$. These rigid displacements have the following structure, provided by the oddness-evenness properties of the solutions $N_l$:

$$N_l(z) = \begin{pmatrix} -b_{1l}z_2 \\ \pm a_{1l} + b_{1l}z_1 \end{pmatrix} + o\left(|z|^{-1}\right), \quad |z_1| \to +\infty,$$
$$N_l(z) = \begin{pmatrix} \pm a_{2l} + b_{2l}z_2 \\ b_{2l}z_2 \end{pmatrix} + o\left(|z|^{-1}\right), \quad |z_2| \to +\infty. \tag{3.12}$$

We shall show below by induction that all $N_l$ $(l > 1)$ conserve the symmetry of directions 1 and 2, i.e. $-b_{1l} = b_{2l}$, $a_{1l} = a_{2l}$. Nevertheless, defining constants $c^q_{il}, d^q_{i,l-1}$ in a special way we can make all $N_l$ tend to zero as $|z| \to +\infty$. Indeed, let for any (for example equal to zero) values of these constants $N_l$ have the asymptotics (3.12). Then we redefine them in the following way:

$$c^{+}_{il} = -a_{il}, \quad c^{-}_{il} = a_{il}, \quad d^{+}_{1,l-1} = -b_{1l},$$
$$d^{-}_{1,l-1} = -b_{1l}, \quad d^{+}_{2,l-1} = b_{2l}, \quad d^{-}_{2,l-1} = b_{2l},$$

i.e.

$$c^{+}_{il} = -c^{-}_{il} = a_{il}, \quad d^{+}_{1,l-1} = -d^{-}_{1,l-1} = (-1)^i b_{1l}. \tag{3.13}$$

Thus we get for the constants $c^q_{il}$, $d^q_{il}$, $g^q_{il}$, $f^q_{il}$, the following equations:

$$f^{-}_{il} = f^{+}_{il}, \quad g^{-}_{il} = -g^{+}_{il}, \quad d^{-}_{il} = d^{+}_{il}, \quad c^{-}_{il} = -^{+}_{il}, \quad 2g^{+}_{il} + 3f^{+}_{il} = 0,$$
$$8c^{+}_{il} + 4d_{il}i^{+} + 2g^{+}_{il} + f^{+}_{il} = 4\delta_{i1}\delta_{l0}, \quad g^{+}_{1,l-2} = g^{+}_{2,l-2}. \tag{3.14}$$

and the relations (3.13).

Consider the case $l = 0$. We get

$$8c^{+}_{i0} + 4d^{+}_{i0} + 2g^{+}_{i0} + f^{+}_{i0} = 4\delta_{i1}, \quad g^{+}_{i0} = -1.5f^{-}_{i0}, \quad g^{+}_{10} = g^{+}_{20}$$

and therefore $f_{10}^{+} = f_{20}^{+}$, $N_0$ is the solution of the problem (3.9) with the 2D vector-valued functions $V_{m0} = c_{m0}^{q}(z_m)\, e_{3-m}$, where $c_{m0}^{q}(z_m) = c_{mi}^{+}$ for $z_m > 0$ and $c_{m0}^{q}(z_m) = c_{mi}^{-}$ for $z_m < 0$, $e_1 = (1,0)^T$; so $N_0 = 0$, $c_{m0}^{q} = 0$.

Consider the case $l = 1$. We get:

$$8c_{i1}^{+} + 4d_{i1}^{+} + 2g_{i1}^{+} + f_{i1}^{+} = 0, \quad g_{i1}^{+} = -1.5f_{i1}^{+}, \quad g_{11}^{+} = g_{21}^{+},$$

and therefore $f_{11}^{+} = f_{21}^{+}$, $N_1$ is the solution of the problem (3.9) with the 2D vector-valued functions $V_{m1} = c_{m1}^{q}(z_m)\, e_{3-m} + d_{m0}^{+}(-1)^m (-z_2, z_1)^T$.

So we pose $N_1 = -d_{10}^{+}(-z_2, z_1)^T (\rho(z_1) + \rho(z_2) - 1)$, $c_{m1}^{q} = 0$, $d_{10}^{+} = d_{20}^{+}$. Now we can calculate all the constants for $l = 0$:

$$4d_{10}^{+} + 2g_{10}^{+} + f_{i0}^{+} = 4, \quad 4d_{20}^{+} + 2g_{20}^{+} + f_{20}^{+} = 0, \quad g_{i0}^{+} = -1.5f_{i0}^{+},$$

$$g_{10}^{+} = g_{20}^{+}, \quad f_{10}^{+} = gf_{20}^{+}, \quad d_{10}^{+} = d_{20}^{-}, \quad g_{10}^{+} = -g_{10}^{-}, \quad f_{10}^{+} = f_{10}^{-},$$

i.e.

$$d_{10}^{+} = 0.5, \quad d_{20}^{+} = -0.5, \quad g_{10}^{+} = 1.5, \quad f_{10}^{+} = -1.$$

Consider the case $l > 1$. We get by induction: $g_{1l}^{+} =_{2l}^{+}$, and therefore $f_{1l}^{+} = f_{2l}^{+}$, $g_{il}^{+} = -1.5f_{il}^{+}$, $c_{il}^{+} = -c_{il}^{+}$, $d_{il}^{+} = d_{il}^{-}$, $g_{il}^{+} = -g_{il}^{-}$, $f_{il}^{+} = f_{il}^{-}$, $8c_{il}^{+} + 4d_{il}^{+} - 2f_{il}^{+} = 0$, and $N_l$ is the solution of the problem (3.9).

Posing the constants $c_{il}^{q} = 0$, $d_{i,l-1}^{q} = 0$, and taking into consideration that $g_{1,l-2}^{+} = g_{2,l-2}^{+}$, $f_{1,l-3}^{+} = f_{2,l-2}^{+}$, we obtain that the problems (3.9) conserve the symmetry in the sense of transpositions of the directions 1 and 2 (i.e. $z_1$ to $z_2$ and the first line to the second one). It means that the limit rigid displacements conserve the same symmetry, i.e. $a_{1l} = a_{2l}$, $b_{1l} = -b_{2l}$, and then (3.13) implies that $c_{1l}^{+} = c_{2l}^{+}d$, $d_{1,l-1}^{+} = d_{2,l-1}^{+}$.

Thus defining $N_l$ up to some rigid displacement we obtain: $c_{il}^{q}, d_{i,l-1}^{q}$ from (3.13) and $f_{i,l-1}^{+} = f_{i,l-1}^{-} = 4c_{i,l-1}^{+} + 2d_{i,l-1}^{+}$, $g_{i,l-1}^{+} - g_{i,l-1}^{-} = -1.5f_{i,l-1}^{+}$, and now we can go to the next value of $l$. At each step we solve the problem (3.9) with the constants $c_{il}^{q}, d_{i,l-1}^{q}$ equal to zero. This problem can be reduced to two auxiliary problems:

$$L_z \bar{U}^q = -L_z \bar{V}^q, \quad z \in \Pi,$$

$$\frac{\partial \bar{U}^q}{\partial \nu_z} = -\frac{\partial \bar{V}^q}{\partial \nu_z}, \quad z \in \partial\Pi, \quad q = 1, 2, \tag{3.15}$$

where

$$\bar{V}^1 = \begin{pmatrix} -2z_1z_2 \\ z_1^2 + \frac{\lambda\left(z_2^2 - \frac{1}{12}\right)}{\lambda+2\mu} \end{pmatrix} \operatorname{sign}(z_1)\,\rho(z_1) + \begin{pmatrix} z_2^2 + \frac{\lambda\left(z_1^2 - \frac{1}{12}\right)}{\lambda+2\mu} \\ -2z_1z_2 \end{pmatrix} \operatorname{sign}(z_2)\,\rho(z_2),$$

$$\bar{V}^2 = \begin{pmatrix} -3z_1^2z_2\frac{3\lambda+4\mu}{\lambda+2\mu}z_2^3 + \frac{11\lambda+12\mu}{4(\lambda+2\mu)}z_2 \\ z_1^3 + \frac{3\lambda}{\lambda+2\mu}z_1\left(z_2^2 - \frac{1}{12}\right) \end{pmatrix} \rho(z_1) +$$

$$+ \begin{pmatrix} z_2^3 + \frac{3\lambda}{\lambda+2\mu}z_2\left(z_1^2 - \frac{1}{12}\right) \\ -3z_2^2z_1\frac{3\lambda+4\mu}{\lambda+2\mu}z_1^3 + \frac{11\lambda+12\mu}{4(\lambda+2\mu)}z_1 \end{pmatrix} \rho(z_2).$$

The problems (2.14) can be reduced to the homogeneous problems

$$L_z\bar{W}^q = 0, \quad z \in \Pi, \qquad \frac{\partial \bar{W}^q}{\partial \nu_z} = 0, \quad z \in \partial\Pi,$$

where the asymptotic behavior of the solution at the infinity coincides with $\bar{V}^q$; i.e. $\bar{W}^q = \bar{U}^q + \bar{V}^q$.

Let

$$\bar{U}^q = \begin{pmatrix} -b^{(q)}z_2 \\ \pm a^{(q)} + b^{(q)}z_1 \end{pmatrix} + o\left(|z|^{-1}\right), \quad z_1 \to +\infty$$

and correspondingly

$$\bar{U}^q = \begin{pmatrix} \pm a^{(q)} + b^{(q)}z_2 \\ -b^{(q)}z_1 \end{pmatrix} + o\left(|z|^{-1}\right), \quad z_2 \to +\infty, \quad q = 1, 2.$$

Therefore

$$N_l = g_{1,l-2}^{+}\bar{U}^1 + f_{1,l-3}^{+}\bar{U}^2 -$$

$$\left(g_{1,l-2}^{+}a^{(1)} + f_{1,l-3}^{+}a^{(2)}\right)\left(e_2 \operatorname{sign}(z_1)\,\rho(z_1) + e_2(z_2)\,\rho(z_2)\right) -$$

$$-\left(g_{1,l-2}^{+}b^{(1)} + f_{1,l-3}^{+}b^{(2)}\right)(-z_2, z_1)^T\left(\rho(z_1) + \rho(z_2)\right),$$

where

$$e_1 = (1,0)\,, e_2 = (0,1)$$

and

$$c_{1l} = -\left(g_{1,l-2}^{+}a^{(1)} + f_{1,l-3}^{+}a^{(2)}\right), \quad d_{1,l-1} = -\left(g_{1,l-2}^{+}b^{(1)} + f_{1,l-3}^{+}b^{(2)}\right).$$

Then

$$N_lV_{1l}(z_1)\,\rho(z_1) + V_{2l}(z_2)\,\rho(z_2) =$$

$$= g_{1,l-2}^{+}\bar{U}^1 + f_{1,l-3}^{+}\bar{U}^2 + g_{1,l-2}^{+}\bar{V}^1 + f_{1,l-3}^{+}\bar{V}^2 = g_{1,l-2}^{+}\bar{W}^1 + f_{1,l-3}^{+}\bar{W}^2.$$

Thus

$$\bar{\mu} = 2\mu\{e(U)\} = 2\mu\delta^{-1}\int_{\Pi} e_z(\sum_{l=0}^{\infty}\delta^l(N_l + V_{1l}(z)\rho(z_1) + V_{2l}(z)\rho(z_2)))dz =$$

$$(15) \qquad 2\mu\sum_{l=0}^{\infty}\delta^{l-1}(g^+_{1,l-2}\int_{\Pi} e_z(\bar{W}^1)dz + f^+_{1,l-3}\int_{\Pi} e_z(\bar{W}^2)dz).$$

We calculate the constants $g^+_{1l}$, $f^+_{1l}$ in the formula (15) by induction:

$$f^+_{10} = -1, \; g^+_{10} = 1.5,$$

$$f^+_{1l} = 4c^+_{1l} + 2d^+_{1l} = -4(g^+_{1,l-2}a^{(1)} + f^+_{1,l-3}a^{(2)}) - 2(g^+_{1,l-1}b^{(1)} + f^+_{1,l-2}b^{(2)}) =$$

$$(16) \quad 3f^+_{1,l-1}b^{(1)} - f^+_{1,l-2}(2b^{(2)} - 6a^{(1)}) - 4f^+_{1,l-3}a^{(2)}, \quad g^+_{1l} = 1.5f^+_{1l}.$$

Here the constants with the negative indices are equal to zero.

The functions $e_z(\bar{W}^q)$ stabilize exponentially at the infinity to $e_z(\bar{V}^q)$ ; if $q = 1$ then the function $e_z(\bar{V}^1) = 0$ for $|z_1| > 1$ or $|z_2| > 1$ and therefore $e_z(\bar{W}^1)$ decays exponentially at the infinity and

$$\langle\, e_z(\bar{W}^1)\,\rangle = \delta^2\int_{\Pi} e_z(\bar{W}^1)dz + O(e^{-\hat{c}/\delta}), \hat{c} > 0.$$

On the contrary ,

$$e_z(\bar{V}^2) = \frac{6(\lambda+\mu)(z_2^2 - 1/4)}{\lambda + 2\mu} \quad for\ |z_1| > 1$$

and analogously

$$e_z(\bar{V}^2) = \frac{6(\lambda+\mu)(z_1^2 - 1/4)}{\lambda + 2\mu} \quad for\ |z_2| > 1\ ;$$

therefore we get:

$$\langle\, e_z(\bar{W}^2)\,\rangle = \delta^2\int_{\Pi_\delta} e_z(\bar{W}^2)dz,$$

where $\Pi_\delta = \{z \in \Pi \;\; and \;\; z \in (-\frac{1}{2\delta}, \frac{1}{2\delta})^2\}$.

So

$$\langle\ e_z(\bar{W}^2)\ \rangle = 8\delta^2 \int_0^{1/2} dz_2 \int_{z_2}^{1/2\delta} dz_1 e_z(\bar{W}^2) =$$

$$8\delta^2 (\int_0^{1/2} dz_2 \int_{z_2}^{1/2\delta} dz_1 (e_z(\bar{W}^2) - \frac{6(\lambda+\mu)(z_2^2 - 1/4)}{\lambda + 2\mu})) +$$

$$\int_0^{1/2} dz_2 \int_{z_2}^{1/2\delta} dz_1 \frac{6(\lambda+\mu)(z_2^2 - 1/4)}{\lambda + 2\mu}) =$$

$$8\delta^2 (\int_0^{1/2} dz_2 \int_{z_2}^{+\infty} dz_1 (e_z(\bar{W}^2) - \frac{6(\lambda+\mu)(z_2^2 - 1/4)}{\lambda + 2\mu})) +$$

$$\int_0^{1/2} dz_2 \int_{z_2}^{1/2\delta} dz_1 \frac{6(\lambda+\mu)(z_2^2 - 1/4)}{\lambda + 2\mu}) + O(e^{-\hat{c}/\delta}) =$$

$$8\delta^2 (I_2 + \frac{6(\lambda+\mu)(\frac{1}{64} - \frac{1}{24\delta})}{\lambda + 2\mu}) + O(e^{-\hat{c}/\delta}) =$$

$$-\frac{2(\lambda+\mu)}{\lambda+2\mu}\delta + (8I_2 + \frac{3(\lambda+\mu)}{4(\lambda+2\mu)})\delta^2 + O(e^{-\hat{c}/\delta}),$$

where

$$I_2 = \int_0^{1/2} dz_2 \int_{z_2}^{+\infty} dz_1 \; (e_z(\bar{W}^2) - \frac{6(\lambda+\mu)(z_2^2 - 1/4)}{\lambda + 2\mu}).$$

Substituting this calculated value of $\langle\ e_z(\bar{W}^2)\ \rangle$ to the expansion (15) and holding the terms of the $\delta^3$ order we obtain:

$$\bar{\mu} = 2\mu\delta^3 (g_{10}^+ \int_\Pi e_z(\bar{W}^1) dz - 2 f_{10}^+ \frac{\lambda+\mu}{\lambda+2\mu}) + O(\delta^4) =$$

$$2\mu\delta^3 (1.5 \int_\Pi e_z(\bar{W}^1) dz + 2\frac{\lambda+\mu}{\lambda+2\mu}) + O(\delta^4) =$$

$$(4\mu\frac{\lambda+\mu}{\lambda+2\mu} + 3\mu \int_\Pi e_z(\bar{W}^1) dz)\delta^3 + O(\delta^4).$$

The Theorem 3.1 is proved.

We can also write the asymptotic expansion of the effective shear modulus:

$$\bar{\mu} = 2\mu(\int_{\Pi} e_z(\bar{W}^1)dz \sum_{l=2}^{\infty} \delta^{l+1} g^+_{1,l-2} +$$

$$(-2\frac{\lambda+\mu}{\lambda+2\mu}\delta + (8I_2 + \frac{3(\lambda+\mu)}{4(\lambda+2\mu)})\delta^2) \sum_{l=2}^{\infty} \delta^{l-1} f^+_{1,l-3}).$$

The justification procedure for the asymptotic expansion is standard: we consider the truncated sum for $N_l$ $\sum_{l=0}^{K} \delta^l N_l$ and the truncated sums $\sum_{l=0}^{K} \delta^l c^q_{il}$ , $\sum_{l=0}^{K} \delta^l d^q_{il}$ , $\sum_{l=0}^{K} \delta^l g^q_{il}$ , $\sum_{l=0}^{K} \delta^l f^q_{il}$ . Their substitution into the equation brings the error $O(\delta^K)$). If we shall multiply $N_l$ by a smooth even function equal to 1 if $|x_1| < 1/3$ and $|x_2| < 1/3$ and vanishing for $|x_1| > 1/2$ or $|x_2| > 1/2$ then the truncated sums will be exactly periodic but this operation will add the error $O(e^{-\hat{c}/\delta})$, $\hat{c} > 0$. The symmetry properties of $N_l$ imply that the average of the truncated solution equals to zero. Then applying the Korn's inequality and the a priori estimate [8, 13, 15] we obtain the estimate for the difference between the exact solution and the truncated solution, and therefore for the exact and approximate values of the effective shear modulus.

**Remark.** An open problem is to find an explicit presentation of the integral $\int_\Pi e_z(u)dz$ (in the formulation of the Theorem 3.1) analogous to the (2.4).

## References


1. N.S. Bakhvalov and G.P. Panasenko, Homogenization: Averaging Processes in Periodic Media, Moscow, Nauka, 1984; English translation: Kluwer, Dordrecht/ Boston/ London, 1989.
2. G.P. Panasenko, Averaging processes in skeletal constructions, *Math. USSR Sbornik*, **122(164)** (2), 220–231 (1983).
3. D. Cioranescu and J. Saint Jean Paulin, Problèmes de Neumann et de Dirichlet dans des structures réticulées de faible épaisseur, *C.R. Acad. Sc. Paris*, t. 303, Serie I, No. 1, 1986.
4. S.M. Kozlov and A.A Malozemov, Hierarchical lattice constructions, *Math. Notes of Academy of Science of USSR*, **51** (2), 53–58 (1992).
5. M.A. Lavrentiev and B.V. Shabat, Methods of Complex Variable Theory, Moscow, Nauka, 1972.

6. H. Le Dret, Modeling of the junction between two rods, *J. Math. Pures. Appl.*, **68**, 356–397 (1989).
7. H. Le Dret, Problèmes Variationnels dans les Multi-domaines: Modélisation des Jonctions et Applications, 1991, Masson, Paris.
8. G. Fichera, Existence Theorems in Elasticity, Handbuch der Physik, Band 6a/2 , Springer-Verlag, Berlin/Heidelberg/New York, 1972.
9. S.M. Kozlov and G.P. Panasenko, Corrections to the strength materials theory for the lattice structures, Publ du Laboratoire d'Analyse Numerique, Universite P.et M. Curie CNRS, R91020, Paris, 1991.
10. O.A. Oleinik and G.A. Yosifian, On the asymptotic behavior at infinity of a solution of linear elasticity equation. *Arch. Rat. Mech. Anal.*, **78** (1), 29–53 (1982).
11. G.P. Panasenko, The principle of splitting of an averaged operator for a nonlinear system of equations in periodic and random skeletal structures, *Soviet Math. Dokl.*, **25** (2), (1982).
12. G.P. Panasenko, Asymptotic solutions of the elasticity theory system of equations for lattice and skeletal structures, *Russian Acad. Sci. Sbornik. Math.*, **75** (1), 85–110 (1993).
13. G.P. Panasenko, Asymptotic solutions of the elasticity system for rods, beam constructions and lattice structures, *VINITI (Zh.V.M.M.F.)*, 05.12.1990, No. 6107-B90, 91 p.
14. G.P. Panasenko, Asymptotic analysis of bar systems, I.*Russian Journ. of Math. Phys.*, **2** (3), 325–352 (1994). II. *Russian Journ. of Math. Phys.*, **4** (1), (1996).
15. G.P. Panasenko, Homogenization of lattice-like domains: L-convergence, *Nonlinear Partial Differential Equations and their Applications*, Ed. H. Brezis, J.L. Lions, College de France Seminar, Vol. XIII, 1996, 23 p.
16. S.M. Kozlov and J. Vucan, Explicit formula for effective thermoconductivity on quadratic lattice structure, *C.R. Acad. Sci. Paris*, **314**, Serie I, 281–286 (1992).

# MULTIPARAMETRIC PROBLEMS OF HOMOGENIZATION THEORY

N.S. BAKHVALOV, M.E. EGLIT
*School of Mechanics and Mathematics, Lomonosov Moscow State University*

Mathematical homogenization theory is used to calculate the effective moduli of composites and mixtures if the ratio $\varepsilon$ of the length scale of inhomogeneity to the macroscopic length scale for the problem is small. In many problems there are additional small parameters $\gamma_i$, besides $\varepsilon$. In this paper structures and materials are considered under the assumptions that certain parameters $\gamma_i$ formed by the ratios of different phases concentrations and moduli, or the ratios of coefficients connected with different properties of a phase are small. The approximate explicit formulae for the effective moduli are proposed and the errors of these formulae are estimated.

## 1 Introduction

Inhomogeneous media are considered in this paper, i.e., composites and mixtures. Let $L$ be the length scale of the problem and $d$ be the typical length of inhomogeneities. The ratio $\varepsilon = d/L$ is supposed to be small. Then the averaged equations can often be obtained that describe a certain homogeneous medium and have solutions close in some sense to solutions of original equations. For periodic media an asymptotic homogenization method to obtain the averaged equations is developed (Bensoussan, Lions, Papanicolaou, 1978; Sanchez-Palencia, 1980; Bakhvalov, Panasenko, 1984) making use of the presence of a small parameter $\varepsilon$.

In many problems there are additional small parameters $\gamma_i$, besides $\varepsilon$. For example the following parameters can be small: the ratios of different phases moduli, the ratios of coefficients determining different properties of a phase, the ratios of inhomogeneity scales in different directions. The averaged equations essentially depend on the relations between small parameters $\varepsilon$ and $\gamma_i$ (Sanchez-Palencia, 1980; Bakhvalov, Panasenko, 1984; Sandrakov 1987, Bakhvalov, Eglit, 1993, 1994, 1995a, 1995b).

For problems with additional small parameters it is sometimes possible to justify the averaged equations for nonperiodical, i.e. random media (Bakhvalov, Eglit, 1993, 1994, 1995a, 1995b).

Construction of averaged equations for periodic media includes solution of the so–called cell problems . They are boundary–value problems for partial differential equations. As a rule, they can be solved only numerically. In some cases analytical approximate solutions to the cell problems and the explicit formulae for effective coefficients can be obtained due to the presence of additional small parameters. The explicit formulae for effective moduli are very useful, especially in optimal design of materials and structures.

The paper gives a brief review of some recent results concerning the effect of different small parameters.

## 2 Structures and composites reinforced by a periodic system of mutually orthogonal plates and bars

Composites under consideration can be imagined as large buildings consisting of identical rectangular rooms filled by soft and weakly conducting material (matrix).

Plates (walls of the rooms) and bars (intersections of walls) are assumed to be relatively thin. Materials can be anisotropic. The properties of matrix, plates, bars and nodes (i.e. intersections of bars) can be different. Moduli of plates, bars and nodes are assumed to be larger then the matrix moduli.

Denote the period along $x_j$ by $\varepsilon_j$, the length of the sample by $L$, the thickness of a plate that is perpendicular to the $x_j$-axis by $d_j$; its relative thickness is $\nu_j = d_j/\varepsilon_j$. It is assumed that $d_j \ll \varepsilon_j \ll L$. The parameter $\nu = \max(d_1, d_2, d_3)/\min(\varepsilon_1, \varepsilon_2, \varepsilon_3)$ will also be used in the following reasoning.

Let us consider the equations

$$\frac{\partial}{\partial x_i} a_{kilj} \frac{\partial u_l}{\partial x_j} = f_k, \quad l, k = 1, \dots, n; \quad i, j = 1, \dots, s. \tag{2.1}$$

They describe, for example, displacements in an elastic composite or stationary heat conduction obeying the Fourier law; $a_{kilj}$ are supposed to be 1-periodic functions of $y_j = x_j/\varepsilon_j$; $n$ is the dimension of the vector $\mathbf{u}$ with components $u_l$ ($l, k \equiv 1$ if $u$ is scalar, e.g. temperature); $s$ is the dimension of the problem. We consider two–dimensional and three–dimensional problems.

Usually $a_{kilj}$ have the following symmetry

$$a_{kilj} = a_{ljki} \quad \forall i, j, k, l. \tag{2.2}$$

In some cases it is more convenient to write the system (2.1) in the form

$$L\mathbf{u} = \varphi, \qquad L\mathbf{u} = \frac{\partial}{\partial x_i}\left(A_{ij}\frac{\partial \mathbf{u}}{\partial x_j}\right), \qquad \varphi = \varphi(x), \tag{2.3}$$

where $A_{ij}$ are matrices with the elements $a_{kilj}$; $A_{ij} = A_{ji}^T$ for all $i, j$ due to (2.2).

Let us denote different parts of a periodicity cell by $\Pi_{j_1\dots j_s}$. In three–dimensional case $\Pi_{100}$, $\Pi_{010}$, $\Pi_{001}$ are the plates orthogonal to the $x_1$ - axis, $x_2$ - axis, and $x_3$ - axis, respectively; $\Pi_{110}$, $\Pi_{101}$, $\Pi_{011}$ are the bars parallel to these coordinate axes; $\Pi_{111}$ is a node; $\Pi_{000}$ is an interior of the room. Let

$$A_{ij} = A_{ij}^{j_1\dots j_s} = \text{const} \quad \text{in} \quad \Pi_{j_1\dots j_s}.$$

The system (2.3) is supposed to be elliptic. Matrices $A_{ij}^{j_1\dots j_s}$ satisfy the conditions

$$0 < Q_{j_1\dots j_s}^- \leq \left(A_{ij}^{j_1\dots j_s}\mathbf{b}_j, \mathbf{b}_i\right)/(\mathbf{b}_j, \mathbf{b}_j) \leq Q_{j_1\dots j_s}^+ < \infty \tag{2.4}$$

for any vectors $\mathbf{b}_j$ if $(\mathbf{b}_j, \mathbf{b}_j) \neq 0$. For elasticity problems the relations (2.4) are valid for vectors $\mathbf{b}_j$ that are columns of a symmetric matrix.

Denote

$$Q_{j_1\dots j_s} = \sqrt{Q_{j_1\dots j_s}^+ Q_{j_1\dots j_s}^-}, \quad \beta = \max_{j_1\dots j_s}\sqrt{Q_{j_1\dots j_s}^+/Q_{j_1\dots j_s}^-}.$$

It follows from (2.4) that

$$\beta^{-1}Q_{j_1\dots j_s} \leq \left(A_{ij}^{j_1\dots j_s}\mathbf{b}_j, \mathbf{b}_i\right)/(\mathbf{b}_j, \mathbf{b}_j) \leq \beta Q_{j_1\dots j_s}.$$

For the elasticity problem it is supposed that the dimension $n$ of the vector $\mathbf{u}$ is equal to $s$, and the following relations are true

$$a_{kilj} = a_{ljki} = a_{iklj} = a_{kijl} \quad \forall i, j, k, l;$$

Since we consider discontinuous $A_{ij}$, the solution (2.3) is understood as generalized one and satisfy the integral identity

$$\int_G (A_{ij}\frac{\partial \mathbf{u}}{\partial x_j}, \frac{\partial \mathcal{M}}{\partial x_i})dx = 0 \quad \forall \quad \mathcal{M} \in H_0^1(G);$$

Here $G$ is the domain where the problem for (2.3) is considered.

If $\varepsilon_j \ll L$, the equations with the constant coefficients $\overline{A}_{ij}$ can be constructed in such a way that their solutions are close to the solutions of the original equations (2.1). Elements $\overline{a}_{kilj}$ of the matrices $\overline{A}_{ij}$ are called the effective moduli of the periodic medium.

To calculate the effective moduli one should find vectors $\mathcal{M}^{qp}$ with the components $m_l^{qp}$ $(q = 1, ..., s; p, l = 1, ..., n)$ satisfying the equations

$$\frac{\partial}{\partial x_i} a_{kilj} \frac{\partial m_l^{qp}}{\partial x_j} = 0, \quad k = 1, \ldots, n \tag{2.5}$$

and the conditions

$$\mathcal{M}^{qp} - x_q \epsilon_p \in \mathbf{M}. \tag{2.6}$$

Here $\epsilon_p$ is a unit vector of coordinate axis $x_p$ ($\epsilon_p$ should be replaced by 1 if $n = 1$), $\mathbf{M}$ is a subspace of functions from $H^{1,loc}(R_s)$ that are periodic over $x_j$ with a period $\varepsilon_j$. The subspace of functions satisfying the condition (2.6) is denoted further by $\mathbf{M}^{qp}$.

Let

$$U(\mathbf{u}, \mathbf{v}) = a_{kilj} \frac{\partial u_l}{\partial x_j} \frac{\partial v_k}{\partial x_i}, \qquad \Lambda(\mathbf{u}, \mathbf{v}) = \langle U(\mathbf{u}, \mathbf{v}) \rangle .$$

Here $< \cdot >$ means averaging over the cell

$$< f >= \frac{1}{\varepsilon_1 \varepsilon_2 \varepsilon_3} \int_0^{\varepsilon_1} \int_0^{\varepsilon_2} \int_0^{\varepsilon_3} f(x_1, x_2, x_3) dx_3 dx_2 dx_1.$$

Coefficients $\overline{a}_{kilj}$ are calculated by the formulae (Bakhvalov, Panasenko, 1984)

$$\overline{a}_{kilj} = \Lambda(\mathcal{M}^{jl}, \mathcal{M}^{ik}).$$

The equations (2.5) are the Euler equations for the functional $\Lambda(\mathcal{M}, \mathcal{M})$;

$$\overline{a}_{qpqp} = \inf_{\mathcal{M} \in \mathbf{M}^{qp}} \Lambda(\mathcal{M}, \mathcal{M}), \tag{2.7}$$

infimum is reached at the functions $\mathcal{M}^{qp}$ that are the solutions of the problems (2.5), (2.6).

According to (2.7) the estimates hold

$$\overline{a}_{qpqp} \leq \inf_{\mathcal{M}(x_q) \in \mathbf{M}^{qp}} \Lambda(\mathcal{M}, \mathcal{M}). \tag{2.8}$$

The right hand sides of these inequalities can be found in quadratures for composites under consideration because the corresponding Euler equations are ordinary equations of special form.

Lower bounds for $\overline{a}_{ijkl}$ can be obtained by different ways.

### *2.1 Effective thermoconductivity of composites with isotropic components*

Consider for example heat conduction in a composite material made up of isotropic components. Let $k_0$, $k_1$, $k_2$ and $k_3$ be the thermoconductivity coefficients for matrix, plates, bars and nodes, respectively. The effective material is orthotropic. In a coordinate system with the axes directed along the bars only the coefficients $\overline{k}_{ii} \equiv \overline{a}_{i1i1}$ can be nonzero. The following inequalities are valid

$$\inf_{\mathcal{M}\in\mathbf{M}^{q1}} \left\langle k\left(\frac{\partial m}{\partial x_q}\right)^2\right\rangle \quad \leq \overline{a}_{q1q1} \equiv \overline{k}_{qq}. \tag{2.9}$$

Solving the Euler equations one obtains the estimates (2.8) and (2.9) for the coefficient $\overline{k}_{11}$ in the form (Bakhvalov, Panasenko, 1984):

$$k_{11}^- \leq \overline{k}_{11} \leq k_{11}^+,$$

where

$$k_{11}^- = \left\langle\left\langle\left\langle (k_{11})^{-1}\right\rangle_1^{-1}\right\rangle_2\right\rangle_3, \qquad k_{11}^+ = \left\langle\langle\langle k_{11}\rangle_2\rangle_3^{-1}\right\rangle_1^{-1}. \tag{2.10}$$

Here $\langle.\rangle_j$ means integration over $y_j$ in the limits $[0,1]$. Analogous estimates can be written for $\overline{k}_{22}$ and $\overline{k}_{33}$.

Fortunately for the composites under consideration the upper and lower estimates (2.10) are close to each other. Calculating $k_{11}^-$, $k_{11}^+$, we obtain the approximate formulae for $\overline{k}_{ii}$ (Bakhvalov, Panasenko, Eglit, 1996).

We write them here for the simplest case. Let all plates have the same relative thickness $\nu_1$. Then

$$\overline{k}_{11} = \overline{k}_{22} = \overline{k}_{33} = k_1\frac{k_0 + 2k_1\nu_1 + k_2\nu_1^2}{k_1 + k_0\nu_1}(1 + O(\Delta)),$$

where $\Delta = (k_1/k_2 + k_2/k_1 + k_2/k_3 + k_3/k_2)\nu_1$.

### *2.2 Effective elastic moduli*

To obtain formulae for effective elastic moduli and for effective thermoconductivity coefficients for locally anisotropic materials we constructed approximate solutions $\overline{\mathcal{M}^{qp}} \in \mathbf{M}^{qp}$ of the problems (2.5), (2.6). We will describe here briefly the solutions in order to determine the elastic moduli.

It is useful to note that at $p = q$ the problems (2.5), (2.6) correspond to a unit extension of a periodicity cell along the $x_q$ - axis at the following conditions: body forces are absent; $\mathbf{u}$ and its derivatives are periodic over the spatial variables $x_j, j \neq q$. At $p \neq q$ they correspond to a unit shear of a periodicity cell along the $x_p$-axis in the planes that are orthogonal to the $x_q$-axis.

For composites under consideration $\overline{\mathcal{M}}^{qp}$ are constructed as follows. At the first step we find $\overline{\mathcal{M}}^{qp}$ by considering $\Pi_{j_1 \dots j_s}$ as separated from each other and solving the problems of their extension or shear. The solutions for plates and bars are constructed under the assumption that the action of the soft interior of a room, i.e. of $\Pi_{0\dots0}$, can be neglected. When finding the solution for $\Pi_{0\dots0}$ we assume that the plates and bars behave as rigid bodies.

At the second step the displacements are spliced specially to make them continuous and satisfy the conditions (2.6).

The following inequalities were proved to be valid for $\overline{\mathcal{M}}^{qp}$

$$|\Lambda(\overline{\mathcal{M}}^{qp}, \varphi)| \leq r^{qp}\sqrt{\Lambda(\varphi, \varphi)} \quad \forall \quad \varphi \in \mathbf{M}$$

where $r^{qp}$ depend on relative thicknesses of plates and typical values of elastic moduli. For example, in two–dimensional case

$$r^{11} = \sqrt{Q_{00}\nu} + \sqrt{\nu_1(Q_{00})^2/Q_{10}} + \sqrt{\nu_2(Q_{00})^2/Q_{01}} +$$

$$\sqrt{\nu_1\nu_2(Q_{00})^2/Q_{11}} + \sqrt{\nu\nu_2 Q_{01}} + \sqrt{\nu_1\nu_2(Q_{01})^2/Q_{11}} + \sqrt{\nu_2 Q_{01}}.$$

Since

$$\Lambda(\mathcal{M}^{qp}, \varphi) = 0 \quad \forall \quad \varphi \in \mathbf{M},$$

if $\mathcal{M}^{qp}$ is a solution of (2.5), (2.6), then

$$|\Lambda(\overline{\mathcal{M}}^{qp} - \mathcal{M}^{qp}, \varphi)| \leq r^{qp}\sqrt{\Lambda(\varphi, \varphi)} \quad \forall \quad \varphi \in \mathbf{M}.$$

With the use of $\varphi = \overline{\mathcal{M}}^{qp} - \mathcal{M}^{qp}$ we obtain

$$\sqrt{\Lambda(\overline{\mathcal{M}}^{qp} - \mathcal{M}^{qp}, \overline{\mathcal{M}}^{qp} - \mathcal{M}^{qp})} \leq r^{qp}.$$

If $r^{qp}$ are small then $\overline{\mathcal{M}}^{qp}$ are the approximate solutions of (2.5), (2.6).

The next step is to separate the main part of $\Lambda(\overline{m}^{jl}, \overline{m}^{ik})$ to write

$$\Lambda(\overline{m}^{jl}, \overline{m}^{ik}) = \hat{a}_{kilj} + \delta_{kilj}$$

with small $\delta_{kilj}$. The approximate formulae for the effective moduli are

$$\overline{a}_{kilj} \sim \hat{a}_{kilj}. \tag{2.11}$$

The relative errors of these approximate formulae depend on $\nu$ and $\theta$ where $\theta$ is defined as follows. Introduce the pliability coefficient $\pi_{j_1\dots j_s}$ of the element $\Pi_{j_1\dots j_s}$ as the ratio of its volume $V_{j_1\dots j_s}$ to typical value of its elastic moduli

$$\pi_{j_1\dots j_s} = V_{j_1\dots j_s}/Q_{j_1\dots j_s}$$

and

$$\theta_{j_1\dots j_s} = \max_{i_1\geq j_1,\dots,i_s\geq j_s,\, i_1+\dots+i_s>j_1+\dots+j_s} \frac{\pi_{i_1\dots i_s}}{\pi_{j_1\dots j_s}}, \qquad \theta = \max_{j_1,\dots,j_s} \theta_{j_1\dots j_s}.$$

For example $\theta_{0\dots0}$ is the ratio of the greatest pliability coefficient to the pliability coefficient of the material inside the room, i.e. in $\Pi_{0\dots0}$.

The exact results are given in the following Theorems 1 and 2 that deal with two–dimensional and three–dimensional problems, respectively.

**Theorem 1.**

*In two–dimensional case $\hat{a}_{kilj}$ are determined by the following formulae*

$$\hat{a}_{k1l1} = (A^{00}_{11}\epsilon_k, \epsilon_l) + ((A^{01}_{11} - A^{01}_{12}(A^{01}_{22})^{-1}A^{01}_{21})\epsilon_k, \epsilon_l)\nu_2. \tag{2.12}$$

$$\hat{a}_{k2l2} = (A^{00}_{22}\epsilon_k, \epsilon_l) + ((A^{10}_{11} - A^{10}_{12}(A^{10}_{22})^{-1}A^{10}_{21})\epsilon_k, \epsilon_l)\nu_1. \tag{2.13}$$

$$\hat{a}_{k1l2} = (A^{00}_{12}\epsilon_k, \epsilon_l), \quad \hat{a}_{k2l1} = (A^{00}_{21}\epsilon_k, \epsilon_l). \tag{2.14}$$

*The error estimates of the approximate formulae (2.11) are as follows*

$$|\overline{a}_{kilj} - \hat{a}_{kilj}| \leq r_{kilj},$$

*where*

$$r_{kiki} = O(q_i(\nu + \theta)), \quad r_{kili} = O(q_i(\sqrt{\nu} + \sqrt{\theta})) \quad \text{at} \quad k \neq l,$$
$$r_{kilj} = O(\sqrt{q_1 q_2}(\sqrt{\nu} + \sqrt{\theta})) \quad \text{at} \quad i \neq j,$$
$$q_1 \equiv Q_{00} + Q_{01}\nu_2, \quad q_2 \equiv Q_{00} + Q_{10}\nu_1.$$

**Theorem 2.**

*In three–dimensional case $\hat{a}_{kilj}$ are determined by the following formulae and analogous ones that can be obtained from them by renumbering of coordinates*

$$\hat{a}_{k1l1} = (A_{11}^{000}\epsilon_k, \epsilon_l) + ((A_{11}^{010} - A_{12}^{010}(A_{22}^{010})^{-1}A_{21}^{010})\epsilon_k, \epsilon_l)\,\nu_2 + \\ ((A_{11}^{001} - A_{13}^{001}(A_{33}^{001})^{-1}A_{31}^{001})\epsilon_k, \epsilon_l)\,\nu_3 + \\ (A_{11}^{011}\epsilon_k + A_{12}^{011}\epsilon_{1k2}^{011} + A_{13}^{011}\epsilon_{1k3}^{011}, \epsilon_l)\nu_2\nu_3, \tag{2.15}$$

$$\hat{a}_{k1l2} = (A_{12}^{000}\epsilon_k, \epsilon_l) + ((A_{12}^{001} - A_{23}^{001}(A_{33}^{001})^{-1}A_{31}^{001})\epsilon_k, \epsilon_l)\nu_1\nu_2, \tag{2.16}$$

*where vectors $\epsilon_{1k2}^{011}$ and $\epsilon_{1k3}^{011}$ are the solutions of the following system*

$$\sum_{j=1}^{3} A_{lj}^{011}\epsilon_{1kj}^{011} = 0, \quad \epsilon_{1k1}^{011} = \epsilon_k.$$

*The error estimates of the approximate formulae (2.11) are*

$$|\overline{a}_{kilj} - \hat{a}_{kilj}| \le r_{kilj},$$

*where*

$$r_{k1k1} = O(\overline{q}_1(\nu + \theta)); \quad r_{k1l1} = O(\overline{q}_1(\sqrt{\nu} + \sqrt{\theta})); \quad \text{at} \quad k \ne l,$$

$$r_{k1l2} = O(\sqrt{\overline{q}_1\overline{q}_2}(\sqrt{\nu} + \sqrt{\theta})),$$

$$\overline{q}_1 \equiv Q_{000} + Q_{010}\nu_2 + Q_{001}\nu_3 + Q_{011}\nu_2\nu_3,$$

$$\overline{q}_2 \equiv Q_{000} + Q_{100}\nu_1 + Q_{001}\nu_3 + Q_{101}\nu_1\nu_3.$$

### *2.3 Effective elastic moduli for locally isotropic materials*

Let the components of matrices $A_{ij}$ be

$$a_{kilj} = \lambda\,\delta_{ik}\delta_{jl} + \mu\,(\delta_{ij}\delta_{kl} + \delta_{il}\delta_{jk}).$$

Due to the symmetry of the cell the effective material is orthotropic in this case. The nonvanishing elastic coefficients are $\hat{a}_{kiki}, \hat{a}_{iikk}$. In two–dimensional case calculation by the formulae (2.12),(2.13), (2.14) gives

$$\hat{a}_{1111} = \lambda_{00} + 2\mu_{00} + 4\frac{\mu_{01}(\lambda_{01} + \mu_{01})}{\lambda_{01} + 2\mu_{01}}\nu_2,$$

$$\hat{a}_{2222} = \lambda_{00} + 2\mu_{00} + 4\frac{\mu_{10}(\lambda_{10} + \mu_{10})}{\lambda_{10} + 2\mu_{10}}\nu_1,$$

$$\hat{a}_{2112} = \hat{a}_{1221} = \hat{a}_{1212} = \hat{a}_{2121} = \mu_{00}, \quad \hat{a}_{2211} = \hat{a}_{1122} = \lambda_{00},$$

the other effective coefficients are equal to zero.

If the moduli and the relative thicknesses of all plates are identical so that

$$\mu_{j_1j_2} = \mu_{j_1+j_2}, \quad \lambda_{j_1j_2} = \lambda_{j_1+j_2} \quad \forall\, j_1, j_2, \quad \nu_k = \nu_1 \quad \forall\, k,$$

the formulae for effective coefficients are simplified to read

$$\hat{a}_{1111} = \hat{a}_{2222} = \lambda_0 + 2\mu_0 + 4\frac{\mu_1(\lambda_1+\mu_1)}{\lambda_1+2\mu_1}\nu_1,$$
$$\hat{a}_{2112} = \hat{a}_{1221} = \hat{a}_{1212} = \hat{a}_{2121} = \mu_0, \quad \hat{a}_{2211} = \hat{a}_{1122} = \lambda_0.$$

In three–dimensional analogous case we have

$$\overline{a}_{1111} \approx \lambda_0 + 2\mu_0 + \frac{4\mu_1(\lambda_1+\mu_1)}{\lambda_1+2\mu_1}(\nu_2+\nu_3) + (2\mu_2 + \frac{\mu_2\lambda_2}{\lambda_2+\mu_2})\nu_2\nu_3,$$
$$\overline{a}_{1212} = \overline{a}_{1221} \approx \mu_0 + \mu_1\nu_3, \qquad \overline{a}_{1122} \approx \lambda_0 + \frac{4\mu_1^2}{\lambda_1+2\mu_1}\nu_1\nu_2.$$

Here $\lambda_1, \mu_1, \lambda_2, \mu_2$ are the Lame coefficients of the plates and bars, respectively. Formulae for the other nonzero coefficients can be obtained by renumbering of the coordinates. For example,

$$\overline{a}_{1313} \approx \mu_0 + \mu_1\nu_2.$$

It is proved that the relative error of the formulae for $\overline{a}_{qpqp}$ in this case is not larger than

$$C(\beta_1, \beta_2)(1 + \mu_1/\mu_2)\nu.$$

Here $C$ is a constant depending on $\beta_i$, $\beta_i = \sqrt{(2\mu_i + 3\lambda_i)/(2\mu_i)^{\operatorname{sign}\lambda_i}}$.

### *2.4 Auxiliary lemmas*

To estimate the errors of approximate solutions of the cell problems the following lemmas were proved.

**Lemma 1.**

*Let $A = ||A_{ij}||$ be a symmetric $2\times 2$ bloc matrix, $A_{11}, A_{22}$ be quadratic matrices and the following relation is valid*

$$0 < 1/\beta \le (A_{ij}\mathbf{b}_j, \mathbf{b}_i)/(\mathbf{b}_j, \mathbf{b}_j) \le \beta < \infty \quad \text{at} \quad (\mathbf{b}_j, \mathbf{b}_j) > 0; \qquad (2.17)$$

*here the summation over repeated indices is implied. Let* $\mathbf{b}_1$ *be a unit vector, and* $\mathbf{b}_2$ *satisfy the equations*

$$A_{21}\mathbf{b}_1 + A_{22}\mathbf{b}_2 = 0.$$

*Then*

$$0 < 1/\beta \leq (A_{ij}\mathbf{b}_j, \mathbf{b}_i) = ((A_{11} - A_{12}(A_{22})^{-1}A_{21})\mathbf{b}_1, \mathbf{b}_1) \leq \beta,$$

$$||\mathbf{b}_2|| = \sqrt{(\mathbf{b}_2, \mathbf{b}_2)} = ||(A_{22})^{-1}A_{21})\mathbf{b}_1|| \leq \sqrt{\beta^2 - 1}.$$

**Lemma 2.**

*Let the condition (2.17) be valid only for vectors that are the columns of a symmetric matrix and in the conditions of Lemma 1 are fulfilled. Then*

$$0 < (A_{ij}\mathbf{b}_j, \mathbf{b}_i) = ((A_{11} - A_{12}(A_{22})^{-1}A_{21})\mathbf{b}_1, \mathbf{b}_1) \leq \beta,$$

$$||\mathbf{b}_2|| = ||(A_{22})^{-1}A_{21})\mathbf{b}_1|| \leq \sqrt{2(\beta^2 - 1)}.$$

**Lemma 3.**

*Let* $\overline{\mathcal{M}}^{ik} \in \mathbf{M}^{ik}$*, and the following inequality be valid*

$$|\Lambda(\overline{\mathcal{M}}^{ik}, \mathcal{M})| \leq r\sqrt{\Lambda(\mathcal{M}, \mathcal{M})} \quad \forall \quad \mathcal{M} \in \mathbf{M}.$$

*Let* $\mathcal{M}^{ik}$ *be a solution to the problem (2.5), (2.6). Then*

$$\sqrt{\Lambda(\overline{\mathcal{M}}^{ik} - \mathcal{M}^{ik}, \overline{\mathcal{M}}^{ik} - \mathcal{M}^{ik})} \leq r, \qquad \overline{a}_{kiki} \leq \Lambda(\overline{\mathcal{M}}^{ik}, \overline{\mathcal{M}}^{ik}) \leq \overline{a}^{ii}_{kk} + r^2.$$

**Lemma 4.**

*Let* $\mathcal{M}^{ik}$ *and* $\mathcal{M}^{jl}$ *satisfy the problem (2.5), (2.6), and* $\overline{\mathcal{M}}^{ik} \in \mathbf{M}^{ik}$*,* $\overline{\mathcal{M}}^{jl} \in \mathbf{M}^{jl}$ *satisfy the conditions*

$$\sqrt{\Lambda(\overline{\mathcal{M}}^{ik} - \mathcal{M}^{ik}, \overline{\mathcal{M}}^{i}k - \mathcal{M}^{i}k)} \leq r^{ik},$$
$$\sqrt{\Lambda(\overline{\mathcal{M}}^{jl} - \mathcal{M}^{jl}, \overline{\mathcal{M}}^{jl} - \mathcal{M}^{jl})} \leq r^{jl}.$$

*Then*

$$|\Lambda(\overline{\mathcal{M}}^{jl}, \overline{\mathcal{M}}^{ik}) - \overline{a}_{kilj}| \leq \sqrt{\Lambda(\overline{\mathcal{M}}^{ik}, \overline{\mathcal{M}}^{ik})}r^{jl} + \sqrt{\Lambda(\overline{\mathcal{M}}^{jl}, \overline{\mathcal{M}}^{jl})}r^{ik}.$$

## 3 Composites reinforced by fibers

Let us consider composites reinforced by fibers (Bakhvalov, Eglit, 1995b). The fibers can be inhomogeneous and curvilinear; they can be located not periodically; the small break of fibers is permitted. Under certain conditions effective characteristics of such composites are some additive functions of the characteristics of fibers. These conditions are the relations that contain the ratios of fiber and matrix moduli, the fiber's diameters and distances between fibers.

We consider the boundary value problem for the equations (2.3) in the domain $G$ under the condition $\mathbf{u}|_{\partial G} = \mathbf{g}$ . As usual the solution is understood in the sense of integral identity

$$\mathbf{u} \in H^1(G), \quad \Lambda_G(\mathbf{u}, \varphi) = (\varphi, \varphi)_G \quad \text{for all} \quad \varphi \in H_0^1(G),$$

here

$$\Lambda_G(\mathbf{u}, \varphi) = \int_G \left( A_{ij} \frac{\partial \mathbf{u}}{\partial x_j}, \frac{\partial \varphi}{\partial x_i} \right) dx, \qquad (\varphi, \varphi)_G = \int_G (\varphi, \varphi)\, dx.$$

Let $V_m$ $(m = 1, ..., M)$ be fibers, $V_0 = G \setminus (\bigcup_{m=1}^{M} V_m)$ be matrix and $\operatorname{dist}(V_i, V_j) \geq \alpha_0 > 0$ for $i$, $j > 0$. We consider the

case of small ratios $\varepsilon$ and $\alpha$ of the typical fiber diameter and distance $\alpha_0$ to the macroscale $L$. Besides we assume that the ratio $\omega$ of the fibers moduli (thermal conductivity, elasticity) to matrix moduli is large. Let $\gamma$ be a volume concentration of fibers.

The main idea is to assume that the influence of matrix on heat conduction and elastic behavior of the composite in the first approximation can be neglected since $\omega$ is large. Then, e.g. heat can propagate only along fibers and the total heat flux equals the sum of fluxes through fibers. On this way it is easy to calculate approximately the effective moduli. More difficult problem is to obtain estimates of the error.

To calculate the effective thermoconductivity coefficients we should first find the relation between the heat flux $\mathbf{q}^m$ along the fiber $V_m$ and the temperature gradient along it. If $V_m$ is homogeneous, straight and oriented along the $x_1$-axis then the temperature $v^{(1)}$ in $V_m$ is given by the relation

$$v^{(1)} = v + N_j^m \frac{\partial v}{\partial x_j},$$

where $v$ is the averaged temperature and $N_j^m$ are linear functions of $x_2$, $x_3$ that can be found from the conditions $q_1^m = 0$, $q_2^m = 0$ for linear $v(x_1)$. Then, for linear $v(x_1)$,

$$q_1^m = k_{1j}\frac{\partial v^{(1)}}{\partial x_j} = c_{11}^m \frac{\partial v}{\partial x_1}.$$

The coefficient $c_{11}^m$ can be called the effective conductivity of $V_m$.

Let $V_m$ be inclined to $x_1$-axis and directed along $x_1'$. Then its effective conductivity tensor components $c_{ij}^m$ equal $c_{11}^m g_{i1} g_{j1}$ where $g_{ij}$ are components of the orthogonal matrix transforming coordinates $x_j'$ into $x_i$.

If the fibers are curvilinear (e.g. circular) they can be transformed into straight ones by a certain nonlinear transformation of coordinates. After that we apply the described algorithm and return to the old variables.

Introduce $\hat{k}_{ij}$ by the relations $\hat{k}_{ij} = 0$ in $V_0$, $\hat{k}_{ij} = \sum_{i=1}^{M} c_{ij}^m$ where $c_{ij}^m = 0$ outside of $V_m$.

Suppose the existence of $\overline{k}_{ij}(x)$ such that

$$\max_{i,j} \left\| \frac{\widehat{k}_{ij}}{\gamma} - \overline{k}_{ij} \right\|_{(-1)} \leq \delta_A \ll 1,$$

As usual, here

$$\|\varphi\|_{(-1)} = \sup_{\varphi \in H_0^1(G)} \frac{(\varphi, \varphi)_G}{\|\nabla \varphi\|_{L_2(G)}}.$$

If the material is homogeneous in average, then $\overline{k}_{ij}$ are the average values of $\widehat{k}_{ij}/\gamma$.

The solution of the original problem is close to the solution of the averaged problem for the homogeneous material with moduli $\overline{k}_{ij}$. The following inequality is proved

$$\int_{G \setminus V_0} |\nabla(\mathbf{u} - \mathbf{v}^{(1)})|^2 \, dx \leq b(R_1 + R_2), \tag{3.1}$$

where

$$R_1 = \varepsilon^2 + \omega^{-1} + \omega\gamma^2\delta_A^2, \qquad R_2 = \varepsilon/(\omega\alpha).$$

There is an additional term $R_3$ in the right part of (3.1) if the fibers have "breaks"; it is equal to a common measure of all breaks.

So the effective moduli $\overline{k}_{ij}$ are the additive functions of $c_{ij}^m$ if $R_1$, $R_2$ and $R_3$ are small.

As an example let us consider a material that can be obtained from a material reinforced by periodic system of homogeneous straight fibers by the following way. Original fibers are replaced by fibers with different diameters not exceeding $O(\varepsilon)$ and moved at a distance not larger than $O(\varepsilon)$ from their original position. Let the ratios of the thermal conductivities of the new and old fibers are inversely proportional to the ratios of their cross-section areas. Then it can be proved that

$$\delta_A^2 = O\left(\varepsilon^2 \ln\frac{1}{\varepsilon} + \frac{1}{\omega^2}\right)$$

and the values of $\overline{k}_{ij}$ for the new material equal those for original one and therefore can be calculated by the formulae given in (Bakhvalov, Panasenko, 1984).

## 4 Parameters related to different properties of each phase

Consider now composites that are characterized by small parameters connected with the weak influence of different physical properties of components. The examples are

i) elastic composites with small ratio $\gamma_1$ of typical values of shear and bulk moduli;

ii) viscous mixtures with small ratio $\gamma_2$ of typical properly scaled viscosity and compressibility coefficients;

iii) thermoconductive materials with small ratio $\gamma_3$ of typical thermoconductivity coefficient and specific heat capacity.

It is found that the effective properties of these materials depend essentially on the relations between the values of additional parameters $\gamma_i$ and the value of dimensionless length of inhomogeneity $\varepsilon$. We studied the process of sound propagation in composites (Sandrakov, 1987; Bakhvalov, Eglit, 1993, 1994, 1995a). Parameters $\gamma_i$ are

$$\gamma_1 = \sqrt{\frac{\mu_0}{\lambda_0}}; \qquad \gamma_2 = \sqrt{\frac{\nu_0}{L\sqrt{\lambda_0/\rho_0}}}; \qquad \gamma_3 = \sqrt{\frac{\chi_0}{L\sqrt{\lambda_0/\rho_0}}}.$$

Here $\lambda_0$, $\mu_0$, $\nu_0$, $\chi_0$ and $\rho_0$ are typical values of the compressibility coefficient, the shear modulus, the viscosity coefficient, the temperature conductivity coefficient and the density, respectively.

The simplest behavior is observed when the structure is very fine, i.e. $\varepsilon$ is much less then $\gamma_i$. At $\varepsilon \ll \gamma_1 \ll 1$ and at $\varepsilon \ll \gamma_2 \ll 1$ the effective compressibility coefficient $\overline{\lambda}$ and density $\overline{\rho}$ are determined by the formulae

$$\overline{\lambda} = \langle 1/\lambda(x)\rangle^{-1}, \qquad \overline{\rho} = \langle \rho(x)\rangle .$$

If $\varepsilon \gg \gamma_1$, $\varepsilon \gg \gamma_2$, $\varepsilon \gg \gamma_3$ and $\gamma_i \ll 1$, then shear elasticity, viscosity and thermoconductivity do not influence on the sound propagation. The sound speed is in general different in different directions and depend not only on the mean density and the compressibility, but on the details of material structure.

If the orders of parameters $\varepsilon$ and $\gamma_2$, $\gamma_3$ are equal then the behavior of the material is the most complicated: it is a material either with memory or with additional internal parameters.

## Acknowledgments

The work was supported by Russian Foundation for Basic Research (the project 96-01-01394) and RFFI-INTAS (the project 95-IN-RU-98)

## References

1. N.S. Bakhvalov and M.E. Eglit, Averaging of the equations of the dynamics of composites of slightly compressible elastic components, *Comp. Maths Math. Phys.*, **33**, 939 (1993).
2. N.S. Bakhvalov and M.E. Eglit, An estimate of the error of averaging the dynamics of small perturbations of very inhomogeneous mixtures, *Comp. Maths Math. Phys.*, **34**, 333 (1994).
3. N.S. Bakhvalov and M.E. Eglit, Homogenization of dynamic problems singularly depending on small parameters, *Proceedings of Second Workshop on Composite Media and Homogenization Theory*, Singapore, World Scientific, 17-35 (1995).
4. N.S. Bakhvalov and M.E. Eglit, Explicit Calculation of Effective Moduli for Composites Reinforced by an Irregular System of fibers, *Doklady Mathematics*, **51**, 46 (1995).
5. N.S. Bakhvalov and G.P. Panasenko, Homogenization. Averaging Processes in Periodic Media, *Mathematical Problems in Mechanics of Composite Materials*, Nauka, Moscow, **1984**, **Kluwer** Academic Publishers, Dordrecht-Boston-London, 1989.

6. N.S. Bakhvalov, G.P. Panasenko and M.E. Eglit, The effective properties of structures and composites with inclusions in the form of walls and rods, *Comp. Maths Math. Phys.*, **36** (12), 1701-1706 (1996).
7. A. Bensoussan, J.L. Lions, G. Papanicolaou, Asymptotic Methods in Periodic Structures, North Holland, Amsterdam, 1978.
8. E. Sanchez-Palencia, Non Homogeneous Media and Vibration Theory, Lecture Notes in Physics, 127, Berlin, Springer-Verlag, 1980.
9. G.M. Sandrakov, Averaging of a linearized system of hydrodynamics with low viscosity and the velocity of sound in mixtures, Preprint No. 178, OVM Akad. Nauk SSSR, Moscow, 1987 (Russian).

# NON-LINEAR DARCY LAW IN A RANDOM POROUS MEDIUM

A.Yu. BELIAEV
*Water Problem Institute, 3 Gubkina Str., 117971 Moscow, Russia*

This paper deals with the homogenization of a sub-differential equation in randomly perforated domain with Dirichlet boundary condition on its numerous "holes". A physical interpretation of this problem corresponds to flows of viscous-plastic liquids through a porous medium. In this asymptotic problem the leading term of the solution satisfies the Darcy law that links the mean value of velocity and the pressure drop. The mean velocity depends on the pressure drop in a very special way due to plasticity of the liquid. Usually there is a vicinity of zero point on the axis of pressure drop where the velocity equals identically zero. It takes place for any periodically perforated domain at least. Here the random checkerboard is considered as an example of a porous structure. In this case the vicinity degenerates, and the mean velocity does not vanish for any positive pressure drop. The explicit asymptotic formula for the Darcy law in the neighborhood of zero point is the main result of this paper.

## Introduction

The idea to conduct the study presented in this paper appeared after we worked with Serguei Kozlov on Darcy's equation in 1993–1995. Actually, the basic points of this contribution have much in common with the ideas of paper [2] and, also, with the approach proposed in paper [7] which was initiated by S. Kozlov.

The Darcy law for the dependence of the mean liquid velocity in a porous medium on the pressure gradient was established experimentally in 1856. Later, theoretical approaches of various levels of rigorousity were proposed to prove this law by means of homogenization of microscopic flows in pores. The first rigorous results with respect to random models of porous media has been obtained in [2]. The homogenization theory provides a way to obtain explicit expressions for the Darcy law in trivial models of porous structure only. All the pores in these structures are supposed to be isolated cylindric pipes. For general periodic structures the problem is reduced to a partial differential equation on the periodicity cell. It can be solved numerically. As for random models, the numerical analysis is too difficult because in this case the "cell" spreads on the whole space, and one can hope to get some asymptotics or estimates for parameters of this law.

Usually for sufficiently slow flows the relation between the liquid velocity and the pressure gradient is practically linear. There is, however, a class of liquids with the non-linear Darcy law, and the nonlinearity is especially significant for slow flows. In particular, oils moving in thin pipes or layers demonstrate this type of behavior. A flow of that liquid through a single cylindric tube takes place if the diameter of the tube is greater than a critical number that depends on the applied pressure gradient. If the diameter is less than the critical value then the liquid cannot move. Fluids of this kind are called viscous-plastic liquids.

For porous flows the effect of this phenomenon is that the liquid subjected to a small pressure gradient can flow through the greatest pores only, and the smaller the gradient the smaller the part of porous space which the liquid might go through. Due to this peculiarity for small pressure gradients the form of the Darcy law is determined by the structure and distribution of the largest pores only. This provides a hope to get explicit formulas for the asymptotic behavior of the Darcy law because the geometric structure of the largest pores could be more simple.

If the pore sizes in the medium are bounded then the curve "mean velocity versus pressure gradient" has a vertical segment with identically zero velocity. In this case, there exists a critical value of the gradient below which any flow is impossible. However, if the distribution law of pore sizes makes pores of arbitrarily large size acceptable then the Darcy law may demonstrate non-trivial behavior in the vicinity of the origin because in this case for any value of the pressure gradient an unbounded sample of the medium may contain a dilute system of sufficiently large pores that is able to conduct the viscous-plastic liquid.

An example of this kind is considered in the paper. Taking the random checkerboard as a model of porous structure we prove an explicit asymptotic formula for the dependence of the averaged liquid velocity $g$ on the pressure gradient $\lambda$ in the vicinity of the origin. The formula reads

$$g \sim \exp\left\{\frac{-C}{\lambda^d}\right\}.$$

where $C$ and $d$ are positive numbers.

The method of the proof appeared from the consideration of the problem relating neither to the Darcy law nor to viscous-plastic liquids. In the papers [12, 8] the authors were interested in the principal eigenvalue of the Laplace operator in random porous domains. The spectrum bottom is specified by the largest pores and clearings of the perforated structure. It depends on their shape and probabilistic properties. This is the reason for similarity of that problem to the Darcy law for viscous-plastic flows.

Dealing with the homogenization of porous flows in 1993, S. Kozlov was interested in the question whether it is possible to extend the results of [12, 8] to other elliptic operators and, in particular, to the Stokes operator. The main trouble was that the methods of [12, 8] were based on probabilistic interpretation of Laplacian as an infinitesimal operator of random diffusion process. So they were not applicable for vector valued operators. In the paper [7] Serguei's idea to re-establish the results of [12, 8] by "non-probabilistic" methods was realized, and this made possible to enlarge the set of resolved problems of this kind. One of them is presented here. The problem of the principal eigenvalue asymptotics for the Stokes operator is still open.

## 1 Posing the problem

For an arbitrary domain $D \subset R^d$ and a real parameter $\lambda \geq 0$ let us define a function $\Pi = \Pi(\lambda, D)$ by the following variational formula:

$$\Pi(\lambda, D) = \sup\left\{\int_D [\lambda\varphi - W(\nabla\varphi)]dx \;:\; \varphi \in C_0^\infty(D)\right\} \tag{1.1}$$

where $W(\xi), \xi \in R^d$, is a given function which grows up as $\xi \to \infty$. We shall be interested in the asymptotic behavior of $\Pi(\cdot, \cdot)$ for the domains $D$ which are supplemented with a perforated structure, i.e. the domains $D$ with a large number of uniformly distributed "holes". The volume fraction of these holes is not assumed to be close neither to 1 or to 0 as their number goes to infinity. To clarify this problem let us consider a couple of examples.

*Example 1.* Let $V$ be a periodic subset of $R^d$, and $Q_r = [-r/2, r/2]^d$ be a family of cubes in $R^d$. Then the sequence $D = D^{(r)} = Q_r \backslash V$, $r \to \infty$, is called a periodically perforated domain.

*Example 2.* Bernoulli's checkerboard is a random structure defined as follows. Let us decompose the integer lattice $Z^d$ into two random subsets by the simplest probabilistic rule, namely, each point $\alpha \in Z^d$ is called either free with the probability $p$, $0 < p < 1$, or occupied with the complementary probability $1 - p$. The states of different vertices are supposed to be independent. Then we put

$$V = \bigcup \left\{ \alpha + [-1/2,\, 1/2]^d : \ \alpha \in Z^d \text{ is occupied} \right\}$$

The Bernoulli's checkerboard is a sequence of random domains $D^{(r)} = Q_r \backslash V$, $r \to \infty$.

For $d = 2$ the problem (1.1) relates to a flow of viscous liquid through a fibrous material. Its fibers occupy 3-dimensional domain $V \times R^1$, and the parameter $\lambda$ stands for the pressure drop per unit fiber's length. The flow occurs in the direction of the fibers, and the solution $\varphi = \varphi^{(r)}(x)$ of the variational problem (1.1) is the unique component of velocity. The function $W(\cdot)$ is called dissipative potential. It describes viscous properties of the liquid. The case $W(\xi) = |\xi|^2/2$ corresponds to linear viscosity. The asymptotics of solutions of (2.1) $\varphi^{(r)}(x)$ was studied in detail for periodic ([1]) and random ([2]) structures. (For the sake of consistency in notations, induce a small parameter $\varepsilon = 1/r$ in stead of $r$ and hold the diameter of the domain $D^{(r)}$ fixed by changing variables $x \to x/\varepsilon$.) In this linear case the following asymptotic relations are valid for structures from examples 1 and 2:

$$\lim_{r \to \infty} \frac{1}{|Q_r|} \Pi(\lambda,\, Q_r \backslash V) = G(\lambda) \tag{1.2}$$

$$\lim_{r \to \infty} \frac{1}{|Q_r|} \int_{Q_r \backslash V} \varphi^{(r)} dx = G'(\lambda) \equiv g(\lambda) \tag{1.3}$$

where $G(\lambda) = k_0 \lambda^2/2$, and the constant $k_0$ depends on the choice of perforating set $V$. For the Bernoulli's checkerboard the non-random limits in (1.2) and (1.3) exist with probability 1, and $k_0$ depends on $p$ only. In accordance with the above physical interpretation the constant $k_0$ is called the scalar permeability of the material. The equality (1.3) is the Darcy law, i.e. a linear relation between the pressure drop and the mean velocity of the flow. The equalities (1.2), (1.3) take place for wide set of potentials $W(\cdot)$. A shape of the function $G(\lambda)$ is not the same for them though, and the Darcy law (1.3) is not necessary linear.

The only subject of this paper is the potential

$$W(\xi) = |\xi| + \frac{1}{2}|\xi|^2 \tag{1.4}$$

This formula was proposed by E.S. Bingham. The first term in (1.4) corresponds to plasticity of the liquid. This property is a distinctive feature of oils flowing in thin layers or pipes. The singularity of $W(\cdot)$ at the point $\xi = 0$ implies that the Euler equation of the variational problem (1.1) is not differential. It could be written as follows

$$\begin{cases} \nabla \cdot S^{(r)} + \lambda = 0, \\ S^{(r)} = \left(|\nabla\varphi^{(r)}|^{-1} + 1\right) \nabla\varphi^{(r)}, \ if \ \ \nabla\varphi^{(r)} \neq 0 \\ |S^{(r)}| \leq 1, \ if \ \ \nabla\varphi^{(r)} = 0 \\ \varphi^{(r)} = 0 \quad on \ \ \partial D^{(r)} \end{cases}$$

It is known that the solution $(S^{(r)}, \varphi^{(r)})$ of this sub-differential equation exists, and the function $\varphi^{(r)} \in W_0^{1,2}(D^{(r)})$ is unique (see [4, 5]). Due to the singularity of $W(\cdot)$ the solution depends on the parameter $\lambda$ crucially. There exists a critical value $\lambda^{(r)} > 0$ such that for any $\lambda \leq \lambda^{(r)}$ the liquid does not flow, i.e. $\varphi^{(r)} \equiv 0$ in $D^{(r)}$, and $\Pi(\lambda, \ D^{(r)}) = 0$. When $r \to \infty$ this peculiarity of Bingham liquid reveals itself by different way for periodic and random structures. For periodically perforated domains from the first Example $\lim \lambda^{(r)} = \lambda^* > 0$, and the limit functions $G(\lambda)$, $g(\lambda)$ equal zero for $\lambda < \lambda^*$. (This statement is a consequence of more general results obtained by A. Bourgeat and A. Mikelic in [3].) At the same time for many random models of perforating set $V$ the sequence $\lambda^{(r)}$ converges to zero, and both $G(\lambda)$ and $g(\lambda)$ are strictly positive for any $\lambda > 0$. In connection with this remark there is an interesting question how those functions behave in the vicinity of the origin $\lambda = 0$. An answer depends on the character of randomness. Here we investigate the case of Bernoulli's checkerboard. The goal of this paper is the following

**Theorem 1.1.** *For the random structure of the Example 2 the following equality holds true with probability 1:*

$$\lim_{\lambda \to 0+} \lambda^d \log G(\lambda) = -\kappa_d \log 1/p \tag{1.5}$$

*where* $\kappa_d = (d\sqrt{\pi})^d \left(\Gamma(1 + d/2)\right)^{-1}$. *The same limit relation is valid for the function* $g(\lambda)$.

An existence of limit functions $G$, $g$ could be established by the same way as it had been provided in [2] for linear case. That is why this part of proof is skipped. The logarithmic asymptotics of $g(\lambda)$ follows from (1.5) due to convexity of $G(\cdot)$, (the latter is a consequence of (1.1) and (1.2)). Indeed, for any $\lambda_1 < \lambda_2$ the following inequalities take place

$$g(\lambda_1) \equiv G'(\lambda_1) \leq \frac{G(\lambda_2) - G(\lambda_1)}{\lambda_2 - \lambda_1} \leq G'(\lambda_2) \equiv g(\lambda_2)$$

Here we put $\lambda_1 = 0$, $\lambda_2 = \lambda$ at first, and $\lambda_1 = \lambda$, $\lambda_2 = (1+\varepsilon)\lambda$ after that. This provides upper and lower estimates for $g(\lambda)$ in terms of $G(\lambda)$ :

$$\frac{G(\lambda)}{\lambda} \leq g(\lambda) \leq \frac{G\left((1+\varepsilon)\lambda\right)}{\varepsilon\lambda}$$

Therefore the logarithmic asymptotics of $g(\lambda)$ and $G(\lambda)$ are the same, and the relation (1.5) is the only thing we have to prove. The proof of this relation is based on a re-scaling method by which the problem can be reduced to a simpler case of perforated domain with low porosity. The same method was used in [6] where the asymptotics of critical values $\lambda^{(r)}$ was studied for Bernoulli's checkerboard. Their asymptotic behavior as $r \to \infty$ in the sense of convergence in probability obeys the relation

$$\lambda^{(r)} \sim \left(\frac{\kappa_d \log 1/p}{d \log r}\right)^{1/d}$$

The re-scaling method itself appeared in [7] with respect to another asymptotic problem that relates to randomly perforated domains. Here the re-scaling method of [6,7] is modified in order to simplify the proof. With that modification the results of [6, 7] could also be established in a more direct way. As in [6,7], in order to justify the re-scaling method, we use here a version of the Poincare inequality. That idea has been proposed by S. Kozlov for the corresponding eigenvalue problem. Actually he was the initiator of the paper [7], but his untimely death interrupted his participation in this work. Logarithmic asymptotics of the same kind could be proved for some other random structures. For instance the formula (1.5) is valid, except of the constant in its right hand side, if the perforating set $V$ is a union of balls distributed with the Poissonian probabilistic rule in $R^d$. The problem of further extensions is still open.

## 2 An estimate of $G(\lambda)$ from below

In this section we will prove the following estimate

$$\liminf_{\lambda\to 0+} \lambda^d \log G(\lambda) \geq -\kappa_d \log 1/p \tag{2.1}$$

To this end we shall use the same method that was applied in [6-8]. The main idea of the method is to replace the set $D^{(r)}$ by a slender domain $\tilde{D}^{(r)} \subset D^{(r)}$ with a simpler structure. The definition (1.1) implies the relation $\Pi(\lambda,\, \tilde{D}^{(r)}) \leq \Pi(\lambda,\, D^{(r)})$. This inequality provides a lower estimate either for $\Pi(\lambda,\, D^{(r)})$ or for the limit function $G(\lambda)$. Let $B_R(x)$ stand for a ball of radius $R$ centered at $x \in R^d$. In order to construct the auxiliary domain $\tilde{D}^{(r)}$ let us fix $R > 1$ and consider a set of balls $B_R(\alpha)$ centered at the vertices $\alpha$ of the re-scaled integer lattice $3R \cdot Z^d$. The radius $R$ and positions of the balls are chosen so that every unit cell of the checkerboard could not intersect any two balls simultaneously. Then we put

$$\tilde{D}^{(r)} = \bigcup_{\alpha} \left\{ B_R(\alpha) \subset Q_r : \ \ \alpha \in 3R \cdot Z^d,\ B_R(\alpha) \cap V = \emptyset \right\}$$

The domain $\tilde{D}^{(r)}$ consists of the identical disjoint balls which do not receive any occupied cell of Bernoulli's checkerboard. Therefore the number $\Pi(\lambda,\, \tilde{D}^{(r)})$ is decomposed into a sum of terms $\Pi(\lambda,\, B_R(\alpha))\mathbf{1}(\alpha)$ where $\mathbf{1}(\alpha)$ is the indicator of the event $\{B_R(\alpha) \cap V = \emptyset\}$. Random variables $\mathbf{1}(\alpha)$ are independent by their definition, and non-random factors $\Pi(\lambda,\, B_R(\alpha))$ are uniform. By the law of large numbers the normalized random number $|Q_r|^{-1}\Pi(\lambda,\, \tilde{D}^{(r)})$ converges to its expectation as $r$ goes to infinity. So we get an estimate

$$G(\lambda) \geq \lim_{r\to\infty} \frac{1}{|Q_r|} \Pi(\lambda,\, \tilde{D}^{(r)}) = \frac{1}{(3R)^d} \Pi(\lambda,\, B_R(0)) E(\mathbf{1}(0)) \tag{2.2}$$

where the notation $E(\cdot)$ stands for the expectation over probabilistic measure corresponding to Bernoulli's checkerboard. The number $\Pi(\lambda,\, B_R(0))$ can be calculated explicitly due to spherical symmetry of the problem (1.1) for a ball. Skipping those simple calculations we present here the answer

$$\Pi(\lambda,\, B_R) = |B_R| f\left( \left( \frac{\lambda R}{d} \right)^d \right)$$

where the function $f$ is defined by the following expression

$$f(s) = \begin{cases} 0, & 0 < s \le 1 \\ \frac{d}{2(d+2)} s^{2/d} - \frac{d}{d+1} s^{1/d} - \frac{1}{(d+1)(d+2)} \frac{1}{s} + \frac{1}{2}, & s \ge 1 \end{cases} \tag{2.3}$$

So $\Pi(\lambda, B_R)$ is equal zero for $\lambda R \le d$ and strictly positive for $\lambda R > d$. The number $E(\mathbf{1}(0))$ in (2.2) equals to probability of the event $\{B_R(0) \cap V = \emptyset\}$. It satisfies upper and lower estimates

$$p^{|B_{R'}|} \le E(\mathbf{1}(0)) \le p^{|B_R|}, \quad R' = R + \sqrt{d}/2 \tag{2.4}$$

Here the radius $R'$ is chosen in such a way that the ball $B_{R'}(0)$ contains all unit cubes intersecting the ball $B_R(0)$. Hence $|B_R|$ and $|B_{R'}(0)|$ are the bounds for the number of the integer vertices $\alpha \in B_R(0)$, and (2.4) follows from the definition of the random checkerboard. Now let us choose the radius $R$ to maximize the right hand side of the inequality (2.2) because this improves the estimate for $G(\lambda)$ from below. For $\lambda R \le d$ that right hand side vanishes due to the factor $\Pi(\lambda, B_R)$, and for large $R$ it goes to zero exponentially in accordance with the upper bound (2.4) of the multiplier $E(\mathbf{1}(0))$. Thus the optimal value of $R$ belongs to the segment $(d/\lambda, \infty)$. Let us take $R$ satisfying the inequalities

$$1 + \varepsilon \le \lambda R/d \le 1 + 2\varepsilon$$

for some $\varepsilon > 0$. Then the estimate follows from (2.2), (2.4):

$$G(\lambda) \ge \frac{|B|}{3^d} f((1+\varepsilon)^d) exp\{-|B|[d(1+2\varepsilon)/\lambda + \sqrt{d}/2]^d \log 1/p\}$$

where $|B| = \kappa_d \cdot d^{-d}$ is the volume of the unit ball in $R^d$. Taking the logarithm of both sides of this inequality and tending $\lambda$ to zero we obtain the following estimate

$$\liminf_{\lambda \to 0+} \lambda^d \log G(\lambda) \ge -(1+2\varepsilon)^d \kappa_d \log 1/p$$

This provides (2.1) due to arbitrariness of $\varepsilon$.

## 3 An estimate for $G(\lambda)$ from above for the case of small porosity $p$

In this section the following estimate will be proved

$$\limsup_{\lambda \to 0+} \lambda^d \log G(\lambda) \le -\kappa_d(\log 1/p - \log C) \tag{3.1}$$

where $C$ is a constant that depends on the dimension only. This upper bound is not exact as the one claimed by theorem 1.1, but its relative error vanishes if $p$ tends to zero. Below we use an isoperimetric inequality for $\Pi(\cdot, \cdot)$. Let us present it by the following

**Lemma 3.1.** *Let $B_R$ be a ball of radius $R$. Then for any domain $D \subset R^d$, $|D| \leq |B_R|$, the number $\Pi(\lambda, D)$ obeys the inequality*

$$\Pi(\lambda, D) \leq \Pi(\lambda, B_R) \equiv |B_R| f\left(\frac{\lambda^d |B_R|}{\kappa_d}\right) \tag{3.2}$$

In other words, the ball is an optimal domain in the sense that it has maximal value of dissipation number $\Pi(\cdot, \cdot)$ among all domains of the same volume. The proof of the lemma will be given below in appendix. We would like to use the results of the percolation theory, and it seems necessary to induce first some definitions. Let $z$ be a fixed odd number. Let us call a set $F \subset Z^d$ *connected* in $Z^d$ if the closed set

$$\tilde{F} = F + [-z/2, z/2]^d$$

is connected in $R^d$. For instance a pair of vertices $\alpha, \beta \in Z^d$ is connected if $\alpha - \beta \in [-z, z]^d$. Such vertices are called *neighbors.* Trough this section the number $z$ could be equal 1 but later we will consider it as an auxiliary large parameter. For any given configuration of free and occupied vertices in $Z^d$ let us define *free cluster* at point $\alpha \in Z^d$ as the maximal connected set $F = F(\alpha) \subset Z^d$ which consists of free vertices only and contains that $\alpha$. If $\alpha$ is occupied then we put $F(\alpha) = \emptyset$. Evidently for any two vertices their free clusters coincide if they have a non-empty intersection. Free clusters of Bernoulli's structure are random subsets of $Z^d$. For sufficiently small value of probability $p$ all of them are finite almost surely. This well-known fact is a consequence of the following

**Lemma 3.2.** *Let $|F(\alpha)|$ stand for the cardinal number of the free cluster $F(\alpha)$, and $P\{\cdot\}$ be the probabilistic measure corresponding to Bernoulli's checker structure. Then for any $s \geq 0$ the following estimate holds true*

$$P\{|F(\alpha)| \geq s\} \leq (C(z)p)^s \tag{3.3}$$

*with a positive constant $C(z) < e(1+2z)^d$.*

The proof of this lemma is given, for instance, in [9]. The estimate for the constant $C$ has another representation there (see formula V.22) because it is done in terms of "the number of the nearest neighbors". The last one is equal to $(2z+1)^d - 1$ for the considered type of connectedness. It follows from lemma 3.2 that if $C(z)p < 1$ then all free clusters are finite with the probability one. This statement is still valid for a wider range of the parameter $p$. To estimate $G(\lambda)$ from above note that the perforated domain $D^{(r)} = Q_r \backslash V$ is decomposed into a set of disjoint domains $\hat{F}(\alpha) = F(\alpha) + [-1/2, 1/2]^d$, $\alpha \in Z^d \cap Q_r$. That is why the number $\Pi(\lambda, D^{(r)})$ can be represented by a sum of terms corresponding to the cluster partition. If all clusters are finite then the limit in (1.2) can be re-written as follows

$$G(\lambda) = \lim_{r\to\infty} \frac{1}{|Q_r|} \sum_{\alpha \in Q_r} \frac{1}{|F(\alpha)|} \Pi(\lambda, \hat{F}(\alpha)) = E\left(\frac{1}{|F(0)|}\Pi(\lambda, \hat{F}(0))\right)$$

In this formula the factor $|F(\alpha)|^{-1}$ takes into account multiplicity of clusters' intersections. Also we have to assume the numbers under expectation sign and under sum symbol equal zero if $F(\alpha) = \emptyset$. The last equality is provided by ergodic properties of Bernoulli's structure. Now let us replace the set $\hat{F}(0)$ by a ball of the same volume. Due to the isoperimetric inequality (3.2), $|F(0)|^{-1}\Pi(\lambda, \hat{F}(0)) \le f\left(\lambda^d |F(0)|/\kappa_d\right)$ where the function $f(\cdot)$ is defined by (2.3). This provides the following inequalities

$$G(\lambda) \le E\left(f\left(\frac{\lambda^d |F(0)|}{\kappa_d}\right)\right) \le \sum_{s=1}^{\infty} f\left(\frac{\lambda^d s}{\kappa_d}\right) \cdot (C(z)p)^s \qquad (3.4)$$

In this sum all terms with the numbers $s \le \lambda^{-d}\kappa_d$ are equal to zero, and the summands with large numbers pay a small contribution due to the exponentially decreasing probabilities. The asymptotic behavior of the right hand side in (3.4) can be investigated by transformation of the sum into an integral over real values of $s$. The result of those standard calculations is given by

**Lemma 3.3.** *For any* $\nu > 0$

$$\lim_{\lambda\to 0+} \lambda^d \log \sum_{s=1}^{\infty} f\left(\frac{\lambda^d s}{\kappa_d}\right) e^{-\nu s} = -\kappa_d \nu$$

Now we put $\log(Cp) = -\nu$ in the inequality (3.4), and the estimate (3.1) follows from lemma 3.3 if $Cp < 1$. For greater values of $p$ it is also valid but becomes ineffective because the trivial bound (i.e. zero) is better.

## 4 An estimate for $G(\lambda)$ from above for any $p < 1$

Here we will prove the following estimate

$$\limsup_{\lambda\to 0+} \lambda^d \log G(\lambda) \le -\kappa' \tag{4.1}$$

with some constant $\kappa'$ in the right hand side. This number is not equal to the desirable constant $\kappa_d \log 1/p$, but it is positive for any $p > 0$.

Let $a$ be some odd number. Consider a partition of $R^d$ into cubes of the size $a$, that is

$$R^d = \bigcup \left\{Q_a(\alpha) : \alpha \in aZ^d\right\}, \quad Q_a(\alpha) \equiv \alpha + [-a/2, a/2]^d$$

We call these cubes *the blocks.* If $a$ is an odd number then each block is a union of $a^d$ unit cubes of the original checkerboard, $a$ is assumed to be a large parameter below. Let us decompose the set of all blocks into the two subsets by the following rule: call the block $Q_a(\alpha)$ *white* if

$$|Q_a(\alpha) \cap V| \le \delta |Q_a(\alpha)| \equiv \delta a^d$$

or *black* if the contrary inequality holds. The parameter $\delta$ will be chosen later. It is assumed small, so the white block is the one containing a relatively small number of the perforating cells. The union $V_1$ of all black blocks is a new Bernoulli's checker structure. It differs from the original structure $V$ by the size and the volume fraction of its cells only. Let us give the following

**Definition 4.1.** *White cluster associated with a point $\alpha \in aZ^d$ is the maximal subset $F(\alpha) \subset aZ^d$ of re-scaled integer lattice such that* (i) *the domain $\hat{F}(\alpha) = F(\alpha) + [-a/2, a/2]^d$ consists of white blocks only, and $\alpha \in F(\alpha)$;* (ii) *the closed domain $\tilde{F}(\alpha) = F(\alpha) - [-za/2, za/2]^d$ is connected in $R^d$.*

In this definition $z$ stands for some odd number. Since now we suppose $z \ge 3$. Henceforth the domain $\hat{F}(\alpha)$ is also called, as well as the discrete set $F(\alpha)$, white cluster, and $\tilde{F}(\alpha)$ is called the cluster's extension. The definition implies that domain $\tilde{F}(\alpha)\backslash\hat{F}(\alpha)$ includes black blocks only. We shall prove that all black blocks outside extended clusters could be excluded from the perforated domain $D^{(r)}$. The number $G(\lambda)$ for this new structure is less than original one, but its logarithmic order is the same for $\lambda \to 0$. In the case if the probability $p_1$ of the event { *"a fixed block is white"* } is sufficiently small, all extended clusters $\tilde{F}(\alpha)$ are bounded a.s., and the desired upper bound could be provided by isoperimetric inequality and exponential estimates for cluster's size like this was done in the previous section. This is the main idea of the re-scaling method. Note that $p_1$ could be, under some certain choice of parameters $a$ and $\delta$, as small as necessary. A different kind of extension was applied in a version of the re-scaling method which was proposed in [7]. All clusters were surrounded there by a "thin" boundary layer instead of "thick" domain $\tilde{F}(\alpha)\backslash\hat{F}(\alpha)$. That way could be applied here too, but the accepted approach implies a certain simplification of the proof. To prove an estimate for $G(\lambda)$ let us induce a cutoff function with a support in the extended clusters.

**Proposition 4.2.** *For any configuration of black and white blocks there exists a continuous function* $\zeta :\ R^d \to [0,\ 1]$ *such that* $\zeta = 1$ *on* $R^d\backslash V_1$, $\zeta = 0$ *outside the union of all extended clusters, and the gradient of that function satisfies everywhere the inequality*

$$|\nabla\zeta| \le \frac{2}{(z-1)a} \tag{4.2}$$

This function $\zeta(x)$ is called a cutoff function. For instance it can be defined by the formula

$$\zeta(x) = \max\left\{0,\ 1 - \frac{2}{(z-1)a}\mathrm{dist}\,(x,\ R^d\backslash V_1)\right\}$$

where dist$(\cdot,\cdot)$ stands for the Euclidean distance. The basic point of the re-scaling method's justification is the following version of the Poincare inequality

**Lemma 4.3.** *Let* $D \subset Q_a = [-a/2,\ a/2]^d$ *be a domain of volume* $|D| > \delta|Q_a|$. *Then for any function* $\varphi :\ Q_a \to R$ *which equals zero on*

*D the following inequalities are valid*

$$\int_{Q_a} |\varphi| dx \le a\Lambda(\delta) \int_{Q_a} |\nabla\varphi| dx \tag{4.3}$$

$$\int_{Q_a} |\varphi|^2 dx \le 2a^2\sqrt{d}\Lambda(\delta) \int_{Q_a} |\nabla\varphi|^2 dx \tag{4.4}$$

*where*

$$\Lambda(\delta) = \frac{\sqrt{d}}{\delta} \int_0^1 \min\left\{\frac{\delta}{t^d}, \frac{1-\delta}{(1-t)^d}\right\} dt$$

The inequality (4.4) was proved in [7], and (4.3) could be established by the same way. Note that factors in the right hand side of (4.3), (4.4) do not depend on the shape of the domain $D$, and their explicit representation in terms of its volume is the only supplement to the usual Poincare inequalities. The important for the further speculations is that the factor $\Lambda(\delta)$ goes to infinity as $\delta \to 0$. Now let us prove the following auxiliary estimate

**Proposition 4.4.** *Let $\varphi \in C_0^\infty(R^d)$ be a function which is equal to zero on the perforating domain $V$, and $\Phi(x) = \zeta(x)\varphi(x)$. Then for any block $Q_a(\alpha) \subset R^d$ the following inequality holds true*

$$\int_{Q_a} [\lambda\varphi - W(\nabla\varphi)]\, dx \le \int_{Q_a} \left[\lambda\Phi - \frac{1-\lambda/\lambda_0}{1+\varepsilon} W(\nabla\Phi)\right] dx \tag{4.5}$$

*where $\lambda_0 = (a\Lambda(\delta))^{-1}$, and $\varepsilon$ is an arbitrary number satisfying the inequality*

$$\frac{1}{z-1} + 8\sqrt{d}\Lambda(\delta)\frac{z}{(z-1)^2} < \varepsilon \tag{4.6}$$

**Proof.** Let $Q_a(\alpha)$ be a black block. In this case

$$\int_{Q_a} |\nabla\Phi| dx \le \int_{Q_a} [\zeta \cdot |\nabla\varphi| + |\varphi| \cdot |\nabla\zeta|]\, dx \le \tag{4.7}$$

$$\int_{Q_a} \left[|\nabla\varphi| + \frac{2}{(z-1)a} a\Lambda(\delta)|\nabla\varphi|\right] dx \le (1+\varepsilon) \int_{Q_a} |\nabla\varphi| dx$$

Here was used the estimate (4.2) for gradient of the cutoff function, and then applied the Poincare inequality (4.3). In the same way the following inequality can be established

$$\int_{Q_a} |\nabla\Phi|^2 dx \le \int_{Q_a} \left[\left(1+\frac{1}{t}\right)\zeta^2|\nabla\varphi|^2 + (1+t)|\nabla\zeta|^2\varphi^2\right] dx \le \tag{4.8}$$

$$\left[1+\frac{1}{t}+(1+t)\frac{4}{(z-1)^2a^2}2a^2\sqrt{d}\Lambda(\delta)\right]\int_{Q_a}|\nabla\varphi|^2 dx$$

where $t > 0$ is some arbitrary number. Posing $t = z - 1$ one can find that the expression inside the square brackets does not exceed $1 + \varepsilon$. Adding of (4.7) and (4.8) we obtain the estimate

$$\int_{Q_a} W(\nabla\Phi)dx \le (1+\varepsilon)\int_{Q_a} W(\nabla\varphi)dx \tag{4.9}$$

It is valid for any white block too because $\Phi \equiv \varphi$ in this case. Then let us apply lemma 4.3 again in order to establish the inequalities

$$\int_{Q_a} (1-\zeta)\varphi dx \le a\Lambda(\delta)\int_{Q_a}|\nabla\varphi|dx \le \frac{1}{\lambda_0}\int_{Q_a} W(\nabla\varphi)dx \tag{4.10}$$

These inequalities hold true either for black or white blocks. Taking into account (4.10) and (4.9) we can write the following chain of relations

$$\int_{Q_a}[\lambda\varphi - W(\nabla\varphi)]\,dx = \int_{Q_a}[\lambda\Phi + \lambda(1-\zeta)\varphi - W(\nabla\varphi)]\,dx \le$$

$$\int_{Q_a}[\lambda\Phi - (1-\lambda/\lambda_0)W(\nabla\varphi)]\,dx \le \int_{Q_a}\left[\lambda\Phi - \frac{(1-\lambda/\lambda_0)}{(1+\varepsilon)}W(\nabla\Phi)\right]dx$$

Thus the proposition is proved.

For sufficiently small $\lambda$ (that is $\lambda < \lambda_0$) the proposition 4.4 yields the inequality

$$\int_{Q_r}[\lambda\varphi - W(\nabla\varphi)]\,dx \le \frac{\lambda}{\lambda'}\int_{Q_r}[\lambda'\Phi - W(\nabla\Phi)]\,dx \tag{4.11}$$

where $\lambda' = \lambda(1+\varepsilon)(1-\lambda/\lambda_0)^{-1}$. It follows by summation of (4.5) over all $\alpha \in aZ^d \cap Q_r$. The estimate (4.11) is valid for any function $\varphi$ which belongs to the domain of the functional in (1.1) with $D = D^{(r)} = Q_r \backslash V$. Each function $\Phi = \zeta\varphi$ has a support inside the domain $D_1^{(r)} = Q_r \cap \left(\cup\{\tilde{F}(\alpha)\backslash V : \ \alpha \in aZ^d\}\right)$. The upper bound of the right hand side in the inequality (4.11) over all functions $\Phi \in C_0^\infty(D_1^{(r)})$ is equal to $\lambda/\lambda'\Pi(\lambda',\, D_1^{(r)})$ with respect to the definition (1.1) of $\Pi(\cdot,\cdot)$. That is why the following estimate holds true

$$\Pi(\lambda,\, D^{(r)}) \le \Pi(\lambda',\, D_1^{(r)})$$

The factor $\lambda/\lambda'$ is dropped out here since it is less than 1. This estimate yields

**Proposition 4.5.** *If the probability $p_1$ of white block is such that $C(z)p_1 < 1$ then for any $\lambda < \lambda_0$ the number $G(\lambda)$ obeys the inequality*

$$G(\lambda) \le E\left(\frac{1}{|\hat{F}|}\Pi(\lambda',\, \tilde{F}\backslash V)\right) \tag{4.12}$$

*where $\tilde{F} = \tilde{F}(0)$ stands for extension of the cluster $\hat{F} = \hat{F}(0)$, and the parameters $\lambda_0$, $\lambda'$, $C(z)$ have been defined above.*

Now we construct an upper bound for the probability $p_1$ of white block. For the sake of simplicity we suppose that the parameters $a$ and $\delta$ are chosen in such a way that $\delta a^d$ is an integer number.

**Proposition 4.6.** *Probability $p_1$ could be estimated from above as follows*

$$p_1 \le \exp\left\{-a^d\left[(1-\delta)\log\frac{1}{p} - \delta - \delta\log\frac{1}{\delta}\right]\right\}$$

**Proof.** By the definition the block $[-a/2,\, a/2]^d$ is white if there are at least $(1-\delta)a^d$ free vertices among $a^d$ points of integer lattice $Z^d$ inside it. The probability $p_1$ of this event satisfies the inequality

$$p_1 \le \binom{N}{M} p^M$$

where the notations $N = a^d$ and $M = (1-\delta)a^d$ have been induced for brevity. The binomial coefficient can be estimated by the Stirling formula as follows

$$\binom{N}{M} = \frac{N!}{M!(N-M)!} \le \frac{N^{N-M}}{(N-M)!} \le \left(\frac{eN}{N-M}\right)^{N-M}$$

Combining both these inequalities and substituting the former notations we get the desired estimate.

It follows from this proposition that for sufficiently small $\delta$ and large $a$ the probability $p_1$ is close to zero. Hence the conditions of the proposition 4.5 are satisfied. In the inequality (4.12) let us replace the domain $\tilde{F}\backslash V$ by a ball of volume $|\tilde{F}|$. Due to lemma 3.1 this cannot break the inequality. So we get the estimate

$$G(\lambda) \leq E\left(\frac{|\tilde{F}|}{|\hat{F}|} f\left(\frac{\lambda'^d |\tilde{F}|}{\kappa_d}\right)\right) \tag{4.13}$$

In order to transform this inequality into the desired estimate it is necessary to find an upper bound for the volume distribution of the extended cluster $\tilde{F}$. It follows from definition 4.1 that

$$|\tilde{F}| \leq z^d |\hat{F}(0)| = (az)^d |F(0)| \tag{4.14}$$

where $|F(0)|$ is the cardinal number of the cluster $F(0) \subset aZ^d$. An estimate for the distribution of the number $|F(0)|$ is given by lemma 3.2. Therefore

$$P\left\{|\tilde{F}| \geq s\right\} \leq P\left\{|F(0)| \geq \frac{s}{(az)^d}\right\} \leq (C(z)p_1)^{(az)^{-d}s} \leq e^{-\nu s}$$

where $\nu = z^{-d}[(1-\delta)\log 1/p - \delta \log 1/\delta - a^{-d} \log C(z)]$, and the probability $p_1$ has been evaluated from above by means of proposition 4.6. At last we obtain from (4.13) the following estimate

$$G(\lambda) \leq z^d \sum_{s=1}^{\infty} f\left(\frac{\lambda'^d s}{\kappa_d}\right) e^{-\nu s} \tag{4.15}$$

For any $p < 1$ there exist parameters $z$, $a$, $\delta$ such that $\nu > 0$. Therefore it is possible to apply lemma 3.3 because $\lambda' \to 0$ as soon as $\lambda \to 0$, and the estimate (4.1) follows. For the constant $\kappa'$ we have got the formula

$$\kappa' = \kappa_d \nu (1+\varepsilon)^{-d}$$

The auxiliary parameters $a$, $\delta$, $z$ could be chosen to optimize the estimate (4.1). The fork formed by (4.1) and (2.1) does not vanish though. For any admissible choice of those parameters the value of $\kappa'$ is less than $z^{-d}\kappa_d \log 1/p$ with $z \geq 3$. The factor $z^{-d}$ had appeared due to the estimate (4.14) where the volume of the cluster $\hat{F}$ is compared with the volume of its extension $\tilde{F}$. Actually the last one is greater, so the estimate (4.14) cannot be improved. Really the loss of exactness in estimates happened in the transformation from (4.12) to (4.13) where the perforated domain $\tilde{F}\backslash V$ was replaced by the cluster extension $\tilde{F}$. In the next section "clearing" of the domain will be made by more delicate means.

## 5 The proof of exact upper estimate

Let us consider a hierarchy of re-scaled checker structures with blocks' sizes $a_1 < a_2 < ... < a_n$. The real parameter $\delta > 0$ and an odd number $z \geq 3$ are supposed to be uniform at all levels of the hierarchy. The criterion of block color and the type of connectedness on re-scaled lattices $a_k Z^d$, $k = 1, ..., n$, are induced like above. The sequence of block sizes is assumed to be a geometric progression $a_k = T^k$, $T > 1$. Let us take this number $T$ being an odd integer. In this case each block $Q_{a_k}(\alpha) = \alpha + [-a_k/2,\, a_k/2]^d$ centered at $\alpha \in a_k Z^d$ can be decomposed into $T^d$ blocks of $(k-1)$-th level. Let us choose the parameters $\delta$, $z$, $T$, $n$ in the following way. At first take a small number $\varepsilon > 0$. After that find $\delta$ satisfying the inequality

$$\delta(\log 1/p + 1 + \log 1/\delta) < \varepsilon \tag{5.1}$$

Then we choose the parameter $z$ satisfying (4.6). At last we take $T$ and $n$ complying with the following inequalities

$$(1 + z/T)^d - 1 < \varepsilon, \quad z^d/n < \varepsilon \tag{5.2}$$

Later we will tend $\varepsilon$ to zero, so the conditions (4.6), (5.1), (5.2) mean that $\delta$ is small number, and $z$, $T$, $n$ are large parameters. Denote by $p_k$, $1 \leq k \leq n$, the probability of event that fixed block $Q_{a_k}(\alpha)$ of the level $k$ is white. It follows from the proposition 4.6 and inequality (5.1) that $\log 1/p_k > T^{kd}(\log 1/p - \varepsilon)$. Hence, for sufficiently small $\varepsilon$ the probabilities $p_k$ are small. The numbers $C(z)p_k$ are close to zero too because $p_k$ tends to zero exponentially with respect to some power of the parameter $T$, and $C(z)$ obeys polynomial estimate with respect to $z$, and $z \ll T$ due to the first estimate (5.2). Since now we suppose that $C(z)p_k < 1$ for all levels of the hierarchy. This assumption provides a.s. finiteness of all white clusters $F_k(\alpha) \subset a_k Z^d$. In particular at the highest level $k = n$ both the cluster $\hat{F} = \hat{F}(0) \subset R^d$ and its extension $\tilde{F}$ are bounded. Therefore the number $G(\lambda)$ obeys the estimate (4.12) of proposition 4.5. Let us construct a set of embedded domains $E_n \subset E_{n-1} \subset ... \subset E_1$ inside this extended cluster $\tilde{F}$ by the following recursive rule for $k = n,\ n-1,\ ...,\ 1$

$$E_n = \tilde{F} \backslash V_n \equiv \hat{F}; \qquad E_k = E_{k+1} \cup \left[\tilde{F} \backslash V_k\right]$$

where $V_k$ stands for the union of all black $k$-blocks. In accordance with this rule each domain $E_k$ consists of white blocks of levels from $k$ to $n$. Non-intersecting domains $F_k = E_k \backslash E_{k+1}$ include those white $k$-blocks from $\tilde{F}$ which are not embedded into $E_{k+1}$. The domain $F_n$ is, up to its closure, consistent with white cluster $\hat{F}$ of $n$-th level centered at the origin. Now we put

$$k_0 = \max\left\{k : \quad |F_k| \leq \frac{1}{n}|E_1|\right\}$$

If $\tilde{F} \neq \emptyset$ the number $k_0$ exists because $|F_1| + ... + |F_n| = |E_1|$. Let us introduce the notations

$$\hat{E} = E_{k_0}; \quad \tilde{E} = \tilde{F} \bigcap \left( \bigcup \left\{ \alpha + \left[-\frac{z a_{k_0}}{2},\ \frac{z a_{k_0}}{2}\right]^d : \quad \alpha \in a_{k_0} Z^d \cap \hat{E} \right\} \right)$$

The domain $\hat{E}$ includes white blocks of levels $k_0,\ ...,\ n$ from $\tilde{F}$, and $\tilde{E}$ is its extension constructed by the surrounding $\hat{E}$ with the boundary layer of thickness $\frac{z-1}{2} a_{k_0}$. We are going to prove that the domain $\tilde{F} \backslash V$ in the right hand side of (4.12) could be replaced by $\tilde{E}$, and the asymptotic behavior of this upper bound would be almost the same. Therefore the isoperimetric estimate (4.13) can be improved as far as the volume of $\tilde{E}$ is less than $|\tilde{F}|$. Later we will prove that in some probabilistic sense the volumes of $\tilde{E}$ and $\hat{F}$ are almost the same asymptotically although the first domain is wider. To begin this program let us formulate the following

**Proposition 5.1.** *Let* $\tilde{F}$ *be an extended cluster of* $n$*-th level and* $\tilde{E}$ *be the domain constructed above. Then for any* $\lambda < \lambda_0$

$$\Pi(\lambda,\ \tilde{F} \backslash V) \leq \Pi(\lambda',\ \tilde{E}) \tag{5.3}$$

*where* $\lambda_0 = (a_n \Lambda(\delta))^{-1}$, $\lambda' = (1+\varepsilon)\lambda(1-\lambda/\lambda_0)^{-1}$.

**Proof.** Let us induce a cutoff function $\zeta : \tilde{F} \to [0,\, 1]$ such that

$$\zeta|_{\hat{E}} = 1; \quad \zeta|_{\tilde{F}\setminus\tilde{E}} = 0; \quad |\nabla\zeta| \le \frac{2}{(z-1)a_{k_0}}$$

It could be constructed by the same way as in proposition 4.2. Then for any $k_0$-block from $\tilde{F}$ and for any function $\varphi \in C_0^\infty(\tilde{F}\setminus V)$ the proposition 4.4 is applicable, and inequality (4.5) holds true with $a = a_{k_0}$. Adding these inequalities over all blocks $Q_a(\alpha)$, $\alpha \in \tilde{F} \cap a_{k_0} Z^d$ we arrive at the relation

$$\int_{\tilde{F}} (\lambda\varphi - W(\nabla\varphi))dx \le \frac{\lambda}{\lambda'} \int_{\tilde{E}} (\lambda'\Phi - W(\Phi))dx$$

where the function $\Phi = \zeta\varphi$ has its support inside $\tilde{E}$. In this inequality the upper bound of the right hand side does not exceed the right hand side of (5.3). As far as $\Pi(\lambda,\, \tilde{F}\setminus V)$ is the supremum of the left hand side over all possible $\varphi$, the estimate (5.3) follows.

Since now let us consider probabilistic properties of the domain $\tilde{E}$. To this end we recall the following version of the Chebyshev inequality

$$P\left\{|\tilde{E}| > s\right\} \le e^{-\nu s} E\left(e^{\nu|\tilde{E}|}\right) \tag{5.4}$$

where $\nu > 0$ is an arbitrary number providing the existence of the expectation in the right hand side. First let us establish the inequality

$$|\tilde{E}| \le (1+2\varepsilon)|E_1| \tag{5.5}$$

by which the probabilistic estimates for $|\tilde{E}|$ could be deduced analyzing the distributions of the set $E_1$. The domain $\hat{E}$ can be represented as follows $\hat{E} = E_{k_0+1} \cup F_{k_0}$. (If $k_0 = n$ then we put $E_{k_0+1} = \emptyset$). Since $\tilde{E}$ is a result of surrounding $\hat{E}$ with a boundary strip, it can also be represented by union of extended domains $E_{k_0+1}$ and $F_{k_0}$. The domain $E_{k_0+1}$ includes large-scaled blocks of sizes $a_{k_0+1} \equiv T a_{k_0}$ or greater. The extending can not change its volume significantly because the thickness $\frac{z-1}{2} a_{k_0}$ of that boundary layer is relatively small. The volume of the extended $(k_0+1)$-block is equal to $a_{k_0+1}^d (1 + z/T)^d$, hence the volume of $E_{k_0+1}$ extended is not greater than $|E_{k_0+1}|(1+z/T)^d$. The extension of the set $F_{k_0}$ can change its volume up to $z^d$ times. But the volume of this set itself is not large because $|F_{k_0}| \le |E_1|/n$ due to definition of $k_0$. Thus

$$|\tilde{E}| \le |E_{k_0+1}|(1+z/T)^d + \frac{z^d}{n}|E_1|$$

and (5.5) follows due to conditions (5.2). To estimate volume distribution we need the following auxiliary

**Proposition 5.2.** *Let $F_{k+1}, ..., F_n$ be some bounded non-intersecting domains in $R^d$ such that each domain $F_l$, $k+1 \le l \le n$, is a union of blocks of l-th level. Denote by $\tilde{F}_n$ an extension of the set $F_n$ :*

$$\tilde{F}_n = \bigcup \left\{ \alpha + \left[ -\frac{za_n}{2}, \frac{za_n}{2} \right]^d : \quad \alpha \in F_n \cap a_n Z^d \right\}$$

*Suppose all domains $F_l$, $k+1 \le l \le n$, are embedded into $\tilde{F}_n$. Denote by $A_-$ and $A_+$ the events { "each $F_l$ includes white l-blocks only."} and { "each set $\tilde{F}_n \backslash (F_n \cup ... \cup F_l)$ consists of black l-blocks only "} correspondingly. We put $H = F_{k+1} \cup ... \cup F_n$. Let $F_k$ be the union of all white k-blocks from the set $\tilde{F}_n \backslash H$. Then for any $\nu > 0$ the following inequality holds true*

$$E\left(e^{\nu|F_k|} \mid A\right) \le E\left(e^{\nu|F_k|}\right) \tag{5.6}$$

*where the conditional expectation is taken with respect to the event $A = A(F_{k+1}, ..., F_n) \equiv A_+ \cap A_-$. In the case $\nu < (1-\delta)\log 1/p - \delta - \delta \log 1/\delta$ an additional estimate is valid*

$$E\left(e^{\nu|F_k|} \mid A\right) \le exp\left\{ \frac{z^d}{a_k^d} |F_n| \log 2 \right\} \tag{5.7}$$

**Proof.** Underline that in this proposition the domains $F_{k+1}, ..., F_n$ are considered as nonrandom fixed sets. Denote all possible configurations of occupied and free vertices in $H \cap Z^d$ by $\Omega_-$, and configurations in $(\tilde{F}_n \backslash H) \cap Z^d$ by $\Omega_+$. Then the probabilistic space of all configurations in $\tilde{F}_n \cap Z^d$ is $\Omega_- \times \Omega_+$. The random variable $\exp\{\nu|F_k|\}$ depends on elements from $\Omega_+$ only. Let $\mathbf{1}_A$ be the indicator of the event $A$. From the definition of the conditional expectation

$$E\left(e^{\nu|F_k|} \mid A\right) = \frac{E\left(e^{\nu|F_k|}\mathbf{1}_A\right)}{E\left(\mathbf{1}_A\right)} = \frac{E_+\left(e^{\nu|F_k|}E_-\left(\mathbf{1}_A\right)\right)}{E\left(\mathbf{1}_A\right)}$$

where $E_-$ and $E_+$ stands for expectations in $\Omega_-$ or $\Omega_+$, respectively. The random variables $exp\{\nu|F_k|\}$ and $E_-(\mathbf{1}_A)$ are monotonic functions on $\Omega_+$ in the sense that the first variable does not enlarge its value, and the second one cannot become smaller when a free point from $(\tilde{F}_n \backslash E_{k+1}) \cap Z^d$ is replaced by occupied vertex. Therefore it is possible to apply FKG inequality to estimate the iterated expectation in the right side of the last equality. (The details appertaining to FKG inequality can be found in [9].) That inequality provides the estimate

$$E_+\left(e^{\nu|F_k|}E_-\left(\mathbf{1}_A\right)\right) \leq E_+\left(e^{\nu|F_k|}\right) \cdot E_+\left(E_-\left(\mathbf{1}_A\right)\right)$$

It is followed by (5.6) because $E_+(\cdot) = E(\cdot)$ for random variables depending on configurations from $\Omega_+$ only. To prove the supplementary statement (5.7) note that domain $\tilde{F}_n \backslash H$ is the union of $k$-blocks, and their number is equal to $m = |\tilde{F}_n \backslash H| \cdot a_k^{-d}$. Colors of these blocks are independent. That is why the unconditional expectation in (5.6) could be found from the formula

$$E\left(e^{\nu|F_k|}\right) = \left[(1-p_k) + p_k \exp\{\nu a_k^d\}\right]^m$$

It follows from proposition 4.6 that $1 - p_k + p_k \exp\{\nu a_k^d\} < 2$ since number $\nu$ obeys the condition declared in the proposition. The obvious estimate $m \leq |F_n| z^d / a_k^d$ completes the proof of the inequality (5.7).

Now let us return to the problem of estimating the term $E\left(\exp\{\nu|\tilde{E}|\}\right)$ in Chebyshev's inequality (5.4). It follows from (5.5) that $E\left(\exp\{\nu\tilde{E}\}\right) \leq E\left(\exp\{\nu'|E_1|\}\right)$, where $\nu' = (1+2\varepsilon)\nu$. For some $k$, $1 \leq k \leq n-1$, let us fix a configuration $A_k$ of the re-scaled structures at the levels $k+1, ..., n$ of the hierarchy. Then the identity

$$E\left(\exp\{\nu'|E_1|\}\right) = E\left[\exp\{\nu'|E_{k+1}|\} E\left(\exp\{\nu'(|F_k| + ... + |F_1|) \mid A_k\right)\right]$$

is valid because the domain $E_{k+1} = F_n \cup ... \cup F_{k+1}$ is fixed since the configuration $A_k$ is given. In the case $k = 1$ it is possible to apply the estimate (5.6) from proposition 5.2 for the conditional expectation in the right hand side of this identity. Indeed, a fixed configuration $A_1$ of checkerboards at the levels 2,...,n extracts the set of configurations at the original checkerboard which corresponds to the event $A = A(F_2, ..., F_n)$ from the proposition 5.2. Then suppose

$\nu' < \log 1/p - \varepsilon$, and the estimate (5.7) is also applicable. Therefore

$$E\left(\exp\{\nu'|E_1|\}\right) \leq E\left(\exp\left\{\nu'|E_2| + \frac{z^d}{a_1^d}|F_n|\log 2\right\}\right)$$

The term in the left hand side can be evaluated by means of induction over $k$ from 1 to $n-1$:

$$E\left(\exp\{\nu'|E_1\}\right) \leq E\left(\exp\{\nu''|F_n|\}\right) \tag{5.8}$$

where $\nu'' = \nu' + z^d\left(a_1^{-d} + a_2^{-d} + \ldots + a_{n-1}^{-d}\right)\log 2 < \nu' + 2z^dT^{-d}\log 2$. The domain $F_n$ is a white cluster of Bernoulli's checkerboard. That is why the distribution of its volume obeys the exponential estimates from lemma 3.2. So (5.8) is followed by the inequality

$$E\left(\exp\{\nu|\tilde{E}|\}\right) \leq \sum_{s=0}^{\infty} \exp\left\{[\nu'' a_n^d - \log\, 1/p_n + \log C(z)]s\right\}$$

The right hand side here is finite if and only if the number inside the square brackets is negative. Taking into account the estimate for $p_n$ from proposition 4.6, we get the following sufficient condition of that finiteness

$$\nu \leq \left(\log 1/p - \varepsilon - 2\frac{z^d}{T^d}\log 2 - \frac{\log C(z)}{a_n^d}\right)(1+2\varepsilon)^{-1} \equiv \nu^*$$

It follows from (5.2) that $2(z/T)^d \log 2 < \varepsilon$ and $a_n^{-d}\log C(z) < \varepsilon$. That is why the number $\nu^*$ for small values of $\varepsilon$ is positive and close to $\log 1/p$. Finally by the Chebyshev inequality (5.4), we obtain

**Proposition 5.3.** *The volume distribution of the domain $\tilde{E}$ obeys the estimate*

$$P\left\{|\tilde{E}| \geq s\right\} \leq C^* e^{-\nu^* s}$$

*where $C^*$ and $\nu^*$ are functions of parameters $\delta$, $a$, $z$, $n$, $\varepsilon$. Under the conditions (5.1), (5.2) the factor $C^*$ is finite for any $\varepsilon < \varepsilon_0(p)$, and $\nu^* = \log 1/p + O(\varepsilon)$ as $\varepsilon \to 0$.*

Now put together estimates from propositions 4.5, 5.1, 5.3 in order to get the inequality

$$G(\lambda) \leq z^d \sum_{s=1}^{\infty} f\left(\frac{\lambda''^d s}{\kappa_d}\right) e^{-\nu^* s}$$

which is satisfied by any $\lambda < \lambda_1$ where

$$\lambda'' = \lambda \frac{(1+\varepsilon)^2}{1-\lambda/\lambda_1} \qquad \lambda_1 = \frac{1}{(2+\varepsilon)\Lambda(\delta)a_n}$$

Due to the lemma 3.3 it is followed by the estimate

$$\limsup_{\lambda\to 0+} \lambda^d \log G(\lambda) \le -\kappa_d \nu^*(1+\varepsilon)^{-2d} = -\kappa_d \log 1/p + O(\varepsilon)$$

This completes the proof of theorem 1.1.

## Appendix A. Isoperimetric lemma

Here we present a proof of the isoperimetric lemma 3.1. An isoperimetric inequality of the same kind is well-known for the function

$$\tilde{\Pi}(\lambda,\, D) = \sup\left\{ \int_D \left(\lambda\varphi^2 - |\nabla\varphi|^2\right) dx \;:\; \varphi \in C_0^\infty(D) \right\}$$

Due to homogeneity of the integrand this function has two possible values, and they are 0 and $+\infty$. The maximum value of $\lambda$ which provides the finite values of $\tilde{\Pi}(\cdot,\,\cdot)$ coincides with the lowest eigenvalue of Laplacian $(-\Delta)$ in $D$ with zero Dirichlet boundary condition on $\partial D$. Therefore the isoperimetric inequality for $\tilde{\Pi}(\cdot,\cdot)$ is equivalent to the statement that, among all domains $D$ of the same volume, ball has the least principal eigenvalue. This proposition was established by Rayleigh in [10], and its proof in modern terms can be found in [11]. To prove lemma 3.1 we will follow this way. Another approach had been used by Mosolov and Miasnikov in the paper [13]. Our proof is applicable for a wider class of convex potentials $W(\cdot)$. For the sake of simplicity let us assume smoothness of $\partial D$. Otherwise one could consider a sequence of embedded domains $D_\varepsilon \to D$, so this is not an important loss of generality. Also let $|D|$ be equal $|B_R|$. For any function $\varphi(x)$ vanishing on $\partial D$ we are going to construct a function $F : B_R \to R^1$ such that it equals zero on the ball's boundary and satisfies inequality

$$\int_D [\lambda\varphi - W(\nabla\varphi)]\, dx \le \int_{B_R} [\lambda F - W(\nabla F)]\, dx \tag{A.1}$$

This inequality provides the claim of lemma due to the definition of $\Pi(\cdot,\cdot)$. Suppose that $\varphi$ is a Morse's function and $\varphi \ge 0$. These assumptions are acceptable since the set of Morse's functions is dense in the natural domain of the functional in variational formula (1.1), and the maximizer $\varphi^{(r)}$ is non-negative. Inside $D$ let us consider a family of subsets $D_t = \{x : \varphi(x) > t\}$ depending on parameter $t$, $0 \leq t \leq t^* \equiv \max\{\varphi(x)\}$. As soon as $\varphi$ is a Morse's function the boundaries $\Gamma_t = \{x : \varphi(x) = t\}$ of $D_t$ are smooth manifolds of dimension $d-1$ for almost all values of $t$. Define a function $r = r(t)$ by equality $|D_t| = |B_{r(t)}|$. Evidently this function is continuous with respect to $t$ and smooth a.e. Induce a function $F$ depending on variable $r = |x|$ by formula $F(r(t)) = t$. Then the following relations hold true

$$|\frac{dr}{dt}| \cdot |\nabla F| = 1, \quad |\nabla F| \cdot \int_{\Gamma_t} \frac{d\sigma}{|\nabla \varphi|} = \frac{d|B_r|}{dr}, \qquad \frac{d|B_r|}{dr} \leq |\Gamma_t| \qquad \text{(A.2)}$$

The second equality is proved by taking the derivative with respect to $r$ of both sides of identity $|D_t| = |B_r|$. The last inequality in (A.2) is well-known minimizing property of ball's surface area. Due to the definition of $F(\cdot)$ the integrals of $\varphi$ or $F$ over the domains $D$ or $B_R$ are equal. Indeed,

$$\int_D \varphi \, dx = \int_0^{t^*} dt \left( t \int_{\Gamma_t} \frac{d\sigma}{|\nabla \varphi|} \right) = \int_0^R F(r) \frac{dB_r}{dr} dr = \int_{B_R} F(|x|) \, dx \qquad \text{(A.3)}$$

Define a function $w : ]0, +\infty[ \to R^1$ by formula

$$|\xi| w \left( \frac{1}{|\xi|} \right) = W(\xi)$$

It is easy to verify that $w(\xi)$ is a convex monotonically decreasing function as soon as $W(\xi)$ is convex and equal to zero for $\xi = 0$. Therefore the following inequalities are valid

$$w \left( \frac{1}{|\nabla F|} \right) \leq w \left( \frac{1}{|\Gamma_t|} \int_{\Gamma_t} \frac{d\sigma}{|\nabla \varphi|} \right) \leq \frac{1}{|\Gamma_t|} \int_{\Gamma_t} w \left( \frac{1}{|\nabla \varphi|} \right) d\sigma$$

Here the first inequality follows from monotonicity of $w(\cdot)$ and relations (A.2). Another one is a consequence of convexity. Due to this estimate we can write

$$\int_D W(\nabla \varphi) \, dx = \int_0^{t^*} dt \left( \int_{\Gamma_t} \frac{W(\nabla \varphi)}{|\nabla \varphi|} d\sigma \right) = \int_0^{t^*} dt \int_{\Gamma_t} w \left( \frac{1}{|\nabla \varphi|} \right) d\sigma \geq$$

$$\int_0^{t^*} |\Gamma_t| w\left(\frac{1}{|\nabla F|}\right) dt \geq \int_0^R |\nabla F| \frac{dB_r}{dr} w\left(\frac{1}{|\nabla F|}\right) dr = \int_{B_R} W(\nabla F)\, dx$$

Combining this estimate with (A.3) we get the inequality (A.1). Thus lemma 3.1 is proved.

## Acknowledgments

This work was supported by Russian fund of Fundamental Researches, grants No. 95-01-01232 and 96-01-01746.

## References

1. J.-L. Lions, Asymptotic Expansions in Perforated Media with a Periodic Structure, *The Rocky Mountain J. of Math.*, **10**, 125–144 (1980).
2. A.Yu. Beliaev and S.M. Kozlov, Darcy Equation for Random Porous Media, *Comm. on Pure and Appl. Math.*, **49**, 1–34 (1996).
3. A. Bourgeat and A. Mikelic, A Note on Homogenization of Bingham Flow trough a Porous Medium, *J. Math. Pures Appliques*, **72**, 405–414 (1993).
4. G. Duvant and J.-L. Lions Les In'equations en M'ecanique et Physique. Dunod. Paris. 1972.
5. Ekeland, I. and Temam, R. Convex Analysis and Variational Problems. North-Holland. Amsterdam. 1976.
6. A.Yu. Beliaev, On critical drop in pressure for flows in porous media, *J. Appl. Math. and Mech. (PMM)*, **61** (6), 1001–1007 (1997).
7. A.Yu. Beliaev and V.V. Yurinsky, Bottom of Laplacian's Spectrum in Poisson Cloud, *Siberian Advances in Math.*, **5** (4), 113–150 (1995).
8. A.-S. Sznitman, On the confinement property of Brownian motion among Poissonian obstacles, *Comm. Pure Appl. Math.*, **44**, 1137–1170 (1991).
9. H. Kesten, Percolation Theory for Mathematicians, Birkhauser, Boston et al., 1982.
10. J.W.S. Rayleigh, The Theory of Sound, McMillan, NY, 1877.

11. I. Chavel, Eigenvalues in Riemannian Geometry, NY, Ac. Press, 1984.
12. M.D. Donsker and S.R.S. Varadhan, Asymptotics for the Wiener Sausage, *Comm. on Pure and Appl. Math.*, **28**, 525–565 (1975).
13. P.P Mosolov and V.P. Miasnikov, Variational methods in the theory of viscous-plastic flows, *J. Applied. Math. and Mech. (PMM)*, **25**, 468–492 (1965).

# DISTRIBUTION OF MINIMUM VALUES OF WEAKLY STOCHASTIC FUNCTIONALS

V. BERDICHEVSKY
*Mechanical Engineering, Wayne State University, Detroit, MI 48202*

Many problems in mechanical engineering and physics are reduced to studying the following question. There is a variational problem to minimize a sum of quadratic and linear functionals. The linear functional is stochastic, therefore the minimum value is random. How to find the probability density function of the minimum value? Such type of problem appears in statistical mechanics of fluid motion, Kosterlitz-Throuless phase transition, brittle-to-ductile transition in solids, reaction of cracked bodies to random excitation, lubrication theory, etc. It is shown in this paper that the probability density function of minimum values is determined by some variational problem. The features of this variational problem are discussed.

## 1 Introduction

The problem under consideration is closely related to the homogenization problem and can be formulated as follows. Consider a quadratic functional

$$I(u) = \frac{1}{2}(Au, u) - (l, u). \tag{1.1}$$

Here $(Au, u)$ and $(l, u)$ stand for quadratic and linear parts of the functional $I(u)$. A typical example is the case of a scalar function $u$ defined in some bounded region $V$ of $n$-dimensional space, and

$$(Au, u) = \int_V \frac{\partial u}{\partial x^i}\frac{\partial u}{\partial x^i} d^n x\ , \quad (l, u) = \int_V \rho(x) u(x) d^n x. \tag{1.2}$$

If $u = 0$ at $\partial V$ then $A$ can be viewed as a positive operator, $A = -\Delta$ ($\Delta$ is the Laplace operator).

Quadratic functional $(Au, u)$ is assumed to be positive.

If $u$ is a vector of $m$-dimensional space $R^m$, then $A$ is a positive symmetric matrix, $l$ is a vector in $R^m$, and $(\cdot, \cdot)$ is the usual scalar product.

Denote the minimizing element of $I(u)$ by $\check{u}$. The minimum value of the functional $I(u)$ is equal to $-1/2\left(A\check{u}, \check{u}\right)$ and always negative. In physical problems, the quadratic part of $I(u)$ has usually the sense of energy while the linear functional represents the external action on the system. In what follows the negative minimum value of $I(u)$ is called energy.

Let the linear functional be stochastic. In the case of the functional (1.2) that means that $\rho(x)$ is a stochastic field. In the problems considered here the randomness comes into play through the dependence of the linear functional on a large number of independent random variables, $r_1, ..., r_N$. The total set $(r_1, ...r_N)$ is denoted by $\mathbf{r}$. So, $I = I(u, \mathbf{r})$, and the negative minimum value of $I$, energy, is a function of $\mathbf{r}$, i.e. it is a random variable. The problem is to find the probability density function of energy.

Here we solve this problem for two cases: in the case when the linear functional is a sum of $N$ small (in the order of $1/N$) linear statistically independent functionals,

$$(l, u) = \frac{1}{N} \sum_{i=1}^{N} (l_i(r_i), u) \tag{1.3}$$

and when each term is much larger, in the order of $1/\sqrt{N}$,

$$(l, u) = \frac{1}{\sqrt{N}} \sum_{i=1}^{N} (l_i(r_i), u). \tag{1.4}$$

In the latter case we assume that the average value of each term of the sum is zero.

Each member of the sum depends only on one random variable. The $i$th random variable has the probability density function, $f_0^i$. Large external actions (1.4) will be discussed in section 13, while here we formulate the results for small excitations (1.3). In this case, the probability density function of energy is

$$f(E) = const \times e^{N \check{S}(E)} \tag{1.5}$$

where $\check{S}$ is the maximum stationary value of the functional

$$S(E, z, u) = Ez + \frac{z}{2}(Au, u) + \frac{1}{N} \sum_{i=1}^{N} \ln \int e^{-z(l_i(r), u)} f_0^i(r) dr. \tag{1.6}$$

The stationary points of the functional (1.6) are sought with respect to real variable $z$ and function $u$. The constant in (1.5) can be found from the normalizing condition,

$$\int f(E)dE = 1.$$

If all linear functionals in (1.3) are the same

$$(l, u) = \frac{1}{N}\sum_{i=1}^{N}(l_0(r_i), u), \tag{1.7}$$

and the random variables have the same probability distribution, $f_0(r)$, then the functional $S(E, z, u)$ takes especially simple form

$$S(E, z, u) = Ez + \frac{z}{2}(Au, u) + \ln\int e^{-z(l_0(r),u)} f_0(r)dr. \tag{1.8}$$

We will discuss mostly this case ; an extension to a more general case (1.3) is straightforward.

It is often convenient to make a change of the variable,

$$u \to v = zu\,,$$

to eliminate the dependence of the last term in (1.8) on $z$ and to make obvious the convexity of $S$ with respect to $z$:

$$S(E, z, v) = Ez + \frac{1}{2z}(Av, v) + \ln\int e^{-(l_0(r),v)} f_0(r)dr. \tag{1.9}$$

The stationary points of the functional (1.9) are denoted by $\beta, \check{v}$. We do not use the symmetric notation $\check{z}$ for the stationary point with respect to $z$ because, in physical problems, $\check{z}$ has usually the sense of the inverse temperature $\beta$. Obviously,

$$\frac{d\check{S}(E)}{dE} = \beta,$$

therefore the function $\check{S}(E) = S(E, \beta, \check{v})$ has the sense of entropy. We call the functional $S(E, z, v)$ the entropy functional.

Formula (1.5) solves, in fact, a problem of large deviations. Indeed, the sum (1.7) converges to the averaged linear functional

$$(\bar{l}, u) = \int (l_0(r), u) f_0(r) dr \tag{1.10}$$

and the most probable value of energy $E$ corresponds to the minimum value of the averaged functional,

$$-\bar{E} = \min_u \left[\frac{1}{2}(Au, u) - (\bar{l}, u)\right]. \tag{1.11}$$

Function $\overset{\vee}{S}(E)$ should have the maximum value at $E = \bar{E}$. If $E - \bar{E}$ is small, in the order of $1/\sqrt{N}$, formula (1.5) yields the central limit theorem. Finite $E - \bar{E}$ corresponds to large deviations of energy from the most probable value.

We call the functionals under consideration weakly stochastic because randomness enters in the linear part. Homogenization problems correspond to strong stochasticity: the randomness of the quadratic part. As will be seen from the derivation the results formulated admit a generalization to the strongly stochastic functionals in some cases.

It is easy to explain why the entropy functional appears in the formula (1.5).

As in any probabilistic problem with independent random variables, it is worth considering the Fourier transform of the probability density

$$Me^{Nz \min_u I(u)}, \tag{1.12}$$

where $M$ is the mathematical expectation, and $z$ is a point on the imaginary axis. It is desirable to find a way to "change the order" of averaging with respect to probabilistic measure and minimization in (1.12). If we can do that, then the original problem is reduced to some other variational problem. A key to performing such a change is the following observation:

Let $u$ and $l$ be vectors in $m$-dimensional space. In this case $A$ is a positive $m \times m$ matrix, and $(\cdot, \cdot)$ is the usual scalar product. It is well known that the integral of the exponent of a quadratic form has the value:

$$\int e^{-\frac{1}{2}(Au,u)+(l,u)} d^m u = e^{\frac{1}{2}(A^{-1}l,l)} \sqrt{\frac{(2\pi)^m}{\det A}}, \tag{1.13}$$

where $A^{-1}$ is the inverse matrix. The remarkable point is that this relation can be written in the from

$$e^{-\min_u\left[\frac{1}{2}(Au,u)-(l,u)\right]} = \int \sqrt{\frac{\det A}{(2\pi)^m}} e^{-\frac{1}{2}(Au,u)+(l,u)} d^n u. \qquad (1.14)$$

We presented $\exp\left[-\min_u I(u)\right]$ as an integral. Now the order of integration and averaging with respect to probabilistic measure can be changed. Doing this in the Fourier representation of the probability density function,

$$f(E) = \frac{N}{2\pi i} \int_{-i\infty}^{+i\infty} M e^{NEz+Nz\min_u I(u)} dz$$

we obtain

$$f(E) = \text{const} \times \int \int e^{NS_m(E,z,v)} d^m v dz, \qquad (1.15)$$

where $S_m$ is an $m$–dimensional truncation of the entropy functional. We have then to find the asymptotics of the integral (1.15), if $N, m \to \infty$. Formula (1.6) corresponds to the case when the asymptotics of the integral (1.15) is determined by the maximum stationary value of $S_m$ as in the steepest descent method. Unfortunately, the steepest descent method cannot be applied directly because it assumes that $m \ll N$, while we are interested in the case $m \gg N$. Thus, we go another way: we establish the identity,

$$f(E) = R(N,E)e^{N\check{S}(E)}, \qquad (1.16)$$

where the prefactor $R$ is determined by some explicit formulas.

It turns out that in many problems $R(N,E)$ does not depend on $E$ in the limit $N \to \infty$, and the formula (1.5) holds.

Weakly stochastic variational problems are of great interest in various problems of physics and mechanics some of which are discussed below. The plan of the paper is as follows. We begin with a simple example supporting formulas (1.5), (1.9) (section 2). In the next five sections some physical problems are considered solutions of which are reduced to a weakly stochastic variational problem. In each problem the entropy functional $S(E,z,u)$ is written down. Some features of the variational problem for the entropy functional and the meaning of its solutions are discussed in section 8. Justification of (1.5) based on the identity (1.16) is derived in section 9. In this section the various forms of the prefactor $R$ are established. Another derivation of (1.16) is given in section 10. In section 11, under some assumptions, the prefactor is shown to be independent on $E$ in the limit $N \to \infty$, and, thus, formula (1.5) holds. The necessity of one of the assumptions is supported by a counterexample in section 12. The last section concerns with the case of large excitations. Appendices contain the derivation of the basic identity (1.16) for the functional (1.3), an outline of the proof of the formula for conditional probability (8.8) and discussion of negative temperature states for a one-dimensional problem which admits a complete investigation.

## 2 An example

We begin with the simplest possible case, a quadratic function $I(u)$ of one real variable,

$$I(u) = \frac{1}{2}u^2 - lu\,,$$

$l$ is a random number of the form

$$l = \frac{1}{N}\sum_{i=1}^{N} l_0(r_i)$$

where $r_i$ are independent random variables. Without loss of generality we may change variables and put $l_0(r) \equiv r$. Then, in accordance with (1.9),

$$S(E,z,v) = Ez + \frac{1}{2z}v^2 + \ln \int e^{-rv} f_0(r)dr. \tag{2.1}$$

The stationary points with the respect to $z$ are determined by the equation

$$\frac{v^2}{2\beta^2} = E \quad \text{or} \quad \beta = \pm\sqrt{\frac{v^2}{2E}}.$$

At these stationary points,

$$S(E,v) = \pm\sqrt{2E}\,|v| + \ln \int e^{-rv} f_0(r)dr. \tag{2.2}$$

To determine $\check{S}(E)$ one has to find the stationary points $\check{v}$ of the function (2.2) at which $S(E,v)$ has the largest value.

Let us compare this statement with the exact formula for the probability density of min $I(u)$, which can be found by the steepest descent method. Indeed, since min $I(u) = -1/2l^2$,

$$f(E) = \left[g\left(\sqrt{2E}\right) + g\left(-\sqrt{2E}\right)\right]\frac{1}{\sqrt{2E}}, \tag{2.3}$$

where $g(\xi)$ is the probability density function of the random variable $l$. For $g(\xi)$ we have

$$g(\xi) = M\delta\left(\xi - \frac{1}{N}\sum l_i\right) = \frac{1}{2\pi i}M\int_{-i\infty}^{i\infty} e^{z(\xi - \frac{1}{N}\sum l_i)}dz =$$

$$= \frac{N}{2\pi i}\int_{-i\infty}^{i\infty} e^{Nz\xi}\left(\int e^{-zr}f_0(r)dr\right)^N dz =$$

$$= \frac{N}{2\pi i}\int_{-i\infty}^{i\infty} e^{NS_1(z,\xi)}dz; \tag{2.4}$$

$$S_1(z,\xi) = z\xi + \ln\int e^{-zr}f_0(r)dr. \tag{2.5}$$

Function $S_1(x,\xi)$ is convex with respect to $z$ on real axis. Denote its minimum value by $S_1(\xi)$. Moving the line of integration in (2.4) to pass the point of minimum and using the steepest descent method, we obtain

$$g(\xi) = \frac{N}{\sqrt{2\pi\, S_1''(\xi)}}e^{NS_1(\xi)}. \tag{2.6}$$

Plugging (2.6) into (2.3) we see that one of the terms, $\exp\left(NS_1\left(\sqrt{2E}\right)\right)$ and $\exp\left(NS_1\left(-\sqrt{2E}\right)\right)$, which correspond to the smaller value of $S_1$, is negligible compared with the other one. The prefactor $(\sqrt{2E}\,S''(\sqrt{2E}))^{-1}$ is not essential since it corresponds to small correction to $S_1$. Thus one has to consider the stationary points of function $S_1(v,\sqrt{2E})$ and $S_1(v,-\sqrt{2E})$ with respect to $v$ and find the stationary point at which the stationary value is the largest. The graphs of two functions $S_1(v,\sqrt{2E})$ and $S_1(v,-\sqrt{2E})$ coincide with the graphs of two branches of the function (2.2). Therefore, the stationary points and the stationary values coincide as well. The only point, $v = 0$, needs special attention since the function (2.2) is not smooth at this point. The point $v = 0$ is the limit of stationary points if $\sqrt{2E} \to <r>$. If $\sqrt{2E} \neq <r>$, $v = 0$ is not a stationary point of smooth function (2.1), thus the variational problem (2.1) yields the same result as the exact formula. Coincidence in the case $\sqrt{2E} = <r>$ we obtain from continuity tending $\sqrt{2E}$ to $<r>$ .

## 3 Statistical mechanics of point vortices

Statistical mechanics of point vortices studies the probabilistic and thermodynamic properties of the Hamiltonian system with the Hamilton function

$$H = \frac{1}{2} \sum_{i \neq j} \gamma_i \gamma_j G(r_i, r_j) \tag{3.1}$$

where $r_1 \ldots r_N$ are points of two – dimensional bounded region $V$, $G(r, r')$ is Green's function of the Laplace operator with zero boundary conditions in this region. The $x$–components of two–dimensional vectors $r_i$ are the generalized coordinates of the system while the $y$–components are the corresponding generalized momenta. It is assumed that the system of point vortices moves ergodically (note that all low–dimensional numerical experiments does not support this hypothesis; it might be, however, that ergodicity takes place approximately for large $N$). Statistical mechanics of point vortices has a long and rich history; a list of references on the subject is given in [3]. All statistical and thermodynamic properties of the ergodic system of vortices can be expressed in terms of probability density function of energy (3.1) if $r_1 \ldots r_N$ are placed randomly and independently in region $V$ [2, 3].

The kinetic energy is the negative minimum value of the functional

$$-H(r_1 \ldots r_N) = \min_u I(u, r)$$

$$I(u) = \frac{1}{2}(Au, u) - (l, u), \quad (Au, u) = \int_V \frac{\partial u}{\partial x^\alpha} \frac{\partial u}{\partial x^\alpha} d^2x, \quad (l, u) = \int_V \omega u d^2x \tag{3.2}$$

$$\omega = \frac{1}{N} \sum_{i=1}^{N} \sigma_i \delta(x - r_i), \qquad \gamma_i = \frac{\sigma_i}{N}. \tag{3.3}$$

Minimum is sought on the set of all functions $u$ taking zero values at the boundary $\partial V$ of the region $V$. Function $u$ is the stream function of the flow, $\omega$ is vorticity. $\delta$-functions in (3.3) yield infinite energy, and the functional $I(u)$ needs a regularization. One way is to remove self–interaction energy and conduct summation in (3.1) over $i \neq j$. Another way is to change $\delta(x)$ by a finite function, $\delta_\varepsilon(x)$, which is equal to $1/\varepsilon$ in a small circle of area $\varepsilon$ centered at the point $x$ and equal to zero outside of this circle. Consider the second way.

The variational problem for the functional (3.2) is a weakly stochastic variational problem. Thus, probability density function of energy is given by formula (1.5) where $\overset{\vee}{S}(E)$ is the maximum stationary value of the functional

$$S_\varepsilon(E, z, \overset{\vee}{v}) = Ez + \frac{1}{2z}\int_V \frac{\partial v}{\partial x^\alpha}\frac{\partial v}{\partial x^\alpha} d^2x + \frac{1}{N}\sum_i \ln \frac{1}{|V|}\int_V e^{-\sigma_i v_\varepsilon(r)} d^2r \tag{3.4}$$

where $v_\varepsilon(x)$ is the function $v(x)$ averaged over the circle vicinity of radius $\varepsilon$ centered at the point $x$. The sum in (3.4) is, in fact, independent on $N$ if there are a few values of $\sigma_i$, say, $\sigma_1, ..., \sigma_k$, which appear with frequencies $c_1, ..., c_k$. In this case,

$$S_\varepsilon(E, z, v) = Ez + \frac{1}{2z}\int_V \frac{\partial v}{\partial x^\alpha}\frac{\partial v}{\partial x^\alpha} d^2x + \sum_\alpha c_\alpha \ln \frac{1}{|V|}\int_V e^{-\sigma_\alpha v_\varepsilon(x)} d^2x. \tag{3.5}$$

The stationary values of the functional $S_\varepsilon$ converges for $\varepsilon \to 0$ to the stationary values of the functional

$$S(E, z, v) = Ez + \frac{1}{2z}\int_V \frac{\partial v}{\partial x^\alpha}\frac{\partial v}{\partial x^\alpha} d^2x + \sum_\alpha c_\alpha \ln \frac{1}{|V|}\int_V e^{-\sigma_\alpha v(x)} d^2x. \tag{3.6}$$

The functional (3.6) was obtained from other reasoning in [2].

## 4 The Kosterlitz-Thouless phase transition

Consider now another problem which is reduced to studying a weakly stochastic functional, so-called Kosterlitz-Thouless phase transition. Originally, this problem was posed by Kosterlitz and Thouless for dislocations in solids [15], but later on it was found to be quite common in various fields. Herein, we discuss this problem for the case of dislocations in linearly elastic solids, but we begin with a more general problem of the internal stresses in solids caused by residual strains.

Let $V$ be a bounded three-dimensional region and $w_i$ are the components of the displacement vector defined in $V$. The components of the total strain tensor are

$$\varepsilon_{ij} = \frac{1}{2}\left(\frac{\partial w_j}{\partial x_j} + \frac{\partial w_i}{\partial x_j}\right). \tag{4.1}$$

Let each material element $x$ have the residual strains $\overset{0}{\varepsilon}_{ij}(x)$. By definition, elastic strains are $\varepsilon_{ij} - \overset{0}{\varepsilon}_{ij}(x)$. The true displacement field minimizes the functional

$$I(w) = \int\limits_V U\, d^3x - \int\limits_{\partial V} f^i w_i d^2x,$$

$$U = \frac{1}{2}A^{ijkl}(\varepsilon_{ij} - \overset{0}{\varepsilon}_{ij}(x))(\varepsilon_{ij} - \overset{0}{\varepsilon}_{ij}(x)),$$

where $A^{ijkl}$ are the elastic moduli.

There is a scalar version of this problem, so-called anti-plane deformation, when only one component of the displacement vector $w_3 \equiv w$ is not zero and depends on $x_1, x_2$ only while the nonzero components of the residual strains are $\overset{0}{\varepsilon}_\alpha \equiv \overset{0}{\varepsilon}_{3\alpha}(x_1, x_2)$ (Greek indices run values 1,2). Then

$$I(w) = \int\limits_V \frac{1}{2}A^{\alpha\beta}\left(\frac{\partial w}{\partial x_\alpha} - \overset{0}{\varepsilon}_\alpha\right)\left(\frac{\partial w}{\partial x_\beta} - \overset{0}{\varepsilon}_\beta\right) d^2x - \int\limits_{\partial V} fw ds. \tag{4.2}$$

For simplicity we focus on this scalar version of the general problem assuming in addition that the elastic body is isotropic: $A^{\alpha\beta} = \mu\delta^{\alpha\beta}$.

If the residual strains are potential,

$$\overset{0}{\varepsilon}_\alpha = \frac{\partial \overset{0}{w}}{\partial x^\alpha},$$

then, for zero external traction $f$, no internal stresses develop: it is clear that the minimum value of the functional (4.2), elastic energy, is zero, and minimum is achieved for $w = \overset{0}{w}$. Nontrivial stress field appears only if the strains $\overset{0}{\varepsilon}_\alpha$ are "incompatible", i.e.

$$\frac{\partial \overset{0}{\varepsilon}_1}{\partial x^2} \neq \frac{\partial \overset{0}{\varepsilon}_2}{\partial x^1}.$$

Incompatibility is measured by the so-called dislocation density,

$$\rho = \frac{\partial \overset{0}{\varepsilon}_1}{\partial x^2} - \frac{\partial \overset{0}{\varepsilon}_2}{\partial x^1}. \tag{4.3}$$

Similar measure appears in the 3D case.

If the residual strains are random we have a problem of the type considered above. Various physical problems yield different types of randomness.

Dislocation is a defect in crystal lattice. Internal stresses caused by dislocations can be found by modeling this crystal lattices by an elastic body with residual strains. The residual strains $\overset{0}{\varepsilon}_\alpha$ are not equal to zero in a thin strip the thickness of which is equal to the interatomic distance. The residual strains $\overset{0}{\varepsilon}_\alpha$ are constant along the strip except the edges where $\overset{0}{\varepsilon}_\alpha$ change smoothly to zero to eliminate the singularities in the stress field. Obviously, the dislocation density $\rho$ is not equal to zero only in these areas which are called the dislocation cores.

The dual variational problem for the functional (4.2) has the same form as the point vortex problem of section 3. Indeed, following the general recipe for construction of the dual variational problem (see, for example, [1]) we have

$$\underline{I} = \min_w I(w) =$$

$$= \min_w \max_{\sigma^\alpha} \left[ \int_V \left[ \sigma^\alpha \left( \frac{\partial w}{\partial x^\alpha} - \overset{0}{\varepsilon}_\alpha \right) - \frac{1}{2\mu} \sigma_\alpha \sigma^\alpha \right] d^2x - \int_{\partial V} f(s) w \, ds \right] =$$

$$= \max_{\sigma^\alpha} \left[ - \int_V \left( \frac{1}{2\mu} \sigma_\alpha \sigma^\alpha + \sigma^\alpha \overset{0}{\varepsilon}_\alpha \right) d^2x \right]. \tag{4.4}$$

In the last variational problem the maximum is sought on the set of all stress field $\sigma^\alpha(x)$ such that

$$\frac{\partial \sigma^\alpha}{\partial x^\alpha} = 0, \qquad \sigma^\alpha \nu_\alpha = f(s) \quad \text{at} \quad \partial V. \tag{4.5}$$

The general solution of (4.5) has the form

$$\sigma^\alpha = e^{\alpha\beta}\frac{\partial u}{\partial x^\beta}, \qquad u|_{\partial V} = \overset{0}{u}(s), \tag{4.6}$$

where $e^{\alpha\beta}$ are the Levi-Civita symbols, $u$ is the stress function, and the boundary values of $u$ is found from the relation

$$\frac{d\overset{0}{u}}{ds} = f(s).$$

Plugging (4.6) in (4.4) we obtain

$$\underline{I} = -\min_u \int_V \left[\frac{1}{2\mu}\frac{\partial u}{\partial x^\alpha}\frac{\partial u}{\partial x^\alpha} + e^{\alpha\beta}\frac{\partial u}{\partial x^\beta}\overset{0}{\varepsilon}_\alpha\right] d^2x =$$

$$= -\min_u \int_V \left[\frac{1}{2\mu}\frac{\partial u}{\partial x^\alpha}\frac{\partial u}{\partial x^\alpha} - \rho u\right] d^2x - \int_{\partial V} e^{\alpha\beta}\nu_\beta \overset{0}{\varepsilon}_\alpha \overset{0}{u}\, ds. \tag{4.7}$$

Here minimum is taken on the set of all functions $u$ having the prescribed values $\overset{0}{u}$ at the boundary. Comparing this problem with the problem of determining the velocity field by known vorticity $\omega$, we see that the stress function $u$ corresponds to the stream function of the vortex flow while the dislocation density $\rho$ corresponds to vorticity $\omega$. Dislocations are analogous to vortex blobs, and for a single dislocation centered at the point $r$

$$\rho = b\delta_\varepsilon(x - r).$$

Dislocations in the antiplane problem may have two signs which are determined by the sign of the constant $b$.

Consider the dislocations positioned at the points $r_1, ..., r_N$. Then $\underline{I}$ is a function of $r_1, ..., r_N$:

$$\underline{I} = \underline{I}(r_1, ..., r_N).$$

Under some assumptions of physical nature, the probability to find dislocations at these positions is

$$f(r_1, ..., r_N) = \frac{1}{Z}e^{-\beta \underline{I}(r_1,...,r_N)} \tag{4.8}$$

where $\beta$ is the inverse temperature ($\beta = 1/T$, $T$ is the absolute temperature). The normalizing factor $Z$ is a function of $\beta$

$$Z(\beta) = \int e^{-\beta \underline{I}(r_1,\dots,r_N)} dr_1 \dots dr_N.$$

Some conclusion on dislocation behavior can be drawn from the calculation of energy of a single dislocations and a pair of dislocations. That can be done explicitly. Energy of two closely positioned dislocations of the same sign is much greater than the energy of two dislocations of the opposite signs. Energy of a single dislocation is also much greater than energy of two close dislocations of the opposite signs (dipole pair). Therefore, typically, if dislocations appear due to thermal fluctuations they appear as dipoles. This changes at high temperature: probability of a pair of dislocations to dissolve, i.e. to be far away, becomes considerable. The explicit estimation gives unrealistically high values of temperature for which dissolving occurs. Kosterlitz and Thouless [15] suggested that the dislocation pairs may dissolve actually for much smaller temperature due to interaction(screening) which decreases $\underline{I}$. Function $Z(\beta)$ may serve as an integral characteristic of dissolving. At the critical inverse temperature of dissolving, $\beta_c$, $Z(\beta)$ should have some steep change.

Calculation of the interaction energy and $\beta_c$ is a nonelementary task. Kosterlitz and Thouless offered a recipe to determine approximately the interaction energy and the critical temperature. The recipe seems quite questionable for me. It is very interesting to compare it with the exact relations following from the approach outlined. Indeed, the function $Z(\beta)$ can be found from (4.8)

$$Z(\beta) = |V|^N \int e^{\beta E} f(E) dE. \tag{4.9}$$

where $f(E)$ is the probability density function of (negative) minimum values of the functional (4.7). The probability density function is sought under condition that all dislocations are distributed homogeneously and independently over $V$. The probability density function is given by formula (1.5) where $S$ is the functional

$$S(E, z, v) = Ez + \frac{1}{2z\mu} \int_V \frac{\partial v}{\partial x^\alpha} \frac{\partial v}{\partial x^\alpha} d^2x + \frac{1}{N} \sum_i \ln \frac{1}{|V|} \int_V e^{-b_i v(r)} dr. \tag{4.10}$$

Comparison of $Z(\beta)$ determined by (4.9) and (4.10) and Kosterlitz-Thouless approach is an interesting open problem.

## 5 Brittle-to-ductile transition

Any material fractures in a brittle way if its temperature is low enough. This means that cracks propagate without deformation of the surrounding material. Any material fractures in a ductile way if its temperature is high enough. This means that considerable material deformation develops in the vicinity of the crack tip, impeding further propagation of the crack. The remarkable phenomenon is that for many materials the transition from brittle to ductile behavior occurs in a very narrow temperature range, on the order on 3 to 5 $^oC$. There is no clear understanding of the mechanism of brittle-to-ductile transition, although some proposals have been made in the pioneering work by J. Rice and R. Thompson [20] and subsequent papers on dislocation generation based models [21, 19, 22, 18, 23, 25, 26] and dislocation mobility – based models [4, 5, 6, 8, 10]. This is one the major challenges in the materials sciences, from both theoretical and practical points of view.

Recently, Khanta, Pope and Vitek [13, 14] put forth an idea that brittle-to-ductile transition is, in fact, a Kosterlitz-Thouless type transition. This means that the simultaneous appearance of a cloud of dislocations in the vicinity of the crack tip is much more probable than the successive appearance of the same dislocations. In other words, the creation of a cloud of dislocations is a collective phenomenon: dislocations help each other to appear.

To evaluate this idea one has to study the variational problem (4.7) complicated by the presence of a crack, a cut in the region $V$ the sides of which are free surfaces. Similarly, a cut appears in the variational problem for the entropy functional $S$ (4.10). This is another interesting open problem.

## 6 Elastic body under action of random external forces

Formulas (1.5), (1.6) determine the probability density function of energy of elastic bodies subjected to action of random external forces. Consider, for example, an elastic beam with the clamped edges bent by random forces applied at the points $r_1, ..., r_N$. The true displacement gives the minimum value to the functional

$$I(u) = \frac{1}{2}\int_0^l D\left(\frac{d^2u}{dx^2}\right)^2 dx - \sum_{i=1}^N \gamma_i u(r_i) \qquad (6.1)$$

where $D$ is the bending rigidity and $\gamma_i$ are the intensities of the external forces. Minimum is sought on the set of functions $u(x)$ obeying the constraints

$$u = \frac{du}{dx} = 0 \quad \text{at} \quad x = 0, l\,. \qquad (6.2)$$

Assuming that all forces are small, in the order of $1/N$,

$$\gamma_i = \frac{\sigma_i}{N},$$

and points of application, $r_i$, are independently distributed over the beam with the probability density $f_0(r)$, we have for the entropy functional

$$S(E,z,v) = Ez + \frac{1}{2z}\int_0^l D\left(\frac{d^2v}{dx^2}\right)^2 dx + \frac{1}{N}\sum_i \ln\int e^{-\sigma_i v(r)} f_0(r)dr. \qquad (6.3)$$

An interesting problem appears if an elastic body contains a crack. The functional $I(u)$ has the form

$$I(w) = \frac{1}{2}\int_V A^{ijkl}\varepsilon_{ij}\varepsilon_{kl}d^3x - \int_{\partial V} f^i w_i d^2x + \gamma|\Sigma|.$$

where $\varepsilon_{ij}$ are given by (4.1), $|\Sigma|$ is the area of the crack and $\gamma$ is the surface energy. Let

$$f^i = \frac{1}{N}\sum_{\alpha=1}^N \varphi^i(x - r_\alpha)$$

and $r_1, ..., r_N$ are statistically independent and have the probability density $f_0(r)$. Then

$$S(E,z,v) = Ez + \frac{1}{2z}\int_V A^{ijkl}\frac{\partial v_i}{\partial x^j}\frac{\partial v_k}{\partial x^l}d^3x +$$

$$+\gamma|\Sigma| + \ln\int e^{-\int_{\partial V}\varphi^i(x-r)v_i(x)d^2x} f_0(r)dr.$$

The minimum is sought over all displacement field $w$ and all possible extension of the crack. If the extension of crack increases $I$ then the crack does not grow. One may expect that small forces can not cause the crack growth. However, some combinations of small forces may make the crack to develop. To find the probability of the crack growth one has to find the stationary points of the functional $S$.

## 7 Homogenization problems

Consider a microinhomogeneous media, for example, a heat conducting body with a periodic microstructure. Let at each cell a source of heat is placed. The positions of heat sources are random and independent at each cell. Then one may expect that formula (1.5) holds where $\check{S}$ is the maximum stationary value of the functional

$$S(E,z,v) =$$

$$= Ez + \frac{1}{2z}\int\limits_V a(x)\frac{\partial v}{\partial x^k}\frac{\partial v}{\partial x^k}d^3x +$$

$$+\frac{1}{N}\sum_i \ln\int e^{-<\varphi(x-r)v(x)>_i} f_0(r)dr. \tag{7.1}$$

where $a(x)$ is a periodic function of $x$ and $\varphi(x)$ is the source intensity. The functional (7.1) can be studied using the variational asymptotic method [1]. This yields the "homogenized" functional $\overline{S}$

$$\overline{S}(E,z,\bar{v}) =$$

$$= Ez + \frac{1}{2z}\int\limits_V \bar{a}(x)\frac{\partial \bar{v}}{\partial x^k}\frac{\partial \bar{v}}{\partial x^k}d^3x + \int\limits_V \left[\ln\int e^{-<\varphi(x-r)>\bar{v}(x)} f_0(r)dr\right] d^3x.$$

For general homogenization problem formula (1.5) should be valid, but the expression for entropy functional is different. One may try to use the identity (1.14) to determine the entropy functional. The difficulty is in randomness of $\det A$. For the cases when $\det A$ is not random or probabilistic properties of $\det A$ are known, the entropy functional can be found from (1.14).

## 8 Some features of stationary points of entropy functional

Entropy functional (1.9) is convex with respect to $v$ for positive $z$ and nonconvex for negative $z$. Therefore, there is the only stationary point for positive $z$, and there might be many for negative $z$. For negative $z$, the variational problem for the entropy functional is a nonlinear eigenvalue problem.

Let us show that entropy $\check{S}(E)$ is negative at each stationary point,

$$\check{S}(E) \leq 0 \tag{8.1}$$

Each stationary point $\beta, \check{v}$ of the entropy functional obeys the equations

$$\frac{1}{2\beta^2}(A\check{v}, \check{v}) = E; \tag{8.2}$$

$$\frac{1}{\beta} A\check{v} = \int l_0(r) e^{-(l_0(r),\check{v})} f_0(r)dr \Big/ \int e^{-(l_0(r),\check{v})} f_0(r)dr. \tag{8.3}$$

Therefore, at a stationary point,

$$\frac{1}{\beta}(A\check{v}, \check{v}) = \int (l_0(r), \check{v})\, e^{-(l_0(r),\check{v})} f_0(r)dr \Big/ \int e^{-(l_0(r),\check{v})} f_0(r)dr = 2\beta E. \tag{8.4}$$

Denote the function of $r$, $(l_0(r), \check{v})$, by $\psi(r)$, and by $J(\psi)$ the functional of $\psi$

$$J(\psi) = \ln \int e^{-\psi(r)} f_0(r)dr$$

Then

$$\check{S}(E) = J(\psi) + \int \psi(r)\, e^{-\psi(r)} f_0(r)dr \Big/ \int e^{-\psi(r)} f_0(r)dr =$$

$$= J(\psi) - \int \psi \frac{\delta J}{\delta \psi} dr \tag{8.5}$$

where $\delta J/\delta\psi$ is the variational derivative of $J$.

For any convex functional $J(\psi)$ the following inequality is true

$$J(\psi_1) - J(\psi_2) \geq \left( \left.\frac{\delta J}{\delta \psi}\right|_{\psi=\psi_2}, \psi_1 - \psi_2 \right) \tag{8.6}$$

see, for example, [24]. Putting in (8.6) $\psi_1 = 0$, $\psi_2 = \psi$, and taking into account that $J(0) = 0$, we see that the expression (8.5) is always negative as claimed.

If a solution of equations (8.2), (8.3) exists for some positive $\beta$, it provides the maximum value of entropy. Indeed, let $\beta_1, \check{v}_1$ and $\beta_2$, $\check{v}_2$ be two solutions of equations (8.2), (8.3). Denote by $\check{S}_1(E)$ and $\check{S}_2(E)$ the corresponding values of entropy. Then the inequality holds

$$\check{S}_1(E) - \check{S}_2(E) \geq 2E(\beta_1 - |\beta_1|). \tag{8.7}$$

If $\beta_1 > 0$, then the right hand side of (8.7) is zero, and we obtain the statement made. To derive (8.7) we note that

$$\frac{1}{\beta_1} A \check{v}_1 = -\frac{\delta J}{\delta \check{v}_1}, \quad \frac{1}{2\beta_1^2}(A \check{v}_1, \check{v}_1) = E;$$

$$\frac{1}{\beta_2} A \check{v}_2 = -\frac{\delta J}{\delta \check{v}_2}, \quad \frac{1}{2\beta_2^2}(A \check{v}_2, \check{v}_2) = E;$$

and, due to convexity of $J(v)$, from 8.6) we have

$$J(\check{v}_1) - J(\check{v}_2) \geq \left(\frac{\delta J}{\delta \check{v}_2}, \check{v}_1 - \check{v}_2\right).$$

Therefore,

$$\check{S}_1(E) - \check{S}_2(E) = 2\beta_1 E + J(\check{v}_1) - (2\beta_2 E + J(\check{v}_2)) =$$

$$= 2E(\beta_1 - \beta_2) + J(\check{v}_1) - J(\check{v}_2) \geq 2E(\beta_1 - \beta_2) + \left(\frac{\delta J}{\delta \check{v}_2}, \check{v}_1 - \check{v}_2\right) =$$

$$= 2E(\beta_1 - \beta_2) - \frac{1}{\beta_2}(A \check{v}_2, \check{v}_1 - \check{v}_2) \geq 2E(\beta_1 - \beta_2) + \frac{1}{\beta_2}(A \check{v}_2, \check{v}_2) -$$

$$-\frac{1}{\beta_2}(A \check{v}_2, \check{v}_1) \geq 2E\beta_1 - \frac{1}{2|\beta_2|}(\alpha(A \check{v}_1, \check{v}_1)) + \frac{1}{\alpha}(A \check{v}_2, \check{v}_2)) =$$

$$= 2E\beta_1 - \frac{E}{|\beta_2|}(\alpha\beta_1^2 + \frac{1}{\alpha}\beta_2^2) = 2E\beta_1 - 2E|\beta_1|.$$

Here we choose the optimum value of an arbitrary constant $\alpha$ to be $|\beta_2/\beta_1|$.

So, if a solution of (8.2), (8.3) exists for positive $\beta$, it is unique, and this is the only solution we need: the values of entropy for negative $\beta$ are smaller the one corresponding to the positive $\beta$.

However, there are interesting cases when the solution for positive $\beta$ does not exist, and only solutions for negative $\beta$ (negative temperature states) should be sought. This happens if the most probable value of energy is zero. Indeed, let the averaged functional $l_0(r)$ be zero:

$$\int l_0(r) f_0(r) dr = 0. \tag{8.8}$$

Then equation (8.3) always have zero solution $\overset{\vee}{v} = 0$ for any $\beta$. Since the solution of (8.3) is unique for positive $\beta$, we have $\overset{\vee}{v} = 0$ for $\beta > 0$. This is inconsistent with equation (8.2) for $E > 0$. We conclude that $\beta$ should be negative if (8.8) holds.

The typical graphs of the dependence of entropy on energy are shown in Fig. 1. If the most probable value of energy, $\bar{E}$, is not zero, then both signs of temperature are possible since $\beta = d\overset{\vee}{S}(E)/dE$. Negative temperature corresponds to large values of energy. If the most probable value of energy, $\bar{E}$, is zero, temperature is always negative. In the example considered in Appendix C all eigenvalues $\beta$ are negative, and the maximum value of $\beta$ corresponds to the maximum value of entropy. It is not known at present whether this is a general feature of the entropy functional.

Some features of the entropy values follow from the inequality (8.7). Let us denote the only positive value of $\beta$ by $\beta_0$, and put all negative values of $\beta$ in decreasing order, $\beta_1 \geq \beta_2 \geq \beta_3 ...$ $(|\beta_1| \leq |\beta_2| \leq |\beta_3| ...)$. The corresponding values of entropy we denote by $S_0(E)$, $S_1(E)$,... . If there is a positive value of $\beta$, then, as we found from (8.7)

$$S_0(E) \geq S_k(E), \quad k = 1, 2, ... .$$

Substituting in (8.7) $S_k$ and $S_m$ instead of $S_1$ and $S_2$ we obtain the estimation

$$S_k(E) \geq S_m(E) - 4E|\beta_k|, \quad m = 0, 1, 2, ... \tag{8.9}$$

If the conjecture $S_k(E) \geq S_m(E)$ for $m > k$, is true, then only the cases $m < k$ in (8.9) yield nontrivial inequalities. They all are consequences of the strongest one,

$$S_k(E) \geq S_0(E) - 4E|\beta_k| \quad k = 1, 2, ... \tag{8.10}$$

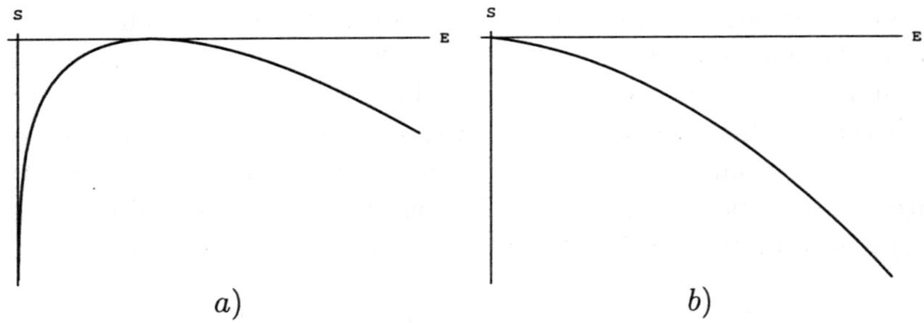


Figure 1: Dependence entropy on energy: a) the most probable value of energy is not zero; b) the most probable value of energy is zero

Similarly we obtain that, if there are no positive $\beta$, the strongest estimate is

$$S_k(E) \geq S_1(E) - 4E|\beta_k| \quad k = 2, ... \tag{8.11}$$

There is a relation which allows one to better understand the meaning of function $\check{v}$. Events $r_1, ..., r_N$ are distributed independently with probability density function $f_0(r)$. One may wonder what is the conditional probability density function of the first event, $\tilde{f}(r)$, if one considers only the events $r_1, ..., r_N$ for which energy takes some given value $E$. It turns out that

$$\tilde{f}(r) = const \times e^{-(l_0(r), \check{v})} f_0(r). \tag{8.12}$$

Conditional probability density functions of $r_2, ..., r_N$ are the same.

A derivation of (8.12) is given in Appendix B.

Formula (8.12) explains the meaning of the stationary point of the entropy functional: the solution of (8.3), $\check{v}$, is the solution of the original variational problem averaged over all events $r_1, ..., r_N$ for which energy has a given value, $E$.

## 9 The basic identity

The proof of the asymptotic formula (1.5) will be based on the following identity

$$f(E) = R\, e^{N \check{S}(E)}$$

$$R = \tilde{M}\left[e^{-\frac{N\beta}{2}(A^{-1}l',l')}\delta\left(\frac{(l',\overset{\vee}{v})}{\beta} + \frac{1}{2}(A^{-1}l',l')\right)\right] \tag{9.1}$$

where $\tilde{M}$ is the mathematical expectation with respect to the auxiliary measure

$$\tilde{f}(r) = e^{-(l_0(r),\overset{\vee}{v})} f_0(r) \Big/ \int e^{-(l_0(r),\overset{\vee}{v})} f_0(r) dr\,, \tag{9.2}$$

the couple $(\beta, \overset{\vee}{v})$ is any solution of equations (8.2), (8.3), $l' = l - \tilde{l}_0$, $\tilde{l}_0$ is the average value of $l$ with respect to the auxiliary measure,

$$\tilde{l}_0 = \int l_0(r)\tilde{f}(r)dr\,, \tag{9.3}$$

and $A^{-1}$ is the inverse operator for $A$, i.e. for any $l$

$$\min_u \left[\frac{1}{2}(Au,u) - (l,u)\right] = -\frac{1}{2}\left(A^{-1}l,l\right).$$

First, we give a derivation of (9.1) based on the identity (1.14). Actually, the author came to (1.5) and (9.1) using this line of reasoning. In the next section, another derivation of (9.1) is presented which is much simpler but could be hardly found before (9.1) was obtained in the long way described in this section.

Consider the following function of complex variable $z$

$$\hat{f}(z) = M\left[e^{z \min\limits_u I(u,r)}\right].$$

By definition of probability,

$$\hat{f}(z) = \int_0^\infty e^{-zE} f(E)dE. \tag{9.4}$$

Thus, function $\hat{f}(z)$ is the Laplace transform of the probability density function of energy, $f(E)$.

Since

$$f(E) \geq 0 \quad \text{and} \quad \int_0^\infty f(E)dE = 1\,,$$

function $\hat{f}(z)$ is analytic in the right half-plane, $\Re z \geq 0$. It tends to zero along each line, $z = x_0 + iy, x_0$ is fixed, $|y| \to \infty$. Formula (9.1) can be inverted (see, for example, [16])

$$f(E) = \frac{1}{2\pi i} \int_{c-i\infty}^{c+i\infty} e^{Ez} \hat{f}(z) dz. \tag{9.5}$$

Consider the scaled function,

$$\hat{f}(Nz) = M\left[e^{Nz \min_u I(u,r)}\right]. \tag{9.6}$$

We expect that for large N

$$\hat{f}(Nz) \approx e^{NB(z)}$$

where $B(z)$ is some function which is independent on $z$. If this is the case, then, by changing the variable of integration $z$ in (9.5) to $Nz$, we arrive at the problem of studying the asymptotics of the integral,

$$f(E) = \frac{N}{2\pi i} \int_{c-i\infty}^{c+i\infty} e^{N(Ez+B(z))} dz.$$

This can be attempted by the steepest descent method. We will proceed this way.

The first step is to study the asymptotic behavior of the function $\hat{f}(Nz)$ (9.6) for $N \to \infty$. A natural desire is to make "a homogenization" in (9.6) by changing the order of minimization with respect to $u$ and averaging with respect to $r$. Fortunately, this is possible in the problems under consideration due to (1.14). We discuss first this identity in more details.

Consider a finite-dimensional variational problem when $I(u,r) = {}^1\!/_2(Au,u) - (l(r),u)$ is a quadratic function of a point $u$ in $m$-dimensional space $R^m$, $u = (u_1, ..., u_m)$, $A$ is a positive $m \times m$ matrix, $(Au,u)$ is the corresponding positive quadratic form. By $(l,u)$ we denote the scalar product of the vector $l$ (which is a function of the event, $r$) and $u$. We introduce also the vector $\zeta = iu$ with pure imaginary components, $\zeta_1 = iu_1, ..., \zeta_m = iu_m$. For complex-valued vectors, the scalar product used here does not contain the complex conjugation, for example: $(l,\zeta) = i(l,u)$.

For any $z$, $\Re z > 0$, the following identity holds

$$\frac{1}{i^m}\sqrt{\frac{z^m \det A}{(2\pi)^m}} \int_{-i\infty}^{i\infty} \cdots \int_{-i\infty}^{i\infty} e^{z\left[\frac{1}{2}(A\zeta,\zeta)-(l,\zeta)\right]} d^m\zeta = e^{-\frac{1}{2}z(A^{-1}l,l)} \tag{9.7}$$

Here $d^m\zeta \equiv d\zeta_1, ...d\zeta_m$ and $A^{-1}$ is the inverse of the matrix $A$. Formula (9.7) (written in terms of variable $u$) can be found, for example, in [7](ch.2, sec.3). As was mentioned in Introduction, (9.7) can be written also in another form:

$$e^{z\min_u\left[\frac{1}{2}(Au,u)-(l,u)\right]} = \frac{1}{i^m}\sqrt{\frac{z^m}{(2\pi)^m}\det A}\int_{-i\infty}^{i\infty}\dots\int_{-i\infty}^{i\infty} e^{z\left[\frac{1}{2}(A\zeta,\zeta)-(l,\zeta)\right]}d^m\zeta. \tag{9.8}$$

This means that the function which is the subject of averaging in (9.6) can be presented in the form of an integral of a function of auxiliary variables $\zeta$. Averaging is an integration over an event, $r$. The order of integration over $\zeta$ and over $r$ can be changed since the integrals converge absolutely. Thus we arrive at the expression:

$$M\left[e^{Nz\min_u I(u,r)}\right] =$$

$$= \frac{1}{i^m}\left(\frac{Nz}{2\pi}\right)^{m/2}\int_{-i\infty}^{i\infty}\dots\int_{-i\infty}^{i\infty} M\left[\sqrt{\det A}e^{Nz\left[\frac{1}{2}(A\zeta,\zeta)-(l,\zeta)\right]}\right]d^m\zeta. \tag{9.9}$$

For weakly stochastic problems, only $(l,\zeta)$ is random. If $(l,\zeta)$ has the form (1.7) then

$$Me^{-Nz(l,\zeta)} = \left[\int e^{-z(l_0(r),\zeta)}f_0(r)dr\right]^N$$

and (9.9) takes the form

$$M\left[e^{Nz\min_u I(u,r)}\right] =$$

$$= \int_{-i\infty}^{i\infty}\dots\int_{-i\infty}^{i\infty}\frac{1}{i^m}\sqrt{\left(\frac{Nz}{2\pi}\right)^m\det A}\, e^{N\left[\frac{z}{2}(A\zeta,\zeta)+\ln\int e^{-z(l_0(r),\zeta)}f_0(r)dr.\right]}d^m\zeta \tag{9.10}$$

For the probability density function of energy we have

$$f(E) = \frac{N}{2\pi i^{m+1}}\int\sqrt{\left(\frac{Nz}{2\pi}\right)^m\det A}\; e^{NS}d^m\zeta dz;$$

$$S = Ez + \frac{z}{2}(A\zeta,\zeta) + \ln\int e^{-z(l_0(r),\zeta)}f(r)dr. \tag{9.11}$$

If a formal change of variables is made, $\zeta \to \zeta/z$ then (9.11) takes the form

$$f(E) = \frac{N}{2\pi i^{m+1}} \int \sqrt{\left(\frac{N}{2\pi z}\right)^m \det A}\, e^{NS} d^m\zeta\, dz;$$

$$S = Ez + \frac{1}{2z}(A\zeta, \zeta) + \ln \int e^{-(l_0(r),\zeta)} f_0(r) dr. \qquad (9.12)$$

In (9.12) the paths of integration for $\zeta$ can be chosen again as the straight lines which are parallel to the imaginary axis. The direct derivation of this fact from (9.11) and (9.12) is cumbersome, and a better way is to choose another starting point instead of the identity (9.7), the analogous identity involving some arbitrary constants $\alpha = (\alpha_1, ..., \alpha_m)$:

$$e^{-\frac{1}{2}z(A^{-1}l,l)} = \frac{1}{i^m}\sqrt{\frac{\det A}{(2\pi z)^m}} \int\limits_{\alpha_1 - i\infty}^{\alpha_1 + i\infty} \cdots \int\limits_{\alpha_m - i\infty}^{\alpha_m + i\infty} e^{\frac{1}{2z}(A\zeta,\zeta) - (l,\zeta)} d^m\zeta. \qquad (9.13)$$

The identity (9.13) follows from (9.7) by the change of variables $\zeta \to \tilde{\zeta} = \alpha + \zeta$ and $l \to \tilde{l} - z^{-1}A\alpha$. In accordance with (9.13), (9.5) and (9.6), the relation (9.12) is true where the paths of integration in $\zeta$-planes are arbitrary straight lines which are parallel to the imaginary axis, while in $z$-plane this path should be in the right half-plane:

$$f(E) = \frac{N}{2\pi i^{m+1}} \int\limits_{c-i\infty}^{c+i\infty} \int\limits_{\alpha_1 - i\infty}^{\alpha_1 + i\infty} \cdots \int\limits_{\alpha_m - i\infty}^{\alpha_m + i\infty} \sqrt{\left(\frac{N}{2\pi z}\right)^m \det A}\, e^{NS} d^m\zeta dz. \qquad (9.14)$$

Relation (9.14) suggests that $f(E)$ can be obtained by the steepest descent method. However, there is a difficulty.

Note that $m$ in (9.14) can be large, in the order of $N$, or even much larger than $N$: the number of basic functions needed for an accurate approximation of the functional $I(u, r)$ may be much larger that $N$. The steepest descent method was developed for the case when $m$ is fixed and $N \to \infty$, and, thus, cannot be applied to (9.14) directly. The fact that the standard steepest descent method does not work for $m >> N$ can be seen from the following example.

Consider a finite-dimensional truncation of the functional,

$$S(u) = \int\limits_0^1 u^2 dx.$$

Let us take the truncation of the form

$$S_m(u) = \frac{1}{m}\sum_{k=1}^{m} u_k^2$$

where $u_k$ are the values of function $u(x)$ at the points $k/m$. It is clear that the integral

$$\int\limits_{R^m} e^{-NS_m(u)} d^m u = \left[\int\limits_{-\infty}^{+\infty} e^{-\frac{N}{m}u^2} du\right]^m = \left(\sqrt{\frac{m}{2\pi N}}\right)^m$$

has different asymptotics depending on the relative order of $N$ and $m$. Slightly complicating this example we may observe that the asymptotics does not necessary relate to the minimum of $S(u)$.

Another difficulty in (9.14) is a study of stationary points of $S$ which is a function of many complex variables.

Thus, we proceed in a different way trying to get an identity for $f(E)$ which contains $e^{N\check{S}}$ as a factor, and evaluating in which cases this factor indeed determines the asymptotics of $f(E)$.

To this end we note that $\alpha$ in (9.14) can be some functions of $z$ and choose $((\beta, \check{v})$ is the stationary point of $S(E, z, \zeta)$ with respect to $z, \zeta)$.[3]

$$\alpha = \frac{z}{\beta}\check{v}, \quad c = \beta.$$

Then, putting $\zeta = \alpha + iv$ ($v$ is a vector in $R^m$) and $z = \beta + iy$, we have

$$S(E, z, \zeta) = Ez + \frac{1}{2z}(A\zeta, \zeta) + \ln\int e^{-(l_0(r), \zeta)} f_0(r)dr =$$

$$= Ez + \frac{1}{2z}\left[\frac{z^2}{\beta^2}(A\check{v}, \check{v}) + 2\frac{z}{\beta}i(A\check{v}, v) - (Av, v)\right] +$$

$$+ \ln\int e^{-(l_0(r), \check{v})\frac{z}{\beta} - i(l_0(r), v) + (l_0(r), \check{v})} \tilde{f}(r)dr + \ln\int e^{-(l_0(r), \check{v})} f_0(r)dr.$$

where $\tilde{f}$ is determined by (9.2). Since

$$(A\check{v}, \check{v}) = 2\beta^2 E\,, \quad \frac{1}{\beta}A\check{v} = \tilde{l}_0\,, \quad (\tilde{l}_0, \check{v}) = 2E\beta$$

[3]In the derivation of this section $\beta$ is assumed to be positive. This is not required in the derivation of the next section.

the functional $S$ takes the form

$$S(E,z,\zeta) = \overset{\vee}{S} + 2Eiy - \frac{1}{2z}(Av,v) + \frac{i}{\beta}(A\overset{\vee}{v},\overset{\vee}{v}) + \ln\int e^{-i(l_0', \frac{y}{\beta}\overset{\vee}{v}+v)}\tilde{f}(r)dr =$$

$$= \overset{\vee}{S} - \frac{1}{2z}(Av,v) + \ln\int e^{-i(l_0', \frac{y}{\beta}\overset{\vee}{v}+v)}\tilde{f}(r)dr.$$

So,

$$f(E) = Re^{N\overset{\vee}{S}} \tag{9.15}$$

where

$$R = \frac{N}{2\pi}\int\limits_{-\infty}^{+\infty}\sqrt{\left(\frac{N}{2\pi z}\right)^m \det A}\; e^{N\left[-\frac{1}{2z}(Av,v)+\ln\int e^{-i(l_0', \frac{y}{\beta}\overset{\vee}{v}+v)}\tilde{f}(r)dr\right]} d^m v\, dy. \tag{9.16}$$

The prefactor can be also put in another form using the transformation,

$$\int\sqrt{\left(\frac{N}{2\pi z}\right)^m \det A}\; e^{-\frac{N}{2z}(Av,v)}\left[\int e^{-i(l_0', \frac{y}{\beta}\overset{\vee}{v}+v)}\tilde{f}(r)dr\right]^N d^m v =$$

$$= \tilde{M}\int\sqrt{\left(\frac{N}{2\pi z}\right)^m \det A}\; e^{-\frac{N}{2z}(Av,v)-Ni(l',\overset{\vee}{v})\frac{y}{\beta}-iN(l',v)} d^m v =$$

$$= \tilde{M}e^{-Ni(l',\overset{\vee}{v})\frac{y}{\beta}-\frac{Nz}{2}(A^{-1}l',l')} \tag{9.17}$$

Since

$$\int\limits_{-\infty}^{+\infty} e^{-Niy\left[\frac{1}{\beta}(l',\overset{\vee}{v})+\frac{1}{2}(A^{-1}l',l')\right]}dy = \frac{2\pi}{N}\delta\left(\frac{1}{\beta}(l',\overset{\vee}{v})+\frac{1}{2}(A^{-1}l',l')\right), \tag{9.18}$$

we arrive at (9.1).

Now we check (9.1) by the direct inspection.

## 10 Another derivation of the basic identity

First we present the probability density function of energy in the form

$$f(E) = M\delta\left(E - \frac{1}{2}(A^{-1}l,l)\right) = \frac{N}{2\pi i}M\int\limits_{-i\infty}^{+i\infty} e^{N\left[Ez-\frac{z}{2}(A^{-1}l,l)\right]}dz =$$

$$= \frac{N}{2\pi i} \int_{-i\infty}^{+i\infty} \int e^{N\left[Ez - \frac{z}{2}(A^{-1}l,l)\right]} f_0(r_1)...f_0(r_N) d^N r\, dz. \tag{10.1}$$

Introducing averaging with respect to measure (9.2), and using (1.7) and the relation

$$\overset{\vee}{S} = E\beta + \frac{1}{2\beta}(A\overset{\vee}{v}, \overset{\vee}{v}) + \ln \int e^{-(l_0, \overset{\vee}{v})} f_0(r) dr =$$

$$= 2E\beta + \ln \int e^{-(l_0, \overset{\vee}{v})} f_0(r) dr \tag{10.2}$$

we can write

$$f(E) = e^{N\overset{\vee}{S}} \frac{N}{2\pi i} \tilde{M} \int_{-i\infty}^{i\infty} e^{N\left[Ez - 2E\beta + (l(r), \overset{\vee}{v}) - \frac{z}{2}(A^{-1}l,l)\right]} dz. \tag{10.3}$$

Changing in (10.3) $l$ by $\tilde{l}_0 + l'$ we have

$$f(E) = e^{N\overset{\vee}{S}} \frac{N}{2\pi i} \tilde{M} \int_{-i\infty}^{i\infty} e^{N\Delta S} dz; \tag{10.4}$$

$$\Delta S = Ez + (\tilde{l}_0, \overset{\vee}{v}) + (l', \overset{\vee}{v}) - \frac{z}{2}(A^{-1}\tilde{l}_0, \tilde{l}_0) - z(A^{-1}\tilde{l}_0, l') - \frac{z}{2}(A^{-1}l', l'). \tag{10.5}$$

Note that equation (8.3) takes a simple form in terms of measure (9.2):

$$\frac{1}{\beta} A\overset{\vee}{v} = \tilde{l}_0. \tag{10.6}$$

Thus, in accordance with (8.2),

$$(\tilde{l}_0, \overset{\vee}{v}) = 2\beta E, \quad (A^{-1}\tilde{l}_0, \tilde{l}_0) = \left(\frac{\overset{\vee}{v}}{\beta}, \tilde{l}_0\right) = \frac{1}{\beta^2}(A\overset{\vee}{v}, \overset{\vee}{v}) = 2E,$$
$$(A^{-1}\tilde{l}_0, l') = \left(\frac{\overset{\vee}{v}}{\beta}, l'\right). \tag{10.7}$$

Due to (10.7) some terms in (10.5) cancel out, and $\Delta S$ is simplified to

$$\Delta S = (l', \overset{\vee}{v}) - z\left(\frac{1}{\beta}(l', \overset{\vee}{v}) + \frac{1}{2}(A^{-1}l', l')\right). \tag{10.8}$$

Integration over $z$ in (10.4) gives $\delta$-function

$$\frac{N}{2\pi i}\int_{-i\infty}^{i\infty} e^{zN\left(\frac{1}{\beta}(l',\overset{\vee}{v})+\frac{1}{2}(A^{-1}l',l')\right)}dz = \delta\left(\frac{1}{\beta}(l',\overset{\vee}{v})+\frac{1}{2}(A^{-1}l',l')\right)$$

and we arrived at (9.1).

## 11 An estimation of the prefactor

Let us find the limit behavior of the prefactor in (9.1),

$$R = \tilde{M}\left[e^{-\frac{N\beta}{2}(A^{-1}l',l')}\delta\left(\frac{1}{\beta}(l',\overset{\vee}{v})+\frac{1}{2}(A^{-1}l',l')\right)\right]. \tag{11.1}$$

To this end we make an $m$-dimensional projection of the space of functions $u$. Then $l'$ becomes an $m$-dimensional vector, while $A^{-1}$ is the positive $m\times m$ matrix. Vector $\xi=\sqrt{N}l'$, as a sum of statistically independent similarly distributed vectors, has Gaussian distribution with the finite second moments

$$B = \tilde{M}\xi\otimes\xi = \tilde{M}l_0\otimes l_0 - (\tilde{M}l_0)\otimes(\tilde{M}l_0). \tag{11.2}$$

Thus, up to small terms of the order $N^{-1/2}$

$$R = \sqrt{\frac{\det B^{-1}}{(2\pi)^m}}\int e^{\frac{\beta}{2}(A^{-1}\xi,\xi)}\delta\left(\frac{1}{\beta\sqrt{N}}(\xi,\overset{\vee}{v})+\frac{1}{2N}(A^{-1}\xi,\xi)\right)e^{-\frac{1}{2}(B^{-1}\xi,\xi)}d^m\xi =$$

$$= \sqrt{\frac{\det B^{-1}}{(2\pi)^m}}\int e^{\frac{1}{2}(C\xi,\xi)}\delta\left(\frac{1}{2N}(A^{-1}\xi,\xi)+\frac{1}{\beta\sqrt{N}}(\xi,\overset{\vee}{v})\right)d^m\xi,$$

where

$$C = B^{-1}+\beta A^{-1}$$

Presenting $\delta$-function in terms of the Fourier integral we have

$$R = \sqrt{\frac{\det B^{-1}}{(2\pi)^m}}\frac{1}{2\pi}\int e^{-\frac{1}{2}((C-\frac{it}{N}A^{-1})\xi,\xi)+\frac{it}{\beta\sqrt{N}}(\xi,\overset{\vee}{v})}d^m\xi\, dt;$$

$$R = \frac{1}{2\pi}\int\sqrt{\frac{\det B^{-1}}{\det\left(C-\frac{it}{N}A^{-1}\right)}}e^{-\frac{1}{2}((C-\frac{it}{N}A^{-1})\overset{\vee}{v},\overset{\vee}{v})\frac{t^2}{\beta^2 N}}\, dt.$$

Changing the variable of integration $t \to t\sqrt{N}$ we put $R$ in the form

$$R = \sqrt{N}\frac{1}{2\pi}\int\sqrt{\frac{\det B^{-1}}{\det\left(C - \frac{it}{\sqrt{N}}A^{-1}\right)}}e^{-\frac{1}{2}((C-\frac{it}{\sqrt{N}}A^{-1})\check{v},\check{v})\frac{t^2}{\beta^2}}\,dt. \tag{11.3}$$

Assuming that there is a limit of $\det B^{-1}/\det C$ for $m \to \infty$, we see that the integral in (11.3) has the finite limit for $N \to \infty$, and, in the first approximation,

$$R = \sqrt{\frac{N\det B^{-1}}{2\pi\,\det C\,(C\check{v},\check{v})}}\,\beta$$

This completes the justification of formula (1.5).

An estimation of the prefactor which does not use the Gaussian distribution of $l'$ can be obtained from (9.16). Changing variables $v \to \sqrt{z}v/\sqrt{N}$, $y \to y/\sqrt{N}$ we have

$$R = \frac{\sqrt{N}}{2\pi}\int\limits_{-\infty}^{+\infty}\sqrt{\frac{\det A}{(2\pi)^m}}\,e^{-\frac{1}{2}(Av,v)+\Phi_N(u)}d^m v\,dy \tag{11.4}$$

$$u = \frac{y}{\beta}\,\check{v} + v\sqrt{z}, \quad z = \beta + i\frac{y}{\sqrt{N}}$$

$$\Phi_N(u) = N\ln\int e^{-\frac{i}{\sqrt{N}}(l'_0,u)}\tilde{f}(r)dr \tag{11.5}$$

Let

$$\Phi_\infty(u) \equiv \lim_{N\to\infty} N\Phi_N(u) = -\frac{1}{2}(Bu,u) \tag{11.6}$$

and

$$\left|e^{\Phi_N(u)-\Phi_\infty(u)} - 1\right| \le \frac{1}{N^\sigma}(u,u)^\kappa, \quad \sigma > 0$$

Then

$$R = \frac{\sqrt{N}}{2\pi}\int\limits_{-\infty}^{+\infty}\sqrt{\frac{\det A}{(2\pi)^m}}\,e^{-\frac{1}{2}(Av,v)-\frac{1}{2}(Bu,u)}d^m v\,dy + R' \tag{11.7}$$

$$R = \frac{\sqrt{N}}{2\pi}\int\limits_{-\infty}^{+\infty}\sqrt{\frac{\det A}{(2\pi)^m}}\,e^{-\frac{1}{2}(Av,v)-\frac{1}{2}(Bu,u)}\left(e^{\Phi_N-\Phi_\infty} - 1\right)d^m v\,dy$$

If the integral

$$\int_{-\infty}^{+\infty} \sqrt{\frac{\det A}{(2\pi)^m}}\,(u,u)^\kappa e^{-\frac{1}{2}(Av,v)-\frac{1}{2}(Bu,u)} d^m v\, dy$$

converges and tends to some limit for $N \to \infty$, $m \to \infty$, and the integral

$$\int_{-\infty}^{+\infty} \sqrt{\frac{\det A}{(2\pi)^m}}\, e^{-\frac{1}{2}(Av,v)-\frac{1}{2}(Bu,u)} d^m v\, dy \tag{11.8}$$

has the limit for $m \to \infty$, $N \to \infty$, then $R' = O(1/N^\sigma)$ and the formula (1.5) is true. These conditions should be checked in each particular case. Note a necessary condition for the convergence of the integral (11.7): the limit

$$\lim_{m\to\infty} \sqrt{\frac{\det A}{\det(A+\beta B)}} \tag{11.9}$$

should exist and be finite.

## 12 A counterexample

The necessity of the existence of the finite limit (11.9) can be seen from the following example. Consider a 1D variational problem: the functional

$$I(u) = \frac{1}{2}\int_0^1 \left(\frac{du}{dx}\right)^2 dx - \int_0^1 \frac{du}{dx} \overset{0}{\varepsilon}(x)dx \tag{12.1}$$

takes its minimum value on the set of all $u(x)$ such that $u(0) = 0$. This is a problem of internal stresses in a beam with one clamped and one free edge, $\overset{0}{\varepsilon}(x)$ has the meaning of the residual strain. Let $\overset{0}{\varepsilon}(x)$ is a piece-wise function taking random independent values on the segments, $\left[\frac{k-1}{N}, \frac{k}{N}\right]$ $k = 1, ..., N$. For simplicity, we assume that $\overset{0}{\varepsilon}(x)$ takes the two values: $\overset{*}{\varepsilon} + \alpha$ with the probability $p_1$ and $\overset{*}{\varepsilon} - \alpha$ with the probability $p_2$ $(p_1 + p_2 = 1)$. The minimum value of the functional (12.1) can be easily found:

$$\underline{I} = -\frac{1}{2}\int_0^1 \overset{0}{\varepsilon}{}^2 dx = -\frac{1}{2}\overset{*}{\varepsilon}{}^2 - \overset{*}{\varepsilon}\int_0^1 \overset{0'}{\varepsilon} dx - \frac{1}{2}\alpha^2 \tag{12.2}$$

$\overset{\circ}{\varepsilon}{}'$ is the piece-wise function taking the values $\pm\alpha$. The random term in (12.2) has the form

$$\xi \equiv \overset{*}{\varepsilon} \int_0^1 \overset{\circ}{\varepsilon}{}' \, dx = \overset{*}{\varepsilon} \frac{1}{N} \sum_{k=1}^{N} \xi_k \tag{12.3}$$

where $\xi_k$ are independent random variables taking the value $\alpha$ with the probability $p_1$ and $-\alpha$ with the probability $p_2$. The probability density function, $f(\xi)$, of the random variable (12.3) can be determined by the steepest descent method from the exact formula

$$f(\xi) = \frac{N}{2\pi i} \int_{-i\infty}^{i\infty} e^{N(\xi z + \ln(p_1 e^{-z\overset{*}{\varepsilon}\alpha} + p_2 e^{z\overset{*}{\varepsilon}\alpha}))} \, dz \tag{12.4}$$

which is obtained in the same way as (2.4) in section 3. We have

$$f(\xi) = \text{const} \times e^{N \check{S}_1(\xi)} \tag{12.5}$$

where $\check{S}_1(\xi)$ is the stationary value of the function

$$S(\xi, z) = \xi z + \ln \left( p_1 e^{-z\overset{*}{\varepsilon}\alpha} + p_2 e^{z\overset{*}{\varepsilon}\alpha} \right)$$

One can check that the exact formula (12.5 )does not coincide with the formula (1.5) written for the functional (12.1). The cause is that the limit (11.9) does not exist in the case under consideration. Indeed, the functional $\Phi_N$ (11.5) has the form (see also Appendix B)

$$\Phi_N(u) = \sum_k \ln \tilde{M} e^{-\frac{i}{\sqrt{N}} \left\langle \frac{du}{dx} \right\rangle_k \left( \overset{0}{\varepsilon} - \tilde{\varepsilon} \right)_k}$$

where $\langle \cdot \rangle_k$ is the space averaging over $k$th segment:

$$\langle \cdot \rangle_k = \frac{1}{1/N} \int_{(k-1)/N}^{k/N} \cdot \, dx \, ,$$

$\tilde{\varepsilon}$ is the average value of $\overset{0}{\varepsilon}$ with respect to the auxiliary measure, $(\overset{0}{\varepsilon} - \tilde{\varepsilon})_k$ is the value $\overset{0}{\varepsilon} - \tilde{\varepsilon}$ at the $k$th segment. For $N \to \infty$

$$\Phi_N(u) \to -\frac{1}{2N} a^2 \sum_{k=1}^{N} \left( \left\langle \frac{du}{dx} \right\rangle_k \right)^2 \equiv -\frac{1}{2}(Cu, u), \quad a^2 = \tilde{M}(\overset{0}{\varepsilon} - \tilde{\varepsilon})^2$$

Function $\varepsilon = du/dx$ is a piece-wise function in this problem, thus

$$(Au, u) = \frac{1}{2N} \sum_{k=1}^{N} \varepsilon_k^2, \quad (Cu, u) = \frac{1}{N} a^2 \sum_{k=1}^{N} \varepsilon_k^2$$

and

$$\frac{\det A}{\det(A + C)} = \frac{(1/N)^N}{\left[\frac{1}{N}(1 + a^2)\right]^N} = e^{-N \ln(1+a^2)}$$

Hence, the prefactor makes the contribution in (1.5) of the same order as $\exp[N \overset{\vee}{S}]$.

## 13 Large excitations

Consider now the case of the linear functional of the form

$$(l, u) = \frac{1}{\sqrt{N}} \sum_{i=1}^{N} (l_0(r_i), u) \tag{13.1}$$

Average value of each member of the sum (13.1) should be zero since the average value of energy is assumed to be finite,

$$\int l_0(r) f_0(r) dr = 0 \tag{13.2}$$

As in the case of negative temperature states, some eigenvalue problem plays the key role in the analysis. For functionals of the form (13.1) this is the problem

$$A\varphi_k = \lambda_k B \varphi_k \tag{13.3}$$

where the positive operator $Bu$ is determined by the condition

$$M(l_0, u)^2 = (Bu, u)$$

We assume that the eigenvectors $\{\varphi_k\}$ form a basis in the space of functions $u$, all eigenvectors are simple, and $\lambda = 0$ is not an eigenvalue. Obviously, all $\lambda_k$ are positive. Eigenfunctions are normalized by the condition

$$(B\varphi_i, \varphi_j) = \delta_{ij}$$

Then

$$(A\varphi_i, \varphi_j) = \lambda_i \delta_{ij}$$

Two cases should be distinguished:

$$\sum_{k=1}^{\infty} \frac{1}{\lambda_k} < +\infty \qquad \text{and} \qquad \sum_{k=1}^{\infty} \frac{1}{\lambda_k} = \infty \tag{13.4}$$

We show that in the first case the probability density function is given by the formula

$$f(E) = \frac{1}{2\pi i} \int_{-i\infty}^{i\infty} \frac{e^{Ez}}{\sqrt{\Phi(z)}} dz \tag{13.5}$$

where $\Phi(z)$ is an analytic function,

$$\Phi(z) = \prod_{k=1}^{\infty} \left(1 + \frac{z}{\lambda_k}\right) \tag{13.6}$$

Moving the contour of integration to the both sides of the negative $x$-axis we transform the integral (13.7) into a fast converging series

$$f(E) = \sum_{\text{odd}\, k=1}^{\infty} \frac{\sqrt{\lambda_k \lambda_{k+1}}}{\pi} e^{-\lambda_{k+1} E} \int_0^{\lambda_{k+1}-\lambda_k} \frac{e^{Ex}}{\sqrt{x(\lambda_{k+1} - \lambda_k - x)}} dx \tag{13.7}$$

In the second cases (13.4), the average value of energy is infinite, as can be observed by direct inspection, and the variational problem should be regularized. We consider the following regularization:

$$I(u) = \lim_{m \to \infty} \left( \frac{1}{2} (Au, u)_m - (l, u)_m + \frac{1}{2} \sum_{k=1}^{m} \frac{1}{\lambda_k} \right) \tag{13.8}$$

where $(Au, u)_m$ and $(l, u)_m$ are $m$-dimensional projections of the corresponding functionals. Then the probability density function of the functional (13.8) is given by (13.5) where function $\Phi(z)$

$$\Phi(z) = \prod_{k=1}^{\infty} \left(1 + \frac{z}{\lambda_k}\right) e^{-z/\lambda_k} \tag{13.9}$$

Derivation of (13.5), (13.6) is based on the identity (9.13). As before, making $m$-dimensional truncation of $I(u)$ we have

$$f(E) = M\delta\left(E + \min_u (I(u)\right) = \frac{1}{2\pi i} \int_{-i\infty}^{i\infty} e^{Ez} M e^{z \min\limits_u (I(u)} dz =$$

$$= \lim_{m\to\infty} \frac{1}{2\pi i} \int\limits_{-i\infty}^{i\infty} e^{Ez} \frac{1}{i^m} \sqrt{\frac{\det A}{(2\pi z)}} M e^{\frac{1}{2\pi z}(A\zeta,\zeta)-(l,\zeta)} d^m\zeta \, dz =$$

$$= \lim_{m\to\infty} \frac{1}{2\pi i} \int\limits_{-i\infty}^{i\infty} e^{Ez} \frac{1}{i^m} \sqrt{\frac{\det A}{(2\pi z)}} e^{\frac{1}{2\pi z}(A\zeta,\zeta)+N\ln\int e^{-\frac{1}{\sqrt{N}}(l_0(r),\zeta)} f_0(r)dr} d^m\zeta \, dz \tag{13.10}$$

Function

$$N \ln \int e^{-\frac{1}{\sqrt{N}}(l_0(r),\zeta)} f_0(r)dr$$

tends for $N \to \infty$ to

$$\frac{1}{2}(B\zeta,\zeta)$$

Substituting the limit value of this function in (13.10) we get

$$f(E) = \lim_{m\to\infty} \frac{1}{2\pi i} \int\limits_{-i\infty}^{i\infty} e^{Ez} \frac{1}{i^m} \sqrt{\frac{\det A}{(2\pi z)}} e^{\frac{1}{2\pi z}(A\zeta,\zeta)+\frac{1}{2}(B\zeta,\zeta)} d^m\zeta \, dz =$$

$$= \frac{1}{2\pi i} \int\limits_{-i\infty}^{i\infty} e^{Ez} \lim_{m\to\infty} \sqrt{\frac{\det A}{\det(A+zB)}} \, dz \tag{13.11}$$

In the basis $\{\varphi_k\}$, the matrix $B$ is the unit matrix while $A$ is a diagonal matrix with the diagonal elements $\lambda_1, \lambda_2, \ldots$. Thus,

$$\lim_{m\to\infty} \frac{\det(A+zB)}{\det A} = \Phi(z)$$

and (13.11) coincides with (13.5). Formula (13.5) with the regularized function $\Phi(z)$ for the case of diverging series $\sum \lambda^{-1}$ can be obtained in the same way.

## 14 Appendices

### Appendix A. The basic identity for the linear functional (1.3)

The basic identity (9.1) has been proven for linear functionals of the form (1.7). Let us show that it remains the same in the case of the linear functionals (1.3).

Euler's equations for the functional $S$ (1.6) are

$$A\check{u} = \frac{1}{N}\sum_{i=1}^{N}\frac{\int e^{-\beta(l_i(r),\check{u})}l_i(r)f_0^i(r)dr}{\int e^{-\beta(l_i(r),\check{u})}f_0^i(r)dr} \qquad (A.1)$$

$$E + \frac{1}{2}(A\check{u},\check{u}) - \frac{1}{N}\sum_{i=1}^{N}\frac{\int e^{-\beta(l_i(r),\check{u})}l_i(r)f_0^i(r)dr}{\int e^{-\beta(l_i(r),\check{u})}f_0^i(r)dr} = 0 \qquad (A.2)$$

Taking the scalar product of (A.1) with $\check{u}$, we see that the last term in (A.2) is equal to $-(A\check{u},\check{u})$. Thus, equation (A.2) can be written also as

$$E = \frac{1}{2}(A\check{u},\check{u}) \qquad (A.3)$$

The stationary value of $S(E, z, u)$ is

$$\check{S}(E) = 2E\beta + \frac{1}{N}\sum_{i=1}^{N}\ln Q_i$$

$$Q_i = \int e^{-\beta(l_i(r),\check{u})}f_0^i(r)dr \qquad (A.4)$$

We introduce an auxiliary probabilistic measure

$$\tilde{f}_0^i(r) = \frac{e^{-\beta(l_i(r),\check{u})}}{Q_i}f_0^i(r) \qquad (A.5)$$

and denote by $\tilde{l}_i$ the functional averaged with respect to this measure,

$$\tilde{l}_i = \int l_i(r)\tilde{f}_0^i(r)dr \qquad (A.6)$$

In these terms equation (A.1) takes a simple form:

$$A\check{u} = \frac{1}{N}\sum_{i=1}^{N}\tilde{l}_i \equiv \tilde{l} \qquad (A.7)$$

Denote by $l_i'$ the difference $l_i - \tilde{l}$ and by $(l', u)$ the functional

$$(l', u) = \frac{1}{N}\sum_{i=1}^{N}(l_i', u)$$

Our goal is to prove that

$$f(E) = E^{N\check{S}}\tilde{M}\left[e^{-\frac{N\beta}{2}(A^{-1}l',l')}\delta((l',\check{u}) + \frac{1}{2}(A^{-1}l,l))\right] \tag{A.8}$$

As in (10.1)

$$f(E) = M\delta(E - \frac{1}{2}(A^{-1}l,l)) = \frac{N}{2\pi i}M\int_{-i\infty}^{i\infty} e^{N(Ez-\frac{z}{2}(A^{-1}l,l))}dz =$$

$$= e^{N\check{S}}\frac{N}{2\pi i}\tilde{M}\int_{-i\infty}^{i\infty} e^{N[Ez-2E\beta+\beta\frac{1}{N}\sum_{i=1}^{N}(l_i,\check{u})-\frac{z}{2}(A^{-1}l,l))}dz =$$

$$= e^{N\check{S}}\frac{N}{2\pi i}\tilde{M}\int_{-i\infty}^{i\infty} e^{N[Ez-2E\beta+\beta(l,\check{u})-\frac{z}{2}(A^{-1}l,l))}dz$$

Here $\tilde{M}$ is the mathematical expectation with respect to the auxiliary measure. Changing $l$ by $\tilde{l}+l'$ we have the following expression in the exponent

$$Ez-2E\beta+\beta(\tilde{l},\check{u})+\beta(l',\check{u})-\frac{z}{2}(A^{-1}l,l)-z(A^{-1}\tilde{l},l')-\frac{z}{2}(A^{-1}l',l') \tag{A.9}$$

From (A.3) and (A.7),

$$(\tilde{l},\check{u}) = (A\check{u},\check{u}) = 2E\,, \quad \frac{z}{2}(A^{-1}\tilde{l},\tilde{l}) = \frac{z}{2}(\check{u},\tilde{l}) = zE$$

$$(A^{-1}\tilde{l},l) = (\check{u},l)$$

and (A.9) simplifies to

$$\beta(l,\check{u}) - z(\check{u},l) - \frac{z}{2}(A^{-1}l,l)$$

Thus,

$$f(E) = e^{N\check{S}}\tilde{M}\int_{-i\infty}^{i\infty} e^{N(\beta(l',\check{u})-z(\check{u},l')-\frac{z}{2}(A^{-1}l',l')}dz =$$

$$= e^{N\check{S}}\tilde{M}\left[e^{N\beta(l',\check{u})}\delta((\check{u},l') + \frac{1}{2}(A^{-1}l',l'))\right]$$

This is equivalent to (A.8).

The modified expression for the prefactor corresponding to (9.16) is

$$R = \frac{N}{2\pi}\int_{-\infty}^{+\infty}\sqrt{\left(\frac{N}{2\pi z}\right)^m \det A}\, e^{N\left[-\frac{1}{2z}(Av,v)+\frac{1}{N}\sum_k \ln\int e^{-i(l'_k\frac{y}{\beta}\check{v}+v)}\tilde{f}_k dr\right]}d^m v dy$$

## Appendix B. An outline of the derivation of (8.12)

Probability density function of the $r_1$ to be in the vicinity A of the point $x$ while energy takes values between $E$ and $E + \Delta E$ is

$$\operatorname{meas}\{r_1, r_2, ..., r_N : E \leq$$

$$-\min_u \left[\frac{1}{2}(Au, u) - \frac{1}{N}(l_0(r_1), u) - \frac{1}{N}\sum_{i=2}^{N}(l_0(r_1), u)\right] \leq$$

$$\leq E + \Delta E,\ r \in A\} \Big/ \operatorname{meas}\left\{r_1, r_2, ..., r_N : E \leq -\min_u I(u, \mathbf{r}) \leq E + \Delta E\right\}$$

If $\Delta E \to 0, A \to x$ then the numerator tends to

$$\text{const} \times e^{N \check{S}_1(E,x,N)} \Delta E\, f_0(x)$$

where $\check{S}_1(E, x, N)$ is the minimum stationary value of the functional

$$S_1(E, z, v, x, N) = Ez + \frac{1}{2z}(Av, v) - \frac{1}{N}(l_0(x), v) + \ln \int e^{-(l_0(r),v)} f_0(r)dr \tag{B.1}$$

while the denominator is const $\times\, e^{N \check{S}_1(E)} \Delta E$. The functional (B.1) depends on small parameter $N^{-1}$ and can be studied by variational-asymptotic method [1]. In first approximation, as it is easy to see,

$$\check{S}_1(E, x, N) = \check{S}_1(E) - \frac{1}{N}(l_0(x), \check{v}) \tag{B.2}$$

There are other terms of order $N^{-1}$ in (B.2), but they do not depend on $x$. Thus, keeping only the terms depending on $x$, we obtain

$$\tilde{f}(x) = const \times e^{-(l_0(x),\check{v})} f_0(x)$$

as claimed.

## Appendix C. Negative temperature states in a 1D problem

Solution of equation (8.3) in case of negative temperatures is a non-linear eigenvalue problem. It is hard to expect an exact solution to this equation. There are, however, exceptions: 1D problems. Here we consider the problem of minimization of the following functional:

$$I(u) = \frac{1}{2}\int_0^1 \left(\frac{du}{dx}\right)^2 dx - \frac{1}{2N}\sum_{i=1}^{N}(u(r_i) - u(r_{N+i})) \qquad \text{(C.1)}$$

At the ends $u(x)$ is assumed to be zero. The minimizing function $\check{u}$ can be interpreted, for example, as the string lateral displacement caused by $2N$ forces of small amplitude $1/2N$ applied at random points; the points are homogeneously and independently distributed over the string; each force has a counter-part with the opposite direction. The average value of the work done by the forces (the linear functional in (C.1)) is zero. Thus, the most probable value of energy is zero, and only negative values of temperature are relevant. The entropy functional has the form

$$S(E, z, v) = Ez + \frac{1}{2z}\int_0^1 \left(\frac{dv}{dx}\right)^2 dx + \frac{1}{2}\ln Q_1 + \frac{1}{2}\ln Q_2 \qquad \text{(C.2)}$$

$$Q_1 = \int_0^1 e^{-v} dx\,, \qquad Q_2 = \int_0^1 e^{v} dx\,,$$

$$v(0) = v(1) = 0 \qquad \text{(C.3)}$$

The Euler equations of the entropy functional are

$$-\frac{1}{\beta}\frac{d^2v}{dx^2} - \frac{1}{2Q_1}e^{-v} + \frac{1}{2Q_2}e^{v} = 0 \qquad \text{(C.4)}$$

$$E = \frac{1}{2\beta^2}\int_0^1 \left(\frac{dv}{dx}\right)^2 dx \qquad \text{(C.5)}$$

We drop here and in what follows the sign $\vee$.

Integrating equation (C.4) over $x$ we obtain that the derivative of $v$ is periodic

$$\frac{dv}{dx}(0) = \frac{dv}{dx}(1) \qquad \text{(C.6)}$$

Therefore, it is reasonable to search for periodic solutions of equation (C.4) with periodic derivatives. If $v(x)$ is such a solution, that, as follows from (C.4), $v(x+c)+b$ is also a solution for any constants $c$ and $b$. Knowing $v(x)$, one can construct a one-parametric family of solutions of the original problem

$$v(x+c) - v(c) \tag{C.7}$$

which are equal to zero at both ends for any constant $c$. Of course, the entropy functional has the same value for each member of the family (C.7). The above mentioned non-uniqueness of the stationary points is caused by the symmetry of the entropy functional under transformations

$$v(x) \to v(x+c) + b$$

on the set of all periodic functions $v(x)$ with the period 1.

Multiplying (C.4) by $dv/dx$ and integrating we have

$$\frac{1}{2b}\left(\frac{dv}{dx}\right)^2 + \frac{1}{2Q_1}e^{-v} + \frac{1}{2Q_2}e^{v} = A = \text{const} \tag{C.8}$$

Here $b \equiv -\beta > 0$. We seek for periodic solutions of (C.8) with periodic derivatives. If $v(x)$ is a solution, then one can make a shift, $v \to v' = v + c$ and choose $c$ in such a way, that $Q_1(v') = Q_2(v')$. Therefore, without loss of generality one can set $Q_1 = Q_2 \equiv Q$. Changing the constant $A$ by the constant $A \equiv aQ$ we have

$$\left(\frac{dv}{dx}\right)^2 = \frac{2b}{Q}(a - \cosh v) \tag{C.9}$$

In the phase plane $(v, dv/dx)$ the integral curves form a family of the closed curves surrounding the origin. The dependence of $v$ on $x$ can be found from (C.9).

$$\int \frac{dv}{\pm\sqrt{a - \cosh v}} = \sqrt{\frac{2b}{Q}}\, x \tag{C.10}$$

The path of integration rotates around the origin a number of times, $n$. For given $n$ and $b$, the value of parameter $a$ is determined by the conditions

$$\oint \frac{dv}{\pm\sqrt{a - \cosh v}} = 4n \int_0^{v_{\max}} \frac{dv}{\sqrt{a - \cosh v}} = \sqrt{\frac{2b}{Q}} \tag{C.11}$$

$$Q = \int_0^1 e^{-v} dx = \int_0^1 \cosh v \, dx = \oint \frac{\cosh v \, dv}{dv/dx} = 4n\sqrt{\frac{2b}{Q}} \int_0^{v_{\max}} \frac{\cosh v \, dv}{\sqrt{a - \cosh v}} \tag{C.12}$$

$$a \equiv \cosh v_{\max}$$

Equation (C.5) provides the link between energy and temperature: from (C.5) and (C.9)

$$E = \frac{1}{2b^2} \int_0^1 \left(\frac{dv}{dx}\right)^2 dx = \frac{1}{bQ}(a - Q) \tag{C.13}$$

The integral appeared can be expressed in terms of elliptic integral $F(t,k)$ and associated Legandre functions [4] $P_\nu^\mu(a)$:

$$\int_0^{v_{\max}} \frac{dv}{\sqrt{a - \cosh v}} = \frac{\pi}{\sqrt{2}} P^0_{-1/2}(a)$$

$$\int_0^{v_{\max}} \frac{\cosh v \, dv}{\sqrt{a - \cosh v}} = \frac{\pi}{\sqrt{2}} P^0_{1/2}(a)$$

$$\int_0^{v_{\max}} \frac{dv}{\sqrt{a - \cosh v}} = \frac{2}{\sqrt{a+1}} \left[ F\left(\frac{\pi}{2}, \sqrt{\frac{a-1}{a+1}}\right) - F\left(t, \sqrt{\frac{a-1}{a+1}}\right) \right]$$

$$\sin^2 t \equiv \frac{a - \cosh v}{a - 1}$$

Finally, for given $E$ and $N$, the values of $b$, $Q$, and $a$ are determined from the system of nonlinear equations:

$$b = 4n^2\pi^2 P^0_{-1/2}(a) P^0_{1/2}(a) \,, \quad Q = P^0_{1/2}(a) \Big/ P^0_{-1/2}(a) \,, \quad E = \frac{1}{b}\left(\frac{a}{Q} - 1\right) \tag{C.14}$$

The corresponding value of entropy is equal to

$$\check{S}(E) = -2Eb + \ln Q$$

Fig. 2 shows the dependence of entropy and inverse temperature on

[4]Note a misprint in 2.464.32 of [9]: the sign on the right hand side must be changed.

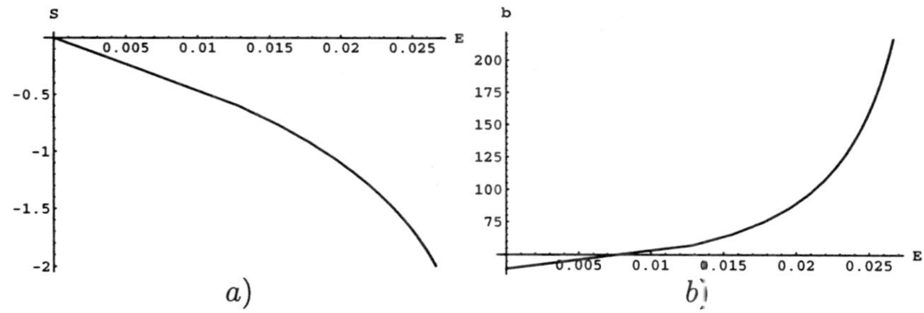


Figure 2: Dependence of entropy (a) and negative inverse temperature (b) on energy for the first branch, $n = 1$

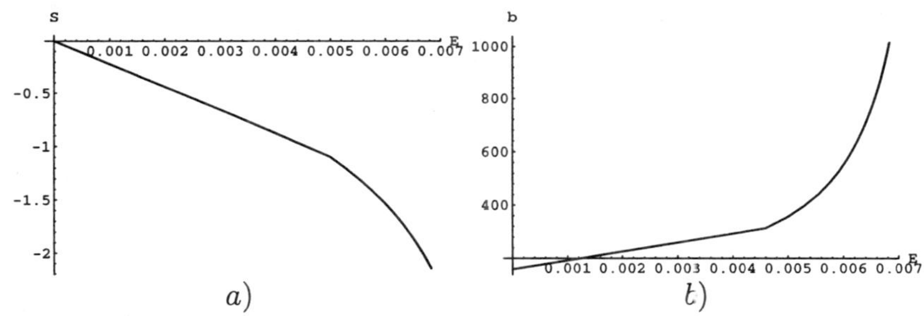


Figure 3: Dependence of entropy (a) and negative inverse temperature (b) on energy for the second branch, $n = 2$

energy for the first branch, $n = 1$. Note that the energy values are bounded above. A numerical value of the upper limit is 0.03. Study of the asymptotics of the solution of equations (C.14) for $a \to \infty$ gives the asymptotic maximum value of energy $1/32n^2$, and very slow approaching of entropy to negative infinity, $S \approx -\ln \ln a$. Thus, the entropy decay is not seen on the Figures. The inverse temperature goes to infinity at the maximum value of energy. Similar graphs for the second branch are plotted in Fig. 3a,b.

The dependence of entropy on energy in vicinity of zero is presented on Fig. 4 for the first five branches. The maximum entropy corresponds to the first branch. Entropy of each next branch is smaller than the

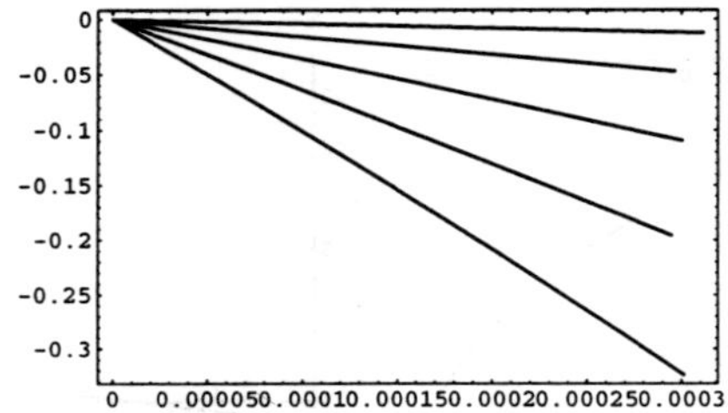


Figure 4: Dependence of entropy on energy for the first five branches

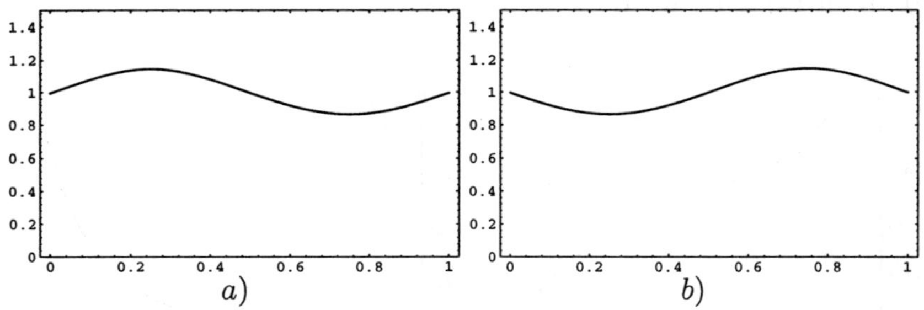


Figure 5: Probability density function of positions of positive (a) and negative (b) force for small $E = 0.000125$

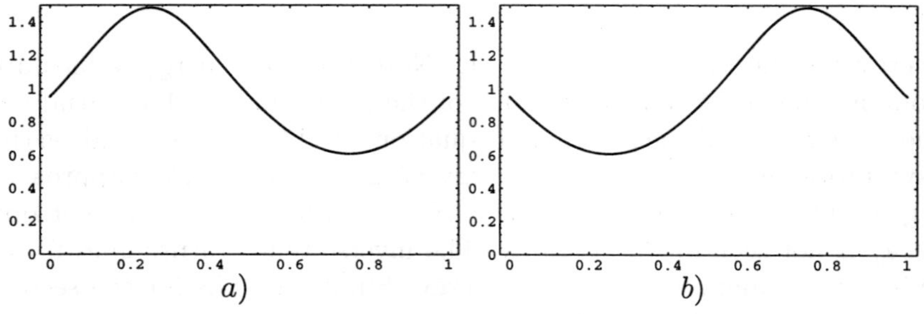


Figure 6: Probability density function of positions of positive (a) and negative (b) force for $E = 0.0012$

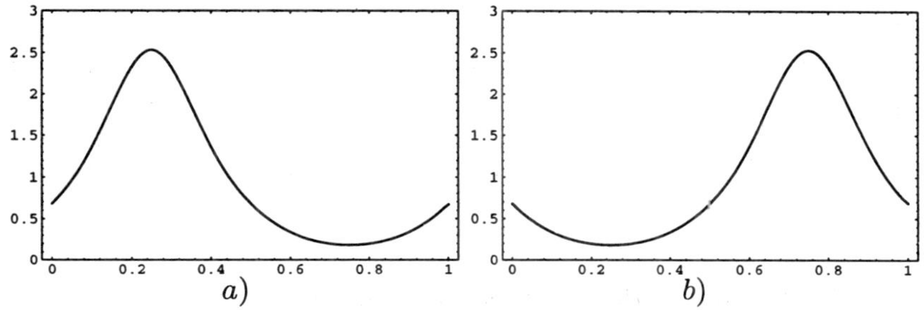


Figure 7: Probability density function of positions of positive (a) and negative (b) force for $E = 0.0075$

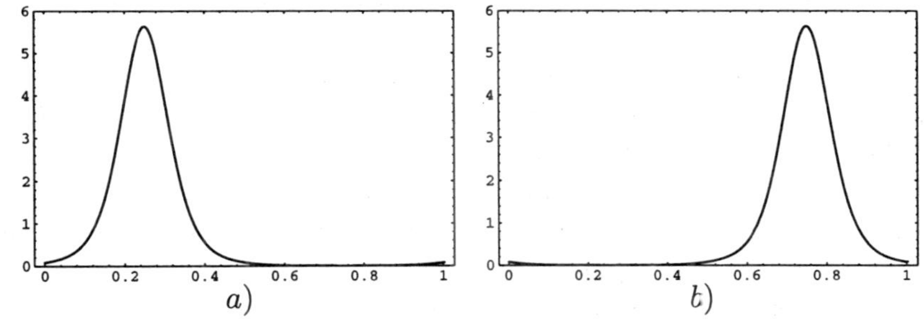


Figure 8: Probability density function of positions of positive (a) and negative (b) force for $E = 0.02$

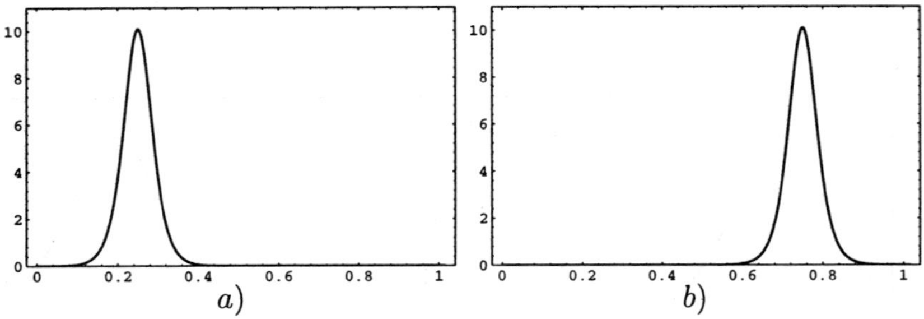


Figure 9: Probability density function of positions of positive (a) and negative (b) force for $E = 0.025$

entropy of previous one.

To provide small values of energy, the points of application of external forces must be almost homogeneously distributed over the beam. The probability density function of distribution of "positive" force is shown in Fig. 5a for $E = 0.000125$. The corresponding distribution of "negative" force is shifted to the right for $x = 0.5$ (Fig. 5b). It is seen formation of zones of localization, the regions where application of the external forces is more probable. The first branch has two localization zones. Of course, both zones can be translated along the string due to the above-mentioned periodicity.

Fig. 6a,b presents probability density functions of points of application of positive and negative forces for the value of energy $E = 0.0012$. Probability density functions for larger values of energy $E = 0.0075, 0.02, 0.025$ are shown in Figs. 7a,b, 8a,b and 9a,b.

Other branches have more localization zones. The $n$th branch has $2n$ localization zones. Qualitatively, the dependence of the size of the zones on energy is the same as for the first branch.

## References


1. V.L. Berdichevsky, Variational Principles of Continuum Mechanics, Nauka, Moscow, 1983.
2. V.L. Berdichevsky, Statistical mechanics of point vortices, *Physical Review E*, **51** (5), 4432–4452 (1995).

3. V. Berdichevsky, Thermodynamics of chaos and order, Addison Wesley Longman, 1997.
4. M. Brede and P. Haasen, *Acta Metall.* **36**, 2003 (1988).
5. G.P. Cherepanov, Nonofracture mechanics approach to dislocation and fracturing, *Appl. Mech. Rev.*, **47** (6), 326–330 (1994).
6. G.P. Cherepanov, A.. Ritcher, V.E. Verijenko, S. Adali, and V. Sutyrin, Dislocation generation and crack growth under monotonic loading, *J. Appl. Phys.*, **78** (10), 6249–6264 (1995).
7. M.V. Fedorjuk, Asymptotics, Integrals and Series, Nauka, Moscow, 1987.
8. W.W. Gerberich, H. Huang, W. Lielinski, and P.G. Marsh, , *Metall. Trans. A*, **24A**, 535 (1993).
9. I.S. Gradstein and I.M. Ryzhik Table of Integrals, Series and Products, Academic Press, 1994, 5th edition.
10. P. B. Hirsch, S.G. Roberts, and J. Samuels, *J. Proc. R. Soc. Land.* **A421**, 25 (1989).
11. M. Khantha, The brittle-to-ductile transition-II: dislocation dynamics and the strain-rate depandence of the transition temperature, *Scripta Metallurgica et Materialia*, **31** (10), 1355–1360 (1994).
12. M. Khantha, D. P. Pope, and V. Vitek, The brittle-to-ductile transition-I: a cooperative dislocation generation instability, *Scripta Metallurgica et Materialia*, **31** (10), 1349–1354 (1994).
13. M. Khantha, D.P. Pope, and V. Vitek, Dislocation screening and the brittle-to-ductile transition: a Kosterlitz-Thouless type instability, *Physical Review Letters*, **73**, 5 (684–687).1994
14. M. Khantha, D. P. Pope, and V. Vitek, Dislocation generation instability and the brittle-to-ductile transition, *Scripta Metallurgica et Materialia*, **A192/193**, 435–442 (1995).
15. J.M. Kosterlitz and D.J. Thouless, Ordering, metastability and phase transcriptions in two-dimentional systems, *J. Phys.*, **6**, 1181 (1973).
16. E. Lucacs, Characteristic functions, Griffin, London, 1970, 2nd edition.
17. D. R. Nelson and B. I. Halperin, *Phys. Rev. B* **19**, 2457 (1979).
18. J.R. Rice and G.E. Beltz, *J. Mech. Phys. Sol.* **42**, 333 (1994).
19. J.R. Rice, G.E. Beltz, and Y. Sun, Peierls Framework for Dislocation Nucleation from a Crack Tip, In A. S. Argon, editor, Topics

in Fracture and fatigue, pp. 1–58. Springer-Verlag, 1992.

20. J.R. Rice and R. Thompson, Ductile versus brittle behaviour of cristals, *Phil. Mag.*, **29**, 73 (1974).
21. J.R. Rice, Dilocation nucleation from a crack tip: an analysis based on the Peierls concept, *J. Mech. Phys. Solids*, **40** (2), 239–271 (1992).
22. G. Schoeck, Dislocation emmision from crack tips, *Phil. Mag. A*, **63**, 111 (1991).
23. R. Thomson, *Solid St. Phys.* **39**, 1 (1986).
24. M.M. Veinberg, Variational Method and Merthod of Monotone Operators in the Theory of Nonlinear Equations, John Wiley and Sons, 1973.
25. G.A. Williams, *Phys. Rev. Lett.* **59**, 1926 (1987).
26. A.P. Young, *Phys. Rev. B* **19**, 1855 (1979).

# ASYMMETRIC STRAIN-STRESS DISTRIBUTION FUNCTION FOR CRYSTAL WITH RANDOM POINT DEFECTS

L. BERLYAND
*Department of Mathematics and Materials Research lab,*
*Penn State University, University Park, PA 16802, USA*

An exactly solvable model of elastic crystal with random point defects is suggested. It is shown that if the defects are distributed according to Poisson's law then the strain-stress distribution function is described by a shifted Cauchy distribution (an asymmetric Lorentz curve). The asymmetry is calculated analytically in terms of the Frenel integral. A simple explanation of the an illusory contradiction between the asymmetry and the distribution's zero mean is presented.

Properties of random media are usually described by some average quantities. In [1, 2] (see also [14, 17] for references to more recent work) a very general approach for such problems was developed. The main objective of this approach is to derive the effective equations for the expected values of the original random fields. In general situation we can hardly hope for more detailed statistics. However, in some cases it is possible and we present here a particular model in which determination of the distribution function of the random field can be done.

The main results presented in this paper were obtained several years ago but at that time they were not published. Short summary of these results appeared in a joint work with experimentalists [12, 13], where the asymmetry predicted by our analytical model was confirmed experimentally and numerically.

I would like to acknowledge with deep gratefulness that my interest to random media problems was stimulated by numerous discussions with S. Kozlov.

## 1 Identical defects

Let us consider a crystal in which the point defects are distributed equiprobably and independently with given density $n = N/V$ ($N$ is the number of defects in the volume $V$).

According to the well-known concepts of the continuum theory of defects [3–5], it is possible to consider crystal as a continuum elastic medium with anisotropy which corresponds to the lattice symmetry. In this medium the point defects are represented as delta-shaped sources of internal stresses. For simplicity we assume that the crystal with no defects is described by isotropic linear elasticity equations [3–5].

Then the displacement and strain produced by a symmetric point defect at the origin are given by the following formulas (see [3–5]):

$$U_i(\vec{r}) = \pm K\Omega_{km}G_{ik,m}(\vec{r}); \varepsilon_{ik}(\vec{r}) = \pm K\Omega_{jm}G_{i)m,(kj} \tag{1.1}$$

where $G_{i)m,(kj} = \frac{1}{2}[G_{im,kj} + G_{km,ij}]$,

$$G_{ik}(\vec{r}) = A\frac{\delta_{ik}}{r} + B\frac{x_i x_k}{r^3} \quad , \quad r = |\vec{r}|. \tag{1.2}$$

is the Green's tensor for isotropic elastic medium in $\mathbb{R}^3$ [5, 6], constants $A$ and $B$ are expressed through Lame's coefficients [3, 6]; dimensionless tensor $\Omega_{jm}$ characterizes the defect symmetry, and coefficient $K$ represents the intensity of the point defect [3, 5]. Index after comma in (1.1) and below denotes differentiation with respect to the corresponding coordinate, and summation is taken with respect to repeating indices. Explicit form of the displacement vector $\vec{U}$ and the strain tensor $\varepsilon_{ik}$ for the defect with spherical symmetry $\Omega_{k\ell} = \delta_{k\ell}$ [4, 5] are given by

$$\vec{U}(\vec{r}) = \pm K\nabla\left(\frac{1}{r}\right), \qquad \varepsilon_{kk}(\vec{r}) = \pm K\frac{1 - 3\cos^2\theta_k}{r^3} \tag{1.3}$$

$$\varepsilon_{ik}(\vec{r}) = \pm 3K\frac{\psi_{ik}}{r^3}, \qquad i \neq k; \quad K = \frac{\Omega_0}{12\pi}\,\frac{1+\nu}{1-\nu}, \qquad i,k = 1,2,3,$$

where $\delta_{k\ell}$ is the Kronecker delta. In (1.1)–(1.3) the sign is defined by the type of the defect ("+" for a vacancy, "−" for an interstitial or impurity), $\theta_k = (\vec{r}, 0X_k)$ is the angle between $\vec{r}$ and $k$-th axis, $\Omega_0$ is the atomic volume, $\nu$ is the dimensionless Poisson's ratio. It is known, that elastic field of a spherical inclusion, which is inserted with tension (i.e. into a spherical hole of the size smaller then the inclusion size) into an isotropic elastic medium which fills $\mathbb{R}^3$ is defined by (1.3). More precisely (1.3) holds sufficiently far from the inclusion (i.e. these formulas are asymptotic, see [3, 4] for details). Then it is possible to model the above problem as an unbounded isotropic elastic medium with randomly distributed inclusions whose radii are much smaller then distances between them.

Without loss of generality we will consider hereafter defects having symmetry of an ellipsoid of revolution [4, 5]:

$$\Omega_{ik} = \Omega_0 \delta_{ik} + \Omega_1 (\ell_i \ell_k - \frac{1}{3}\delta_{ik}) \tag{1.4}$$

where $\vec{\ell}$ is the unit vector in the direction of the ellipsoid's axis, parameter $\Omega_1$ determines the magnitude of the deviatoric part of $\Omega_{ik}$. We are interested in the case of small concentration of defects. Then the resulting strain is an additive quantity and the desired distribution function can be determined as follows:

$$I(\varepsilon_{ik}) = \lim_{\substack{V \to \infty \\ N \to \infty \\ n \to N/V}} \frac{1}{(4nV)N} \times$$
$$\times \int\limits_{(S_0)^N} \int\limits_{(V)^N} \delta \left[ \varepsilon_{ik} - \sum_{j=1}^{N} \varepsilon_{ik}(\vec{r}_j, \vec{\ell}_j) \right] d\vec{r}_1 d\vec{\ell}_1 \ldots d\vec{r}_N d\vec{\ell}_N, \tag{1.5}$$

where $\varepsilon_{ik}(\vec{r}, \vec{\ell})$ is given by the formulas (1.1), (1.3).

In (1.5) the integration is taken over all configurations $\vec{r}_1, \ldots \vec{r}_N$ and all orientations $\vec{\ell}_j$ on the unit sphere $S_0$, which produce fixed resulting strain $\varepsilon_{kk}$. Search over all configurations is done in accordance with the above assumptions that the defects are placed independently and equiprobably, which leads to the Poisson's distribution of the defects positions in the thermodynamic limit ($n = N/V$ remains constant as $V \to \infty$,).

The difficulty in the calculation of the integral (1.5) is caused by the fact that the integral in (1.5) has infinite multiplicity and the integrand has singularity at the origin (see (1.1)). That is why instead of calculating the distribution function $I(\varepsilon_{ik})$ we will find the corresponding characteristic function.

For the sake of definiteness we take $i = k = 1$ in (1.5) and choose the "$-$" sign in (1.1) (interstitial). We further expand $\delta$-function into Fourier integral and change the order of integration. Then the expression under the limit in (1.5) takes the form

$$\frac{1}{2\pi}\int_{-\infty}^{\infty} e^{i\varepsilon_{11}t}dt\left\{\frac{1}{4nV}\int_{S_0} d\ell\int_V \exp\left[itK\Omega_{jk}G_{1k,1j}\right]dV\right\}^N . \tag{1.6}$$

Further let us write the internal integral in the following form

$$1-\frac{n}{N}\frac{1}{4\pi}\int_{S_0} d\ell\int_V \{1-\exp[itK\Omega_{jk}G_{1k,1j}]\}\,dV = 1-\frac{n}{N}\,J_V. \tag{1.7}$$

Then the problem is reduced to the calculation of $J = \lim\limits_{V\to\infty} J_V$.

We pass to the limit in the polar coordinates (which corresponds to the spherical shape of a large crystal specimen) and separate real and imaginary parts. Then we get

$$\begin{aligned} J = {} & \frac{1}{4\pi}\int_{S_0} d\ell\int_0^{\infty} r^2\,dr\int_0^{2\pi} d\varphi\int_0^{\pi}\sin\theta d\theta\{1-\cos[tK\Omega_{jk}G_{1k,1j}]\} - \\ & - i\frac{1}{4\pi}\int_{S_0} d\ell\int_0^{\infty} r^2\,dr\int_0^{2\pi} d\varphi\int_0^{\pi}\sin\theta d\theta\{\sin[tK\Omega_{jk}G_{1k,1j}]\} \\ = {} & J_1 - i\ J_2. \end{aligned} \tag{1.8}$$

It is easy to see that $r^3 G_{ik,\ell m}$ does not depend on $r$. Therefore we introduce the notation

$$g_{ik,\ell m}(\theta,\varphi) = r^3 G_{ik,\ell m}. \tag{1.9}$$

To calculate $J_1$ we note that $\int_0^{\infty} r^2 dr[1-\cos(br^{-3})]$ after substitution $x = r^{-3}$ is reduced to the well- known improper integral ([10])

$$\int_0^{\infty}\frac{(1-\cos bx)dx}{x^2} = \frac{\pi|b|}{2} \tag{1.10}$$

and therefore after elementary transformations we obtain

$$J_1 = |t|\,K\,\gamma\,, \tag{1.11}$$

where

$$\gamma = \frac{1}{4\pi}\int_{S_0} d\vec{\ell}\int_0^{2\pi} d\varphi\int_0^{\pi}\sin\theta d\theta\Omega_{jk}g_{1k,1j}(\theta,\varphi). \tag{1.12}$$

To calculate $J_2$ we observe that the integral

$$\frac{1}{4\pi}\int_{S_0} d\vec{\ell}\int_0^{R} r^2 dr\int_0^{2\pi} d\varphi\int_0^{\pi}\sin\theta d\theta \times \sin[tK\Omega_{jk}g_{1k,1j}r^{-3}]$$

converges if $R \to \infty$. Indeed, since

$$\frac{1}{4\pi}\int_{S_0} \ell_i \ell_k d\vec{\ell} = \frac{1}{3}\delta ik,$$

$$g_{11,11} + g_{12,12} + g_{13,13} = (3\cos^2\theta - 1)\,\text{const}$$

$$\int_0^{\pi} (3\cos^2\theta - 1)\sin\theta d\theta = 0, \tag{1.13}$$

the integral of the first term in the expansion in the powers of $r^{-3}$ is zero:

$$\frac{1}{4\pi}\int_0^{2\pi} d\varphi \int_0^{\pi} \sin\theta d\theta \int_{S_0} d\vec{\ell}\,[\Omega_{jk} g_{1k}, g_{1j}] = 0. \tag{1.14}$$

Therefore

$$J_2 = t\,K\,\alpha, \tag{1.15}$$

$$\alpha = \frac{1}{4\pi}\int_{S_0} d\vec{\ell} \int_0^{\infty} r^2 dr \int_0^{2\pi} d\varphi \int_0^{\pi} \sin\theta d\theta [\Omega_{jk} g_{1k,1j}(\theta,\varphi)]. \tag{1.16}$$

Substituting (1.11), (1.15) into (1.8) and using (1.5) – (1.7), we obtain

$$\begin{aligned} I(\varepsilon_{11}) &= \lim_{N\to\infty} \frac{1}{2\pi}\int_{-\infty}^{\infty} e^{i\varepsilon_{11}t}\left[1 - \frac{n|t|K\gamma}{N} + i\,\frac{n\,t\,K\alpha}{N}\right]^N dt = \\ &= \frac{1}{2\pi}\int_{-\infty}^{\infty} e^{i\varepsilon_{11}t} e^{-nK\gamma|t| + inK\alpha t} dt = \\ &= \frac{1}{\pi}\,\frac{n\,K\,\gamma}{(\varepsilon_{11} - nK\alpha)^2 + (nK\gamma)^2} \end{aligned} \tag{1.17}$$

Thus, the desired function is the Cauchy's distribution (the Lorentz's curve) centered at $nK\alpha$ and with width $nK\gamma$ at the semiheight ( i.e.the semiwidth is $nK\gamma$ ).

The constants $K, \gamma, \alpha$ are defined in (1.1), (1.12) and (1.16) respectively, $n$ is the density of defects.

Note that the constants $\gamma$ and $\alpha$ depend on the tensor $\Omega_{ik}$ only, i.e. depend only on the geometry (shape) of the defects.

**Remark 1**. To simplify calculations we consider hereafter only the case of defects with spherical symmetry $\Omega_{ik} = \delta_{ik}$. However, general case of ellipsoidal symmetry can be analyzed in the same way and all the results are analogous (but, more cumbersome).

In view of this remark, the field of each defect is defined by formulas (1.3) and the expressions (1.8) have a simpler form:

$$J_1 = 2\pi \int_0^\infty r^2 dr \int_0^\pi \{1 - \cos[tK(1-3\cos^2\theta)r^{-3}]\} \sin\theta d\theta$$
$$J_2 = 2\pi \int_0^\infty r^2 dr \int_0^\pi \sin[(1-3\cos^2\theta)tKr^{-3}] \sin\theta d\theta \tag{1.18}$$

After substitution $x = r^{-3}$, $z = \cos\theta$ we get

$$J_1 = \frac{4\pi}{3} \int_0^1 dz \int_0^\infty \frac{1-\cos[tK(1-3z^2)x]}{x^2} dx$$
$$J_2 = \frac{4\pi}{3} Kt \int_0^\infty \frac{dx}{x^2} \int_0^1 \sin[x(1-3z^2)]dz \tag{1.19}$$

It remains to refer to (1.10) to obtain

$$\gamma^0 = \frac{\delta\,\pi^2}{9\sqrt{3}}\,, \tag{1.20}$$

$$\alpha^0 = \frac{4\pi}{3} \int_0^\infty \frac{dx}{x^2} \int_0^1 \sin[x(1-3z^2)]dz \simeq \frac{4\pi}{3} \cdot 0.159714\ldots, \tag{1.21}$$

i.e. $\gamma^0$ can be calculated explicitly and $\alpha^0$ can be expressed, for example, through the Frenel's integral. Its numerical value can be obtained by computing the integral (1.21).

We proceed now with determination of the distribution of the shear component $\varepsilon_{12}$ of the strain tensor. We first observe that due to (1.3) and Remark 1, the integral $J$ defined in (1.8) can be written in the following form

$$J = \int_0^{2\pi} d\varphi \int_0^\pi \sin\theta \, d\theta \int_0^\infty r^2 dr \{1 - \exp[it\, 3K\, \psi_{12}(\theta,\varphi) r^{-3}]\} \tag{1.22}$$

Writing $\int_0^\pi \sin\theta\, d\theta$ as the sum $\int_0^{\pi/2} \sin\theta\, d\theta + \int_{\pi/2}^\pi \sin\theta\, d\theta$, we decompose (1.22) into two integrals and make the change of variable $\theta' = \pi - \theta$, $r' = -r$ in the second term. Then we obtain

$$J = \int_0^{2\pi} d\varphi \int_0^{\pi/2} \sin\theta d\theta \int_{-\infty}^\infty r^2 dr \{1 - \exp[it3K\psi_{12}(\theta,\varphi)r^{-3}]\}$$

$$= \int_0^{2\pi} d\varphi \int_0^{\pi/2} \sin\theta \, d\theta \int_{-\infty}^\infty r^2 dr [1 - \cos[it\, 3K\, \psi_{12}(\theta,\varphi) r^{-3}]]\} \tag{1.23}$$

After elementary transformations, which are analogous to (1.19), we obtain

$$J = |t|\lambda^0 , \qquad \lambda^0 = \frac{4\pi}{3} K \tag{1.24}$$

And, thus, the strain distribution has the form of Lorentz's curve (Cauchy distribution) centered at the origin (unlike (1.17)!) with semi-width $nK\lambda^0$

$$I(\varepsilon_{12}) = \frac{1}{\pi} \frac{n K \lambda^o}{\varepsilon_{12}^2 + (nK\lambda^0)^2} \tag{1.25}$$

## 2 Strain and stress distributions for defects of two different kinds

We now assume that in the density of defects (number of defects per unit volume) with "$-$" sign in (1.1) or (1.3) is $n_1$ and with "$+$" sign is $n_2 = n - n_1$.

It means that each defect with probability $n_1/n$ is an interstitial and with probability $n_2/n$ it is a vacancy. Taking into account Remark 1, the integral, which defines the characteristic function of desired distribution (and is analogous to the integral (1.7)), can be written as follows:

$$\begin{aligned} 1 &- \frac{1}{V} \frac{n_1}{n} \int_V \{1 - \exp[it\, K(1 - 3\cos^2\theta) r^{-3}]\} dV - \\ &- \frac{1}{V} \frac{n_2}{n} \int_V \{1 - \exp[it\, K(3\cos^2\theta - 1) r^{-3}]\} dV = \\ &= 1 - \frac{n_1}{N} \int_{2V} \{1 - \exp[it\, K(1 - 3\cos^2\theta) r^{-3}]\} dV - \\ &- \frac{n_1 - n_2}{N} \int_V \{1 - \exp[it\, K(1 - 3\cos^2\theta) r^{-3}]\} dV \end{aligned} \tag{2.1}$$

Here the integration over the volume $V$ $(0 < r < R)$ of two exponents with opposite signs is replaced by the integration over the doubled volume $(-R < r < R)$. Calculation of the limit $V \to \infty$ in (2.1) reduces to determination of the constants $\alpha^0$ and $\gamma^0$ (see (1.19) – (1.21)).

Therefore from (2.1), taking into account the integration over the doubled volume (i.e. in $r$ from $-\infty$ to $\infty$)), we get (analogously to (1.17))

$$I(\varepsilon_{11}) = \frac{1}{2\pi} \int_{-\infty}^{\infty} e^{i\varepsilon_{11}t} e^{-n2\gamma^0|t| + i\alpha^0(n_1 - n_2)t} dt =$$

$$= \frac{1}{\pi} \frac{n\,K\,2\gamma^0}{[\varepsilon_{11} - K\,\alpha^0(n_1 - n_2)]^2 + (n\,K\,2\gamma^0)^2} \tag{2.2}$$

Thus in the presence of the defects of the two kinds, the distribution is centered at the point $(n_1 - n_2)\,K\,\alpha^0$, i.e. it can be viewed as a shift of the Cauchy distribution which was originally symmetric (with respect to the origin).

For ellipsoidal defects the distribution is also defined by formula (2.2), with $\alpha$ and $\gamma$ defined in (1.12) and (1.16) in place of $\alpha^0$ and $\gamma^0$ respectively.

It should be noted that in an improbable but theoretically possible case $n_1 = n_2$, the distribution (2.2) becomes symmetric (around the origin).

It is well known ([3]), that the trace $\varepsilon_{ii}(r)$ the strain tensor of the dilation centers in isotropic medium is zero, i.e. $\varepsilon_{ii}(r) = 0$ for each positive $r$.

Therefore the Hook's law can be written in a particulary simple form $\sigma_{ik} = 2G\,\varepsilon_{ik}$; $i, k = 1, 2, 3$ where $G$ is the shear modulus. Hence the distribution functions (1.17), (1.25), and (2.2) describe the distribution of the corresponding components of the stress tensor (up to the multiplicative constant $2G$).

## 3 Tail and first moment of the distribution

The Lorentz's form of the strain and stress distribution was obtained above under the assumption that defects can be located arbitrary close to the observation point $r = 0$. However, from physical consideration it is obvious that there exists the minimal distance $a$, whose order of magnitude is determined by the period of the crystal lattice so that there are no defects in the sphere $S_a$ of radius $a$ centered at $r = 0$ ($a$ determines the maximal strain $\sim K\,a^{-3}$). Therefore, in order to refine the distribution curve in the range of maximal strains, we replace the integration over $V$ by the integration over $V_a = V \backslash S_a$ in the inner integral in (1.6). Then we obtain the distribution

$$I_a(\varepsilon) = \frac{1}{2\pi} \int_{-\infty}^{\infty} e^{i\varepsilon t} \varphi_a(t)dt \tag{3.1}$$

where

$$\varphi_a(t) = \exp\left\{-n \int_{V_a} [1 - \exp[itK(1 - 3\cos^2\theta)r^{-3}]]dV\right\} \tag{3.2}$$

Expanding the inner exponent in the Taylor series and differentiating (taking into account (1.13)), we find $\varphi_a'(0) = 0$, $\varphi_a''(0) = -C_2 n\, K^2 a^{-3}$, $C_2 = \frac{4\pi}{3}\int_0^1 (3z^2-1)^2 dz$. Consequently, the first and the second moments of the distribution $I_a(\varepsilon)$ are given by

$$M_1(I_a) = 0\,, \qquad M_2(I_a) = C_2 n\, K^2 a^{-3}. \tag{3.3}$$

It is well-known that the mean $M_1(I)$ of the Cauchy distribution (1.17) is, in general a divergent integral. However, it is possible to regularize it in a natural way. Define

$$M_1(I) := \lim_{a \to 0} M_1(I_a). \tag{3.4}$$

Note that the definition (3.4) corresponds to the physics of the problem ($a \ll 1$). Thus, although the mode (maximum) of the distribution (1.17) is located at the point $\varepsilon = n\, K\, \alpha$, the mean value of this distribution is equal to zero. It means that the most probable value of the strain differs from the mean. In Section 4 we shall provide an explanation of this very interesting fact.

We now expand the external exponent in (3.2) into the Taylor series and, taking into account the inequality

$$(1 - \exp[itK(1 - 3\cos^2\theta) r^{-3}]) < 2,$$

we find the error estimate

$$|\varphi(t) - \varphi_a(t)| < \frac{8\pi^3}{3}\,\mu\,. \tag{3.5}$$

Here the dimensionless small parameter $\mu$ is the ratio of the atomic volume $a^3$ and the specific volume $1/n$. The above mentioned assumption of small concentration of the defects quantitatively can be characterized by the condition $\mu \ll 1$.

Combining (3.5) and the inverse limit theorem for characteristic functions ([9]), we conclude that the distribution function $I_a$ approximates the distribution function $I$ when $\mu$ is sufficiently small.

Taking into account estimate (3.5), we find the shape of the curve $I_a$, which approximately describes the strain distribution in case when the critical distance $a$, maximal $\theta$ and minimal $\xi$ values of strain are introduced into the (refined) model.

From (3.3) we have

$$\frac{q}{\pi}\int_{\xi}^{\theta}\frac{x\,dx}{(x-p)^2+q^2}=0,\quad \frac{q}{\pi}\int_{\xi}^{\theta}\frac{x^2\,dx}{(x-p)^2+q^2}=C_2 n\,K^2 a^{-3}\,,$$
$$p=(n_1-n_2)K\,\alpha^0,\qquad q=n\,K\,\gamma^0. \tag{3.6}$$

From (3.6) and the normalization condition

$$\frac{q}{\pi}\int_{\xi}^{\theta}\frac{dx}{(x-p)^2+q^2}=1,$$

we obtain the system which determines $\theta$ and $\xi$:

$$\theta-\xi=\pi C_2\,K\,a^{-3},\quad \frac{(\theta/q-c)^2+1}{(\xi/q-c)^2+1}=e^{-2\pi C},\quad C=\frac{p}{q}=\frac{n_1-n_2}{n_1+n_2}\,\frac{\alpha^0}{\gamma^0}\,. \tag{3.7}$$

It follows from the first equality (3.7), that the order of the critical strain is $K/a^3$. It corresponds to the strain value, produced by a defect, located the interatomic distance ((lattice spacing) apart from the observation point. Thus, the desired curve $I_a$ has the form, which is shown in Fig. 1. Note that the mean is zero due to the asymmetry in the cut off for the left and right tails which compensates the shift of the mode.

Using analogy with magnetic resonance in paramagnetics [11], one can say that the obtained shift in the distribution mode is not caused by the mean magnetization (such shift would depend on the sample shape, type of the lattice and it would result in the nonzero first moment $M_1$). It is rather related to fluctuations of the density of magnetic moment over the distance of order $n^{-1/3}$. In particular, for a spherical sample and cubic lattice the shift of the distribution mode, caused by the mean magnetization, is equal to zero [11].

## 4 Why is the distribution shifted?

To provide a heuristic explanation of the nature of the positiveness of the distribution mode $p$ (shift of the distribution curve), we consider the field of one defect in a ball $B_R$ of a large radius $R$.

It follows from the formula (1.1), that $\varepsilon_{11}>0$ if $(1-3\cos^2\theta)>0$, and $\varepsilon_{11}<0$ if $(1-3\cos^2\theta)<0$. It is easy to see that the volume of the cone $|\cos\theta|<1/\sqrt{3}$ is larger, than the volume of the cone

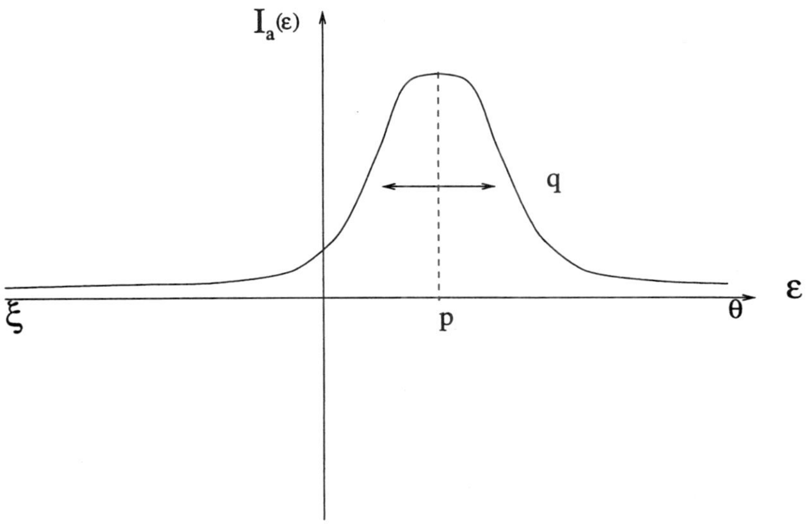


Figure 1: The distribution $I_a(\epsilon)$

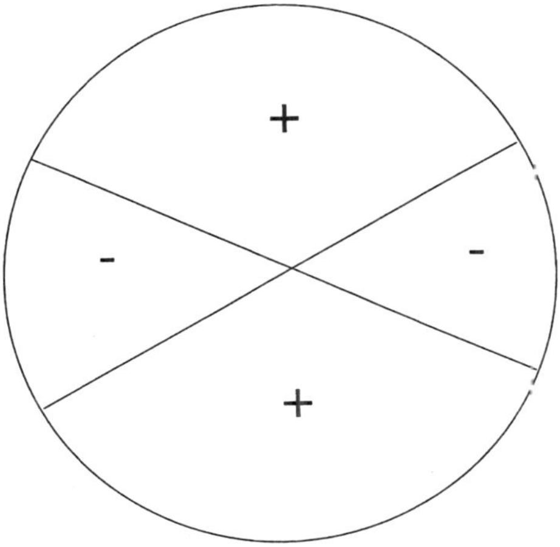


Figure 2:

$|\cos\theta| > 1/\sqrt{3}$ (see Fig. 2). Thus, if we choose a point in $B_R$ at random then a positive value of $\varepsilon$ at this point is more probable, then a negative one. On the other hand, integral of the expression $(1-3\cos^2\theta)$ over a sphere or spherical layer ( see (1.13)) is zero, which provides the distribution's mean equal to zero.

## 5 Comparison with the statistics of gravitational field arising from a random distribution of stars

One of the classical problems of stellar mechanics is concerned with the analysis of the force $F$ acting on a star, which is a member of a stellar system (see [16]). In particular the stationary distribution $W(|\mathbf{F}|)$ of the force $\mathbf{F}$ acting on a star, per unit mass, due to gravitational attraction of the neighboring stars was studied in [16, 15] (see also references in [16]). Instead of the points defects one considers randomly distributed stars and the strain field of each defect in the above consideration corresponds to the gravitational field of each star.The difference, which appears to be crucial, is in the fact that the gravitational field decays as $1/r^2$ whereas the strain field decays as $1/r^3$. The resulting distribution function $W(|\mathbf{F}|)$ is known as the Holtzmark distribution [15] and expressed through the function $H(\beta)$ defined by the following improper integral:

$$H(\beta) = \frac{2}{\pi\beta}\int_0^\infty e^{[-(\frac{x}{\beta})^{3/2}]} x \sin x dx \qquad (5.1)$$

The asymptotic behavior of the distribution $W(|\mathbf{F}|)$ can be obtained from the following formulas ( see [16] for details):

$$H(\beta) = \frac{4\beta^2}{3\pi} + O(\beta^4)\ ,\ (\beta \to 0) \qquad (5.2)$$

and

$$H(\beta) = (\frac{15}{8})(\frac{2}{\pi})^{1/2}\beta^{-5/2} + O(\beta^{-4})\ ,\ (\beta \to \infty) \qquad (5.3)$$

Then

$$W(|\mathbf{F}|) \sim (\frac{4}{3\pi Q_H^3})|\mathbf{F}| \ \ (|\mathbf{F}| \to 0) \qquad (5.4)$$

and

$$W(|\mathbf{F}|) \sim (\frac{15}{8})(\frac{2}{\pi})^{1/2} Q_H^{3/2} |\mathbf{F}|^{-5/2} \ \ (|\mathbf{F}| \to \infty), \qquad (5.5)$$

where $Q_H$ is a normalizing factor (called normal field [16]). So in this case there is no explicit algebraic formula for the resulting distribution as oppose to the case of the point defects in the elastic crystal and the best what one can do is to find asymptotic behavior of the distribution.

## Acknowledgments

I would like to thank B.N. Provotorov, A.K. Khitrin and Yu.A. Kosevich for useful discussions. The preparation of this work for publication was supported in part by NSF grant DMS-9622927.

## References

1. S.M.Kozlov, Averaging of random operators, *Matem. Sbornik*, **109** (2), 188–203 (1979).
2. G. Papanicolaou and S. Varadan, Boundary value problems with rapidly oscillating random coefficients, (Colloquia Mathematica Societatis Janos Bolyai 27, Radon Fields, Esztergom (Hungary), 1979, pp. 835–873.
3. A.M. Kosevich, Physical mechanics of real crystals, Kiev, Naukova Dumka, 1981, 327 p. (in Russian).
4. J.D. Eshelby, The Continuum Theory of Lattice Defects, in *Progress in Solid State Physics*, **3**, 79–144 (1956), New York.
5. J.D. Eshelby, The determination of the elastic field of an ellipsoidal inclusion, and related problems. in Proc. Roy. Soc. (London), Ser. A, 241, 376, 376–390 (1957).
6. I.M. Lifshitz and L.N. Rozentsveig, On the theory of elastic properties of polycrystals, *JETF*, **16** (11), 966–980 (1946). (in Russian).
7. V.M. Entov and R.L. Salganik, The Prandtl model of Brittle fracture, *Mechanics of Solids*, **3** (6), 79–89 (1968).
8. Yu.A. Rozanov, Stochastic processes, Moscow, Nauka, 1971, 286 p. (in Russian).
9. B.V. Gnedenko, The theory of probability (translated from Russian by B. D. Seekler (3rd ed.)), New York, Chelsea Pub. Co., 1966, 459 p.
10. I. S. Gradshtein and I. M. Ryzhik, Table of Integrals, Series, and Products, Correcte and Enlarged Edition Prepared by Alan Jef-

frey Incorporating, 4-th edition, prepared by Yu.V. Geronimus and M.Yu. Tseytlin Translated from Russian by Scripta Technika Inc. and edited by Alan Jeffrey New York, Acad. Press, 1980.

11. A. Abragam and M. Goldman, Nuclear magnetism, order and disorder, *International series of monographs on physics*, Claredan press, Oxford, 1982, 626 p.
12. L.V. Berlyand, N.V. Chukanov, and V.A. Dubovitsky, Exactly solvable random model and IR spectroscopy of a strained defective lattice, *Chemical Physics letters*, **181** (5), 450–453 (1991).
13. N.V. Chukanov, L.V. Berlyand, B.L. Korsunskii and V.A. Dubovitsky, Reactivity of crystals with defects, *Russian Journal of Physical Chemistry*, **64** (7), 970–973 (1990).
14. V.V. Zhikov, S.M. Kozlov and O.A. Oleinik, Homogenization of Differential Operators, Springer-Verlag, Berlin, New York, 1993.
15. J. Holtsmark, Ann. d. Phys. V. 58, 577 (1919), Phys., Zeitz, V. 20, 162 (1919), V. 25, 73 (1924).
16. S. Chandrasekhar, Stochastic Problems in Physics and Astronomy, *Reviews of Modern Physics*, V. 15, Nox 1, Jan. 1943.
17. L.V. Berlyand and K. Golden, Exact Result for the Effective Conductivity of a Continuum Percolation Model Phys, Rev. B., **50** (4), 2114–2117 (1994).
18. V.L. Berdichevsky, Variational principles of continuum mechanics, Nauka, 1983 (in Russian).

# OPTIMAL DESIGN FOR UNCERTAIN LOADING CONDITION

A. CHERKAEV, E. CHERKAEVA
*Department of Mathematics, the University of Utah,*
*Salt Lake City UT 84112, USA*

Optimal design problems are usually formulated as problems of minimization of the energy, stored in the design under a prescribed loading. Solutions of these problems are unstable to perturbations of the loading. We suggest a new min-max formulation of the optimal design problem which has a stable solution. The loading is not prescribed, but only a set of admissible loadings is given. The stable optimal design problem is formulated as minimization of the stored energy of the project under the most unfavorable loading. This most dangerous loading is one that maximizes the stored energy over the class of admissible functions. The problem is reduced to minimization of Steklov eigenvalues. Several stable solutions of various optimal design problems are demonstrated; among them are the optimal structure of a material stable to variations in uniaxial loading, the optimal specific stiffness of an uncertainly loaded beam, and the stable design of an optimal wheel.

## 1 Introduction

In a rather extensive literature on optimal design, major attention is paid to optimization of structures that are subject to a fixed loading. However, the typical situation for the practical use of an optimal design is different: acting forces are either varying in time, or varying from one sample to another, or are unpredictable. This motivates a reformulation of the problem to account possible variations and uncertainties in loading.

One can foresee a significant change in the reformulated design if the loading is not completely known. Indeed, the optimality requirement forces the structure to concentrate its resistivity against applied loading, since its abilities to resist other loadings are limited. This high sensitivity to the loading restricts the applicability of most optimal designs.

The goal of this paper is to formulate a stable optimal design problem with solution insensitive to small variations in the loadings. We show, that this stable reformulation of the problem leads to optimization of Steklov eigenvalues [2, 9]. Such formulation makes the problem similar to the well investigated problem of optimization of the principle eigenfrequency of the vibration of a structure [10, 11]. We introduce a concept of the compliance of a whole structure that is similar to the compliance of the material. The structure compliance is the energy that corresponds to "the worst" admissible loading, or to the maximal energy produced by a loading from a set of admissible loadings. Then we formulate the optimal design problem for the project which minimizes that maximum, or minimizes the compliance. We end up with the min-max problem that leads to minimization of a non-smooth functional. The approach we use, is similar to one developed for the problem of optimization of boundary excitations in nondestructive testing and electrical tomography [4, 5] and to structural optimization [12]. Several optimal design problems are discussed, and their stable solutions are demonstrated.

## 2 The problem formulation

**The problem of optimal design.** The overall compliance of an elastic structure is characterized by the mechanical work produced by an applied loading. This work is equal to the total energy stored in the loaded structure. It is found from the following variational problem

$$H(\boldsymbol{p}, \boldsymbol{f}) = \min_{\boldsymbol{\sigma} \in \Sigma} \left\{ \int_\Omega W(\boldsymbol{p}, \boldsymbol{\sigma}) \right\}, \tag{2.1}$$

where $W$ is the (doubled) elastic energy

$$W(\boldsymbol{p}, \boldsymbol{\sigma}) = \boldsymbol{\sigma} : \boldsymbol{S}(\boldsymbol{p}, \boldsymbol{x}) : \boldsymbol{\sigma}. \tag{2.2}$$

$\boldsymbol{\sigma}$ is the stress tensor, the set $\Sigma$ is (see (2.1, 2.2))

$$\Sigma = \left\{ \boldsymbol{\sigma} : \nabla \cdot \boldsymbol{\sigma} = 0 \ \text{ in } \Omega, \ \boldsymbol{\sigma} = \boldsymbol{S}^{-1} (\nabla u)^s, \ \boldsymbol{\sigma} = \boldsymbol{\sigma}^T, \ \boldsymbol{n} \cdot \boldsymbol{\sigma} = \boldsymbol{f} \ \text{ on } \partial\Omega \right\} \tag{2.3}$$

$\boldsymbol{f}$ is the vector of applied boundary forces and $\boldsymbol{u}$ is the vector of deflection, $(\nabla \boldsymbol{u})^s = \boldsymbol{\epsilon}$ is the strain field, that is given by a symmetrized part of the gradient of $\boldsymbol{u}$, $(\nabla \boldsymbol{u})^s = (\nabla \boldsymbol{u} + (\nabla \boldsymbol{u})^T)/2$. $\boldsymbol{S}(\boldsymbol{p}, \boldsymbol{x})$ is the tensor of elastic compliance: a fourth order symmetric positive tensor, which depends on the point $\boldsymbol{x}$ in $\Omega$ and on the structural parameter

$\boldsymbol{p}$ that defines the material's properties. The symbol (:) denotes the contraction by two indices.

The stored energy $H$ is a quadratic functional of the loading $\boldsymbol{f}$ that depends on the layout of the material's properties $\boldsymbol{p}$, called the design variables.

Consider the typical problem of optimal design: minimize $H$ with respect to layout $\boldsymbol{p}$:

$$\min_{\boldsymbol{p}\in\mathcal{P}} H(\boldsymbol{p},\boldsymbol{f}), \tag{2.4}$$

where $\mathcal{P}$ is the admissible set of design variables. There are many possible settings for the set $\mathcal{P}$ : it can be defined as the set of effective moduli of the composite [8], or it could describe the shape of the body, the thickness of a thin structure, and so on.

**Instabilities in the optimal design problems.** The following example demonstrates the instability of the optimal structure and suggests ways of reformulating the problem in order to stabilize the design.

**Problem 1.** Suppose that a square domain $a\,b\,c\,d$ filled with a composite material, is loaded by a uniaxial loading. Suppose for simplicity, that the composite is assembled from the material with unit compliance tensor $\boldsymbol{S}_1 = \boldsymbol{I}$ (the Poisson ratio is equal to zero and the Young modulus is equal to one) and from the void with infinite compliance: $\boldsymbol{S}_2 = \infty$. Suppose also, that the fractions $m_1$ of the material and $m_2$ of the void are equal to one half each:

$$m_1 = m_2 = \frac{1}{2}. \tag{2.5}$$

Let the domain be loaded by a prescribed loading

$$\boldsymbol{f}_0 = \begin{cases} \boldsymbol{i}_1 & \text{on} \quad a\,b \\ 0 & \text{on} \quad b\,c \\ -\boldsymbol{i}_1 & \text{on} \quad c\,d \\ 0 & \text{on} \quad a\,d \end{cases}. \tag{2.6}$$

The optimal design is homogeneous. The loading $\boldsymbol{f}_0$ creates a stress field $\boldsymbol{\sigma}_1$,

$$\boldsymbol{\sigma}_1 = \begin{pmatrix} 1 & 0 \\ 0 & 0 \end{pmatrix}, \quad \text{or} \quad \boldsymbol{\sigma}_1 = \boldsymbol{i}_1 \otimes \boldsymbol{i}_1 \tag{2.7}$$

inside the domain.

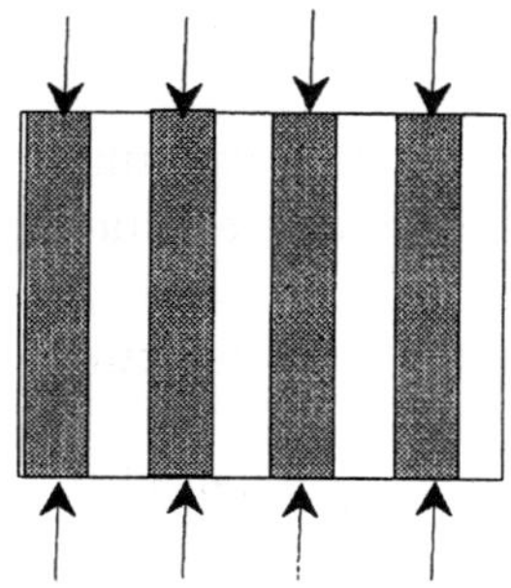

Figure 1: The optimal composite under the homogeneous axial loading.

The problem is to find the composite that minimizes the energy of the project under the loading $\boldsymbol{f}_0$. Obviously, the best structure is a simple laminate, with layers oriented along the loading (see Fig.1).

The effective compliance $s_{1111}$ in the direction $\boldsymbol{i}_1$ of the loading is equal to the harmonic mean of the (unit) material's complaince $s_m = 1$ and the (infinite) complaince of the void $s_v = \infty$:

$$s_{1111} = \left( \frac{m_1}{s_m} + \frac{m_2}{s_v} \right)^{-1} = 2 \tag{2.8}$$

The minimal energy and the problem cost are:

$$W(\boldsymbol{\sigma}_1) = \boldsymbol{\sigma}_1 : \boldsymbol{S}_* : \boldsymbol{\sigma}_1 = s_{1111}\boldsymbol{\sigma}_{11}^2 = 2, \quad H(\boldsymbol{\sigma}_1) = 2 \tag{2.9}$$

This solution, however, is not satisfactory from a common sense viewpoint. Indeed, the laminate structure is extremely unstable, and its compliance tensor is singular. The laminate structure cannot resist any loading but the prescribed one. Its compliance is infinitely large for all other loadings. Simply speaking, the structure falls apart under any infinitesimally small applied stress that has either shear component, or a component along the axis $\boldsymbol{i}_2$.

*Remark.* The described instability is typical for the projects that are designed to optimally resist to a prescribed loading, at the expense of the resistivity in other directions.

**Formulation of the problem of stable optimal design.** Let us consider a problem of energy optimization of an elastic body $\Omega$ loaded by unknown forces $\boldsymbol{f}$ applied on the boundary $\partial\Omega$. In this paper, we focus on the dependence of the optimal project on the loading that belongs to a set $\mathcal{F}$: $\boldsymbol{f} \in \mathcal{F}$. We define the compliance $\Lambda$ of a structure as the maximum of compliances upon all admissible loadings

$$\Lambda(\boldsymbol{p}) = \max_{\boldsymbol{f} \in \mathcal{F}} H(\boldsymbol{p}, \boldsymbol{f}), \tag{2.10}$$

and we formulate the problem of the optimal design against the "worst" loading:

$$\min_{\boldsymbol{p} \in \mathcal{P}} \Lambda = \min_{\boldsymbol{p} \in \mathcal{P}} \left\{ \max_{\boldsymbol{f} \in \mathcal{F}} H(\boldsymbol{p}, \boldsymbol{f}) \right\}. \tag{2.11}$$

The energy $H$ is a quadratic functional of $||\boldsymbol{f}||$ To impose constraints on acting forces, we formulate a problem for a design that offers a minimal compliance in a class of loadings.

**Integral constraints for the loading.** Let the set of the loadings $\mathcal{F}$ be characterized by an integral constraint. It is convenient to consider the constraints as a quadratic form of the loading: this form leads to rather simple equations and possesses a needed generality and flexibility. Suppose that an unknown loading by normal forces $\boldsymbol{f} \in \mathcal{F}$ is constrained as following:

$$\mathcal{F} = \left\{ \boldsymbol{f} : \oint_{\partial\Omega} \boldsymbol{f} \cdot \Psi(S) \boldsymbol{f} \leq 1 \right\},$$

where $\Psi(S)$ is a positively defined weight matrix,

$$\Psi(S) > 0, \quad \forall\, S \in \partial\Omega.$$

The introduced here weight function $\Psi$ expresses a priori assumptions about the unknown loading. For instance, the case when all loadings are equally possible, corresponds to $\Psi = const(S)$.

The compliance of the design, introduced in (2.10), is given by the solution of the problem of maximization of the stored in the design energy with respect to the applied loadings $\boldsymbol{f} \in \mathcal{F}$:

$$\Lambda = \max_{\boldsymbol{f}} H(\boldsymbol{p}, \boldsymbol{f}) = \max_{\boldsymbol{f}} \min_{\boldsymbol{\sigma}} \frac{\int_\Omega W(\boldsymbol{p}, \boldsymbol{\sigma})}{\oint_{\partial\Omega} \boldsymbol{f} \cdot \Psi(S) \boldsymbol{f}}. \tag{2.12}$$

The problem (2.12) is an eigenvalue problem. The value $\Lambda(\boldsymbol{p})$ corresponds to the first eigenfunction or to the set of the eigenfunctions, that generate the most "dangerous" loading(s) from the considered class. Hence we formulate the stable optimal design problem as a problem of eigenvalue optimization:

$$J_* = \min_{\boldsymbol{p}\in\mathcal{P}} \Lambda(\boldsymbol{p}) = \min_{\boldsymbol{p}\in\mathcal{P}} \max_{\boldsymbol{f}\in\mathcal{F}} \min_{\boldsymbol{\sigma}\in\Sigma} \frac{\int_\Omega W(\boldsymbol{p},\boldsymbol{\sigma})}{\oint_{\partial\Omega} \boldsymbol{f}\cdot\Psi\boldsymbol{f}}. \tag{2.13}$$

## 3 Eigenvalue problem

**Saddle point case.** The question of whether or not the multiple eigenvalue case is taking place, depends on the power of the control. It the control $\boldsymbol{p}$ is "weak", that is if the control cannot change the sequence of eigenvalues, then we are dealing with a saddle point situation. In this case, the minimal upon the control $\boldsymbol{p}$ eigenvalue corresponds to a unique eigenfunction $\boldsymbol{f}(\boldsymbol{p})$. The example below illustrates this situation.

In this case the functional $\Lambda$ 2.12 is a saddle function of the arguments, and the operations of max with respect to $\boldsymbol{f}$ and min with respect to $\boldsymbol{\sigma}$ can be switched. Then varying the functional, we find the Euler equations for the most dangerous loading. Let us find this loading. Variation of 2.12 with respect to $\boldsymbol{f}$ gives:

$$\delta\Lambda_{\boldsymbol{f}} = -\left(\oint_{\partial\Omega} \boldsymbol{f}\cdot\Psi(S)\boldsymbol{f}\right)^{-1} (\boldsymbol{u} - \Lambda\Psi\boldsymbol{f})\,\delta\boldsymbol{f},$$

which implies the relation pointwise between the optimal loading and the boundary deflection

$$\boldsymbol{f}(S) = \frac{1}{\Lambda}\Psi^{-1}\boldsymbol{u}(S), \quad \forall S\in\partial\Omega.$$

It is also easy to see that the stationary condition corresponds to the *maximum* not the minimum of the functional using the second variation technique.

The problem of the most dangerous loading $\boldsymbol{f}_0$ becomes an eigenvalue problem

$$\frac{1}{\Lambda} = \min_{\boldsymbol{\sigma}} \frac{\int_\Omega W(\boldsymbol{p},\boldsymbol{\sigma})}{\oint_{\partial\Omega} \boldsymbol{u}\cdot\Psi(S)^{-1}\boldsymbol{u}}. \tag{3.1}$$

The cost $\Lambda$ corresponds to the minimal eigenvalue, given by the Rayleigh ratio 3.1, and the most "dangerous" loading corresponds to the first eigenfunction of this problem.

*Remark.* One can consider also the problem of the most "favorable" loading, that is

$$\Lambda_- = \min_{\boldsymbol{f}} \min_{\boldsymbol{\sigma}} \frac{\int_\Omega W(\boldsymbol{p}, \boldsymbol{\sigma})}{\oint_{\partial\Omega} \boldsymbol{f} \cdot \Psi \boldsymbol{f}}.$$

However, $\Lambda_-$ is zero. Clearly, the spectrum of the operator is clustered at zero. A minimizing sequence is formed from often oscillating forces.

**Euler equations.** The Euler equations (with respect to $\boldsymbol{\sigma}$) are

$$\begin{aligned} &\nabla \cdot \boldsymbol{\sigma} = 0, \quad \boldsymbol{\sigma} = \boldsymbol{S}^{-1}(\boldsymbol{p}) : (\nabla \boldsymbol{u})^s && \text{in } \Omega, \\ &\boldsymbol{u} = \Lambda \Psi \boldsymbol{\sigma} \cdot \boldsymbol{n} && \text{on } \partial\Omega. \end{aligned} \tag{3.2}$$

They describe the vibration of the body with inertial elements concentrated on $\partial\Omega$.

The problem admits the following physical interpretation: the optimal loading forces are equal to a distribution of inertial elements (concentrated masses) on the boundary component $\partial\Omega$. The specific inertia is described by the tensor $\Psi$, so it could include the resistance to the turning as well. The vibration of such loaded system excites the forces that are proportional to the deflection $\boldsymbol{u}$. The compliance is proportional to the eigenfrequency of vibrations. One can see that the introduced quantity $\Lambda$ characterizes the domain or the structure itself, it represents the maximum of possible stored energy under any loading from the class $\mathcal{F}$.

These equations form an eigenvalue problem that possesses infinitely many solutions. We pick up the pair $\{\Lambda_1, \ \boldsymbol{\sigma}_1\}$ that corresponds to the maximal eigenvalue $\Lambda_1 = \max\{\Lambda_k\}$.

The problem (3.1) with unit matrix $\Psi$ is called the Steklov eigenvalue problem, which considers the ratio of integrals of different dimensionality. The corresponding Euler equation (3.2) has an eigenvalue in the boundary condition. Similar optimality conditions were derived in [5] for the optimal boundary sources in electrical tomography problem.

**Problem 2. Stable optimal design of a loaded beam.** The problems for beams and bending plates admit the loading distributed

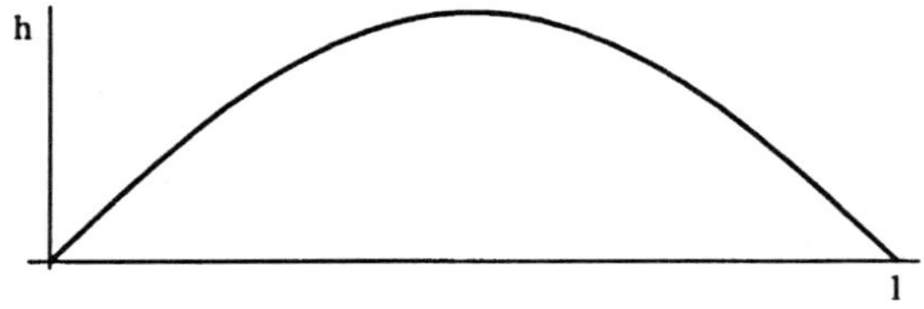


Figure 2: The stiffness $p(x)$ of the optimal beam.

in the whole domain of the definition: on the interval in the case of the beam, and in the plane domain in the case of the bending plate or shell. In these problems, the loaded surface $\partial\Omega$ coincide with the domain $\Omega$ itself.

Consider an elastic beam whose energy density is

$$W = p(w'')^2 - 2fw,$$

where $p \geq 0$ is a material's stiffness, that can be varied from point to point and subject to the integral constraint

$$\int_0^l p\,dx = V \tag{3.3}$$

which expresses the limits on resources; $f$ is the intensity of the normal loading, subject to the constraint

$$\int_0^l f^2\,dx = 1.$$

Consider an optimization problem of choosing a stiffness $p(x)$ that maximally resists to the most dangerous loading $f$:

$$\min_{p \geq 0, p \in (3.3)} \max_f \min_w \mu, \quad \mu = \frac{\int_0^l (p(w'')^2 - 2fw)\,d\,x}{\int_0^l (f^2)\,d\,x}.$$

The stationary conditions are:

$$\delta\, w: \quad (p\, w'')'' - f = 0,\ \forall x \in (0,\ 1), \quad pw''|_{x=0} = pw''|_{x=l} = 0, \tag{3.4}$$
$$\delta\, f: \quad f + \frac{w}{\mu} = 0,\ \forall x \in (0,\ 1), \tag{3.5}$$
$$\delta\, p: \quad (w'')^2 = \gamma,\ \forall x \in (0,\ 1), \tag{3.6}$$

where $\gamma$ is the Lagrange multiplier for the constraint (3.3). This system admits a solution

$$w = -\mu f = -\gamma x(l-x)/2,$$
$$f = \frac{\sqrt{\gamma}}{\mu} x(l-x)/2,$$
$$p_0 = \frac{1}{\mu}\frac{x}{24}(x-l)\left((x-\frac{l}{2})^2 - \frac{5l}{4}\right).$$

Accounting for the constraints, we get

$$\mu = \frac{l^5}{5!V},$$
$$p(x) = \frac{5V}{4l^5} x(l-x)\left(5l^2 - (2x-l)^2\right),$$
$$w(x) = -\frac{1}{V}\sqrt{\frac{l^5}{5!}}\frac{x(l-x)}{2}.$$

The optimal stiffness $p(x)$ of the beam is shown on Fig. 2. Interestingly, that the optimal solution is found analytically.

**A min-max problem for a partly known loading.** The developed formulation can be generalized to the case of a partly known loading. Suppose, that the applied loading $\boldsymbol{f}$ is composed of the known component $\boldsymbol{f}_0$ and of some unknown component $\boldsymbol{f}_1$ which describes an uncertainty in the loading. The constraint has the form

$$\mathcal{F}_1 = \left\{ \boldsymbol{f}_1 : \ \oint_{\partial\Omega} \boldsymbol{f}_1 \cdot \Psi(S) \boldsymbol{f}_1 \le 1 \right\},$$

or

$$\oint_{\partial\Omega} (\boldsymbol{f} - \boldsymbol{f}_0) \cdot \Psi(S)(\boldsymbol{f} - \boldsymbol{f}_0) \le 1.$$

The problem is no longer homogeneous. The functional is

$$I = \min_{\boldsymbol{p}\in\mathcal{P}} \Lambda(\boldsymbol{p}) = \min_{\boldsymbol{p}\in\mathcal{P}} \max_{\boldsymbol{f}_1\in\mathcal{F}_1} \min_{\boldsymbol{\sigma}\in\Sigma} \frac{\int_\Omega W(\boldsymbol{p},\boldsymbol{\sigma})}{\oint_{\partial\Omega}(\boldsymbol{f}-\boldsymbol{f}_0)\cdot\Psi(S)(\boldsymbol{f}-\boldsymbol{f}_0)}.$$

Depending on $\mathcal{P}$ and $\mathcal{F}_1$, we can encounter two different situations: minimization of a single or multiple maximal eigenvalues $\Lambda(\boldsymbol{p})$. In the first case, the stationary conditions with respect to $\boldsymbol{f}$ and $\boldsymbol{\sigma}$ lead to the Euler equation:

$$\begin{aligned}
&\nabla\cdot\boldsymbol{S}^{-1}(\nabla\,\boldsymbol{u})^s = 0 \quad \text{in } \Omega,\\
&\boldsymbol{u} - \Lambda\Psi\cdot\boldsymbol{S}^{-1} : (\nabla\,\boldsymbol{u})^s\cdot\boldsymbol{n} = \boldsymbol{f}_0 \quad \text{on } \partial\Omega.
\end{aligned}$$

They describe the inhomogeneous boundary value problem, which has a unique solution. The optimality condition

$$\frac{\partial W(\boldsymbol{p},\boldsymbol{\sigma})}{\partial \boldsymbol{p}}\,\delta\,\boldsymbol{p} \geq 0,$$

has different forms depending on the type of used control and of possible isoperimetric restrictions.

Discussed below problems 3 and 4 illustrate the problem of a design with partly known loading. The occurrence of multiple eigenvalues in the problem 3 switches in the problem 4 to a situation when the problem with a single maximal eigenvalue becomes one with multiple eigenvalues, if we change slightly the set of design parameters $\mathcal{P}$ or the set of possible variations in the loading $\mathcal{F}_1$.

## 4 Multiple eigenvalues

**Eigenvalue optimization.** We return to the discussion of the project that minimizes the functional $\Lambda$, or minimizes the stored energy in the most unfortunate situation. The problem has the form (2.13). The specific effect of the min-max problem is the possibility of appearance of multiple eigenvalues. The mechanism of this phenomenon is the following. Minimization of the maximal eigenvalue likely leads to the situation when its value meets the second eigenvalue of the problem. In this case, both eigenvalues must be minimized together, until their common value reaches the third eigenvalue, and so on. The multiplicity means that two or more loadings give the same value of the problem.

We will bring below an example demonstrating this phenomenon: the resistance of the structure to five different loadings in this example of the stable optimal design is the same. The second example of that section illustrates the situation when the eigenvalue could either be multiple or stay single, depending on the set of loadings $\mathcal{F}$ and the class of admitted materials $\mathcal{P}$, constraining controls $\boldsymbol{p} \in \mathcal{P}$. This is a problem with a partly known loading, the Euler equation of which for one most dangerous loading was derived in the previous section. Similar min-max problem with multiple eigenvalues was considered in [4] for nondestructive testing of the worst possible damage by applying optimal boundary currents.

There is an extended literature on eigenvalue optimization. It was understood in a different setting: the maximization of the fundamental frequency. We refer to the recent review papers [6, 11] and references therein.

**Optimal composite structures.** Consider the following problem of structural optimization. A domain made of a two-phase composite material of an arbitrary structure is loaded by an uncertain loading $f_0$. We want to find the most resistant structure of the composite, that is to minimize the functional (2.10). Here $\boldsymbol{p}$ is a vector of parameters that defines the tensor $\boldsymbol{S}_*$ of the effective compliance of the composite. For definiteness, consider the two-dimensional elasticity problem.

We do not know a priori, how many loadings should be taken in consideration. But clearly, it is sufficient to enlarge the set of admissible composites to those which minimize the sum of elastic energies caused by any number of different loadings. These composites are described in the paper by Avellaneda [1]: in two-dimensional elasticity, they form the class of the so-called matrix laminates of the third rank (see Fig.3).

The effective property tensors of these composites admit an analytical expression through their structural parameters. To describe the class of the effective tensors of these anisotropic structures, we use the natural tensor basis

$$\boldsymbol{e}_1 = \begin{pmatrix} 1 & 0 \\ 0 & 0 \end{pmatrix}, \quad \boldsymbol{e}_2 = \begin{pmatrix} 0 & 0 \\ 0 & 1 \end{pmatrix}, \quad \boldsymbol{e}_3 = \frac{1}{\sqrt{2}} \begin{pmatrix} 0 & 1 \\ 1 & 0 \end{pmatrix}. \qquad (4.1)$$

Any stress and strain matrices are represented as vectors in their basis, and the effective compliance $\boldsymbol{S}_*$ of matrix laminates is given by the

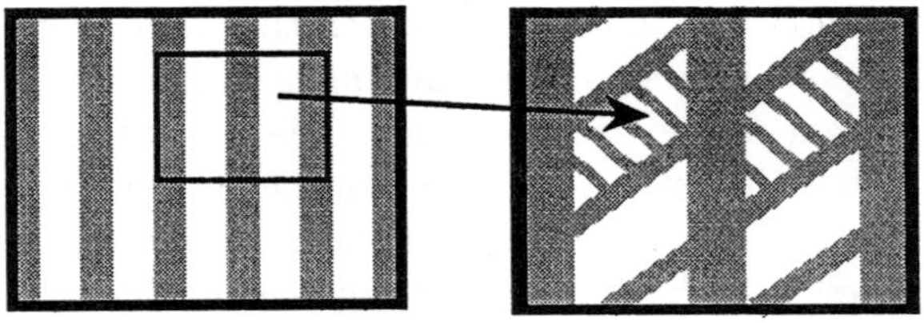

Figure 3: The schematic picture of the composite of the third rank.

$3 \times 3$ matrix (see [8])

$$\boldsymbol{S}_* = \boldsymbol{S}_1 + m_2 \left( (\boldsymbol{S}_2 - \boldsymbol{S}_1)^{-1} + \frac{2m_1}{E_1} \boldsymbol{N} \right)^{-1}, \tag{4.2}$$

where $\boldsymbol{S}_1$ and $\boldsymbol{S}_2$ are the compliance matrices of the first and the second materials, $m_1$ and $m_2$ are the volume fractions, $E_1$ is the Young modulus of the first material (which forms the envelope). The matrix $\boldsymbol{N}$ depends on the structural parameters: on the angles $\theta_i$ between the tangent to the laminates and the axis $\boldsymbol{i}_1$, and on the relative thickness $\alpha_i$ (see Fig.3):

$$\boldsymbol{N} = \textstyle\sum_i^3 \alpha_i \begin{pmatrix} \cos^4 \theta_i & \cos^2 \theta_i \sin^2 \theta_i & \cos^3 \theta_i \sin \theta_i \\ \cos^2 \theta_i \sin^2 \theta_i & \sin^4 \theta_i & \cos \theta_i \sin^3 \theta_i \\ \cos^3 \theta_i \sin \theta_i & \cos \theta_i \sin^3 \theta_i & \cos^2 \theta_i \sin^2 \theta_i \end{pmatrix},$$

$$\textstyle\sum_i^3 \alpha_i = 1, \quad \alpha_i \geq 0. \tag{4.3}$$

$\alpha_i$ and $\theta_i$ form the control vector $\boldsymbol{p}$.

**Problem 1 revisited. Unstable design for a uniaxial loading.** Discussing the instabilities of the optimal project in the problem 1 below, we considered the optimization problem

$$\min_{\boldsymbol{p} \in \mathcal{P}} H(\boldsymbol{p}, \boldsymbol{f}_0), \tag{4.4}$$

where $\boldsymbol{f}_0$ is given by (2.6), and the set $\mathcal{P}$ constrains the parameters of the composite $\alpha_i$ and $\theta_i$. The solution of the problem is a laminate, that is easily found from (4.2, 4.3). It corresponds to the parameters $\alpha_1 = 1, \theta_1 = 0, \alpha_2 = \alpha_3 = 0$. This structure is shown in Fig.1, below we discussed the instabilities of this solution. Indeed, the compliance tensor $\boldsymbol{S}_*$ of a third rank composite becomes ($m_1 = m_2 = 1/2$)

$$\boldsymbol{S}_* = \boldsymbol{S}_1 + \frac{1}{2}(\boldsymbol{N})^{-1}. \tag{4.5}$$

For the optimal choice of the parameters $\alpha_i$, $\theta_i$ the matrix $\boldsymbol{N}$ (see (4.3)) has two zero eigenvalues, and the two eigenvalues of $\boldsymbol{S}_*$ corresponding to the shear loading and the loading in the direction $\boldsymbol{i}_2$, are infinite (see (4.2)). Hence the compliance of the structure is infinitely large for any loading that has a projection on these two eigenvectors.

**Problem 3. Stable design for a uniaxial loading.** Now we reformulate the design problem (4.4) to obtain a stable project. Suppose that the loading is not exactly known. Namely, the loading field $\boldsymbol{\sigma}$ can take one of the following six values $\boldsymbol{\sigma}_1 + \boldsymbol{\tau}_i$, $i = 1, ...6$, where $\boldsymbol{\sigma}_1$ is given by (2.7) and

$$\boldsymbol{\tau}_{1,2} = \pm r\boldsymbol{e}_1 = \begin{pmatrix} \pm r & 0 \\ 0 & 0 \end{pmatrix}, \quad \boldsymbol{\tau}_{3,4} = \pm r\boldsymbol{e}_2 = \begin{pmatrix} 0 & 0 \\ 0 & \pm r \end{pmatrix},$$
$$\boldsymbol{\tau}_{5,6} = \pm r\boldsymbol{e}_3 = \begin{pmatrix} 0 & \pm r/\sqrt{2} \\ \pm r/\sqrt{2} & 0 \end{pmatrix}. \tag{4.6}$$

Here, $r > 0$ is a real parameter.

Assume, in addition, that $r$ is smaller than the magnitude of the "main" loading, which in our example is equal to one. The six loadings are viewed as small perturbations of the main loading, that correspond to all linearly independent directions of the symmetric tensor $\boldsymbol{\sigma}$. In spite of the smallness of $r$, the perturbation of the functional (4.4) is infinitely large, if $\boldsymbol{S}_*$ is optimally chosen. This characterizes the instability of the optimal project to those perturbations.

Let us reformulate the optimization problem. We are looking for a structure of a composite that minimizes the *maximum* of compliances $H(\boldsymbol{p}, \boldsymbol{\sigma}_1 + \boldsymbol{\tau}_i)$ upon all considered loadings.

$$\min_{\boldsymbol{p}=\{\alpha_i,\ \theta_i\}} \left\{ \max_{i=1..6} H(\boldsymbol{p}, \boldsymbol{\sigma}_1 + \boldsymbol{\tau}_i) \right\}. \tag{4.7}$$

The obtained min-max problem asks for the minimal compliance in the case of the "most dangerous" loading. To construct the solution of the optimization problem, we introduce a variable $z$ that is greater than any of $H(\boldsymbol{p}, \boldsymbol{\sigma}_1 + \boldsymbol{\tau}_i)$,

$$z \geq H(\boldsymbol{p}, \boldsymbol{\sigma}_1 + \boldsymbol{\tau}_i), \quad i = 1, ...6. \tag{4.8}$$

The problem (4.7) can be formulated as follows (see [7]):

$$\min_{\boldsymbol{p}} \left\{ z + \sum_i^6 \lambda_i^2 (z - H(\boldsymbol{p}, \boldsymbol{\sigma}_1 + \boldsymbol{\tau}_i)) \right\}, \tag{4.9}$$

where $\lambda_i^2$ are the non-negative Lagrange multipliers by the constraints (4.8). The Lagrange multiplier is equal to zero, if this relation is satisfied as a strong inequality, and is non-zero, if it is satisfied as an equality [7]:

$$\lambda_i^2 = \begin{cases} = 0 & \text{if} \quad z > H(\boldsymbol{p}, \boldsymbol{\sigma}_1 + \boldsymbol{\tau}_i) \\ > 0 & \text{if} \quad z = H(\boldsymbol{p}, \boldsymbol{\sigma}_1 + \boldsymbol{\tau}_i) \end{cases}. \tag{4.10}$$

The problem requires minimization of the weighted sum of energies of the 'dangerous' loadings $(\boldsymbol{\tau}_i)$, $i \in I$ . Here $I$ is the set of such 'dangerous' loadings. Other loadings lead to the smaller energies $H(\boldsymbol{p}, \boldsymbol{\sigma}_1 + \boldsymbol{\tau}_j)$: $H(\boldsymbol{p}, \boldsymbol{\sigma}_1 + \boldsymbol{\tau}_j) < H(\boldsymbol{p}, \boldsymbol{\sigma}_1 + \boldsymbol{\tau}_i)$, if $i \in I$, $j \notin I$, and therefore to $\lambda_j = 0$. This leads to the equalities

$$\begin{aligned} z &= H(\boldsymbol{p}, \boldsymbol{\sigma}_1 + \boldsymbol{\tau}_i), \quad \text{if } i \in I, \\ z &> H(\boldsymbol{p}, \boldsymbol{\sigma}_1 + \boldsymbol{\tau}_i), \quad \text{if } i \notin I \end{aligned}. \tag{4.11}$$

Applying to the problem (4.9), we argue that the set of dangerous loadings in this example consists of five elements, $I = \{1, 3, 4, 5, 6\}$ :

$$H(\boldsymbol{p}, \boldsymbol{\sigma}_1 + \boldsymbol{\tau}_1) > H(\boldsymbol{p}, \boldsymbol{\sigma}_1 + \boldsymbol{\tau}_2), \tag{4.12}$$

$$\begin{aligned} H(\boldsymbol{p}, \boldsymbol{\sigma}_1 + \boldsymbol{\tau}_3) &= H(\boldsymbol{p}, \boldsymbol{\sigma}_1 + \boldsymbol{\tau}_4), \\ H(\boldsymbol{p}, \boldsymbol{\sigma}_1 + \boldsymbol{\tau}_5) &= H(\boldsymbol{p}, \boldsymbol{\sigma}_1 + \boldsymbol{\tau}_6). \end{aligned} \tag{4.13}$$

The inequality (4.12) is explained by the observation that an additional loading, if codirected with the main load, will either increase or decrease its magnitude independently of the composite structure. Clearly, the energy of the more intensive loading is greater.

The symmetry of the loadings # 3 and #4 and of the loadings # 5 and # 6 together with the symmetry of the set of admissible structural tensors $\mathcal{P}$ suggests that the "twin" loadings lead to the same cost of the problem. In other words, the same project $\boldsymbol{p}$ minimizes both

$H(\boldsymbol{p}, \boldsymbol{\sigma}_1+\boldsymbol{\tau}_3)$ and $H(\boldsymbol{p}, \boldsymbol{\sigma}_1+\boldsymbol{\tau}_4)$, keeping them equal to each other; the same for the other pair of loadings. To achieve the equalities (4.13), we require the symmetry of the would be optimal tensor $\boldsymbol{S}_*$ (see (4.2)):

$$\alpha_1 = 1-a, \quad \alpha_2 = \alpha_3 = a/2, \quad \theta_1 = 0, \quad \theta_2 = -\theta_3 = \theta, \tag{4.14}$$

where $a$ and $\theta$ are two parameters. Physically, we require the orthotropy of $\boldsymbol{S}_*$.

Under the conditions (4.14), the matrix $\boldsymbol{N}$ (see (4.3)) takes the form

$$\boldsymbol{N} = (1-a)\begin{pmatrix} 1 & 0 & 0 \\ 0 & 0 & 0 \\ 0 & 0 & 0 \end{pmatrix} + a\begin{pmatrix} \cos^4\theta & \cos^2\theta\sin^2\theta & 0 \\ \cos^2\theta\sin^2\theta & \sin^4\theta & 0 \\ 0 & 0 & \cos^2\theta\sin^2\theta \end{pmatrix} \tag{4.15}$$

and, from (4.5)

$$\boldsymbol{S}_* = \begin{pmatrix} 1 & 0 & 0 \\ 0 & 1 & 0 \\ 0 & 0 & 1 \end{pmatrix} +$$

$$+ \frac{1}{2a(1-a)}\begin{pmatrix} a & -a\cot\theta & 0 \\ -a\cot\theta & T & 0 \\ 0 & 0 & (1-a)\csc^2\theta\sec^2\theta \end{pmatrix}, \tag{4.16}$$

$$T = \tfrac{1}{8}(8 - 5\,a + 4\,a\,\cos 2\,\theta + a\,\cos 4\,\theta\,)\,\csc^4\theta. \tag{4.17}$$

*Remark.* Note, that the matrix becomes singular when $a \to 0$, which corresponds to unstable design.

**Optimal parameters.** The described class of symmetric composites is defined by two parameters $\theta$ and $a$. The symmetry of the project eliminates the necessity to compare the loadings except from those with numbers 1, 3, 5. It turns out that these loadings are equally "dangerous":

$$H(a, \theta; \boldsymbol{\sigma}_1 + \boldsymbol{\tau}_1) = H(a, \theta; \boldsymbol{\sigma}_1 + \boldsymbol{\tau}_3) = H(a, \theta; \boldsymbol{\sigma}_1 + \boldsymbol{\tau}_5). \tag{4.18}$$

Two equalities (4.18) allow to compute the optimal values of $\theta$ and $a$. One can easily see that the problem is always solvable. The optimal

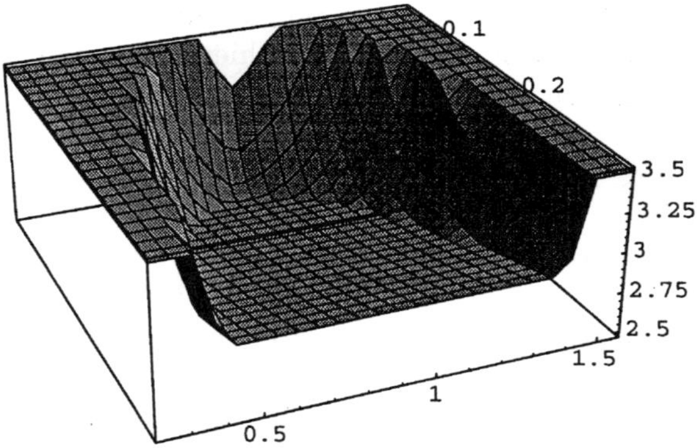


Figure 4: The graph of the function $J(a, \theta)$.

values of the parameters $\theta$ and $a$ correspond to the solution of the min-max problem:

$$J(a,\theta) = \min_{a,\theta}\left\{\max\left\{H(a,\theta;\boldsymbol{\sigma}_1+\boldsymbol{\tau}_1),\right.\right.$$
$$\left.\left. H(a,\theta;\boldsymbol{\sigma}_1+\boldsymbol{\tau}_3),\ H(a,\theta;\boldsymbol{\sigma}_1+\boldsymbol{\tau}_5)\right\}\right\}. \tag{4.19}$$

Note, that the project (4.17) is not optimal for any single loading but it is optimal for the set of them. The solution provides an example of a mixed strategy in the game: loadings versus design.

**Illustration.** Set $r = 0.1$. The graph of the function $J(a,\theta)$ is presented on Fig.4. The magnified neighborhood of the minimum is shown on Fig.5, where the viewpoint is changed for better visibility. The optimal values of the parameters are $\theta = 0.889,\ a = 0.0496,\ J = 2.483$. We see that the compliance is bigger than the compliance of the structure optimal for a single load. On the other hand, the found structure is stable to all loadings, unlike the original design.

The picture of the optimal structure is shown on the Fig.3. Note that a part of the material is removed from the laminates that resist

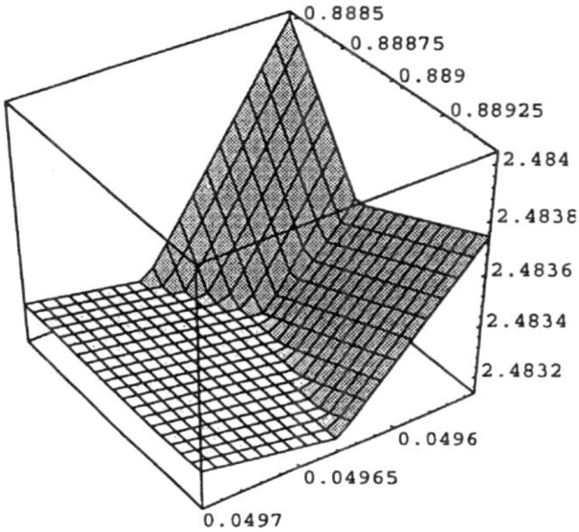


Figure 5: The magnified neighborhood of the minimum of the functional.

the main load. This material is placed in "reinforcements" that reduce the compliance in all directions.

**The sensitivity.** The non-smoothness of the min-max formulation leads to the nonlinear 'sensitivity' to the perturbations. Particularly, the optimal design may stay the same independently of sufficiently small perturbations of the loading. To demonstrate this, consider the following problem.

**Problem 4**. Consider the previous optimization problem, but suppose that both materials in the mixture have finite stiffness. The structure for this problem belongs to the same class of the third-rank laminates. The scheme of the solution is quite similar to the discussed case. However, there is an important difference. The maximal compliance in any direction is no longer an infinity, but it is estimated (see [8]) by the arithmetic mean of the compliances of the materials. Assume that the compliance matrices are equal to $s_1\boldsymbol{I}$ and $s_2\boldsymbol{I}$, and the volume fractions are equal to one half. The compliance of the laminate material is

$$\boldsymbol{S}_{lam} = \begin{pmatrix} \frac{2s_1s_2}{s_1+s_2} & 0 & 0 \\ 0 & \frac{s_1+s_2}{2} & 0 \\ 0 & 0 & \frac{s_1+s_2}{2} \end{pmatrix}. \tag{4.20}$$

If the perturbation parameter $r$ is small enough, then the maximum of compliance can correspond to one loading only.

Indeed, the energy of the loading $\boldsymbol{\sigma}_1$ in the initial laminate structure $\boldsymbol{p}_1$ is equal to

$$W_1(\boldsymbol{p}_1) = (1+r)^2\frac{2s_1s_2}{s_1+s_2}. \tag{4.21}$$

For each of the loadings $\boldsymbol{\tau}_i$, $i = 3, ..., 6$ there is a structure $\boldsymbol{p}_i$ which minimizes its energy. The energy $W_i(\boldsymbol{p}_1)$ of the loadings $\boldsymbol{\tau}_i$, $i = 3, ..., 6$ in the initial laminated structure is clearly greater than the energy $W_i(\boldsymbol{p}_i)$ corresponding to the optimal structure $\boldsymbol{p}_i$. Therefore $W_i(\boldsymbol{p}_i)$ is bounded as

$$W_i(\boldsymbol{p}_i) \le W_i(\boldsymbol{p}_1) = \frac{2s_1s_2}{s_1+s_2} + r^2\frac{s_1+s_2}{2}, \quad i = 3, ..., 6. \tag{4.22}$$

If the upper bound of $W_i$ is still smaller than $W_1$: $W_i(\boldsymbol{p}_1) < W_1(\boldsymbol{p}_1)$, $i = 3, ..., 6$, then the laminate is optimal, because $W_1$ reaches its minimum on it. The corresponding condition for $r$ is

$$0 \le r < \frac{8s_1s_2}{(s_1-s_2)^2}. \tag{4.23}$$

If $r$ satisfies this inequality, then the laminate structure remains optimal. This means that the design is insensitive to sufficiently small but final perturbations of the loading. If (4.23) holds, then the set of dangerous loadings consists of the first loading only.

**Symmetry. The optimal wheel.** The next example of the multiplicity of the "most dangerous" loadings demonstrates the appearance of symmetric projects in a min-max optimal design problem.

**Problem 5.** Consider the problem of a design of an optimal wheel. A circular domain is loaded by a non-axisymmetric loading that consists of a pair of radial forces applied to the rim and to the hub. These forces can move circumferential, which corresponds to the revolution of the

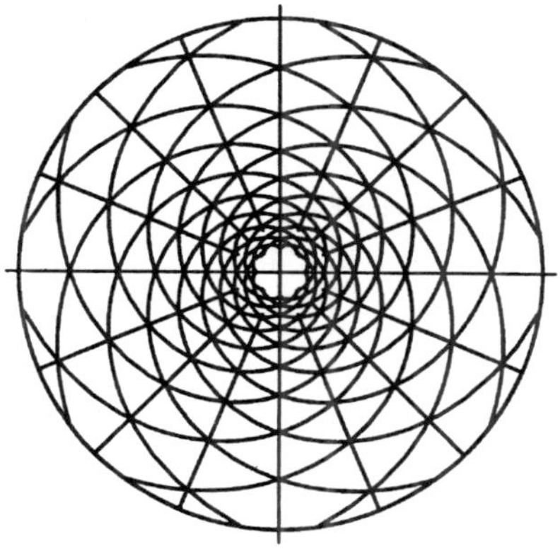

Figure 6: The cartoon of the optimal structure of the wheel.

wheel. If a loading $\boldsymbol{f}(S)$ is admissible, then any shifted loading $\boldsymbol{f}(S+\theta)$ is admissible too. Here, $S$ is the circumferential coordinate and $\theta$ is an arbitrary real number.

Consider an optimal design problem. Suppose, that it is required to minimize the maximal compliance of the wheel in a class of forces. The design which minimizes the maximal compliance is obviously axisymmetric even if a particular loading is not. The symmetry comes from the min-max requirement of the equal resistance to all forces $\boldsymbol{f}(S+\theta)$: the project is independent of the angle $\theta$.

The optimal axisymmetric layout of the composite properties $\boldsymbol{S}_*(\rho)$ in any particular point $\rho$ minimizes the integral over $\theta$ of the energy distribution. The solution locally is again the third rank laminate, symmetric with respect to angular coordinate $\theta$. The properties of the structure vary with the radius.

In the large, it can be represented as a periodic system of radii and two symmetric spirals (see Fig.6). The period of the spirals is infinitesimal, and the thickness of the materials varies with radius.

## Acknowledgements

We are thankful to Leonid Gibiansky for valuable comments. The support from the National Science Foundation through the grant DMS-9625129 and from the United States - Israel Binational Science Foundation through the grant No 9400349 is gratefully acknowledged.

## References

1. M. Avellaneda, Optimal bounds and microgeometries for elastic two-phase composites, *SIAM J. Appl. Math.*, **47**, 1216–1228 (1987).
2. C. Bandle, Isoperimetric inequalities and applications, Pitman Publishing Program, 1980.
3. A.V. Cherkaev, Stability of optimal structures of elastic composites, in: Topology Design of Structures, M.Bendsoe and C.A. Mota Soares, eds., Kluwer, 1992, 547-558.
4. E. Cherkaeva and A. Cherkaev, Bounds for detectability of material damage by noisy electrical measurements, in: Structural and multidisciplinary optimization, N. Olhoff and G. I. N. Rozvany eds, Pergamon, 1995, 543-548.
5. E. Cherkaeva , Optimal source control and resolution in non-destructive testing, *J. Structural Optimization,* **13** (1), 12–16 (1997).
6. S.J. Cox and M.L. Overton , On the optimal design of columns against buckling, *SIAM J. Math. Anal.*, **23**, 287–325 (1992).
7. V.F Demjanov and V.N. Malozemov, Introduction to minimax, Moscow, Nauka, 1972.
8. L.V. Gibiansky and A.V. Cherkaev, Microstructures of composites of extremal rigidity and exact bounds on the associated energy density, Ioffe Physico-Technical Institute, Academy of Sciences of USSR, Report N. 1115, Leningrad, 1987, translation in: Topics in the mathematical modelling of composite materials, A. Cherkaev and R.V. Kohn, eds., Birkhauser, 1997, 273-317.
9. J. R. Kuttler, Bounds for Stekloff eigenvalues, *SIAM J. Numer. Anal.*, **19** (1), 121–125 (1982).
10. G.I.N. Rozvany Structural design via optimality criteria. Kluwer Academic Publishers, Dordrecht, The Netherlands, 1989.

11. A.P. Seyranian, E. Lund and N. Olhoff, Multiple eigenvalues in structural optimization problems, *J. Structural Optimization*, **8**, 207–227 (1994).
12. J. Zowe, M. Kocvara and M.P. Bendsoe, Free material optimization via mathematical programming, *Math. Programming*, **79** (1–3) B, 445–446 (1997).

# EFFECTIVE PROPERTIES OF A PLANE TWO-PHASE ELASTIC COMPOSITES: COUPLED BULK-SHEAR MODULI BOUNDS

L.V. GIBIANSKY
*Princeton Material Institute, Princeton University, Princeton, NJ 08544, USA.*

The paper is concerned with the bounds on the effective properties of planar isotropic elastic composites. Specifically, coupled bulk-shear moduli bounds are presented that improve Hashin-Shtrikman-Walpole uncoupled bounds. They are given by the combination of the Hashin-Shtrikman-Walpole bulk modulus bounds, and the two new curves (hyperbolas) in the bulk-shear moduli plane. The bounds are identical to those obtained earlier by Cherkaev and Gibiansky[1]. However, the derivation is new and the form of the bounds is simpler. Derivation involves an analysis of quasi-affine quadratic forms of generalized fourth-order strain and stress tensors associated with eighth-order rotationally invariant tensors. Open questions, attainability of the bounds, and generalization to the three-dimensional problem are discussed.

## 1 Introduction

The paper is concerned with the long-standing problem of the bounds on the effective properties of plane elastic composites comprised of two isotropic phases. We aim to achieve two goals.

First, we provide a new derivation of the coupled bulk-shear moduli bounds originally obtained by Cherkaev and Gibiansky [1]. The new procedure utilizes recent advances in the translation method and provides guidance in generalization of the result to the three-dimensional problem. The form of the bounds that we obtain is simple and easy to use. The bounds improve upon the classical Hashin-Shtrikman-Walpole (HSW) bounds in a sense that they are coupled and allow one to use bulk modulus measurements to tighten the shear modulus bounds, and vice versa. The method allows the universal approach to the well-ordered and badly-ordered pairs of the materials (proper definitions and details are given in the body of the paper). In the process of deriving the bounds, we found all quasiaffine quadratic forms of generalized (fourth-order) stress and strain tensors associated with eighth-order rotationally invariant tensors. Obtained results are readily applied to the effective moduli of a transversely isotropic three-dimensional composite with cylindric inclusions of an arbitrary cross-section. Generalization to the full three-dimensional problem is also discussed.

Second, we use the problem to demonstrate the simplicity and the power of the translation method. The method was originated independently by Lurie and Cherkaev,[2]$^-$[5] and by Murat and Tartar[6, 7, 8]. It was developed and applied to a variety of problems by Gibiansky and Cherkaev [9, 10], Kohn and Milton [11], Milton and Kohn[12], Milton [13]$^-$[15], Cherkaev and Gibiansky [1, 16], Gibiansky and Milton [17], Gibiansky and Torquato [18, 19], Avellaneda *et al*[20]. The list is by no means complete since we are not aiming to give a comprehensive review of the extensive literature. Here we mention only works that directly contributed to the procedure applied in this paper. Additional references can be found in the cited papers. We combine ideas and achievements of the previous works on the subject to present a simple, well-structured, and powerful approach to bounding the effective properties of composite materials.

The rest of the paper is organized as follows: In Section 2 we introduce the basic equations and notations of the planar elasticity, review the previously known bounds, and state the new form of the Cherkaev and Gibiansky [1] coupled bulk-shear moduli bounds. In Section 3 we outline the translation method procedure in application to the plane elasticity problem. In Section 4 we derive the coupled bulk-shear moduli bounds. In Section 5 we summarize and discuss the results of the paper and generalization to the three-dimensional problem.

## 2 Statement of the problem and the main results

### *2.1 Equations of the planar elasticity*

We deal with the plane problem of elasticity. The state of the body is described by the local relations

$$\nabla \cdot \boldsymbol{\sigma} = 0, \quad \boldsymbol{\epsilon} = \frac{1}{2}[\nabla \boldsymbol{u} + (\nabla \boldsymbol{u})^T], \quad \boldsymbol{\sigma} = \boldsymbol{\sigma}^T, \quad \boldsymbol{\sigma} = \boldsymbol{C} : \boldsymbol{\epsilon}, \tag{2.1}$$

where $\boldsymbol{u}$ is the displacement vector, $\boldsymbol{\epsilon}$ and $\boldsymbol{\sigma}$ are the strain and stress tensors, respectively, and $\boldsymbol{C}$ is the stiffness tensor. Here the symbol :

denotes the contraction with respect to two indices, i.e.,

$$\boldsymbol{a}:\boldsymbol{b}=\sum_{i=1}^{2}\sum_{j=1}^{2}a_{ij}b_{ji};$$

$$\boldsymbol{a}=\boldsymbol{D}:\boldsymbol{b}\quad\text{if}\quad a_{ij}=\sum_{k=1}^{2}\sum_{l=1}^{2}D_{ijkl}b_{lk},\quad i,j=1,2. \tag{2.2}$$

It is convenient to introduce the following orthonormal basis in the space of the second-order tensors

$$\begin{aligned}\boldsymbol{a}_1&=(\boldsymbol{ii}+\boldsymbol{jj})/\sqrt{2}, &\quad \boldsymbol{a}_2&=(\boldsymbol{ii}-\boldsymbol{jj})/\sqrt{2},\\ \boldsymbol{a}_3&=(\boldsymbol{ij}+\boldsymbol{ji})/\sqrt{2}, &\quad \boldsymbol{a}_4&=(\boldsymbol{ij}-\boldsymbol{ji})/\sqrt{2},\\ &\boldsymbol{a}_k:(\boldsymbol{a}_\ell)^T=\delta_{k\ell},\end{aligned} \tag{2.3}$$

where $\boldsymbol{i}$ and $\boldsymbol{j}$ are the unit vectors of the Cartesian $x_1$ and $x_2$ axes, $\delta_{k\ell}$ is the Kronecker symbol, and $\boldsymbol{ij}$ represents the direct tensor product of $\boldsymbol{i}$ and $\boldsymbol{j}$. These tensors have the following matrix form

$$\begin{aligned}\boldsymbol{a}_1&=\frac{1}{\sqrt{2}}\begin{pmatrix}1&0\\0&1\end{pmatrix}, &\quad \boldsymbol{a}_2&=\frac{1}{\sqrt{2}}\begin{pmatrix}1&0\\0&-1\end{pmatrix},\\ \boldsymbol{a}_3&=\frac{1}{\sqrt{2}}\begin{pmatrix}0&1\\1&0\end{pmatrix}, &\quad \boldsymbol{a}_4&=\frac{1}{\sqrt{2}}\begin{pmatrix}0&1\\-1&0\end{pmatrix}\end{aligned} \tag{2.4}$$

in the Cartesian basis $\boldsymbol{i}$, $\boldsymbol{j}$. The subspace of the symmetric second-order tensors admits the orthonormal basis $\boldsymbol{a}_1$, $\boldsymbol{a}_2$, $\boldsymbol{a}_3$. Symmetric strain and stress tensors have the following representation in this basis

$$\boldsymbol{\epsilon}=\sum_{i=1}^{3}\epsilon_i\boldsymbol{a}_i\;,\quad \boldsymbol{\sigma}=\sum_{i=1}^{3}\sigma_i\boldsymbol{a}_i. \tag{2.5}$$

Here the coefficients $\epsilon_i=\boldsymbol{\epsilon}:\boldsymbol{a}_i$ , $\sigma_i=\boldsymbol{\sigma}:\boldsymbol{a}_i$ are given by the equations

$$\begin{aligned}\epsilon_1&=\frac{1}{\sqrt{2}}(\epsilon_{11}+\epsilon_{22}), &\quad \epsilon_2&=\frac{1}{\sqrt{2}}(\epsilon_{11}-\epsilon_{22}), &\quad \epsilon_3&=\sqrt{2}\epsilon_{12},\\ \sigma_1&=\frac{1}{\sqrt{2}}(\sigma_{11}+\sigma_{22}), &\quad \sigma_2&=\frac{1}{\sqrt{2}}(\sigma_{11}-\sigma_{22}), &\quad \sigma_3&=\sqrt{2}\sigma_{12},\end{aligned} \tag{2.6}$$

and $\epsilon_{ij}$, $\sigma_{ij}$ are the elements of the tensors $\boldsymbol{\epsilon}$ and $\boldsymbol{\sigma}$ in the Cartesian basis $\boldsymbol{i}$, $\boldsymbol{j}$.

In the basis (2.3), the Hook's law for an isotropic material has the form [21]

$$\sigma_1 = 2K\epsilon_1\ , \qquad \sigma_2 = 2\mu\epsilon_2\ , \qquad \sigma_3 = 2\mu\epsilon_3. \tag{2.7}$$

Here $K = K(\boldsymbol{x})$ and $\mu = \mu(\boldsymbol{x})$ are the bulk and shear moduli, respectively, of the plane elasticity (plane-strain) problem at the point $\boldsymbol{x} = (x_1, x_2)$. The plane-strain bulk modulus $K$ is expressed as

$$K = \kappa + \mu/3, \tag{2.8}$$

in terms of the bulk modulus $\kappa$ of the three-dimensional theory [22]. The plane-strain shear modulus $\mu$ does not differ from the transverse shear modulus used in the three-dimensional theory. If the composite is a thin plate, then we consider the plane-stress problem. In this case the plane-stress bulk modulus $K$ is expressed as [22]

$$K = \frac{9k\mu}{3k + 4\mu}. \tag{2.9}$$

Again, the plane-stress shear modulus $\mu$ does not differ from the shear modulus used in the three-dimensional theory.

The isotropic stiffness tensor $\boldsymbol{C}(K, \mu)$ is represented by the diagonal matrix

$$\boldsymbol{C}(K, \mu) = \begin{pmatrix} 2K & 0 & 0 \\ 0 & 2\mu & 0 \\ 0 & 0 & 2\mu \end{pmatrix} \tag{2.10}$$

in the basis (2.3). The elastic energy density can be written either as a quadratic form of strains

$$W_\epsilon(\boldsymbol{\epsilon}) = \boldsymbol{\epsilon} : \boldsymbol{C} : \boldsymbol{\epsilon}, \tag{2.11}$$

or as a quadratic form of stresses

$$W_\sigma(\boldsymbol{\sigma}) = \boldsymbol{\sigma} : \boldsymbol{S} : \boldsymbol{\sigma}, \tag{2.12}$$

where

$$\boldsymbol{S} = \boldsymbol{C}^{-1} = \begin{pmatrix} \frac{1}{2K} & 0 & 0 \\ 0 & \frac{1}{2\mu} & 0 \\ 0 & 0 & \frac{1}{2\mu} \end{pmatrix} \tag{2.13}$$

is the compliance tensor, and we omit the factor 1/2 in (2.11) and (2.12).

Following Cherkaev and Gibiansky,[1] we introduce the non-symmetric matrix $\boldsymbol{\zeta} = \boldsymbol{\nabla u}$ of the gradient of the displacement vector $\boldsymbol{u}$

$$\boldsymbol{\zeta} = \boldsymbol{\nabla u}, \quad \boldsymbol{\zeta} = \sum_{i=1}^{4} \zeta_i \boldsymbol{a}_i, \quad \zeta_i = \boldsymbol{\zeta} : \boldsymbol{a}_i, \quad i = 1, 2, 3, 4. \tag{2.14}$$

Note that the first three components of the tensors $\boldsymbol{\zeta}$ and $\boldsymbol{\epsilon}$ coincide, i.e.,

$$\zeta_i = \epsilon_i, \quad i = 1, 2, 3, \tag{2.15}$$

and the fourth component of $\boldsymbol{\zeta}$ ($\zeta_4 \neq \epsilon_4 = 0$) does not affect the elasticity equations. Specifically, the elastic energy density (2.11) as a function of the tensor $\boldsymbol{\zeta}$ becomes

$$W_\zeta(\boldsymbol{\zeta}) = \boldsymbol{\zeta} : \boldsymbol{C}' : \boldsymbol{\zeta}, \tag{2.16}$$

where the $(4 \times 4)$ matrix $\boldsymbol{C}'$ has the diagonal form

$$\boldsymbol{C}' = \begin{pmatrix} 2K & 0 & 0 & 0 \\ 0 & 2\mu & 0 & 0 \\ 0 & 0 & 2\mu & 0 \\ 0 & 0 & 0 & 0 \end{pmatrix} \tag{2.17}$$

in the basis (2.3).

### *2.2 Homogenization*

Let us consider a composite that is a double-periodic plane structure. The element of periodicity is divided into two parts with the volume fractions $f_1$ and $f_2 = 1 - f_1$, occupied by two isotropic materials with elastic moduli $(K_1, \mu_1)$ and $(K_2, \ \mu_2)$.

It is desired to study the homogenization problem, i.e. the problem of describing the medium's effective properties. It is well known [22]–[24] that the average behavior of a mixture is described by the homogenized elasticity equations

$$\begin{aligned} \langle \boldsymbol{\epsilon} \rangle &= \frac{1}{2}\left[\boldsymbol{\nabla}\langle \boldsymbol{u} \rangle + (\boldsymbol{\nabla}\langle \boldsymbol{u} \rangle)^T\right], \quad \langle \boldsymbol{\sigma} \rangle = \langle \boldsymbol{\sigma} \rangle^T, \\ \langle \boldsymbol{\sigma} \rangle &= \boldsymbol{C}_* : \langle \boldsymbol{\epsilon} \rangle, \quad \nabla \cdot \langle \boldsymbol{\sigma} \rangle = 0, \end{aligned} \tag{2.18}$$

where the angular brackets denote averaging of the bracketed quantity over the periodic cell. The tensor $\boldsymbol{C}_*$ that relates the average stress to the average strain, is by definition the *effective stiffness tensor*.

The effective stiffness tensor $\boldsymbol{C}_*$ depends on the phase properties, phase volume fractions, and the geometrical structure of the composite, independent of the loading. Note that any homogeneous (periodic or random) composite is equivalent to some periodic structure with respect to the effective elasticity tensor. The assumption of periodicity is not very restrictive; it is imposed only for the sake of simplicity of the description.

The elastic energy density $W^*$ stored in the composite is known to be equal to

$$W_\epsilon^*(\boldsymbol{\epsilon}_0) = \boldsymbol{\epsilon}_0 : \boldsymbol{C}_* : \boldsymbol{\epsilon}_0 = \min_{\substack{\boldsymbol{\epsilon} : \langle \boldsymbol{\epsilon} \rangle = \boldsymbol{\epsilon}_0 \\ \boldsymbol{\epsilon} = [\nabla \boldsymbol{u} + (\nabla \boldsymbol{u})^T]/2}} \langle \boldsymbol{\epsilon} : \boldsymbol{C} : \boldsymbol{\epsilon} \rangle. \tag{2.19}$$

Here minimum is taken over all admissible fields $\boldsymbol{\epsilon}$ that are the strain fields with given mean value $\boldsymbol{\epsilon}_0$ that satisfy compatibility conditions [21, 25]. For the conjugate functional of the complimentary energy we have

$$W_\sigma^*(\boldsymbol{\sigma}_0) = \boldsymbol{\sigma}_0 : \boldsymbol{S}_* : \boldsymbol{\sigma}_0 = \inf_{\substack{\boldsymbol{\sigma} : \langle \boldsymbol{\sigma} \rangle = \boldsymbol{\sigma}_0 \\ \boldsymbol{\sigma} = \boldsymbol{\sigma}^T, \ \nabla \cdot \boldsymbol{\sigma} = 0}} \langle \boldsymbol{\sigma} : \boldsymbol{S} : \boldsymbol{\sigma} \rangle, \tag{2.20}$$

where the effective compliance tensor $\boldsymbol{S}_*$ is determined as $\boldsymbol{S}_* = \boldsymbol{C}_*^{-1}$ and infimum is taken over all admissible stress fields.

### *2.3 The known bounds*

The problem of bounds on elastic moduli has a long history. Several bounds were found for the three-dimensional problem and all of them had natural two-dimensional analogies; isotropy in the planar elasticity corresponds to transversal isotropy in the three-dimensional problem for the structures with cylindrical arrangement of phases. In this section we briefly summarize classical bounds and give the new form of the Cherkaev and Gibiansky [1] coupled bulk-shear moduli bounds.

The first bounds of the effective moduli $K_*$, $\mu_*$ were found by Hill [26, 27]. They have a form

$$K_h \ \leq K_* \leq \ K_a, \quad \mu_h \ \leq \mu_* \leq \ \mu_a, \tag{2.21}$$

where

$$K_h = \left[\frac{f_1}{K_1} + \frac{f_2}{K_2}\right]^{-1}, \qquad K_a = f_1 K_1 + f_2 K_2,$$

$$\mu_h = \left[\frac{f_1}{\mu_1} + \frac{f_2}{\mu_2}\right]^{-1}, \qquad \mu_a = f_1 \mu_1 + f_2 \mu_2, \tag{2.22}$$

are the harmonic and arithmetic means of the phase bulk and shear moduli, respectively.

Hashin and Shtrikman [28] suggested new variational principles which allowed them to find new sharp bounds on the elastic moduli of an isotropic mixture made from isotropic materials. Originally, the Hashin-Shtrikman bounds were formulated for isotropic three dimensional mixtures; however, later Hashin [29] applied the same method to obtain bounds for the transversal isotropic composites with cylindrical inclusions as well. The above problem is exactly the case we will study here.

The phases were supposed to be "well-ordered," i.e.,

$$K_2 \leq K_1, \qquad \mu_2 \leq \mu_1. \tag{2.23}$$

The obtained bounds have the form

$$K_{2*} \leq K_* \leq K_{1*}, \quad \mu_{2*} \leq \mu_* \leq \mu_{1*}, \tag{2.24}$$

where

$$K_{1*} = f_1 K_1 + f_2 K_2 - \frac{f_1 f_2 (K_1 - K_2)^2}{f_1 K_2 + f_2 K_1 + \mu_1}, \tag{2.25}$$

$$K_{2*} = f_1 K_1 + f_2 K_2 - \frac{f_1 f_2 (K_1 - K_2)^2}{f_1 K_2 + f_2 K_1 + \mu_2}, \tag{2.26}$$

$$\mu_{1*} = f_1 \mu_1 + f_2 \mu_2 - \frac{f_1 f_2 (\mu_1 - \mu_2)^2}{f_1 \mu_2 + f_2 \mu_1 + y_{1*}}, \quad y_{1*} = \frac{K_1 \mu_1}{K_1 + 2\mu_1}, \tag{2.27}$$

$$\mu_{2*} = f_1 \mu_1 + f_2 \mu_2 - \frac{f_1 f_2 (\mu_1 - \mu_2)^2}{f_1 \mu_2 + f_2 \mu_1 + y_{2*}}, \quad y_{2*} = \frac{K_2 \mu_2}{K_2 + 2\mu_2}. \tag{2.28}$$

The Hashin-Shtrikman bounds are more restrictive than the Hill's ones. Moreover, they describe the minimal rectangle in the bulk-shear moduli plane which contains the set of all possible values of these moduli. It is proved that two opposite corner points A and C

$$A = (K_{1*},\ \mu_{1*}), \quad C = (K_{2*},\ \mu_{2*}) \tag{2.29}$$

of this rectangle (see Fig. 1a) that correspond to the maximal or minimal values of both moduli are attainable by special microstructures [30]–[33] (i.e., there exist microstructures with such properties).

Walpole [34] considered the opposite case of "badly-ordered" phases when

$$K_1 \le K_2, \quad \mu_2 \le \mu_1. \tag{2.30}$$

By using similar variational method he obtained the bounds on the effective moduli of an isotropic mixture. The two-dimensional Walpole's bounds have a form (see Fig.1b)

$$K_{2*} \le K_* \le K_{1*}, \qquad \mu_{3*} \le \mu_* \le \mu_{4*}, \tag{2.31}$$

where

$$\mu_{3*} = f_1\mu_1 + f_2\mu_2 - \frac{f_1 f_2(\mu_1 - \mu_2)^2}{f_1\mu_2 + f_2\mu_1 + y_{3*}}, \quad y_{3*} = \frac{K_1\mu_2}{K_1 + 2\mu_2}, \tag{2.32}$$

$$\mu_{4*} = f_1\mu_1 + f_2\mu_2 - \frac{f_1 f_2(\mu_1 - \mu_2)^2}{f_1\mu_2 + f_2\mu_1 + y_{4*}}, \quad y_{4*} = \frac{K_2\mu_1}{K_2 + 2\mu_1}. \tag{2.33}$$

Note that the cases (2.23) and (2.30) cover all possible relations between the elastic moduli of two original materials because one always can order the materials so that

$$\mu_2 \le \mu_1. \tag{2.34}$$

All the mentioned bounds provide independent ranges for both bulk and shear moduli variations. They bound the set of these moduli by a rectangular domain in the $(K_*, \mu_*)$ plane (see Fig.1a,b). Milton and Phan-Thien [35] and Berryman and Milton [36] used a different variational method to obtain *coupled* bounds for the effective properties of isotropic three dimensional composites; Kublanov and Milton [37] in their unpublished paper applied the same method to the plane problem we are interested in here. They obtained the first coupled bulk–shear moduli bounds. In other words, the rectangular domain of the permitted values of the effective moduli was replaced by a smaller domain. For the sake of brevity we will not provide their formulas here, and refer the reader to the paper [1] since the original work [37] is not published.

Cherkaev and Gibiansky [1] obtained the coupled bulk-shear moduli bounds that are sharper than all the previously mentioned ones. Here we give the new simpler statement of their results:

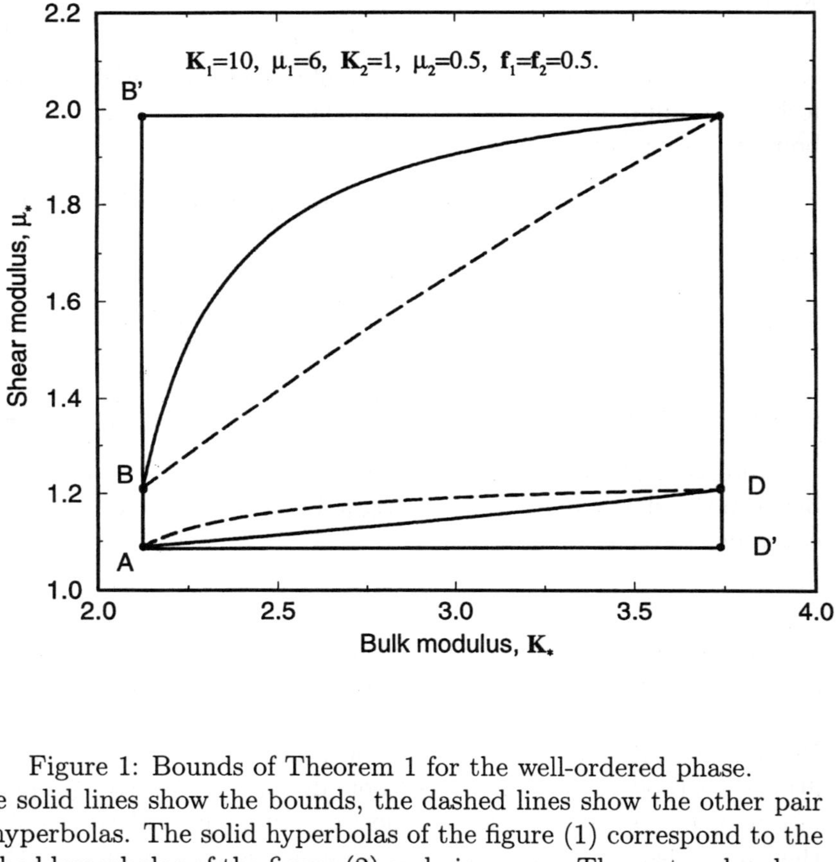


Figure 1: Bounds of Theorem 1 for the well-ordered phase. The solid lines show the bounds, the dashed lines show the other pair of hyperbolas. The solid hyperbolas of the figure (1) correspond to the dashed hyperbolas of the figure (2) and vice versa. The rectangles show the Hashin-Shtrikman-Walpole bounds.

**Theorem 1:** The pair $(K_*, \mu_*)$ of the effective bulk and shear moduli of any elastic isotropic composite lies in the set restricted in the bulk-shear moduli plane $(K_*, \mu_*)$ by two vertical lines $K_* = K_{1*}$ and $K_* = K_{2*}$ and by the pair of the hyperbolas

$$Hyp[(K_{1*}, \mu_{1*}), (K_{2*}, \mu_{4*}), (K_h, \mu_1)],$$

$$Hyp[(K_{2*}, \mu_{2*}), (K_{1*}, \mu_{3*}), (K_1, \mu_1)],$$

if the phases are well-ordered and satisfy (2.23), or by the pair of the hyperbolas

$$Hyp[(K_{1*}, \mu_{1*}), (K_{2*}, \mu_{4*}), (K_2, \mu_2)],$$

$$Hyp[(K_{2*}, \mu_{2*}), (K_{1*}, \mu_{3*}), (K_h, \mu_2)],$$

if the phases are badly-ordered and satisfy (2.30).

Here $K_{1*}$, $K_{2*}$ are given by (2.25), (2.26), $\mu_{1*}$, $\mu_{2*}$, $\mu_{3*}$, $\mu_{4*}$, are given by (2.27), (2.28), (2.32), (2.33), and $K_h$ is the arithmetic mean (2.22) of the phase bulk moduli. Moreover, $Hyp[(x_1, y_1), (x_2, y_2), (x_3, y_3)]$ denotes the segment of the hyperbola in the $(x, y)$ plane that can be presented in the parametric form as

$$x = \gamma x_1 + (1-\gamma)x_2 - \frac{\gamma(1-\gamma)(x_1 - x_2)^2}{\gamma x_2 + (1-\gamma)x_1 - x_3},$$

$$y = \gamma y_1 + (1-\gamma)y_2 - \frac{\gamma(1-\gamma)(y_1 - y_2)^2}{\gamma y_2 + (1-\gamma)y_1 - y_3}, \quad \gamma \in [0,1]. \quad (2.35)$$

Such segment passes through the points $(x_1, y_1)$ (when $\gamma = 1$) and $(x_2, y_2)$ (when $\gamma = 0$), and when extended passes through the point $(x_3, y_3)$ (when $\gamma = \infty$). Excluding the parameter $\gamma$, one obtains the relation

$$\frac{(y - y_2)(y_3 - y_1)}{(y_1 - y_2)(y_3 - y)} = \frac{(x - x_2)(x_3 - x_1)}{(x_1 - x_2)(x_3 - x)} \quad (2.36)$$

that describes this hyperbola in the $(x, y)$ plane.

The form of the bounds presented by Theorem 1 is simpler than the original formulation [1]. Its form is very similar to the conductivity-shear modulus bounds obtained by Gibiansky and Torquato [19].

The results are illustrated by Fig. 1, 2 . In Figure 1 we present the bounds for the well-ordered phases with the values of the parameters

$$K_1 = 10, \quad \mu_1 = 6, \quad K_2 = 1, \quad \mu_2 = 0.5, \quad f_1 = f_2 = 0.5. \quad (2.37)$$

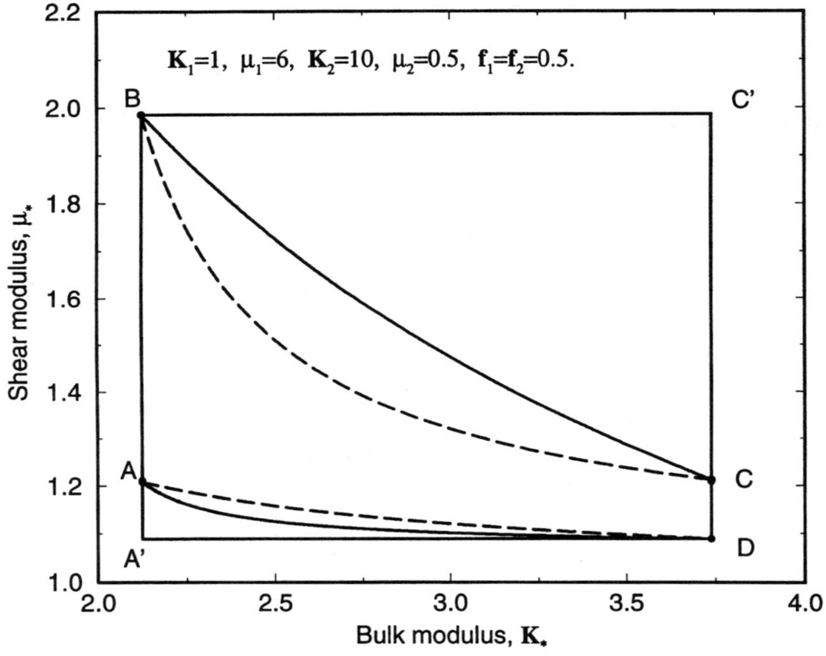


Figure 2: Bounds of Theorem 1 for the badly-ordered phases. The solid lines show the bounds, the dashed lines show the other pair of hyperbolas. The solid hyperbolas of the figure (1) correspond to the dashed hyperbolas of the figure (2) and vice versa. The rectangles show the Hashin-Shtrikman-Walpole bounds

The Hashin-Shtrikman bounds (2.24) are given by the rectangle $AB'CD'$, the solid curves show the bounds, the dashed curves correspond to the other two hyperbolas mentioned in Theorem 1. Similar results for the badly-ordered phases with the moduli

$$K_1 = 1, \quad \mu_1 = 6, \quad K_2 = 10, \quad \mu_2 = 0.5, \quad f_1 = f_2 = 0.5, \tag{2.38}$$

are presented in Figure 2.

## 3 The translation method

To prove Theorem 1 we use the translation method. The procedure that we describe here is a blend of ideas of the original paper,[1] the idea of "embedding" the problem into the higher-order problem,[13, 14] and the geometrical interpretation of translation bounds by Gibiansky and Milton [17] and Gibiansky and Torquato [18, 19]. Several steps are essential in the implementation of the translation method. In this section we discuss these steps in detail. Description of Sections 3.1 and 3.2 closely follows the paper [18].

### *3.1 Functionals*

The method is based on bounding from below the relevant energy functional $I$ that for the problem under study has the general form

$$I = \sum_{i=1}^{N} W_\epsilon^*(\epsilon_0^{(i)}), \tag{3.1}$$

[see relations (2.19)-(2.20)]. This functional is equal to the sum of $N$ values of the elastic energy stored in the element of periodicity of a composite which is exposed to $N$ elastic fields with fixed mean values. Here $N = 1$ for the bulk modulus bounds, $N = 2$ for the shear modulus bounds, and $N = 3$ for the coupled bulk-shear moduli bounds. Note that the energy also can be treated in terms of the stress fields. The energy functional is used because its value is equal to the energy stored by an equivalent homogeneous medium in the uniform field. The equivalent medium is characterized by the tensor of the effective properties, and the uniform external field coincides with the mean value of the field in the composite. Clearly, the lower bound on the functional

(3.1) provides bounds on the effective moduli of interest. Moreover, results by Milton [38] (see also Francfort and Milton [39]) evidence that sharp bounds on such energy functionals lead to optimal bounds on the effective moduli.

Let us now specify the functionals of the type (3.1) that attain minimal values at the boundary of the set of pairs $(K_*,\ \mu_*)$ and yield our coupled bounds.

It is instructive to begin with functionals for the bulk modulus bounds and shear modulus bounds before presenting the coupled bounds. To obtain bounds on the bulk modulus, the composite is exposed to an external hydrostatic strain $\boldsymbol{\epsilon}_h = \epsilon_h \boldsymbol{a}_1$ or stress $\boldsymbol{\sigma}_h = \sigma_h \boldsymbol{a}_1$ field. Indeed, the energy of an isotropic composite under the action of these fields is proportional to the effective bulk modulus $K_*$ according to

$$I_\epsilon(\boldsymbol{\epsilon}_h) = W_\epsilon^*(\boldsymbol{\epsilon}_h) = \boldsymbol{\epsilon}_h : \boldsymbol{C}_* : \boldsymbol{\epsilon}_h = 2K_*(\epsilon_h)^2, \tag{3.2}$$

or to its inverse value $1/K_*$ according to

$$I_\sigma(\boldsymbol{\sigma}_h) = W_\sigma^*(\boldsymbol{\sigma}_h) = \boldsymbol{\sigma}_h : \boldsymbol{S}_* : \boldsymbol{\sigma}_h = \frac{1}{2K_*}(\sigma_h)^2 \tag{3.3}$$

(see relations (2.10) and (2.13)). The minimization (by changing the microstructure) of the energy stored in the hydrostatic strain field [i.e. the functional (3.2)] is equivalent to the minimization of the bulk modulus. By contrast, the minimization of the energy stored in the hydrostatic stress field [see (3.3)] leads to maximization of this modulus. As we will see further, we need to express the elastic energy in terms of $\boldsymbol{\zeta}$ (the gradient of the displacement vector) instead of the strain fields $\boldsymbol{\epsilon}$, i.e.,

$$I_\epsilon(\boldsymbol{\epsilon}_h) = I_\zeta(\boldsymbol{\zeta}_h) = \boldsymbol{\zeta}_h : \boldsymbol{C}'_* : \boldsymbol{\zeta}_h, \tag{3.4}$$

where we assume that $\boldsymbol{\epsilon}_h : \boldsymbol{a}_i = \boldsymbol{\zeta}_h : \boldsymbol{a}_i, \quad i = 1, 2, 3.$

In a similar manner, one can get the bounds on the shear modulus of a composite by bounding the energy stored in the composite exposed to shear-type trial strain or stress fields $\boldsymbol{\epsilon}_s = \epsilon_s \boldsymbol{a}_2$ or $\boldsymbol{\sigma}_s = \sigma_s \boldsymbol{a}_2$, i.e.,

$$I_\epsilon(\boldsymbol{\epsilon}_s) = W_\epsilon^*(\boldsymbol{\epsilon}_s) = \boldsymbol{\epsilon}_s : \boldsymbol{C}_* : \boldsymbol{\epsilon}_s, \tag{3.5}$$

or

$$I_\sigma(\boldsymbol{\sigma}_s) = W_\sigma^*(\boldsymbol{\sigma}_s) = \boldsymbol{\sigma}_s : \boldsymbol{S}_* : \boldsymbol{\sigma}_s. \tag{3.6}$$

In this way we obtain the bounds on any of two shear moduli of the composite which is anisotropic in general. To ensure isotropy of the composite, one need to consider the reaction of the composite on two orthogonal shear fields. Hence, to estimate the shear modulus of the isotropic composite one should minimize the functional equal to the sum of the values of the energy stored by the medium under the action of two trial orthogonal shear strain fields $\boldsymbol{\epsilon}_s^{(1)} = \epsilon_s \boldsymbol{a}_2$, $\boldsymbol{\epsilon}_s^{(2)} = \epsilon_s \boldsymbol{a}_3$, or stress fields $\boldsymbol{\sigma}_s^{(1)} = \sigma_s \boldsymbol{a}_2$, $\boldsymbol{\sigma}_s^{(2)} = \sigma_s \boldsymbol{a}_3$, i.e.,

$$I_{\epsilon\epsilon}(\boldsymbol{\epsilon}_s^{(1)}, \boldsymbol{\epsilon}_s^{(2)}) = \boldsymbol{\epsilon}_s^{(1)} : \boldsymbol{C}_* : \boldsymbol{\epsilon}_s^{(1)} + \boldsymbol{\epsilon}_s^{(2)} : \boldsymbol{C}_* : \boldsymbol{\epsilon}_s^{(2)} = 4\mu_* \epsilon_s^2, \tag{3.7}$$

(see (2.10)), or

$$I_{\sigma\sigma}(\boldsymbol{\sigma}_s^{(1)}, \boldsymbol{\sigma}_s^{(2)}) = \boldsymbol{\sigma}_s^{(1)} : \boldsymbol{S}_* : \boldsymbol{\sigma}_s^{(1)} + \boldsymbol{\sigma}_s^{(2)} : \boldsymbol{S}_* : \boldsymbol{\sigma}_s^{(2)} = \frac{1}{\mu_*} \sigma_s^2, \tag{3.8}$$

(see (2.13)). The functional (3.7) can be re-expressed in terms of $\boldsymbol{\zeta}$ as in the relation (3.4), i.e.

$$I_{\epsilon\epsilon}(\boldsymbol{\epsilon}_s^{(1)}, \boldsymbol{\epsilon}_s^{(2)}) = I_{\zeta\zeta}(\boldsymbol{\zeta}_s^{(1)}, \boldsymbol{\zeta}_s^{(2)}) = \boldsymbol{\zeta}_s^{(1)} : \boldsymbol{C}'_* : \boldsymbol{\zeta}_s^{(1)} + \boldsymbol{\zeta}_s^{(2)} : \boldsymbol{C}'_* : \boldsymbol{\zeta}_s^{(2)} = 4\mu_* \epsilon_s^2, \tag{3.9}$$

where $\boldsymbol{\epsilon}_s^{(1)} : \boldsymbol{a}_i = \boldsymbol{\zeta}_s^{(1)} : \boldsymbol{a}_i$, and $\boldsymbol{\epsilon}_s^{(2)} : \boldsymbol{a}_i = \boldsymbol{\zeta}_s^{(2)} : \boldsymbol{a}_i$, $\quad i = 1, 2, 3$.

In order to get the coupled bulk-shear moduli bounds, we combine the functionals mentioned above. We have a choice between stress and strain trial fields for the bulk modulus problem, and between the stress and strain fields for the shear modulus problem. The following functionals should be considered for the coupled bulk-shear modulus bounds:

$$I_{\sigma\sigma\sigma}(\boldsymbol{\sigma}_h, \boldsymbol{\sigma}_s^{(1)}, \boldsymbol{\sigma}_s^{(2)}) = W_\sigma^*(\boldsymbol{\sigma}_h) + W_\sigma^*(\boldsymbol{\sigma}_s^{(1)}) + W_\sigma^*(\boldsymbol{\sigma}_s^{(2)}), \tag{3.10}$$

$$I_{\zeta\zeta\zeta}(\boldsymbol{\zeta}_h, \boldsymbol{\zeta}_s^{(1)}, \boldsymbol{\zeta}_s^{(2)}) = W_\zeta^*(\boldsymbol{\zeta}_h) + W_\zeta^*(\boldsymbol{\zeta}_s^{(1)}) + W_\zeta^*(\boldsymbol{\zeta}_s^{(2)}), \tag{3.11}$$

$$I_{\sigma\zeta\zeta}(\boldsymbol{\sigma}_h, \boldsymbol{\zeta}_s^{(1)}, \boldsymbol{\zeta}_s^{(2)}) = W_\sigma^*(\boldsymbol{\sigma}_h) + W_\zeta^*(\boldsymbol{\zeta}_s^{(1)}) + W_\zeta^*(\boldsymbol{\zeta}_s^{(2)}), \tag{3.12}$$

$$I_{\zeta\sigma\sigma}(\boldsymbol{\zeta}_h, \boldsymbol{\sigma}_s^{(1)}, \boldsymbol{\sigma}_s^{(2)}) = W_\zeta^*(\boldsymbol{\zeta}_h) + W_\sigma^*(\boldsymbol{\sigma}_s^{(1)}) + W_\sigma^*(\boldsymbol{\sigma}_s^{(2)}). \tag{3.13}$$

The lower bound of each of these functionals gives some component of the boundary. This is illustrated by Fig.3, where the directions of the steepest descend of all of these functionals are shown in the $(K_*, \mu_*)$ plane. Let us assume that we start with some microstructure that possesses the effective properties $(K_*, \mu_*)$ and we place the origin of the $(K_*, \mu_*)$ plane at this point. If one changes the microstructure with the aim to decrease the energy $W_\zeta^*(\boldsymbol{\zeta}_h)$, one moves horizontally to the left, that corresponds to a lower $K_*$. Decreasing the energy $W_\sigma^*(\boldsymbol{\sigma}_h)$ allows one to move horizontally to the right, that corresponds to a larger $K_*$. Similarly, the minimization of the functional $W_\zeta^*(\boldsymbol{\zeta}_s^{(1)}) + W_\zeta^*(\boldsymbol{\zeta}_s^{(2)})$ moves one vertically down in the plane, that corresponds to a lower $\mu_*$; minimization of the functional $W_\sigma^*(\boldsymbol{\sigma}_s^{(1)})+W_\sigma^*(\boldsymbol{\sigma}_s^{(2)})$ moves one vertically up in this plane, that corresponds to a larger $\mu_*$.

The functionals (3.10)-(3.13) move one in any fixed direction in the $(K_*, \mu_*)$ plane. Therefore, by bounding these functionals from below one can find bounds on the set of pairs of the values $(K_*, \mu_*)$ of composites of all possible microstructures. For example, consider the functional (3.11). For any fixed amplitudes $\epsilon_h^2$ and $\epsilon_s^2$, the bounds on the functional (3.11) show how far one can move in the direction $\boldsymbol{v} = -2\epsilon_h^2 \boldsymbol{i} - 4\epsilon_s^2 \boldsymbol{j}$, where $\boldsymbol{i}$ and $\boldsymbol{j}$ are the unit vectors of the $K_*$-axis and $\mu_*$-axis, respectively (see Fig.3). By changing the ratio $\epsilon_h^2/\epsilon_s^2$ one can direct $\boldsymbol{v}$ at any point within the third quadrant of the $(K_*, \mu_*)$ plane. This means that the functional (3.11) can provide the part of the boundary having a normal directed into the third quadrant (like $(\boldsymbol{CD})$ in Fig.3). Similarly, the functional (3.13) provides the part of the boundary having a normal directed into the second quadrant of the $(K_*, \mu_*)$ plane (like $(\boldsymbol{DA})$ in Fig.3). The functional (3.12) corresponds to the part of the boundary having a normal directed into the fourth quadrant of the $(K_*, \mu_*)$ plane (like $(\boldsymbol{BC})$ in Fig.3), and the functional (3.10) provides the part of the boundary having a normal directed into the first quadrant of the plane $(K_*, \mu_*)$ (like $(\boldsymbol{AB})$ in Fig.3).

### *3.2 Translation method*

The key element of the translation method is the translation inequality that we will derive in this section. Consider a two-phase composite with the local constitutive relation

$$\boldsymbol{j}(\boldsymbol{x}) = \boldsymbol{D}(\boldsymbol{x}) \cdot \boldsymbol{e}(\boldsymbol{x}) \tag{3.14}$$

at a point $\boldsymbol{x}$. Here $\boldsymbol{j}$ is a generalized "flux", $\boldsymbol{e}$ is a generalized "gradient", and $\boldsymbol{D}$ is some local property, generally a tensor, equal to $\boldsymbol{D}_1$ in

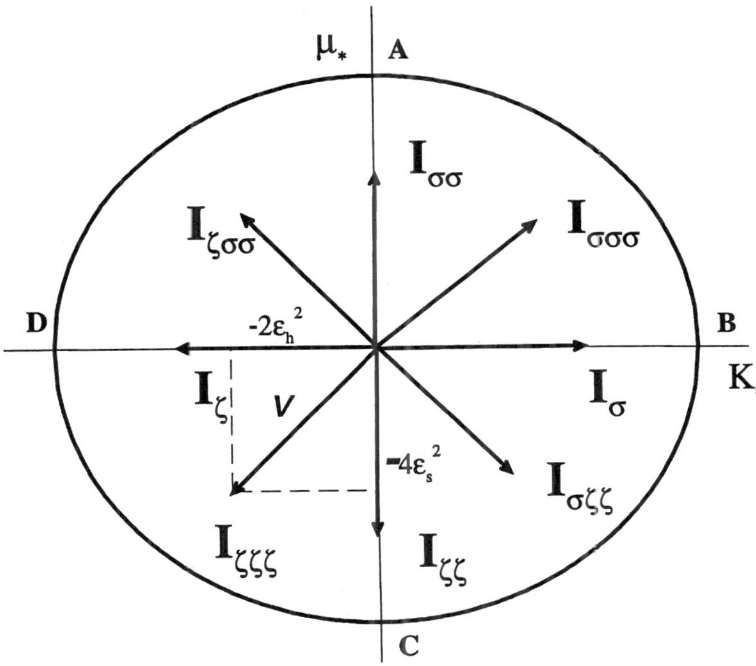


Figure 3: Directions of the steepest descend of various functionals described in the text.
For example, $\boldsymbol{v}$ represents the direction of the steepest descend of the functional $I_{\zeta\zeta\zeta}$; $-2\epsilon_h^2$ and $-4\epsilon_s^2$ are the components of the vector $\boldsymbol{v}$ in the $K_*$ and $\mu_*$ directions. The ellipse demarcates schematically the boundary of the set of the admissible pairs $(K_*, \mu_*)$

phase 1 and $\boldsymbol{D}_2$ in phase 2. For example, in the elasticity problem, $\boldsymbol{j}$, $\boldsymbol{e}$, and $\boldsymbol{D}$ represent the stress, strain, and stiffness tensor, respectively. In the present problem of coupled bounds, $\boldsymbol{D}$ is actually an eighth-order "super-tensor" associated with one of the quadratic forms (3.10)-(3.13), as discussed below. The effective tensor $\boldsymbol{D}_*$ is defined by the relation

$$\langle \boldsymbol{j} \rangle = \boldsymbol{D}_* \cdot \langle \boldsymbol{e} \rangle, \tag{3.15}$$

or equivalently by the averaged energy expression

$$\langle \boldsymbol{e} \cdot \boldsymbol{D} \cdot \boldsymbol{e} \rangle = \langle \boldsymbol{e} \rangle \cdot \boldsymbol{D}_* \cdot \langle \boldsymbol{e} \rangle, \tag{3.16}$$

where $\boldsymbol{e}$ and $\boldsymbol{j}$ is the solution of the corresponding system of differential equations for the medium with the properties $\boldsymbol{D}(\boldsymbol{x})$. It can be alternatively defined by the variational principle

$$\boldsymbol{e}_0 \cdot \boldsymbol{D}_* \cdot \boldsymbol{e}_0 = \inf_{\substack{\boldsymbol{e}:\ \langle \boldsymbol{e} \rangle = \boldsymbol{e}_0, \\ \boldsymbol{e} \in EK}} \langle \boldsymbol{e} \cdot \boldsymbol{D}(\boldsymbol{x}) \cdot \boldsymbol{e} \rangle, \tag{3.17}$$

where $EK$ is an appropriate space of generalized gradients.

Now consider a "translated" medium with the local property tensor

$$\boldsymbol{D}'(\boldsymbol{x}) = \boldsymbol{D}(\boldsymbol{x}) - \boldsymbol{T}, \tag{3.18}$$

where $\boldsymbol{T}$ is a constant *translation* tensor chosen in such a way that
**(i)** $\boldsymbol{D}'(\boldsymbol{x})$ is positive semi-definite and
**(ii)** the quadratic form associated with $\boldsymbol{T}$ is *quasiconvex*,[40] i.e., such that

$$\langle \boldsymbol{e} \cdot \boldsymbol{T} \cdot \boldsymbol{e} \rangle - \langle \boldsymbol{e} \rangle \cdot \boldsymbol{T} \cdot \langle \boldsymbol{e} \rangle \geq 0, \qquad \text{for any} \quad \boldsymbol{e} \in EK. \tag{3.19}$$

The effective properties of this medium can be defined via

$$\boldsymbol{e}_0 \cdot \boldsymbol{D}'_* \cdot \boldsymbol{e}_0 = \inf_{\substack{\boldsymbol{e}:\ \langle \boldsymbol{e} \rangle = \boldsymbol{e}_0, \\ \boldsymbol{e} \in EK}} \langle \boldsymbol{e} \cdot (\boldsymbol{D}(\boldsymbol{x}) - \boldsymbol{T}) \cdot \boldsymbol{e} \rangle \tag{3.20}$$

[c.f. with (3.17)]. Let $\boldsymbol{e}'(\boldsymbol{x})$ be a solution of the variational problem (3.17) and let us use this field as a trial field for the variational problem (3.20). This yields

$$\inf_{\substack{\boldsymbol{e}:\ \langle \boldsymbol{e} \rangle = \boldsymbol{e}_0, \\ \boldsymbol{e} \in EK}} \langle \boldsymbol{e} \cdot (\boldsymbol{D}(\boldsymbol{x}) - \boldsymbol{T}) \cdot \boldsymbol{e} \rangle \leq \langle \boldsymbol{e}' \cdot \boldsymbol{D}(\boldsymbol{x}) \cdot \boldsymbol{e}' \rangle - \langle \boldsymbol{e}' \cdot \boldsymbol{T} \cdot \boldsymbol{e}' \rangle$$

$$\leq \; \boldsymbol{e}_0 \cdot \boldsymbol{D}_* \cdot \boldsymbol{e}_0 - \boldsymbol{e}_0 \cdot \boldsymbol{T} \cdot \boldsymbol{e}_0. \tag{3.21}$$

Here we took into account the quasiconvexity of the quadratic form with the matrix $\boldsymbol{T}$ and the equality (3.17) for the field $\boldsymbol{e} = \boldsymbol{e}'$. Hence, the effective properties of the comparison and original media are related by

$$\boldsymbol{D}_* - \boldsymbol{T} \geq \boldsymbol{D}'_*. \tag{3.22}$$

Now use of the well-known harmonic mean bound [22] yields

$$(\boldsymbol{D}_* - \boldsymbol{T}) \; \geq \; \boldsymbol{D}'_* \; \geq \; [f_1(\boldsymbol{D}_1 - \boldsymbol{T})^{-1} + f_2(\boldsymbol{D}_2 - \boldsymbol{T})^{-1}]^{-1}, \tag{3.23}$$

or

$$(\boldsymbol{D}_* - \boldsymbol{T})^{-1} \; \leq \; f_1(\boldsymbol{D}_1 - \boldsymbol{T})^{-1} + f_2(\boldsymbol{D}_2 - \boldsymbol{T})^{-1}, \tag{3.24}$$

for any matrix $\boldsymbol{T}$ of a quasiconvex quadratic form such that

$$\boldsymbol{D}(\boldsymbol{x}) - \boldsymbol{T} \geq 0 \qquad \text{for any} \;\; \boldsymbol{x}. \tag{3.25}$$

For two-phase composites, the restriction (3.25) means

$$\boldsymbol{D}_1 - \boldsymbol{T} \geq 0, \qquad \boldsymbol{D}_2 - \boldsymbol{T} \geq 0. \tag{3.26}$$

The equation (3.24) is the basic inequality of the translation method. In fact, it is valid for any number of phases when written in the form

$$(\boldsymbol{D}_* - \boldsymbol{T})^{-1} \quad \leq \quad \langle (\boldsymbol{D}(\boldsymbol{x}) - \boldsymbol{T})^{-1} \rangle, \tag{3.27}$$

where $\boldsymbol{T}$ is subjected to (3.25). The essential point is that one wants to choose $\boldsymbol{T}$ so as to optimize the bound, to make it as restrictive as possible for the effective property tensor $\boldsymbol{D}_*$.

One can transform the bounds by using the so-called Y – transformation [15, 16]

$$\boldsymbol{Y}(f_1, \boldsymbol{D}_1, \boldsymbol{D}_2, \boldsymbol{D}_*) = -f_1\boldsymbol{D}_2 - f_2\boldsymbol{D}_1 \tag{3.28}$$
$$+ f_1 f_2(\boldsymbol{D}_1 - \boldsymbol{D}_2) \cdot (f_1\boldsymbol{D}_1 + f_2\boldsymbol{D}_2 - \boldsymbol{D}_*)^{-1} \cdot (\boldsymbol{D}_1 - \boldsymbol{D}_2).$$

Henceforth, we will omit the first three arguments of the Y-transform and will denote it simply as $\boldsymbol{Y}(\boldsymbol{D}_*)$. If the matrix $(\boldsymbol{D}_1 - \boldsymbol{D}_2)$ is not degenerate, then the bound (3.24) can be presented in a surprisingly simple form [15, 16]

$$\boldsymbol{Y}(\boldsymbol{D}_*) + \boldsymbol{T} \geq 0. \tag{3.29}$$

Note that the bound (3.29) in terms of the Y-transforms does not depend on the volume fraction. All volume-fraction information is "hidden" in the definition of the Y-transformation. Note some of the remarkable properties of this transformation

$$\begin{aligned} \boldsymbol{Y}(f_1, \boldsymbol{D}_1, \boldsymbol{D}_2, \boldsymbol{D}_i) &= -\boldsymbol{D}_i, \ i = 1, 2, \\ \boldsymbol{Y}(f_1, \boldsymbol{D}_1^{-1}, \boldsymbol{D}_2^{-1}, \boldsymbol{D}_*^{-1}) &= \boldsymbol{Y}^{-1}(f_1, \boldsymbol{D}_1, \boldsymbol{D}_2, \boldsymbol{D}_*), \\ \boldsymbol{Y}(f_1, \boldsymbol{D}_1 - \boldsymbol{T}, \boldsymbol{D}_2 - \boldsymbol{T}, \boldsymbol{D}_* - \boldsymbol{T}) &= \boldsymbol{Y}(f_1, \boldsymbol{D}_1, \boldsymbol{D}_2, \boldsymbol{D}_*) + \boldsymbol{T}, \end{aligned} \quad (3.30)$$

that are helpful in simplifying the bounds.

The parameters of the symmetric matrix $\boldsymbol{T}$ should be chosen in order to make the bounds (3.29) the most restrictive ones. In the problem under study, the matrix $\boldsymbol{D}_1 - \boldsymbol{D}_2$ may be degenerate, i.e. some of the eigenvectors and eigenvalues of the matrices $\boldsymbol{D}_1$ and $\boldsymbol{D}_2$ coincide. Indeed, for any material, the stiffness matrix $\boldsymbol{C}'$ has one eigenvalue equal to zero (see (2.17)). One can find an appropriate form of the bounds (3.24) for this case as well [1] but here we will not go into details. As we will see, in our problem all of the matrices in the matrix inequality (3.29) are block-diagonal if the component materials and the composite are isotropic which is the case. For the block of the matrix (3.29) that gives the bounds, the difference $\boldsymbol{D}_1 - \boldsymbol{D}_2$ is not degenerate and one can use the bound in the form (3.29).

### *3.3 "Embedding" into the higher-order problem*

In this section we briefly discuss how to make a correspondence between the energy-type functionals (3.10)-(3.13) and the general problem (3.14) with the "super-tensor" $\boldsymbol{D}$ and the associated energy (3.16). We discuss the problem on the example of the functional (3.10). The other functionals (3.11)-(3.13) can be studied in a similar fashion.

There are two possible ways of establishing such a correspondence. Specifically, one can view the energy functional (3.10) as a quadratic form

$$I_{\sigma\sigma\sigma} = \begin{pmatrix} \boldsymbol{\sigma}^{(1)} \\ \boldsymbol{\sigma}^{(2)} \\ \boldsymbol{\sigma}^{(3)} \end{pmatrix}^T \begin{pmatrix} \boldsymbol{S} & \mathbf{o} & \mathbf{o} \\ \mathbf{o} & \boldsymbol{S} & \mathbf{o} \\ \mathbf{o} & \mathbf{o} & \boldsymbol{S} \end{pmatrix} \begin{pmatrix} \boldsymbol{\sigma}^{(1)} \\ \boldsymbol{\sigma}^{(2)} \\ \boldsymbol{\sigma}^{(3)} \end{pmatrix} \quad (3.31)$$

associated with a large $N$-by-$N$ block-diagonal compliance matrix ($N = 3$ for the functional (3.31)). Each of the diagonal elements of this matrix is, in turn, a three-by-three matrix that represents the compliance tensor $\boldsymbol{S}$ in the basis (2.3). One can then study the properties of the corresponding general problem (3.14) with the "super-tensor" $\boldsymbol{D}$ being such special $3N$-by-$3N$ matrix. The advantage of this approach is its simplicity. However, it requires a great deal of intuition and experience later, in defining the optimal choice of the translation parameters in the $3N$-by-$3N$ translation matrix $\boldsymbol{T}$. Such approach to the problem was adopted in the original paper [1] [see also papers by Lurie and Cherkaev [3]–[5] who first introduced this approach for the conductivity problem]. It leads to the bounds of Theorem 1.

Alternatively, one can construct an equivalent higher-order problem by using the approach suggested for the elasticity problem by Milton [14][see also Tartar,[8] Avellaneda *et al.*[41] who used similar idea for the conductivity problem]. He viewed elasticity problems (corresponding to different external fields in (3.10)) as elements of the higher-order problem (3.14) with the fourth-order tensors of generalized strain and stress fields and the stiffness matrix that is an eighth-order tensor. The advantages of such "embedding" the elasticity problem into the problem with the eighth-order generalized stiffness tensor is a more systematic choice of the translation matrix $\boldsymbol{T}$. Specifically, Milton [13, 14] suggested to use *rotationally invariant* (i.e., isotropic) translation matrices $\boldsymbol{T}$. Then the choice of the translation matrix $\boldsymbol{T}$ is reduced to determining all rotationally invariant eighth-order tensors associated with quasiaffine (or quasiconvex) quadratic forms of the generalized strain or stress tensors. Milton [14] proved Hashin-Shtrikman-Walpole shear modulus bounds by using the translation method procedure that involved the analyses of an isotropic eighth-order tensors. We use similar approach in the next section to prove Theorem 1.

## 4 Coupled bulk-shear moduli bounds

### *4.1 Generalized Hooke's law*

Let us fix some Cartesian basis $\boldsymbol{i}$, $\boldsymbol{j}$ and an associated basis $\boldsymbol{a}_1$, $\boldsymbol{a}_2$, $\boldsymbol{a}_3$ in the space of the second-order symmetric tensors. Consider the following fourth-order "generalized stress" tensor

$$\boldsymbol{\Sigma} = \boldsymbol{a}_1\boldsymbol{\sigma}^{(1)} + \boldsymbol{a}_2\boldsymbol{\sigma}^{(2)} + \boldsymbol{a}_3\boldsymbol{\sigma}^{(3)}. \tag{4.1}$$

The tensors $\boldsymbol{a}_1$, $\boldsymbol{a}_2$, $\boldsymbol{a}_3$, and $\boldsymbol{\sigma}^{(1)}$, $\boldsymbol{\sigma}^{(2)}$, $\boldsymbol{\sigma}^{(3)}$ are symmetric. Therefore, the elements $\Sigma_{ijkl}$ of the fourth-order tensor $\boldsymbol{\Sigma}$ satisfy the symmetry conditions

$$\Sigma_{ijkl} = \Sigma_{jikl} = \Sigma_{ijlk} \tag{4.2}$$

However, the tensor $\boldsymbol{\Sigma}$ is less symmetric than the fourth-order elastic tensors, because

$$\Sigma_{ijkl} \neq \Sigma_{klij}. \tag{4.3}$$

The tensor $\boldsymbol{\Sigma}$ can be written in the basis $\boldsymbol{a}_1$-$\boldsymbol{a}_3$ in the form

$$\boldsymbol{\Sigma} = \begin{pmatrix} \sigma_h^{(1)} & \sigma_{s1}^{(1)} & \sigma_{s2}^{(1)} \\ \sigma_h^{(2)} & \sigma_{s1}^{(2)} & \sigma_{s2}^{(2)} \\ \sigma_h^{(3)} & \sigma_{s1}^{(3)} & \sigma_{s2}^{(3)} \end{pmatrix}, \tag{4.4}$$

where the super-index distinguishes between the fields, and the sub-index denotes the components of the field according to

$$\sigma_h^{(i)} = \boldsymbol{\sigma}^{(i)} : \boldsymbol{a}_1, \qquad \sigma_{s1}^{(i)} = \boldsymbol{\sigma}^{(i)} : \boldsymbol{a}_2, \qquad \sigma_{s2}^{(i)} = \boldsymbol{\sigma}^{(i)} : \boldsymbol{a}_3. \tag{4.5}$$

Similar to the combination of the stress fields, the following fourth-order tensor of the "generalized strain" field can be considered

$$\boldsymbol{\mathcal{E}} = \boldsymbol{a}_1\boldsymbol{\epsilon}^{(1)} + \boldsymbol{a}_2\boldsymbol{\epsilon}^{(2)} + \boldsymbol{a}_3\boldsymbol{\epsilon}^{(3)} = \begin{pmatrix} \epsilon_h^{(1)} & \epsilon_{s1}^{(1)} & \epsilon_{s2}^{(1)} \\ \epsilon_h^{(2)} & \epsilon_{s1}^{(2)} & \epsilon_{s2}^{(2)} \\ \epsilon_h^{(3)} & \epsilon_{s1}^{(3)} & \epsilon_{s2}^{(3)} \end{pmatrix}. \tag{4.6}$$

The Hooke's law of the form

$$\boldsymbol{\epsilon}^{(i)} = \boldsymbol{S} : \boldsymbol{\sigma}^{(i)}, \quad i = 1, 2, 3, \tag{4.7}$$

for each of the fields involved in the construction of the tensors $\boldsymbol{\Sigma}$ and $\boldsymbol{\mathcal{E}}$ results in the following relation (which we call generalized Hooke's law) among the generalized strain and stress tensors

$$\boldsymbol{\mathcal{E}} = \boldsymbol{D}_{\sigma\sigma\sigma} :: \boldsymbol{\Sigma}, \tag{4.8}$$

where the eighth-order elastic tensor $\boldsymbol{D}_{\sigma\sigma\sigma}$ is given by

$$\boldsymbol{D}_{\sigma\sigma\sigma} = \boldsymbol{a}_1\boldsymbol{S}\boldsymbol{a}_1 + \boldsymbol{a}_2\boldsymbol{S}\boldsymbol{a}_2 + \boldsymbol{a}_3\boldsymbol{S}\boldsymbol{a}_3, \tag{4.9}$$

and a contraction-type operation " :: " is defined as

$$\boldsymbol{D}_{\sigma\sigma\sigma} :: \boldsymbol{\Sigma} = \{D_{\sigma\sigma\sigma}\}_{ijklmnop}\Sigma_{ponm} \tag{4.10}$$

(summation from one to two is assumed over the repeated indices). Indeed, with such a definition we have

$$\boldsymbol{a}_i \boldsymbol{\epsilon}^{(i)} = \boldsymbol{D}_{\sigma\sigma\sigma} :: [\boldsymbol{a}_i \boldsymbol{\sigma}^{(i)}] = \boldsymbol{a}_i [\boldsymbol{S} : \boldsymbol{\sigma}^{(i)}], \quad i = 1, 2, 3 \tag{4.11}$$

(compare with (4.7)). Now it is obvious that

$$I_{\sigma\sigma\sigma}(\boldsymbol{\sigma}^{(1)}, \boldsymbol{\sigma}^{(2)}, \boldsymbol{\sigma}^{(3)}) = \boldsymbol{\Sigma} :: \boldsymbol{D}_{\sigma\sigma\sigma} :: \boldsymbol{\Sigma}. \tag{4.12}$$

Similarly, one can obtain such a "super-tensor" formulation for any of the functionals (3.11)-(3.13). The only specific feature is that in all of these problems strain tensors should be replaced by the gradient of the displacement tensors, similar to how it was done by Cherkaev and Gibiansky [1]. This is convenient since the quasiaffine combinations of the gradients are easier to find than the quasiconvex combinations of the strain tensors.

Specifically, for the functional (3.13) the generalized "flux" is actually the combination of the stress and the gradients of the displacements fields

$$\boldsymbol{j} = \boldsymbol{a}_1 \boldsymbol{\sigma} + \boldsymbol{a}_2 \boldsymbol{\zeta}^{(1)} + \boldsymbol{a}_3 \boldsymbol{\zeta}^{(2)}, \tag{4.13}$$

and the generalized gradient is given by

$$\boldsymbol{e} = \boldsymbol{a}_1 \boldsymbol{\zeta} + \boldsymbol{a}_2 \boldsymbol{\sigma}^{(1)} + \boldsymbol{a}_3 \boldsymbol{\sigma}^{(2)}. \tag{4.14}$$

Then the "generalized Hooke's law" is given by $\boldsymbol{j} = \boldsymbol{D}_{\zeta\sigma\sigma} :: \boldsymbol{e}$, where the tensor $\boldsymbol{D}_{\zeta\sigma\sigma}$ is equal to

$$\boldsymbol{D}_{\zeta\sigma\sigma} = \boldsymbol{a}_1 \boldsymbol{C}' \boldsymbol{a}_1 + \boldsymbol{a}_2 \boldsymbol{S} \boldsymbol{a}_2 + \boldsymbol{a}_3 \boldsymbol{S} \boldsymbol{a}_3, \tag{4.15}$$

and associated energy $I_{\zeta\sigma\sigma} = \boldsymbol{e} :: \boldsymbol{D}_{\zeta\sigma\sigma} :: \boldsymbol{e}$ is equal to the functional (3.13).

The problem for the functional (3.11) is a "dual" to the one considered in detail in this section. It is described by the generalized Hooke's law

$$\boldsymbol{\Sigma} = \boldsymbol{D}_{\zeta\zeta\zeta} :: \boldsymbol{Z}, \quad \boldsymbol{D}_{\zeta\zeta\zeta} = \boldsymbol{a}_1 \boldsymbol{C}' \boldsymbol{a}_1 + \boldsymbol{a}_2 \boldsymbol{C}' \boldsymbol{a}_2 + \boldsymbol{a}_3 \boldsymbol{C}' \boldsymbol{a}_3, \tag{4.16}$$

and the energy

$$I_{\zeta\zeta\zeta} = \boldsymbol{Z} :: \boldsymbol{D}_{\zeta\zeta\zeta} :: \boldsymbol{Z}, \tag{4.17}$$

where

$$\boldsymbol{\Sigma} = \boldsymbol{a}_1\boldsymbol{\sigma}^{(1)} + \boldsymbol{a}_2\boldsymbol{\sigma}^{(2)} + \boldsymbol{a}_3\boldsymbol{\sigma}^{(3)}, \quad \boldsymbol{\mathcal{Z}} = \boldsymbol{a}_1\boldsymbol{\zeta}^{(1)} + \boldsymbol{a}_2\boldsymbol{\zeta}^{(2)} + \boldsymbol{a}_3\boldsymbol{\zeta}^{(3)}. \quad (4.18)$$

Finally, the problem associated with the functional (3.12) is "dual" to the one associated with the functional (3.13).

Summarizing the discussion of this section, each of the functionals (3.10)-(3.13) can be viewed as a quadratic form associated with some eighth-order tensor. Our next goal is to find rotationally invariant eighth-order tensors that generate quasiaffine quadratic forms of the generalized strain or stress tensors. This is non-trivial algebraic problem that we decompose into two steps. In the next section we find a convenient representation of the rotationally invariant eighth-order tensors. Then we will establish conditions that guarantee quasiaffine property of the corresponding quadratic forms.

### *4.2 Rotationally invariant eighth-order tensors*

Eighth-order tensors can be viewed as linear operators acting on the space of the fourth-order tensors. Thus, to find the rotationally invariant eighth-order tensors we need to study the rotation properties of the appropriate fourth-order tensors.

Let us study the transformation of the elements of the fourth-order tensor due to a rotation of the Cartesian axes. Unless stated otherwise, we assume in this section that the fourth-order tensors are of a $\boldsymbol{\Sigma}$-type, i.e., satisfy the symmetry properties (4.2). We follow the approach used by Avellaneda *et al.*[20] for the anisotropic elastic tensors. Namely, consider two different systems of axes with orthonormal basis vectors $\boldsymbol{i}, \boldsymbol{j}$ and $\boldsymbol{i}', \boldsymbol{j}'$ and let $\varphi$ denotes an angle between $\boldsymbol{i}$ and $\boldsymbol{i}'$, i.e.,

$$\begin{aligned} \boldsymbol{i}' &= \boldsymbol{i}\ \cos\varphi + \boldsymbol{j}\sin\varphi, \\ \boldsymbol{j}' &= -\boldsymbol{i}\sin\varphi + \boldsymbol{j}\cos\varphi, \end{aligned} \quad (4.19)$$

or, in the matrix form:

$$\begin{pmatrix} \boldsymbol{i}' \\ \boldsymbol{j}' \end{pmatrix} = \begin{pmatrix} \cos\varphi & \sin\varphi \\ -\sin\varphi & \cos\varphi \end{pmatrix} \begin{pmatrix} \boldsymbol{i} \\ \boldsymbol{j} \end{pmatrix}. \quad (4.20)$$

We associate the tensor basis $\boldsymbol{a}_1', \boldsymbol{a}_2', \boldsymbol{a}_3'$ with the vectors $\boldsymbol{i}', \boldsymbol{j}'$ in the same way as (2.3) associates the basis $\boldsymbol{a}_1, \boldsymbol{a}_2, \boldsymbol{a}_3$ with the vectors $\boldsymbol{i}, \boldsymbol{j}$.

It is easy to verify that the two bases are connected by the relationships

$$\begin{aligned} \boldsymbol{a}_1' &= \boldsymbol{a}_1, \\ \boldsymbol{a}_2' &= \quad \cos 2\varphi\ \boldsymbol{a}_2 + \sin 2\varphi\ \boldsymbol{a}_3, \\ \boldsymbol{a}_3' &= -\sin 2\varphi\ \boldsymbol{a}_3 + \cos 2\varphi\ \boldsymbol{a}_3, \end{aligned} \tag{4.21}$$

or, in the matrix form,

$$\boldsymbol{a}' = \boldsymbol{R} : \boldsymbol{a}, \tag{4.22}$$

where $\boldsymbol{R} = \boldsymbol{R}(\varphi)$ is the rotation matrix

$$\boldsymbol{R} = \boldsymbol{R}(\varphi) = \begin{pmatrix} 1 & 0 & 0 \\ 0 & \cos 2\varphi & \sin 2\varphi \\ 0 & -\sin 2\varphi & \cos 2\varphi \end{pmatrix}, \tag{4.23}$$

$\boldsymbol{a}' = (\boldsymbol{a}_1', \boldsymbol{a}_2', \boldsymbol{a}_3')$, and $\boldsymbol{a} = (\boldsymbol{a}_1, \boldsymbol{a}_2, \boldsymbol{a}_3)$. Under this change of basis, the matrix $\boldsymbol{\Sigma}$ transforms to a new matrix

$$\boldsymbol{\Sigma}' = \boldsymbol{R} : \boldsymbol{\Sigma} : \boldsymbol{R}^T. \tag{4.24}$$

We are looking for the decomposition of the space of the fourth-order tensors into several orthogonal subspaces. One can study rotation properties of the $\boldsymbol{\Sigma}$-type tensors, as it was done by Avellaneda *et al.* [20] for the elastic tensors, or use more formal approach that is used in quantum mechanics [see Messiah,[42] and also Milton[13]]. The result is that the space of the $\boldsymbol{\Sigma}$-type tensors can be decomposed into the following six orthogonal rotationally invariant subspaces (presented below in the tensor basis $\boldsymbol{a}_1$-$\boldsymbol{a}_3$):

- the one-dimensional subspaces $\Lambda_1$, $\Lambda_2$, and $\Lambda_3$ of the tensors of the form

$$\boldsymbol{\Sigma}^{(1)} = q_1 \boldsymbol{a}_1 \boldsymbol{a}_1 = \begin{pmatrix} q_1 & 0 & 0 \\ 0 & 0 & 0 \\ 0 & 0 & 0 \end{pmatrix}, \tag{4.25}$$

$$\boldsymbol{\Sigma}^{(2)} = q_2(\boldsymbol{a}_2\boldsymbol{a}_2 + \boldsymbol{a}_3\boldsymbol{a}_3) = \begin{pmatrix} 0 & 0 & 0 \\ 0 & q_2 & 0 \\ 0 & 0 & q_2 \end{pmatrix}, \tag{4.26}$$

$$\boldsymbol{\Sigma}^{(3)} = q_3(\boldsymbol{a}_2\boldsymbol{a}_3 - \boldsymbol{a}_3\boldsymbol{a}_2) = \begin{pmatrix} 0 & 0 & 0 \\ 0 & 0 & q_3 \\ 0 & -q_3 & 0 \end{pmatrix}, \tag{4.27}$$

  respectively;

- the two-dimensional subspace $\Lambda_4$, $\Lambda_5$, and $\Lambda_6$ of the tensors of the form

$$\boldsymbol{\Sigma}^{(4)} = \boldsymbol{a}_1(q_4\boldsymbol{a}_2 + q_5\boldsymbol{a}_3) = \begin{pmatrix} 0 & q_4 & q_5 \\ 0 & 0 & 0 \\ 0 & 0 & 0 \end{pmatrix}, \tag{4.28}$$

$$\boldsymbol{\Sigma}^{(5)} = (q_6\boldsymbol{a}_2 + q_7\boldsymbol{a}_3)\boldsymbol{a}_1 = \begin{pmatrix} 0 & 0 & 0 \\ q_6 & 0 & 0 \\ q_7 & 0 & 0 \end{pmatrix}, \tag{4.29}$$

$$\boldsymbol{\Sigma}^{(6)} = q_8(\boldsymbol{a}_2\boldsymbol{a}_2 - \boldsymbol{a}_3\boldsymbol{a}_3) + q_9(\boldsymbol{a}_2\boldsymbol{a}_3 + \boldsymbol{a}_3\boldsymbol{a}_2) = \begin{pmatrix} 0 & 0 & 0 \\ 0 & q_8 & q_9 \\ 0 & q_9 & -q_8 \end{pmatrix}, \tag{4.30}$$

respectively.

Under the rotation by the angle $\varphi$, the tensor elements of the subspaces $\Lambda_1$, $\Lambda_2$, and $\Lambda_3$ transform as scalars

$$q_1' = q_1, \qquad q_2' = q_2, \qquad q_3' = q_3, \tag{4.31}$$

where prime denotes the elements in the rotated basis. The elements of the subspaces $\Lambda_4$ and $\Lambda_5$ transform as vectors but with the rotation by the double angle, i.e.,

$$\begin{gathered} \begin{pmatrix} q_4' \\ q_5' \end{pmatrix} = \begin{pmatrix} \cos 2\varphi & \sin 2\varphi \\ -\sin 2\varphi & \cos 2\varphi \end{pmatrix} \begin{pmatrix} q_4 \\ q_5 \end{pmatrix}, \\ \begin{pmatrix} q_6' \\ q_7' \end{pmatrix} = \begin{pmatrix} \cos 2\varphi & \sin 2\varphi \\ -\sin 2\varphi & \cos 2\varphi \end{pmatrix} \begin{pmatrix} q_6 \\ q_7 \end{pmatrix}. \end{gathered} \tag{4.32}$$

The elements of the subspace $\Lambda_6$ transform as a second-order tensor with the rotation by the double angle,

$$\begin{pmatrix} q_8' & q_9' \\ q_9' & q_8' \end{pmatrix} = \begin{pmatrix} \cos 2\varphi & \sin 2\varphi \\ -\sin 2\varphi & \cos 2\varphi \end{pmatrix} \begin{pmatrix} q_8 & q_9 \\ q_9 & q_8 \end{pmatrix} \begin{pmatrix} \cos 2\varphi & -\sin 2\varphi \\ \sin 2\varphi & \cos 2\varphi \end{pmatrix}. \tag{4.33}$$

In turn, the relation (4.33) can be further simplified. Specifically, the pair $(q_8, q_9)$ transforms as a vector but with the quadrupled angle of rotation, i.e.,

$$\begin{pmatrix} q_8' \\ q_9' \end{pmatrix} = \begin{pmatrix} \cos 4\varphi & \sin 4\varphi \\ -\sin 4\varphi & \cos 4\varphi \end{pmatrix} \begin{pmatrix} q_8 \\ q_9 \end{pmatrix}. \tag{4.34}$$

It follows that the quadratic form of the fourth-order tensor $\boldsymbol{\Sigma}$ associated with any rotationally invariant symmetric eighth-order tensor $\boldsymbol{T}_\Sigma$ can be written as a sum

$$\boldsymbol{\Sigma} :: \boldsymbol{T}_\Sigma :: \boldsymbol{\Sigma} = \begin{pmatrix} q_1 \\ q_2 \\ q_3 \end{pmatrix} \begin{pmatrix} a_{11} & a_{12} & a_{13} \\ a_{12} & a_{22} & a_{23} \\ a_{13} & a_{23} & a_{33} \end{pmatrix} \begin{pmatrix} q_1 \\ q_2 \\ q_3 \end{pmatrix} \tag{4.35}$$
$$+b_{11}(q_4^2 + q_5^2) + 2b_{12}(q_4 q_6 + q_5 q_7) + b_{22}(q_6^2 + q_7^2) + d(q_8^2 + q_9^2)$$

Note that the energy associated with the isotropic compliance tensor $\boldsymbol{D}_{\sigma\sigma\sigma}$ (see (4.9)) can be written as

$$\boldsymbol{\Sigma} :: \boldsymbol{D}_{\sigma\sigma\sigma} :: \boldsymbol{\Sigma} = \frac{1}{2K}(q_1^2 + q_6^2 + q_7^2) + \frac{1}{2\mu}[q_4^2 + q_5^2 + 2(q_2^2 + q_3^2 + q_8^2 + q_9^2)]. \tag{4.36}$$

Therefore, for the eighth-order isotropic elastic tensor $\boldsymbol{D}_{\sigma\sigma\sigma}$ the coefficients $a_{ij}$, $b_{ij}$, and $d$ are given by

$$a_{11} = b_{22} = \frac{1}{2K}, a_{22} = a_{33} = d = \frac{1}{\mu},$$

$$b_{11} = \frac{1}{2\mu}, \quad a_{12} = a_{13} = a_{23} = b_{12} = 0. \tag{4.37}$$

In the same way, one can find the irreducible rotationally invariant subspaces of the $\boldsymbol{\mathcal{E}}$-type tensors. More precisely, one need to study the generalized gradient of the displacements tensor

$$\boldsymbol{\mathcal{Z}} = \boldsymbol{a}_1 \boldsymbol{\zeta}^{(1)} + \boldsymbol{a}_2 \boldsymbol{\zeta}^{(2)} + \boldsymbol{a}_3 \boldsymbol{\zeta}^{(3)}, \tag{4.38}$$

because it will be easier to find the quasiaffine quadratic form of this tensor than the quasiconvex quadratic form of the tensor $\boldsymbol{\mathcal{E}}$. Any such tensor can be presented as a three-by-four matrix in the basis $\boldsymbol{a}_1$-$\boldsymbol{a}_4$. Since the gradient of the displacement is not always a symmetric matrix, the tensor $\boldsymbol{\mathcal{Z}}$ has even weaker symmetry properties than the tensor $\boldsymbol{\Sigma}$. Namely,

$$\mathcal{Z}_{ijkl} = \mathcal{Z}_{jikl}, \tag{4.39}$$

but in general

$$\mathcal{Z}_{ijkl} \neq \mathcal{Z}_{ijlk}, \quad \mathcal{Z}_{ijkl} \neq \mathcal{Z}_{klij}. \tag{4.40}$$

Therefore, the space of the fourth-order $\boldsymbol{\mathcal{Z}}$-type tensors (with the symmetry properties (4.39)) is richer than the set of the $\boldsymbol{\Sigma}$-type tensors.

This leads to a richer space of the eighth-order rotationally invariant tensors (i.e., linear operators acting on the space of the fourth-order $\mathcal{Z}$-type tensors) and harder calculations. Similar analysis also can be performed for the generalized fields that correspond to the other two functionals (3.12) and (3.13). The result is the general representation of the quadratic forms similar to (4.35) associated with the corresponding eighth-order rotationally invariant tensors.

### *4.3 Quasiaffine functions*

Our next step is to look for the rotationally invariant eighth-order tensors associated with the *quasiaffine* quadratic forms of the generalized stress field $\boldsymbol{\Sigma}$. In this section we show how to find such tensors.

To prove that the quadratic form associated with the eighth-order tensor $\boldsymbol{T}_\Sigma$ is quasiaffine, we need to establish the relation

$$\langle \boldsymbol{\Sigma} :: \boldsymbol{T}_\Sigma :: \boldsymbol{\Sigma} \rangle - \langle \boldsymbol{\Sigma} \rangle :: \boldsymbol{T}_\Sigma :: \langle \boldsymbol{\epsilon} \rangle = 0, \tag{4.41}$$

for all generalized stress fields $\boldsymbol{\Sigma}$. Following the standard procedure we use the Plancherel's formula to rewrite the equality (4.41) as the equality in the Fourier space,

$$\sum_{\boldsymbol{k}\neq 0} \hat{\boldsymbol{\Sigma}}(\boldsymbol{k}) :: \boldsymbol{T}_\Sigma :: \hat{\boldsymbol{\Sigma}}(\boldsymbol{k}) = 0, \tag{4.42}$$

in which $\hat{\boldsymbol{\Sigma}}(\boldsymbol{k})$ are the Fourier components of generalized stress tensor, and $\boldsymbol{k}$ is the usual wave-vector in the Fourier space. The equilibrium constraints (2.1) can be presented in the Fourier space as algebraic equalities

$$\hat{\boldsymbol{\sigma}}^{(i)}(\boldsymbol{k}) = [\hat{\boldsymbol{\sigma}}^{(i)}(\boldsymbol{k})]^T, \quad \boldsymbol{k} \cdot \hat{\boldsymbol{\sigma}}^{(i)}(\boldsymbol{k}) = 0, \quad i = 1, 2, 3, \tag{4.43}$$

where $\hat{\boldsymbol{\sigma}}^{(i)}(\boldsymbol{k})$ are the Fourier coefficients of the stress fields $\boldsymbol{\sigma}^{(i)}$.

In other words the integral equality (4.41) with the differential constraints (2.1) in the real space is now reduced to the algebraic equality (4.42) with the algebraic constraints (4.43) in the Fourier space. We are aiming to find the relations among the parameters $a_{ij}$, $b_{ij}$, $d$ that guarantee the quasiaffine property (4.42).

To find these relations we first fix $\boldsymbol{k}$ and choose a basis of the Cartesian coordinate system with the unit vectors $\boldsymbol{v}_1$, $\boldsymbol{v}_2$ with $\boldsymbol{v}_2$ being parallel to $\boldsymbol{k}$. The equalities (4.43) are satisfied if and only if

$$\hat{\boldsymbol{\sigma}}^{(i)}(\boldsymbol{k}) = \sqrt{2}\lambda_i(\boldsymbol{k})\boldsymbol{v}_1\boldsymbol{v}_1, \qquad i = 1, 2, 3, \tag{4.44}$$

where $\lambda_i(\boldsymbol{k})$ are some constants. Therefore, the tensor $\hat{\boldsymbol{\Sigma}}(\boldsymbol{k})$ has the form

$$\hat{\boldsymbol{\Sigma}}(\boldsymbol{k}) = \begin{pmatrix} \lambda_1 & \lambda_1 & 0 \\ \lambda_2 & \lambda_2 & 0 \\ \lambda_3 & \lambda_3 & 0 \end{pmatrix}, \tag{4.45}$$

in the basis (2.3) associated with the Cartesian basis $\boldsymbol{i} = \boldsymbol{v}_1$, $\boldsymbol{j} = \boldsymbol{v}_2$. One can now check that

$$\hat{\boldsymbol{\Sigma}}(\boldsymbol{k}) :: \boldsymbol{T}_\Sigma :: \hat{\boldsymbol{\Sigma}}(\boldsymbol{k}) = \begin{pmatrix} \lambda_1 \\ \lambda_2/2 \\ -\lambda_3/2 \end{pmatrix}^T \begin{pmatrix} a_{11} & a_{12} & a_{13} \\ a_{12} & a_{22} & a_{23} \\ a_{13} & a_{23} & a_{33} \end{pmatrix} \begin{pmatrix} \lambda_1 \\ \lambda_2/2 \\ -\lambda_3/2 \end{pmatrix} \tag{4.46}$$
$$+ b_{11}\lambda_1^2 + b_{22}(\lambda_2^2 + \lambda_3^2) + 2b_{12}\lambda_1\lambda_2 + d(\lambda_2^2 + \lambda_3^2)/4,$$

where $a_{ij}$, $b_{ij}$, and $d$ characterize an isotropic tensor $\boldsymbol{T}_\Sigma$. To guarantee (4.42), this expression should be identically equal to zero for any values of the variables $\lambda_1$, $\lambda_2$, and $\lambda_3$, thus leading to the requirements

$$\begin{aligned} a_{11} + b_{11} &= 0, \\ a_{12} + 2b_{12} &= 0, \\ a_{13} = a_{23} &= 0, \\ a_{22} + 4b_{22} + d &= 0, \\ a_{33} + 4b_{22} + d &= 0. \end{aligned} \tag{4.47}$$

The system (4.47) of six linear equations for ten unknowns has a set of the solutions

$$\begin{aligned} &a_{11} = -t_1, \quad a_{12} = -2t_3, \quad a_{13} = a_{23} = 0, \quad a_{22} = a_{33} = 2(t_2 - t_4), \\ &b_{11} = t_1, \quad b_{12} = t_3, \quad b_{22} = -t_2, \quad d = 2(t_2 + t_4), \end{aligned} \tag{4.48}$$

that depend on four parameters $t_1$, $t_2$, $t_3$, and $t_4$. This choice guarantees the quasiaffine property of the quadratic form associated with the matrix $\boldsymbol{T}_\Sigma$ for any values of the parameters $t_i$. Since the translation matrix $\boldsymbol{T}_\Sigma$ is isotropic, the result does not depend on the choice of the basis, and on the direction of the wave-vector $\boldsymbol{k}$.

Similarly one can obtain quasiaffine quadratic forms of the generalized gradient of the displacements tensor $\boldsymbol{\mathcal{Z}}$. Calculations are more involved, but the result is similar: there is a four-parameter family of the rotationally invariant eighth-order tensors associated with the quasiaffine quadratic form of the $\boldsymbol{\mathcal{Z}}$-type tensors. The same is true for the "mixed" problems with the arguments being fourth-order tensors combined of the stress and strain fields. This can be shown in a very similar way and hence we omit it.

### *4.4 Coupled bulk-shear moduli bounds*

Here we derive the bulk-shear moduli bounds that follow from the analysis of the functional (3.10) in the form (4.12). We need to interpret the general bounds (3.29) and find the optimal values of the parameters that satisfy (3.26) and lead to the sharpest bounds. As we found in the previous section, the rotationally invariant eighth-order tensors for the problem under study can be characterized by the three-by-three matrix $\boldsymbol{A} = a_{ij},\ i,j = 1,2,3$, the two-by-two matrix $\boldsymbol{B} = b_{ij},\ i,j = 1,2$, and the parameter $d$. The table below provides the values of these parameters for the tensors $\boldsymbol{D}_i$, $i = 1,2$ (representing the eighth-order isotropic phase compliance tensors), the effective compliance tensor $\boldsymbol{D}_*$, the $Y$-tensor $\boldsymbol{Y}_* = \boldsymbol{Y}(f_1, \boldsymbol{D}_1, \boldsymbol{D}_2, \boldsymbol{D}_*)$, and the translation matrix $\boldsymbol{T}_\Sigma$.

In this table the quantities $y_K(K_*)$ and $y_\mu(\mu_*)$ are the $Y$-transforms of the effective bulk and shear moduli, respectively. They are given by the scalar relations

$$y_K(K_*) = -f_1K_2 - f_2K_1 + \frac{f_1f_2(K_1 - K_2)^2}{f_1K_1 + f_2K_2 - K_*}, \tag{4.49}$$

$$y_\mu(\mu_*) = -f_1\mu_2 - f_2\mu_1 + \frac{f_1f_2(\mu_1 - \mu_2)^2}{f_1\mu_1 + f_2\mu_2 - \mu_*}, \tag{4.50}$$

deduced from the tensor definition of the $Y$-transformation (3.28).

Table 1. Parameters of the eighth-order rotationally invariant tensors $\boldsymbol{D}_i$, $i = 1, 2, *$, $\boldsymbol{Y}_* = \boldsymbol{Y}(f_1, \boldsymbol{D}_1, \boldsymbol{D}_2, \boldsymbol{D}_*)$, and $\boldsymbol{T}_\Sigma$.

| | $\boldsymbol{A}$ | $\boldsymbol{B}$ | $d$ |
|---|---|---|---|
| $\boldsymbol{D}_i$ | $\begin{pmatrix} \frac{1}{2K_i} & 0 & 0 \\ 0 & \frac{1}{\mu_i} & 0 \\ 0 & 0 & \frac{1}{\mu_i} \end{pmatrix}$ | $\begin{pmatrix} \frac{1}{2\mu_i} & 0 \\ 0 & \frac{1}{2K_i} \end{pmatrix}$ | $\frac{1}{2\mu_i}$ |
| $\boldsymbol{Y}_*$ | $\begin{pmatrix} \frac{1}{2y_K(K_*)} & 0 & 0 \\ 0 & \frac{1}{y_\mu(\mu_*)} & 0 \\ 0 & 0 & \frac{1}{y_\mu(\mu_*)} \end{pmatrix}$ | $\begin{pmatrix} \frac{1}{2y_\mu(\mu_*)} & 0 \\ 0 & \frac{1}{2y_K(K_*)} \end{pmatrix}$ | $\frac{1}{2y_\mu(\mu_*)}$ |
| $\boldsymbol{T}_\Sigma$ | $\begin{pmatrix} -t_1 & -2t_3 & 0 \\ -2t_3 & 2(t_2 - t_4) & 0 \\ 0 & 0 & 2(t_2 + t_4) \end{pmatrix}$ | $\begin{pmatrix} t_1 & t_3 \\ t_3 & -t_2 \end{pmatrix}$ | $2(t_2 + t_4)$ |

The tensor inequality

$$\boldsymbol{D}_1 - \boldsymbol{D}_2 \geq 0 \tag{4.51}$$

for any two isotropic eighth-order $\boldsymbol{D}_{\sigma\sigma\sigma}$ tensors $\boldsymbol{D}_1$ and $\boldsymbol{D}_2$ is equivalent to the system

$$\boldsymbol{A}_1 - \boldsymbol{A}_2 \geq 0, \quad \boldsymbol{B}_1 - \boldsymbol{B}_2 \geq 0, \quad d_1 - d_2 \geq 0, \tag{4.52}$$

where $\boldsymbol{A}_i$, $\boldsymbol{B}_i$, and $d_i$ are the parameters that describe the tensor $\boldsymbol{D}_i$.

From the general bounds of Section 3.2 we deduce that

$$\boldsymbol{Y}_* + \boldsymbol{T}_\Sigma \geq 0, \quad \text{if} \quad \boldsymbol{D}_1 - \boldsymbol{T}_\Sigma \geq 0, \quad \boldsymbol{D}_2 - \boldsymbol{T}_\Sigma \geq 0. \tag{4.53}$$

Applying these general formulas to our case we obtain

$$\begin{pmatrix} \frac{1}{2y_K(K_*)} - t_1 & -2t_3 \\ -2t_3 & \frac{1}{2y_\mu(\mu_*)} + 2(t_2 - t_4) \end{pmatrix} \geq 0, \tag{4.54}$$

if

$$\begin{pmatrix} \frac{1}{2\mu_i} - t_1 & -t_3 \\ -t_3 & \frac{1}{2K_i} + t_2 \end{pmatrix} \geq 0, \quad i = 1, 2, \tag{4.55}$$

$$\begin{pmatrix} \frac{1}{2K_i} + t_1 & 2t_3 \\ 2t_3 & \frac{1}{\mu_i} - 2(t_2 - t_4) \end{pmatrix} \geq 0, \quad i = 1, 2, \tag{4.56}$$

$$\frac{1}{\mu_i} - 2(t_2 + t_4) \geq 0, \quad i = 1, 2. \tag{4.57}$$

One can compare these bounds and the bounds derived by Cherkaev and Gibiansky [1] [see Eqs. (7.77), (7.65)-(7.68) of their paper] to see that they are identical.

Now we want to find the best choice of the parameters (subject to the restrictions (4.55)-(4.57)) in order to optimize the bound (4.54). We use the geometrical interpretation of the translation bounds that was suggested by Gibiansky and Milton [17].

Let us first assign

$$t_4 = \frac{1}{2\mu_{\max}} - t_2, \tag{4.58}$$

where $\mu_{\max} = \mu_1$ is the maximal phase shear modulus. Thus both of the inequalities (4.57) are satisfied (one with the equality sign), and the inequalities (4.56) result in

$$\left(\frac{1}{\mu_i} + \frac{1}{\mu_{\max}} - 4t_2\right)\left(\frac{1}{2K_i} + t_1\right) - 4t_3^2 \geq 0, \quad i = 1, 2. \tag{4.59}$$

By calculating the determinant of the matrix (4.54) we arrive at the bound

$$\left(\frac{1}{y_\mu(\mu_*)} - \frac{1}{\mu_{\max}} + 4t_2\right)\left(\frac{1}{2y_K(K_*)} - t_1\right) - 4t_3^2 \geq 0. \tag{4.60}$$

Clearly, the inequality (4.60) confines the point $\left(y_K^{-1}(K_*), y_\mu^{-1}(\mu_*)\right)$ to lie above the positive branch of the hyperbola

$$\left(\frac{1}{y_\mu(\mu_*)} - \frac{1}{\mu_{\max}} + 4t_2\right)\left(\frac{1}{2y_K(K_*)} - t_1\right) - 4t_3^2 = 0 \tag{4.61}$$

in the $\left(y_K^{-1}(K_*), y_\mu^{-1}(\mu_*)\right)$ plane (see Fig. 4, the dashed curve). The precise form of the hyperbola depends on the choice of the translation parameters $t_i$. Let us now study the properties of such hyperbolas.

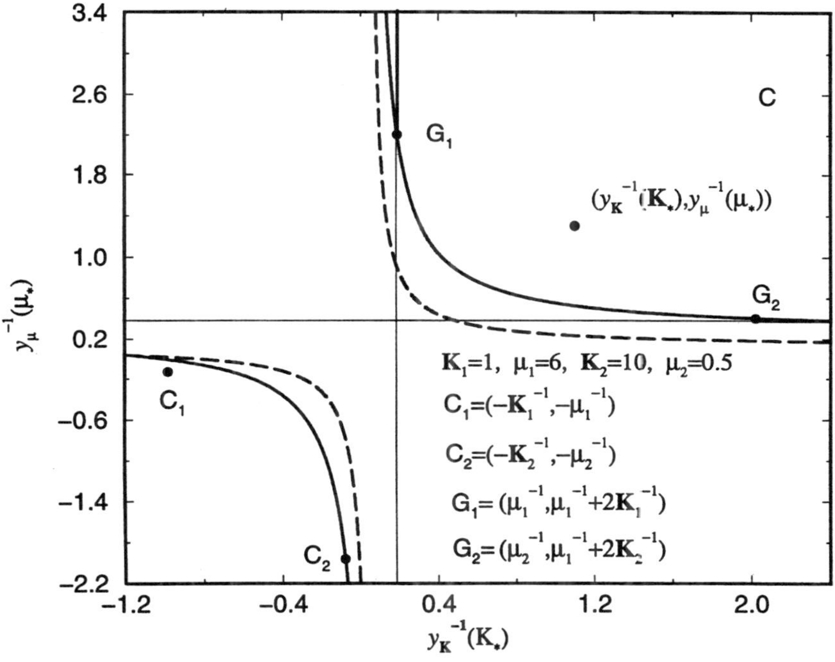


Figure 4: Bounding hyperbolas in the $\left(y_K^{-1}(K_*), y_\mu^{-1}(\mu_*)\right)$ plane. The dashed curves correspond to the hyperbola (4.61) for some admissible values of the parameters $t_i$; the solid curves show the optimal bound; the thin lines correspond to the Walpole bounds (4.65), (4.67).

First, the inequalities (4.59) (that are very similar to the bound (4.60)), show that both points

$$C_1 = (-K_1^{-1}, -\mu_1^{-1}), \qquad C_2 = (-K_2^{-1}, -\mu_2^{-1}) \tag{4.62}$$

lie below the negative branch of the hyperbola (4.61). Second, one can check by the straightforward calculation that the inequalities (4.55) are equivalent to the requirement

$$\left(\frac{2}{K_i} + 4t_2\right)\left(\frac{1}{2\mu_i} - t_1\right) - 4t_3^2 \geq 0. \tag{4.63}$$

Comparing (4.63) and (4.60), one can conclude that the points

$$G_1 = \left(\frac{1}{\mu_1}, \frac{1}{\mu_{\max}} + \frac{2}{K_1}\right), \qquad G_2 = \left(\frac{1}{\mu_2}, \frac{1}{\mu_{\max}} + \frac{2}{K_2}\right) \tag{4.64}$$

lie above the positive branch of the hyperbola (4.61).

By changing the parameters $t_1$, $t_2$, and $t_3$ one can move and resize the hyperbola (4.61). In fact, there is a one to one correspondence between the translation parameters $t_1$, $t_2$, $t_3$ and hyperbolas (4.61) in the $(y_K^{-1}(K_*), y_\mu^{-1}(\mu_*))$ plane. Changing the parameters $t_1$, $t_2$, $t_3$ is equivalent to moving and resizing this hyperbola. The restrictions on the values of the parameters can be interpreted as the restrictions on the position of the bounding hyperbola. Our discussion allow us to formulate the following result:

**Lemma**: For any elastic isotropic two-phase composite with the effective moduli $K_*$ and $\mu_*$, the point $\left(y_K^{-1}(K_*), y_\mu^{-1}(\mu_*)\right)$ lies above the positive branch of any hyperbola (4.61) such that

- the points $C_1$ and $C_2$ (see (4.62)) lie below the negative branch of this hyperbola;
- the points $G_1$ and $G_2$ lie above the positive branch of this hyperbola

in the $\left(y_K^{-1}(K_*), y_\mu^{-1}(\mu_*)\right)$ plane.

Figure 4 depicts the characteristic points $C_i$ and $G_i$ and the bounds that can be obtained by applying the Lemma. The simplest of these bounds is the bulk modulus bound (the thin vertical line in Fig. 4)

$$\frac{1}{y_K(K_*)} \geq \frac{1}{\mu_{\max}} = \frac{1}{\mu_1}, \tag{4.65}$$

which is equivalent to the Hashin-Shtrikman-Walpole upper bulk modulus bound $K_* \leq K_{1*}$. It corresponds to the values

$$t_1 = \frac{1}{2\mu_1}, \qquad t_3 = 0 \tag{4.66}$$

of the parameters $t_1$ and $t_3$ independently of the value of the parameter $t_2$. The other obvious result is the shear modulus bound (the thin horizontal line in Fig. 4)

$$\frac{1}{y(\mu_*)} \geq \frac{1}{\mu_{\max}} + \frac{2}{\kappa_{\max}}, \tag{4.67}$$

which is equivalent to either Hashin-Shtrikman ($\mu_* \leq \mu_{1*}$) or Walpole ($\mu_* \leq \mu_{4*}$) upper shear modulus bounds depending on whether the phases are well- or badly-ordered. This corresponds to the values

$$t_2 = -\frac{1}{2K_{\max}}, \qquad t_3 = 0 \tag{4.68}$$

of the parameters $t_2$ and $t_3$ independently of the choice of $t_1$.

One cannot obtain better bounds from the system (4.54)-(4.57) if the phases are well-ordered. However, for the badly – ordered phases the coupled bound is given by the hyperbola $Hyp[G_1, G_2, C_1]$ or $Hyp[G_1, G_2, C_2]$ (depending of which of them lies lower) as can be seen in Fig.4. One can check by the direct calculation [1] that the bound is given by the hyperbola $Hyp[G_1, G_2, C_2]$ in the $(y_K^{-1}(K_*), y_\mu^{-1}(\mu_*))$ plane.

Now we need to transform the bounds into the plane of the actual moduli, not their inverse Y-transforms. First we mention that the superposition of the inversions $1/y_K(K_*)$, and $1/y_\mu(\mu_*)$ and the inverse Y-transformations

$$K_* = f_1K_1 + f_2K_2 - \frac{f_1f_2(K_1 - K_2)^2}{f_1K_2 + f_2K_1 + y_K}, \tag{4.69}$$

$$\mu_* = f_1\mu_1 + f_2\mu_2 - \frac{f_1 f_2(\mu_1-\mu_2)^2}{f_1\mu_2 + f_2\mu_1 + y_\mu}, \tag{4.70}$$

are fractional – linear functions. Therefore, hyperbolas in the $(y_K^{-1}(K_*), y_\mu^{-1}(\mu_*))$ plane correspond to the hyperbolas in the $(K_*, \mu_*)$ plane. Any such hyperbola can be defined by three points that it passes through. Hence, in order to transform the results into the plane of the actual moduli, we need to study the correspondence between the characteristic points on the boundary hyperbolas. We note that

$$\begin{aligned} y_K(K_{i*}) &= \mu_i, \qquad y_K(K_i) = -K_i, \qquad y_\mu(\mu_i) = -\mu_i, \qquad i = 1,2; \\ y_K(0) &= \kappa_h; \qquad y_\mu(\mu_{i*}) = y_{i*}, \qquad i = 1,2,3,4, \end{aligned} \tag{4.71}$$

where the values $K_{1*}$, $K_{2*}$, $\kappa_h$, and $\mu_{i*}$ $i = 1..4$ are defined by the equations (2.25), (2.26), (2.22), (2.27), (2.28), (2.32), and (2.33). Therefore, the hyperbola $Hyp[G_1, G_2, C_2]$ in the $(y_K^{-1}(K_*), y_\mu^{-1}(\mu_*))$ plane corresponds to the hyperbola $Hyp[(K_{1*}, \mu_{1*}), (K_{2*}, \mu_{4*}), (K_2, \mu_2)]$ in the $(K_*, \mu_*)$ plane. Thus, we proved one of the bounds of Theorem 1 for the badly-ordered phases.

Calculations for the other bounds are similar and we omit them.

## 5 Discussion and generalization

In this section we summarize the results, address some questions that remain unresolved, discuss the optimality of the bounds of Theorem 1 and generalization to the three-dimensional problem.

### 5.1 *Summary of the results*

In this work we applied the translation method to prove the coupled bulk-shear moduli bounds of Theorem 1. Embedding the elasticity problem into a more general framework that involves fourth-order tensors of generalized strain and stress fields related via the eighth-order elasticity tensor, proved to be a very fruitful idea. This approach allowed us to choose appropriate translations and derive tight bounds on the elastic moduli of isotropic composites. The bounds coincide with the ones previously derived [1]. However, they are formulated in a much simpler and geometrically clear form.

The bounds of Theorem 1 are illustrated in Fig.1 for the well-ordered phases (2.37) and in Fig.2 for the badly-ordered phases (2.38).

The new bounds are tighter than the uncoupled Hashin-Shtrikman-Walpole bounds presented by the rectangles $AB'CD'$ and $A'BC'D$ in Fig. 1 and 2, respectively. The curves on these figures show all four hyperbolas mentioned in Theorem 1. The bounds are given by the solid curves. The dashed curves show the pair of the hyperbolas that are not the bounds for such a choice of the parameters but may become bounds if one changes the well-ordered phases into the badly-ordered ones, or vice versa. Note that one can reformulate Theorem 1 in the following way:

**Theorem 1a:** The pair $(K_*, \mu_*)$ of the effective bulk and shear moduli of any elastic isotropic composite lies in the set restricted by two vertical lines $K_* = K_{1*}$ and $K_* = K_{2*}$ and by the outermost pair of four hyperbolas

$$Hyp[(K_{1*}, \mu_{1*}), (K_{2*}, \mu_{4*}), (K_h, \mu_1)],$$

$$Hyp[(K_{2*}, \mu_{2*}), (K_{1*}, \mu_{3*}), (K_1, \mu_1)],$$

$$Hyp[(K_{1*}, \mu_{1*}), (K_{2*}, \mu_{4*}), (K_2, \mu_2)],$$

$$Hyp[(K_{2*}, \mu_{2*}), (K_{1*}, \mu_{3*}), (K_h, \mu_2)],$$

in the bulk-shear moduli plane $(K_*, \mu_*)$ .

It is interesting to compare the coupled bulk-shear moduli bounds of Theorem 1 and the bounds on the elastic moduli of plane isotropic polycrystals. Unlike the coupled bounds for the two-phase composite, any attempts to find coupled bounds on the elastic moduli of an isotropic polycrystal failed. The best bounds [20, 43] on the bulk and shear moduli of an isotropic polycrystal represent a rectangle in the $(K_*, \mu_*)$ plane, similar to the Hashin-Shtrikman-Walpole rectangles in Fig. 1, 2. Surprisingly, these bounds are optimal in the polycrystal case! For each point within the bounding rectangle, Avellaneda *et al.*[20, 43] found microstructure that corresponds to that point, although the laminate composites of infinite rank were needed to attain some of the points.

This means that one can change the bulk and shear moduli of the isotropic polycrystal independently, by changing the microstructure. It is true even if the polycrystal possesses the extremal bulk or shear modulus. On the contrary, one can change the bulk and shear moduli of an isotropic two-phase composite independently only when these moduli are far from being optimal. For the extremal composite with the bulk-shear moduli pair being on the boundary of the admissible set, these moduli are interrelated.

### *5.2 Optimality of the bounds*

Figures 1 and 2 reveal that the bounding hyperbolas pass through the points

$$A = (\kappa_{1*}, \mu_{1*}), \qquad C = (\kappa_{2*}, \mu_{2*}), \tag{5.1}$$

and the points

$$B = (\kappa_{1*}, \mu_{3*}), \qquad D = (\kappa_{2*}, \mu_{4*}) \tag{5.2}$$

in both cases of the well- and badly-ordered phases. The points $A$ and $C$ correspond to the Hashin-Shtrikman bounds for the well-ordered materials. They are attainable, i.e., there exist isotropic composites with the pair of the elastic moduli that correspond to these points [30]–[33]. The points $B$ and $D$ correspond to the Walpole bounds for the badly-ordered phases. On the contrary to the Hashin-Shtrikman points $A$ and $C$, the structures that correspond to the Walpole points $B$ and $D$ have not been found in spite of the long history of the problem. Moreover, it is unclear whether such microstructures exist, or the bounds of Theorem 1 can be further improved.

Note that in terms of the $Y$-transforms $y_K(K_*)$ and $y_\mu(\mu_*)$, the bounds of Theorem 1 are independent of the phase volume fractions. Namely, the following statement is equivalent to Theorem 1:

**Theorem 2:** The pair $(y_K(K_*), y_\mu(\mu_*))$ of the $Y$-transforms of the effective bulk and shear moduli of any elastic isotropic composite lies in the set of the $(y_K(K_*), y_\mu(\mu_*))$ plane restricted by two vertical lines

$$y_K(K_*) = \mu_1 \quad \text{and} \quad y_K(K_*) = \mu_2,$$

and the hyperbolas

$$Hyp[(\mu_1, y_{1*}), (\mu_2, y_{4*}), (0, -\mu_1)], \quad Hyp[(\mu_2, y_{2*}), (\mu_1, y_{3*}), (-K_1, -\mu_1)],$$

for the well-ordered phases (2.23), or by the hyperbolas

$$Hyp[(\mu_1, y_{1*}), (\mu_2, y_{4*}), (-K_2, -\mu_2)], \quad Hyp[(\mu_2, y_{2*}), (\mu_1, y_{3*}), (0, -\mu_2)],$$

for the badly-ordered phases (2.30). Recall that the points $y_{1*}$, $y_{2*}$, $y_{3*}$, $y_{4*}$ are given by

$$y_{1*} = \frac{K_1\mu_1}{K_1+2\mu_1}, \quad y_{2*} = \frac{K_2\mu_2}{K_2+2\mu_2}, \quad y_{3*} = \frac{K_1\mu_2}{K_1+2\mu_2}, \quad y_{4*} = \frac{K_2\mu_1}{K_2+2\mu_1}. \tag{5.3}$$

There are evidence that the bounds of Theorem 2 are optimal, whereas the bound of Theorem 1 are not attainable, at least by any high-rank laminate composite. This idea sounds controversial, since Theorems 1 and 2 are equivalent if the phase volume fractions are given. However, the bounds of Theorem 2 are independent of the phase volume fractions. Different parts of the set restricted by Theorem 2 in the $(y_K(K_*), y_\mu(\mu_*))$ plane may correspond to the composites with various phase volume fractions. Together, for all possible phase volume fractions, they may cover the whole set.

Specifically, we make a conjecture that for any point of the set described by Theorem 2 there exist composite that correspond to this point *if a composite is allowed to have any phase volume fractions.* The following facts support this hypothesis. First of all, the Hashin-Shtrikman points $(\mu_1, y_{1*})$ and $(\mu_2, y_{2*})$ in the $(y_K(K_*), y(\mu_*))$ plane are attainable, as it trivially follows from the attainability of their images (5.1) in the $(K_*, \mu_*)$ plane. Moreover, one can find the microstructures that correspond to the points

$$(\mu_1, y_\mu(\mu_{5*}))\,, \qquad (\mu_2, y_\mu(\mu_{6*}))\,, \tag{5.4}$$

on the vertical sides of the set in the $(y_K(K_*), y_\mu(\mu_*))$ plane. Here

$$\mu_{5*} = \left[\sqrt{(K_{1*}^{-1}+\mu_h^{-1})(K_{1*}^{-1}+\mu_{q1}^{-1})} - K_{1*}^{-1}\right]^{-1}, \tag{5.5}$$

$$\mu_{6*} = \left[\sqrt{(K_{2*}^{-1}+\mu_h^{-1})(K_{2*}^{-1}+\mu_{q2}^{-1})} - K_{2*}^{-1}\right]^{-1}, \tag{5.6}$$

and

$$\mu_{qi} = f_1\mu_1 + f_2\mu_2 - \frac{f_1f_2(\mu_1-\mu_2)^2}{f_1\mu_2+f_2\mu_1+K_i}, \qquad i=1,2. \tag{5.7}$$

These are the isotropic polycrystals made of the square symmetric matrix laminate composites [9, 10, 32] of the second rank with the phase 1 being a matrix and the phase 2 being inclusion (the first point (5.4)), or vice versa (the second point (5.4)). By mixing these composites and the composites that correspond to the Hashin-Shtrikman points (5.1) on can attain the segments

$$y_K(K_*) = \mu_1, \qquad y_\mu(\mu_*) \in [y_{1*}, y_\mu(\mu_{5*})] \tag{5.8}$$

$$y_K(K_*) = \mu_2, \qquad y_\mu(\mu_*) \in [y_{2*}, y_\mu(\mu_{6*})] \tag{5.9}$$

One can check that

$$\lim_{f_2=0} y_\mu(\mu_{5*}) = y_{3*}, \qquad \lim_{f_2=1} y_\mu(\mu_{5*}) = y_{1*},$$
$$\lim_{f_2=0} y_\mu(\mu_{6*}) = y_{2*}, \qquad \lim_{f_2=1} y_\mu(\mu_{6*}) = y_{4*}. \tag{5.10}$$

5.10 Therefore, the entire vertical segment of the boundary

$$y_K(K_*) = \mu_1, \qquad y_\mu(\mu_*) \in [y_{1*}, y_{3*}], \tag{5.11}$$

can be attained in the limiting case when the phase 2 volume fraction is close to zero. Similarly, the vertical segment

$$y_K(K_*) = \mu_2, \qquad y_\mu(\mu_*) \in [y_{2*}, y_{4*}] \tag{5.12}$$

is attainable when the phase 1 volume fraction vanishes.

Our conjecture is also supported by Erker and Milton [44]. We briefly review their results here because they are still unpublished. Specifically, Erker and Milton [44] wrote a computer program that generated (randomly) a laminate composite of a high rank, and calculated the effective moduli of this (anisotropic) structure. They then took this anisotropic material and constructed an isotropic polycrystal from it with extremal bulk and shear moduli. They used unpublished results of Avellaneda and Milton [43] [see also Avellaneda *et al.*[20]] which give a precise characterization of the possible bulk and shear moduli of two–dimensional elastic polycrystals constructed from a single crystal. In this way they generated a point in the $(K_*, \mu_*)$ plane that corresponded to some random isotropic microstructure (laminate of high rank). The program kept track of the phase volume fractions as well. Then the bulk and shear moduli of the resulting isotropic composite were used to plot the corresponding point in the $(y_K(K_*), y_\mu(\mu_*))$ plane. Millions of microstructures were examined and marked in the $Y$-plane together with the bounds of Theorem 2. The results revealed that the whole admissible set in the $Y$-plane was filled with the numerically found points. However, all the attempts to restrict the range of admissible phase volume fractions led to the appearance of the "white spots" near the Walpole points (5.2). This is another evidence that the bounds of Theorem 1 cannot be attained by the high-rank laminate composites in the $(K_*, \mu_*)$ plane, but may be optimal in the corresponding $Y$-plane if the phase volume fractions are not fixed.

It is important to note that all the known methods such as the translation, Hashin-Shtrikman, analytical, and Beran-type methods, give the bounds that are independent of the phase volume fractions in terms of the $Y$-transforms. Therefore, none of the existing methods in their present form can improve upon the bounds of Theorems 1 and 2, if our conjecture is correct. New methods may be needed to improve upon the bounds of Theorem 1. These methods should be capable of obtaining bounds that explicitly depend on the phase volume fractions even if written in the $Y$-terms.

Alternatively, new types of structures may be needed to close the gap between the bounds and the set in the $(K_*, \mu_*)$ plane attainable by the high-rank laminate composites. The optimal structures should be capable of achieving such unusual combinations of the effective moduli as the maximal bulk modulus and a small shear modulus, or the minimal bulk modulus and a large shear modulus.

### *5.3 Three-dimensional problem*

The ideas presented in this paper admit generalization to the three-dimensional problem. However, the realization is much more difficult. In three dimensions, there exist five linear independent shear fields instead of two for the planar elasticity. Therefore, the functional of the type (3.7) and (3.8) responsible for the shear modulus bounds should include five energies. The four functionals of the type (3.10)-(3.13) responsible for the coupled bounds, should include six energies (one for the bulk modulus and five for the shear modulus). Again, one has a choice of the stress or strain energy for the bulk modulus part, and five stress or five strain fields for the shear modulus part.

Unfortunately, the shears in three dimensions do not have any nice basis. Still, one can find the orthonormal basis $\boldsymbol{a}_1$-$\boldsymbol{a}_6$ with the tensor $\boldsymbol{a}_1$ being the identity tensor, and $\boldsymbol{a}_2$-$\boldsymbol{a}_6$ being mutually orthogonal trace-free tensors. Then the eighth-order elastic tensors

$$\mathcal{C} = \sum_{i=1}^{6} \boldsymbol{a}_i \boldsymbol{C} \boldsymbol{a}_i, \qquad \mathcal{S} = \mathcal{C}^{-1} = \sum_{i=1}^{6} \boldsymbol{a}_i \boldsymbol{S} \boldsymbol{a}_i, \tag{5.13}$$

relate the fourth order generalized strain and stress tensors

$$\boldsymbol{\mathcal{E}} = \sum_{i=1}^{6} \boldsymbol{a}_i \boldsymbol{\epsilon}^{(i)}, \quad \boldsymbol{\Sigma} = \sum_{i=1}^{6} \boldsymbol{a}_i \boldsymbol{\sigma}^{(i)}, \tag{5.14}$$

via the Hooke's law $\boldsymbol{\Sigma} = \mathcal{C} :: \boldsymbol{\mathcal{E}}$, $\boldsymbol{\mathcal{E}} = \mathcal{S} :: \boldsymbol{\Sigma}$. Similarly to the two-dimensional problem, one can study the rotation properties of the generalized stress tensors and the generalized strain tensors to find irreducible rotationally invariant subspaces of the $\boldsymbol{\Sigma}$-type (or $\boldsymbol{\mathcal{E}}$-type) tensors, and the corresponding set of rotationally invariant eighth-order tensors. It is a technically difficult work, but it can be done by using the ideas developed in quantum physics [13, 42].

The next problem is to find isotropic eighth-order tensors that correspond to the quasiaffine combinations of the generalized strains or stresses (5.14). However, there are no such quadratic forms! The systems similar to (4.47) have only trivial (zero) solutions in three-dimensions.

It is sufficient for the purpose of the translation method to have quasiconvex quadratic forms. However, it is much more difficult to find such combinations since the conditions (4.47) much be replaced by the condition that the matrix of second partial derivatives of the quadratic form (4.46) over the parameters $\lambda_i$ is positive definite. One such quasiconvex quadratic form of the generalized strain tensor was found by Milton [14]. He guessed the form of an isotropic eighth-order tensor that depends on several free parameters. Then he found the restrictions on these parameters that guarantee quasiconvexity of the corresponding quadratic form. This way he was able to derive the Hashin-Shtrikman-Walpole bulk and shear moduli bounds. However, the quasiconvex quadratic form that he found does not lead to the coupled bounds.

We tried to use the other approach that worked in the two–dimensional problem. Namely, we considered the strain energy as a quadratic form of the gradient of the displacement vectors, and studied the fourth-order tensor of the generalized gradient of the displacements.

We found that the corresponding eighth-order rotationally invariant tensors depend on at most twenty three free parameters. The restriction similar to the system (4.47) has a non-trivial solution that depends on eight free parameters. Thus, we found an eight-parameter family of the translation matrices that correspond to quasiaffine quadratic forms. However, all the attempts to obtain the coupled bulk-shear moduli bounds, or even prove the Hashin-Shtrikman shear modulus bounds by using these new quasiaffine quadratic forms of the generalized gradient of the displacements, failed. This shows that the three-dimensional problem is much more difficult to deal with than the simpler two-dimensional version. However, the success of Milton [14] who found the quasiconvex quadratic form and used it to prove the Hashin-Shtrikman-Walpole bounds, evidence that this is the right way to attack the problem of the coupled bounds.

## Acknowledgments

The author thanks Andrei Cherkaev and Graeme Milton for helpful discussions, and gratefully acknowledges the support by AFOSR grant F49620-96-1-0182.

## References

1. A.V. Cherkaev and L.V. Gibiansky, *J.Mech.Phys.Solids* **41**, 937 (1993).
2. K.A. Lurie and A.V. Cherkaev, *J. Opt. Th. Appl.* **42**, 305 (1984).
3. K.A. Lurie and A.V. Cherkaev, *Proc.R.Soc.Edinburgh A* **99**, 71 (1984).
4. K.A. Lurie and A.V. Cherkaev, *Proc.R.Soc.Edinburgh A* **104**, 21 (1986).
5. K.A. Lurie and A.V. Cherkaev, *Advances' in Mechanics (Poland)*, **9**, 3 (1986), (translation in *Topics in the mathematical modeling of composite materials*, A.V. Cherkaev, R.V. Kohn Eds., Birkhauser, 175, 1997).
6. F. Murat and L. Tartar, in *Les Methodes de l'Homogeneisation: Theorie et Applications en Physique, Coll. de la Dir. des Etudes et Recherches de Electr. de France, Eyrolles, Paris*, 319 (1985), (translation in *Topics in the mathematical modeling of compos-*

*ite materials*, A.V. Cherkaev, R.V. Kohn Eds., Birkhauser, 139, 1997).

7. L. Tartar, in *Computing methods in applied sciences and engineering*, (Lecture Notes in Math. 704, Springer - Verlag), 364 (1978), (translation in *Topics in the mathematical modeling of composite materials*, A.V. Cherkaev, R.V. Kohn Eds., Birkhauser, 9, 1997).
8. L. Tartar, *Ennio de Giorgi Colloquium, P.Kree,ed., Pitman Research Notes in Math. 125*, 168 (1985).
9. L.V. Gibiansky and A.V. Cherkaev, Preprint 914. Phys.-Tech. Inst. Russian Acad. Sci., St. Petersburg (1984), (translation in *Topics in the mathematical modeling of composite materials*, A.V. Cherkaev, R.V. Kohn Eds., Birkhauser, 95, 1997).
10. L.V. Gibiansky and A.V. Cherkaev, Preprint 1115. Phys.-Tech. Inst. Russian Acad. Sci., St. Petersburg (1987), (translation in *Topics in the mathematical modeling of composite materials*, A.V. Cherkaev, R.V. Kohn Eds., Birkhauser, 273 1997).
11. R.V. Kohn and G.W. Milton, in *Homogenization and Effective Moduli of Materials and Media, J. Ericksen et al. eds., Springer-Verlag*, 97 (1986).
12. G.W. Milton and R.V. Kohn, *J. Mech. Phys. Solids*, **36**, 597 (1988).
13. G.W. Milton, *Comm. of Pure and Appl. Math.* **XLIII**, 63 (1990).
14. G.W. Milton, *Continuum Models and Discrete Systems, ed. by G. A. Maugin, vol. 1*, 60 (1990).
15. G.W. Milton, in *Composite Media and Homogenization Theory*, ed. by G. Dal Maso and G. Dell-Antonio, vol. 5, *Progress in Non-linear Differential Equations and their Applications*, series ed. H. Brezis, Birkhauser, Boston, 223 (1991).
16. A.V. Cherkaev and L.V. Gibiansky, *Proc.R.Soc.Edinburgh A* **122**, 93 (1992).
17. L.V. Gibiansky and G.W. Milton, *Proc.R.Soc.London A* **440**, 163 (1993).
18. L.V. Gibiansky and S. Torquato, *Phyl. Trans. R. Soc. London* A, **353**, 243 (1995).
19. L.V. Gibiansky and S. Torquato, *Proc.R.Soc.London A* **452**, 253 (1996).
20. M. Avellaneda, A.V. Cherkaev, L.V. Gibiansky, G.W. Milton, and M. Rudelson, *J. Mech. Phys. Solids* **44**(7), 1179 (1996).

21. R.J. Atkin and N. Fox, *An introduction to the theory of elasticity*, London: Longman (1990).
22. R.M. Christensen, *Mechanics of Composite Materials* (Wiley-Interscience, New-York, 1979).
23. M. Beran, *Nuovo Cimento* **38**, 771 (1965).
24. E. Sanchez - Palencia, *Nonhomogeneous Media and Vibration Theory* (Lecture Notes in Physics 127, Springer - Verlag, 1980).
25. J. Willis, in *Mechanics of Solids, H.G. Hopkins and M.J. Sewell eds, Pergamon Press, Oxford* 653 (1982).
26. R. Hill, *Proc. Phys. Soc. London A*, **65**, 349 (1952).
27. R. Hill, *J. Mech. Phys. Solids* **11**, 357 (1963).
28. Z. Hashin and S. Shtrikman, *J. Mech. Phys. Solids*, **11**, 127 (1963).
29. Z. Hashin, *J. Mech. Phys. Solid*, **13**, 119 (1965).
30. K.A. Lurie and A.V. Cherkaev, *Preprint 895, Ioffe Physico-Technical Institute, Leningrad 1984*).
31. A.N. Norris, *Mech. Mat.* **4**, 1 (1985).
32. G. Francfort and F. Murat, *Arch. Rational Mech. Anal.*, **94**, 307 (1986).
33. G.W. Milton, in *Homogenization and Effective Moduli of Materials and Media*, Ericksen, J.L., Kinderlehrer, D., Kohn, R., and Lions, J.-L. eds., Springer-Verlag, New-York, 150 (1986).
34. L. Walpole, *J. Mech. Phys. Solids* **14**, 151 and 289 (1966).
35. G.W. Milton, N. Phan-Thien, *Proc.R.Soc.Lond.*, A **380**, 305 (1982).
36. J.G. Berryman and G.W. Milton, *J. Phys. D: Appl. Phys.* **21**, 87 (1988).
37. L. Kublanov and G.W. Milton, unpublished manuscript (1991).
38. G.W. Milton, *Comm. Pure Appl. Math.* **XLVII**, 959 (1994).
39. G.A. Francfort, G.W. Milton, *Comm. Pure Appl. Math* **17**, 257 (1994).
40. B. Daçorogna, *Weak continuity and weak lower semicontinuity of non-linear functionals* (Lecture Notes in Math. 922, Springer-Verlag 1982).
41. M. Avellaneda, A.V. Cherkaev, K.A. Lurie, and G.W. Milton, *J. Appl. Phys.*, **63**, 4989 (1988).
42. A. Messiah, *Quantum mechanics.* Vol. 2, (North-Holland, Amsterdam, 536-581, 1964).

43. M. Avellaneda, and G.W. Milton, in preparation (1997).
44. J. Erker and G.W. Milton, unpublished work (1992).

# HOMOGENIZATION OF THE LAPLACE EQUATION IN A PARTIALLY PERFORATED DOMAIN

W. JÄGER
*Institut für Angewandte Mathematik, Universität Heidelberg,*
*Im Neuenheimer Feld 294, W-6900 Heidelberg, Germany*

A. MIKELIĆ
*U.M.R. CNRS 5585, Laboratoire d'Analyse Numérique, Bât. 101,*
*Université Lyon 1, 43, bd. du Onze Novembre, 69622*
*Villeurbanne Cedex, France*

We consider the Laplace equation in a bounded domain consisting of a porous medium, a nonperforated domain and an interface between them. The homogeneous Dirichlet condition is prescribed on the boundaries of the perforations. We suppose the porous medium having a periodic structure with a period $\varepsilon$ and consider the limit $\varepsilon \to 0$. The effective behaviors inside the porous medium and the nonperforated domain are obtained and in addition the boundary conditions to be imposed at the interface between two media are determined. Using the boundary layers describing interaction between two very different media we estimate the $L^2$-norm of the difference between the solution $u^\varepsilon$ of the $\varepsilon$-problem and the effective behavior terms and show that it is bounded from above by $C\varepsilon^{\frac{1}{2}}$. Finally, we calculate the weak $L^2$-limit of the renormalized function $u^\varepsilon/\varepsilon^2$ and weak $M(\Omega)$-limit of the renormalized $\nabla(u^\varepsilon/\varepsilon^2)$.

## Dedication

My [1] first memories about Serguei do not concern the person, they are connected with his articles. Before meeting him for the first time I had a knowledge of his fundamental results in stochastic homogenization. We met for the first time in Trieste, Italy, during the "Workshop on Composite Media and Homogenization" in January 1990. This conference went on for 3 weeks and we spent a lot of time together. I mentioned him getting the Darcy law from the Stokes system in a random porous medium. He found the problem amusing and we thought of working on it. For the second time we met in the Spring of 1991 and afterwards several times in Spring and Summer 1992. Together with Alain Bourgeat, we worked on homogenization of the equations

[1]Written by Andro Mikelić

describing two-phase flows in a stochastic porous medium. It was a successful marriage of Serguei's stochastic homogenization results with our PDE's techniques. The results are in the joint paper [2] which appeared in 1995. Obtaining the Darcy law for flows through random porous media led to the beautiful paper of Beliaev and Kozlov in 1996. Even if he collaborated mostly with his colleagues from Moscow we kept quite intense contacts. For the last time I saw Serguei in June 1994, when he organized a small workshop on asymptotic methods in Marseille.

Let me also mention that he visited Heidelberg a number of times and had close contacts with the local stochastic homogenization group (H. Rost, K. Oelschläger). W. Jäger knew him very well.

His death under tragic circumstances is a great loss to the scientific community, especially to us working on homogenization.

The mathematical paper which follows corresponds to obtaining the effective behavior at the interface between a perforated and a homogeneous medium for the solutions of the Poisson equation under the Dirichlet condition at the boundaries of perforations. This was an important step in getting the effective laws on a rigorous way at the interface between a porous medium and a free fluid flow.[3] This material corresponds to the contents of the preprint,[4] though it has never been published, it has been quoted a number of times by numerous authors.

We dedicate this paper to Serguei as a homage from Heidelberg and Lyon.

## 1 Statement of the problem and of the results

### *1.1 Introduction*

The behavior of the solutions of the boundary value problems for elliptic equations in perforated domains, as the characteristic size $\varepsilon$ of

[2]A. Bourgeat, S.M. Kozlov, A. Mikelić, Effective Equations of Two-Phase Flow in Random Media, *Calculus of Variations and Partial Differential Equations*, **3**, 385–406 (1995).

[3]W. Jäger, A. Mikelić, On the Boundary Conditions at the Contact Interface between a Porous Medium and a Free Fluid, Annali della Scuola Normale Superiore di Pisa, Classe di Scienze, 1996.

[4]W. Jäger, A. Mikelić: Homogenization of the Laplace Equation in a Partially Perforated Domain, preprint no. 157, Equipe d'Analyse Numérique Lyon-St Etienne, Septembre 1993.

perforations tends to zero, has been studied in many papers and we refer to [1, 4, 5, 6] and to the references therein. However, the most of the literature ignores the study of the boundary effects that occur near the boundary. The existing results concern the behavior in the neighborhood of the closed flat boundary, pioneering results and methods can be found in [4], pp. 50-80. The detailed exposition of the results obtained during the last decade can be found in [5].

On the contrary, rigorous results concerning the effective behavior of solutions of the partial differential equations on the contact surfaces between the perforated domain and the nonperforated one (or another perforated domain but with different geometric structure) do not exist, to our knowledge. There are some results of that type for the differential operators with different oscillating coefficients in two domains separated by a plane (see [1], Ch. 9), but that situation is considerably simpler.

The main difficulty comes from the appearance of the boundary layers in the neighborhoods of the contact surfaces.

The principal objective of this work is the systematic study of the effective behavior of a solution $u^\varepsilon$ of the problem

$$-\Delta u^\varepsilon = F^\varepsilon \quad \text{in } \Omega^\varepsilon; \tag{1.1}$$

$$u^\varepsilon = 0 \quad \text{on } \partial\Omega^\varepsilon; \tag{1.2}$$

as $\varepsilon \to 0$. Here

$$F^\varepsilon = \begin{cases} \varepsilon^\gamma f & \text{in } \Omega_1 \text{ (the nonperforated domain);} \\ f & \text{in } \Omega_1 \text{ (the porous medium).} \end{cases} \tag{1.3}$$

The problem (1.1)–(1.3) with $\gamma = 0$ arises in study of the effective behavior of a porous, but conducting, medium in an electric field. In that situation we search the electric potential satisfying the Laplace equation and the constant Dirichlet boundary condition on the boundaries of the conductor. Another related model is a heating or cooling of a porous medium with a fixed constant temperature on the surfaces of the pores. Similarly, we find physical models corresponding to $\gamma = 2$. Since the most interesting values of $\gamma$ are $\gamma = 0$ and $\gamma = 2$, we restrict ourselves to the case $\gamma \geq 0$, but, in fact, our convergence result is valid for $\gamma > -1/2$ with somewhat weaker order of convergence $O\left(\varepsilon^{\gamma+1/2}\right)$.

Our main results are presented in the Section 1.3 for rectangular domain $\Omega$. The auxiliary and homogenized problems are introduced and studied and the main convergence theorems are stated. The sections 2.1–2.3 are devoted to the proofs of convergence theorems.

The final section is an appendix collecting various results concerning the auxiliary problems, being necessary for the convergence proofs.

It should be noted that study of the problem (1.1)–(1.3) is one step towards homogenization of the Stokes system in a partially perforated domain. However, it is more complicated since the Stokes operator contains the pressure gradient.

The results on the Laplace equation in a partially perforated domain with the Neumann type conditions instead of (1.2) can be found in [2].

### *1.2 Notations, assumptions and auxiliary results*

Consider a bounded domain $\Omega = ]0,L]^{n-1} \times ]-L,L] \subset R^n$, $L > 0$, consisting of a subdomain $\Omega_1 = ]0,L]^{n-1} \times ]0,L]$ and a porous medium $\Omega_2 = ]0,L]^{n-1} \times ]-L,0]$. We assume that the structure of the porous medium is periodic and generated by translations of a cell $Z^\varepsilon = \varepsilon Z$, where Z is a standard cell, $Z = ]0,1[^n$. $Z$ contains an open set $Z^* \subset Z$, $\partial Z^* \in C^\infty$, being a union of the finite number of components homeomorphic to a unit ball.

Let $Y^* = Z \setminus \bar{Z}^*$ and let $\chi$ be the characteristic function of $Y^*$. We set $\chi^\varepsilon(x) = \chi(x/\varepsilon)$, $x \in R^n$, and define $\Omega_2^\varepsilon = \{x | \, x \in \Omega_2, \chi^\varepsilon(x) = 1\}$ (the fluid part of the porous medium). Furthermore, $\Sigma = ]0,L[^{n-1} \times \{0\}$ is the surface separating porous and nonperforated medium and $\Omega^\varepsilon = \Omega_1 \cup \Sigma \cup \Omega_2^\varepsilon$.

We study the problem (1.1)–(1.3) with $f \in C_0^\infty(\Omega)$ and $f$ not being identically zero on $\Sigma$. Obviously, it has the unique solution $u^\varepsilon \in H_0^1(\Omega^\varepsilon) \cap W^{2,\infty}(\Omega^\varepsilon), \forall \varepsilon > 0$.

We study the limit $\varepsilon \to 0$. In order to formulate the results corresponding to the different values of $\gamma$ it is necessary to introduce the auxiliary problems connected with the periodic structure.

The first auxiliary problem corresponds to the study of $u^\varepsilon$ in the interior of $\Omega_2^\varepsilon$:

$$\begin{cases} -\Delta_y w = 1 \text{ in } Y^*; \\ \quad w = 0 \text{ on } \partial Z^*, \quad w \text{ is } Z-\text{periodic.} \end{cases} \tag{1.4}$$

The unique solubility of the problem (1.4) is well-known (see e.g. [6]). Furthermore, $C^\infty$-regularity of $\partial Z^*$ implies $C^\infty_{loc}(R^n)$-regularity of the solution $w$.

We define $w^\varepsilon(x) = w\,(x/\varepsilon)\,, x \in \Omega^\varepsilon$ and extend $w^\varepsilon$ by zero to $\Omega \setminus \Omega^\varepsilon$. Then we have

$$\|w^\varepsilon\|_{L^q(\Omega)} = C, \quad \forall q \in [1,+\infty]; \tag{1.5}$$

$$\|\nabla w^\varepsilon\|_{L^q(\Omega)^n} = \frac{C}{\varepsilon} \quad \forall q \in [1,+\infty]; \tag{1.6}$$

$$\|w^\varepsilon(\cdot,0)\|_{L^q(\Sigma)} = \frac{C}{\varepsilon} \quad \forall q \in [1,+\infty]; \tag{1.7}$$

$$\|\frac{\partial}{\partial x_n} w^\varepsilon(\cdot,0)\|_{L^q(\Sigma)} = C \quad \forall q \in [1,+\infty]; \tag{1.8}$$

Furthermore,

$$w^\epsilon \rightharpoonup \int_{Y^*} w(y)dy = m \tag{1.9}$$

weakly in $L^q(\Omega)$, $\forall q \in [1,\infty[$, and weakly* in $L^\infty(\Omega)$.

The next auxiliary problem corresponds to the boundary layer around $\Sigma$.

Let $Z^- = \bigcup_{k=1}^\infty \{Y^* - k\vec{e}_n\}$, $\Pi = \bigcup_{k=1}^\infty \{\partial Z^* - k\bar{e}_n\}$, $S = ]0,1[^{n-1} \times \{0\}$, $Z^+ = ]0,1[^{n-1} \times ]0,+\infty[$, $Z_{BL} = Z^- \cup S \cup Z^+$ and $\tilde{y} = (y_1,\dots,y_{n-1})$. We consider the problem

$$-\Delta_y \omega = 0 \qquad \text{in } Z^+ \cup Z^-; \tag{1.10}$$

$$\omega(\tilde{y},+0) - \omega(\tilde{y},-0) = \omega(\tilde{y},0) \qquad \text{on } S; \tag{1.11}$$

$$\frac{\partial}{\partial y_n}\omega(\tilde{y},+0) - \frac{\partial}{\partial y_n}\omega(\tilde{y},-0) = \frac{\partial}{\partial y_n}\omega(\tilde{y},0) \text{ on } S; \tag{1.12}$$

$$\omega = 0 \text{ on } \bigcup_{k=1}^\infty \{\partial Z^* - k\vec{e}_n\}, \qquad \omega \text{ is } \tilde{y}\text{- periodic.} \tag{1.13}$$

In the Corollary A1 of the Appendix A the existence and uniqueness of the solution $\omega \in C^\infty_{loc}(Z^+ \cup Z^-)$ of (1.10)–(1.13) is established. Furthermore, it is proved that there exist constants $\gamma_0 \in ]0,1[$ and $C_0$ such that

$$e^{\gamma_0|y_n|}\nabla_y\omega \in L^2(Z^+ \cup Z^-)^n, \quad e^{\gamma_0|y_n|}\omega \in L^2(Z^-)^n,$$

and

$$\|\omega(\cdot,y_n) - C - 0\|_{L^2(]0,1[^{n-1})} \le Ce^{-\gamma_0|y_n|}, \quad y_n > 0. \tag{1.14}$$

The constant $C_0$ is given by

$$C_0 = \lim_{a\to\infty} \int\limits_{[0,1]^{n-1}} \omega(\tilde{y}, a)d\tilde{y} = \int\limits_{[0,1]^{n-1}} \omega(\tilde{y}, +0)d\tilde{y}.$$

We define $\omega^\varepsilon = \omega(x/\varepsilon), x \in \Omega^\varepsilon$ and extend $\omega^\varepsilon$ by zero to $\Omega \setminus \Omega^\varepsilon$. Then we have

$$\|\omega^\varepsilon - C_0 H(x_n)\|_{L^q(\Omega)} = C\varepsilon^{1/q}, \qquad \forall q \in [1, +\infty], \tag{1.15}$$

$$\|\nabla\omega^\varepsilon\|_{L^q(\Omega_1\cup\Omega_2)^n} = C\varepsilon^{1/q-1}, \quad \forall q \in [1, +\infty], \tag{1.16}$$

where $H$ is the Heaviside function.

The second auxiliary problem connected with $\Sigma$ is

$$-\Delta_y \beta = 0 \quad \text{in } Z^+ \cup Z^-; \tag{1.17}$$

$$\beta(\tilde{y}, +0) - \beta(\tilde{y}, -0) = 0 \quad \text{on } S; \tag{1.18}$$

$$\frac{\partial}{\partial y_n}\beta(\tilde{y}, +0) - \frac{\partial}{\partial y_n}\beta(\tilde{y}, -0) = -1 \text{ on } S; \tag{1.19}$$

$$\beta = 0 \text{ on } \cup_{k=1}^{\infty} \{\partial Z^* - k\vec{e}_n\}, \qquad \beta \text{ is } \tilde{y}\text{–periodic.} \tag{1.20}$$

In the Corollary A2 of the Appendix A the existence and uniqueness of the $\beta \in C^\infty_{loc}(Z^+\cup Z^-)\cap H^1_{loc}(Z_{BL})$ being the solution of (1.17)–(1.20) is established. Furthermore, there are constants $\gamma_0 \in ]0, 1[$ and

$$C_\beta = \lim_{a\to\infty} \int\limits_{[0,1]^{n-1}} \beta(\tilde{y}, a)d\tilde{y} = \int\limits_{[0,1]^{n-1}} \beta(\tilde{y}, +0)d\tilde{y}$$

such that

$$e^{\gamma_0|y_n|}\nabla_y\beta \in L^2(Z_{BL})^n, \quad e^{\gamma_0|y_n|}\beta \in L^2(Z^-)$$

and

$$\|\beta(\cdot, y_n) - C_\beta\|_{L^2(]0,1[^{n-1})} \leq Ce^{\gamma_0|y_n|}, \quad y_n > 0. \tag{1.21}$$

Now we introduce $\beta^\varepsilon(x) = \varepsilon\beta(x/\varepsilon),\ x \in \Omega^\varepsilon$, and extend $\beta^\varepsilon$ by zero to $\Omega \setminus \Omega^\varepsilon$. Then $\beta^\varepsilon$ satisfies

$$\frac{\partial}{\partial y_n}\beta(\tilde{x}, +0) - \frac{\partial}{\partial y_n}\beta(\tilde{x}, -0) = -1 \text{ on } \Sigma \tag{1.22}$$

and we have

$$\|\beta^\varepsilon - \varepsilon C_\beta H(x_n)\|_{L^q(\Omega)} = C\varepsilon^{1+1/q}, \quad \forall q \in [1, +\infty]; \tag{1.23}$$

$$\|\nabla \beta^\varepsilon\|_{L^q(\Omega^\varepsilon)^n} = C\varepsilon^{1/q}, \quad \forall q \in [1, +\infty]. \tag{1.24}$$

The final auxiliary problem connected with $\Sigma$ is

$$\Delta_y \zeta_i = \frac{\partial}{\partial y_n}(\beta - C_\beta H(y_n)) \text{ in } Z_{BL}, \tag{1.25}$$

$$\zeta_i = 0 \text{ on } \cup_{k=1}^\infty \{\partial Z^* - k\vec{e}_n\}, \quad \zeta_i \text{ is } y\text{–periodic}, \tag{1.26}$$

where $i = 1, \dots, n$ and $\beta$ is the solution of (1.17)–(1.20).

The Corollary A3 of the Appendix A gives the existence and uniqueness of $\zeta_i \in C^\infty_{loc}(Z^+ \cup Z^-) \cap H^1_{loc}(Z_{BL})$ being the solution of (1.25)–(1.26). Furthermore, there are constants $\gamma_0 \in ]0, 1[$ and

$$C_{i,\zeta} = \lim_{a\to\infty} \int_{[0,1]^{n-1}} \zeta_i(\tilde{y}, a) d\tilde{y}, \quad 1 \le i \le n;$$

$$C_{i,\zeta} = \int_{[0,1]^{n-1}} \zeta_i(\tilde{y}, 0) d\tilde{y}, \quad 1 \le i \le n-1,$$

such that

$$e^{\gamma_0|y_n|}\nabla_y \zeta_i \in L^2(Z_{BL})^n, \quad e^{\gamma_0|y_n|}\zeta_i \in L^2(Z^-),$$

and

$$\|\zeta_i(\cdot, y_n) - C_{i,\zeta}\|_{L^2(]0,1[^{n-1})} \le Ce^{-\gamma_0|y_n|}, \quad y_n > 0. \tag{1.27}$$

Now we introduce $\zeta_i^\varepsilon(x) = \varepsilon^2 \beta(x/\varepsilon)$, $x \in \Omega^\varepsilon$ and extend $\zeta_i^\varepsilon$ by zero to $\Omega \setminus \Omega^\varepsilon$. Then $\zeta_i^\varepsilon$ satisfies

$$\|\zeta_i^\varepsilon - \varepsilon^2 C_{i,\zeta} H(x_n)\|_{L^q(\Omega)} = C\varepsilon^{2+1/q}, \quad \forall q \in [1, +\infty]; \tag{1.28}$$

$$\|\nabla \zeta_i^\varepsilon\|_{L^q(\Omega^\varepsilon)^n} = C\varepsilon^{1+1/q}, \quad \forall q \in [1, +\infty]. \tag{1.29}$$

Now we consider the "counterflow-type" problems in $\Omega_1$ (and in $\Omega$) caused by approaching of $\omega$, $\beta$, and $\zeta_i$ to nonzero constants.

The first auxiliary problem in $\Omega_1$ corrects the force:

$$\begin{cases} -\Delta v = f & \text{in } \Omega_1; \\ v = 0 & \text{on } \partial\Omega_1. \end{cases} \tag{1.30}$$

Obviously, there exists the unique solution $v \in C^{\infty}_{loc}(\Omega_1) \cap W^{2,\infty}(\Omega_1) \cap H^1_0(\Omega_1)$ of (1.30).

However, we need more regularity at the corners $(0,0)$ and $(L,0)$. We get it by extending $v$ in an odd way to $]-L,0[^{n-1} \times ]0,L[$. Due to the homogeneous Dirichlet boundary condition we easily prove that it is an $H^2$–extension and we get $-\Delta\tilde{v} = \tilde{f}$ in $]-L,L[^{n-1} \times ]0,L[$, where $\tilde{v}$ and $\tilde{f}$ are the odd extensions of $v$ and $f$, respectively. Since $f \in C^{\infty}_0(\Omega_1)$, we easily establish $C^{\infty}$–regularity of $\tilde{v}$ at the corner $(0,0)$. The $C^{\infty}$–regularity at the other corners is established analogously.

In particular, we have

$$V(\tilde{x}) = \frac{\partial}{\partial x_n} v(\tilde{x},+0) \in H^k_0(\Sigma), \quad \forall k \in n. \tag{1.31}$$

The second auxiliary problem in $\Omega_1$ corrects the effects of the stabilization of $\omega$ to $C_0$ in $\Omega_1$:

$$\begin{cases} -\Delta u = 0 \quad \text{in } \Omega_1; \\ u = f(\tilde{x},0) \text{ on } \Sigma; \quad u = 0 \text{ on } \partial\Omega_1 \setminus \bar{\Sigma}. \end{cases} \tag{1.32}$$

Taking into account that $f \in C^{\infty}_0(\Omega)$ implies $u \in C^{\infty}_0(\partial\Omega_1)$ we easily get the existence of the unique solution $u \in C^{\infty}(\bar{\Omega}_1) \cap H^1_0(\Omega_1)$ of (1.32).

The third auxiliary problem in $\Omega_1$ is connected with values of derivatives of $v$ on $\Sigma$:

$$\begin{cases} -\Delta\bar{V} = 0 \quad \text{in } \Omega_1; \\ \bar{V} = V(\tilde{x}) \quad \text{on } \Sigma, \quad \bar{V} = 0 \quad \text{on } \partial\Omega_1 \setminus \bar{\Sigma}. \end{cases} \tag{1.33}$$

(1.31) and the regularity of $\partial v(\tilde{x},+0)/\partial x_n$ imply the existence of the unique solution $\bar{V} \in W^{1,q}(\Omega_1), \forall q \in [1,+\infty[$ of (1.33). In order to get more regularity we repeat the above extension argument and establish $W^{k,q}(\Omega_1)$–regularity of $\bar{V}$, $\forall k \in N$ and $\forall q \in [1,+\infty[$.

Our final class of the problems involves the partial derivatives of $\bar{V}$ on $\Sigma$.

By the above established regularity results we have

$$\nabla_{\tilde{x}} V(\tilde{x}) \in H^k_0(\Sigma)^{n-1}, \quad \forall k \in N. \tag{1.34}$$

Hence the problem

$$\begin{cases} -\Delta U_i = 0 & \text{in } \Omega_1, \\ U_i = \dfrac{\partial}{\partial x_n}\bar{V}(\tilde{x}) & \text{on } \Sigma, \\ U_i = 0 & \text{on } \partial\Omega_1 \setminus \bar{\Sigma}, \quad 1 \le i \le n-1. \end{cases} \tag{1.35}$$

has a unique solution $U_i \in W^{k,q}(\Omega_1)$, $\forall k \in N$, $\forall q \in [1, +\infty[$.

Finally we consider

$$\begin{cases} -\Delta \bar{U} = 0 & \text{in } \Omega_1; \\ \bar{U} = \dfrac{\partial}{\partial x_n}\bar{V}(\tilde{x}, +0) \equiv \tilde{V}(\tilde{x}) \text{ on } \Sigma, & U = 0 \text{ on } \partial\Omega_1 \setminus \bar{\Sigma}. \end{cases} \tag{1.36}$$

As above, we get $\bar{U} \in W^{k,q}(\Omega_1), \forall k \in N, \forall q \in [1, +\infty[$.

Now we state two simple results to be used as basic tools in our convergence proofs.

**Proposition 1.1** *Let $\varphi \in H^1(\Omega_2^\varepsilon)$ such that $\varphi = 0$ on $\partial\Omega_2^\varepsilon \setminus \Omega_2$. Then we have*

$$\|\varphi\|_{L^2(\Sigma)} \leq C\varepsilon^{1/2}\|\nabla_x \varphi\|_{L^2(\Omega_2^\varepsilon)^n}; \tag{1.37}$$

$$\|\varphi\|_{L^2(\Omega_2^\varepsilon)} \leq C\varepsilon\|\nabla_x \varphi\|_{L^2(\Omega_2^\varepsilon)^n}. \tag{1.38}$$

**Proof.**

Let $\tilde{\varphi}(y) = \varphi(\varepsilon y)$, $y \in Y_k^*$. Then $\tilde{\varphi} \in H^1(Y_k^*)\, \forall k$ and $\varphi = 0$ on $S_k$. Therefore by the trace theorem and the Poincaré inequality

$$\int_S |\tilde{\varphi}(\tilde{y}, 0)|^2 d\tilde{y} \leq C \int_{Y^*} |\nabla_y \tilde{\varphi}|^2 dy.$$

Change of variables gives:

$$\int_{\varepsilon S_k} |\tilde{\varphi}(\tilde{x}, 0)|^2 \varepsilon^{1-n} d\tilde{x} \leq C \int_{\varepsilon Y^*} |\varepsilon^2 \nabla_y \tilde{\varphi}|^2 \varepsilon^{-n} dx.$$

Now after summation over $k$ we get:

$$(\int_\Sigma |\tilde{\varphi}(\tilde{x}, 0)|^2 \varepsilon^{1-n} d\tilde{x})^{1/2} \leq C\varepsilon^{1/2}(\int_{\Omega_2^\varepsilon} |\nabla_x \tilde{\varphi}(x)|^2 dx)^{1/2}. \tag{1.39}$$

and (1.37) is proved.

Estimate (1.38) is well-known (see e.g. [6]).

The next result concerns solubility of the Laplace equation with nonsmooth data:

**Proposition 1.2** *Let $G \in H^{-1}(\Omega_1)$ and $\xi \in L^2(\Sigma)$, $\xi = 0$ on $\partial\Omega_1/\Sigma$. Then the problem*

$$\begin{cases} -\Delta B = G & \text{in } \Omega_1; \\ B = \xi & \text{on } \partial\Omega_1 \end{cases} \tag{1.40}$$

*has a unique solution* $B \in L^2(\Omega_1)$ *and*

$$\|B\|_{L^2(\Omega_1)} \leq C\{\|G\|_{H^{-1}(\Omega_1)} + \|\zeta\|_{L^2(\Sigma)}\}. \tag{1.41}$$

**Proof.**

It is a proof by transposition which is a simple generalization of Theorem 6.2 page 182 from [3] to convex polygonal domains.

### *1.3 Statements of the main results*

The purpose of this section is to state our results on asymptotic behavior of $u^\varepsilon$ as $\varepsilon \to 0$ for $\gamma \geq 0$.

In order to investigate behavior of solution of the problems (1.1)–(1.2), as $\varepsilon \to 0$, we need to extend $u^\varepsilon$ to the whole $\Omega$. Denote by the symbol $\tilde{}$ the extension of a function to $\Omega/\Omega^\varepsilon$ by zero. It is well-known that extension by zero preserves the $L^2$-norm and for $z^\varepsilon \in H_0^1(\Omega^\varepsilon)$ we have $\tilde{z}^\varepsilon \in H_0^1(\Omega^\varepsilon)$ with the equal $H_0^1$-norms.

We meet some difficulties in choosing the right scaling for $u^\varepsilon$. Our choice is to consider $u^\varepsilon/\varepsilon^2$ and the motivation comes from the homogenization in porous medium with the homogeneous Dirichlet boundary conditions, since in simpler situations boundary effects enter as boundary layers, being exponentially small in the interior of $\Omega_2^\varepsilon$.

It should be noted that we can derive the a priori estimates for $u^\varepsilon$. They are given in the following proposition.

**Proposition 1.3.** *Let* $u^\varepsilon \in H_0^1(\Omega^\varepsilon)$ *be a solution of (1.1)–(1.2). Then we have*

$$\|\nabla u^\varepsilon\|_{L^2(\Omega)^n} \leq Ce^{\min\{1,\gamma\}}; \tag{1.42}$$

$$\|u^\varepsilon\|_{L^2(\Omega_2^\varepsilon)} \leq Ce^{1+\min\{1,\gamma\}}; \tag{1.43}$$

$$\|u^\varepsilon\|_{L^2(\Omega_1)} \leq Ce^{\min\{3/2,\gamma\}}. \tag{1.44}$$

We skip the proof of Proposition 1.3 (it is simple) because we are not going to use the estimates (1.42)–(1.44).

We expect our results to depend strongly on $\gamma$. For sufficiently large $\gamma$ it is reasonable to expect $u^\varepsilon/\varepsilon^2$ to be dominated or at least balanced by its values in porous medium. In that sense we have Theorem 1.1.

**Theorem 1.1.** *Let $\gamma \geq 3/2$ and let $u^\varepsilon \in H_0^1(\Omega^\varepsilon)$ be a solution of (1.1)–(1.2). Then we have*

$$\|\frac{u^\varepsilon}{\varepsilon^2} - C_0 u - \varepsilon^{\gamma-2} v - \{\omega^\varepsilon - C_0\} f\|_{L^2(\Omega_1)} \leq Ce^{1/2}, \qquad (1.45)$$

$$\|\frac{u^\varepsilon}{\varepsilon^2} - \{\omega^\varepsilon + w^\varepsilon\} f\|_{L^2(\Omega_2^\varepsilon)} \leq Ce, \qquad (1.46)$$

*where $w^\varepsilon$ is defined by (1.4)–(1.8), $\omega^\varepsilon$ by (1.10)–(1.16), $C_0$ by (1.15), $v$ by (1.30), and $u$ by (1.32).*

The preceding theorem can be generalized to yield the weak limits of renormalized $u^\varepsilon/\varepsilon^2$.

**Corollary 1.1.** *Let $\gamma > 2$ and let $u^\varepsilon \in H_0^1(\Omega^\varepsilon)$ be a solution of (1.1)–(1.2). Then we have*

$$\frac{\tilde{u}^\varepsilon}{\varepsilon^2} \rightharpoonup \chi_{\Omega_1} C_0 u + \chi_{\Omega_2} m f \quad \textit{weakly in } L^2(\Omega); \qquad (1.47)$$

$$\nabla \frac{\tilde{u}^\varepsilon}{\varepsilon^2} \rightharpoonup \chi_{\Omega_1} C_0 \nabla u + \chi_{\Omega_2} m \nabla f + (C_0 - m) f \vec{e}_n \delta_\Sigma \quad \textit{weakly in } H^{-1}(\Omega)^n; \qquad (1.48)$$

$$\nabla\{\frac{\tilde{u}^\varepsilon}{\varepsilon^2} - \tilde{w}^\varepsilon f(1 - H(x_n))\} \rightharpoonup \chi_{\Omega_1} C_0 \nabla u + (C_0 - k_\Sigma) f \vec{e}_n \delta_\Sigma \quad \textit{weakly* in } M(\Omega)^n; \qquad (1.49)$$

*where $m$ is given by (1.9), $\delta_\Sigma$ is the Dirac measure concentrated on $\Sigma$ and $k_\Sigma = \int_\Sigma w(\tilde{y}, 0) d\tilde{y}$.*

**Corollary 1.2.** *Let $\gamma = 2$ and let $u^\varepsilon \in H_0^1(\Omega^\varepsilon)$ be a solution of (1.1)–(1.2). Then we have*

$$\frac{\tilde{u}^\varepsilon}{\varepsilon^2} \rightharpoonup \chi_{\Omega_1}(C_0 u + v) + \chi_{\Omega_2} m f \quad \textit{weakly in } L^2(\Omega); \qquad (1.50)$$

$$\nabla \frac{\tilde{u}^\varepsilon}{\varepsilon^2} \rightharpoonup \chi_{\Omega_1} \nabla(C_0 u + v) + \chi_{\Omega_2} m \nabla f + (C_0 - m) f \vec{e}_n \delta_\Sigma \quad \textit{weakly in } H^{-1}(\Omega)^n; \qquad (1.51)$$

$$\nabla\{\frac{\tilde{u}^\varepsilon}{\varepsilon^2} - \tilde{w}^\varepsilon f(1 - H(x_n))\} \rightharpoonup \chi_{\Omega_1} \nabla(C_0 u + v) + (C_0 - k_\Sigma) f \vec{e}_n \delta_\Sigma \quad \textit{weakly* in } M(\Omega)^n. \qquad (1.52)$$

**Corollary 1.3.** *Let $3/2 < \gamma \leq 2$ and let $u^\varepsilon \in H_0^1(\Omega^\varepsilon)$ be a solution of (1.1)-(1.2). Then we have*

$$\frac{\tilde{u}^\varepsilon}{\varepsilon^2} - \varepsilon^{\gamma-2} v H(x_n) \rightharpoonup \chi_{\Omega_1} C_0 u + \chi_{\Omega_2} m f \quad \textit{weakly in } L^2(\Omega); \quad (1.53)$$

$$\nabla \frac{\tilde{u}^\varepsilon}{\varepsilon^2} - \varepsilon^{\gamma-2} v H(x_n) \rightharpoonup \chi_{\Omega_1} C_0 \nabla u + \chi_{\Omega_2} m \nabla f + (C_0 - m) f \vec{e}_n \delta_\Sigma$$
$$\textit{weakly in } H^{-1}(\Omega)^n; \quad (1.54)$$

$$\nabla \{ \frac{\tilde{u}^\varepsilon}{\varepsilon^2} - \varepsilon^{\gamma-2} v H(x_n) - \tilde{w}^\varepsilon f(1 - H(x_n)) \} \rightharpoonup \chi_{\Omega_1} C_0 \nabla u$$
$$+ (C_0 - k_\Sigma) f \vec{e}_n \delta_\Sigma \quad \textit{weakly* in } M^(\Omega)^n. \quad (1.55)$$

A careful analysis of the proof of Theorem 1.1 shows that for $\gamma < 3/2$ estimates (1.45) and (1.46) break down due to the boundary layer caused by $\partial v\left(\tilde{x}, +0\right)/\partial x_n$ and being different from zero. Elimination of this boundary layer improves situation for a limited range of $\gamma$.

**Theorem 1.2.** *Let $1 \leq \gamma < 3/2$ and let $u^\varepsilon \in H_0^1\left(\Omega^\varepsilon\right)$ be a solution of (1.1, 1.2).*

*Then we have*

$$\left\| \tfrac{u^\varepsilon}{\varepsilon^2} - C_0 u - \varepsilon^{\gamma-2} v - \{\omega^\varepsilon - C_0\} f - \varepsilon^{\gamma-2} V\left(\beta^\varepsilon - \varepsilon C_\beta\right) - \right.$$
$$\left. - \varepsilon^{\gamma-1} C_\beta \bar{V} \right\|_{L^2(\Omega_1)} \leq C \varepsilon^{\frac{1}{2}}; \quad (1.56)$$

$$\left\| \tfrac{u^\varepsilon}{\varepsilon^2} - \{\omega^\varepsilon + w^\varepsilon\} f - \varepsilon^{\gamma-2} V \beta^\varepsilon \right\|_{L^2(\Omega_2^\varepsilon)} \leq C\varepsilon; \quad (1.57)$$

*where $V$ is defined by (1.31), $\bar{V}$ by (1.33), $\beta^\varepsilon$ by (1.17)-(1.24) and other quantities are defined as in Theorem 1.1.*

It is easy to see that for function $\tilde{u}^\varepsilon/\varepsilon^2 - \varepsilon^{\gamma-2}\beta^\varepsilon V$ applies Corollary 1.3. For $\gamma = 1$ we get a bit different weak limit.

**Corollary 1.4.** *Let $\gamma = 1$ and let $u^\varepsilon \in H_0^1\left(\Omega^\varepsilon\right)$ be a solution of (1.1)(1.2). Then we have*

$$\frac{\tilde{u}^\varepsilon}{\varepsilon^2} - \varepsilon^{\gamma-2} v H(x_n) \rightharpoonup \chi_{\Omega_1} C_0 u + \chi_{\Omega_2} m f \quad \textit{weakly in } L^2(\Omega); \quad (1.58)$$

$$\nabla \frac{\tilde{u}^\varepsilon}{\varepsilon^2} - \varepsilon^{\gamma-2} v H(x_n) \rightharpoonup \chi_{\Omega_1} C_0 \nabla u + \chi_{\Omega_2} m \nabla f + (C_0 - m) f \vec{e}_n \delta_\Sigma$$
$$\textit{weakly in } H^{-1}(\Omega)^n; \quad (1.59)$$

$$\nabla\{\frac{\tilde{u}^\varepsilon}{\varepsilon^2} - \varepsilon^{\gamma-2}vH(x_n) - \tilde{w}^\varepsilon f(1-H(x_n))\} \rightharpoonup \chi_{\Omega_1} C_0 \nabla u$$
$$+(C_0 - k_\Sigma) f \vec{e}_n \delta_\Sigma \quad \textit{weakly* in } M(\Omega)^n \tag{1.60}$$

The homogenization theorem covering the case $0 \leq \gamma < 1$ can now be obtained by correcting the volume and boundary terms created by the boundary layer's correction constructed in Theorem 1.2.

**Theorem 1.3.** *Let $0 \leq \gamma < 1$ and let $u^\varepsilon \in H_0^1(\Omega^\varepsilon)$ be a solution of (1.1)(1.2). Then we have*

$$\Big\| \tfrac{u^\varepsilon}{\varepsilon^2} - C_0 u - \varepsilon^{\gamma-2} v - \{\omega^\varepsilon - C_0\} f - \varepsilon^{\gamma-2} V(\beta^\varepsilon - \varepsilon C_\beta)$$
$$-2\varepsilon^{\gamma-2} \sum_{i=1}^{n-1} \tfrac{\partial V}{\partial x_i}(\tilde{x})(\varsigma_i^\varepsilon - \varepsilon^2 C_{i,\varsigma}) - \varepsilon^{\gamma-1} C_\beta \bar{V} - 2\varepsilon^\gamma \sum_{i=1}^{n-1} C_{i,\varsigma} U_i$$
$$-\varepsilon^\gamma C_\beta^2 \bar{U} - \varepsilon^{\gamma-1} C_\beta \tilde{V}(\beta^\varepsilon - \varepsilon C_\beta) \Big\|_{L^2(\Omega_1)} \leq C\varepsilon^{\frac{1}{2}}, \tag{1.61}$$
$$\| \tfrac{u^\varepsilon}{\varepsilon^2} - \{\omega^\varepsilon - w^\varepsilon\} f - \varepsilon^{\gamma-2} V \beta^\varepsilon \|_{L^2(\Omega_2^\varepsilon)} \leq C\varepsilon \tag{1.62}$$

*where $\varsigma_i^\varepsilon$ is defined by (1.25)–(1.29), $U_i$ is defined by (1.35), $\tilde{V}, \bar{U}$ are defined by (1.36) and other quantities are defined as in Theorem 1.2.*

It is interesting to calculate weak limit of $\gamma = 0$. We have

**Corollary 1.5.** *Let $\gamma = 0$ and let $u^\varepsilon \in H_0^1(\Omega^\varepsilon)$ be a solution of (1.1)(1.2). Then we have*

$$\frac{\tilde{u}^\varepsilon}{\varepsilon^2} - \varepsilon^{-1} vH(x_n) \rightharpoonup \chi_{\Omega_1}(C_0 u + C_\beta \bar{V}) + \chi_{\Omega_2} m f \quad \textit{weakly in } L^2(\Omega); \tag{1.63}$$
$$\nabla\{\frac{\tilde{u}^\varepsilon}{\varepsilon^2} - \varepsilon^{-1} vH(x_n)\} \rightharpoonup \chi_{\Omega_1} \nabla(C_0 u + C_\beta \bar{V}) \chi_{\Omega_2} m \nabla f + (C_0 - m) f \vec{e}_n \delta_\Sigma$$
$$\textit{weakly in } H^{-1}(\Omega); \tag{1.64}$$
$$\nabla\{\frac{\tilde{u}^\varepsilon}{\varepsilon^2} - \varepsilon^{-1} vH(x_n) - \tilde{w}^\varepsilon f(1-H(x_n))\} \rightharpoonup \chi_{\Omega_1} \nabla(C_0 u + C_\beta \bar{V})$$
$$+(C_0 - k_\Sigma) f \vec{e}_n \delta_\Sigma \qquad \textit{weakly* in } M(\Omega). \tag{1.65}$$

## 2 Proofs of the Theorems

### *2.1 Proof of Theorem 1.1 and Corollaries 1.1–1.3*

Let us introduce function $\varphi_\varepsilon$ by

$$\varphi_\varepsilon(x) = \frac{u^\varepsilon}{\varepsilon^2} - C_0 u H(x_n) - \varepsilon^{\gamma-2} v H(x_n)$$
$$-\{\omega^\varepsilon - C_0 H(x_n)\} f - \{1 - H(x_n)\} \omega^\varepsilon f. \qquad (2.1)$$

By a straightforward calculation we conclude the following auxiliary result.

**Lemma 2.1.** $\varphi_\varepsilon \in H_0^1(\Omega^\varepsilon)$.

Now we are ready to prove Theorem 1.1.

**Proof of Theorem 1.1.**

We start with weak formulation corresponding to (1.1)–(1.2) :

$$\int_{\Omega^\varepsilon} \nabla \frac{u^\varepsilon}{\varepsilon^2} \nabla \varphi = \int_{\Omega_1} \varepsilon^{\gamma-2} f\varphi + \int_{\Omega_2^\varepsilon} \varepsilon^{-2} f\varphi, \qquad \forall \varphi \in H_0^1(\Omega^\varepsilon). \qquad (2.2)$$

Due to Lemma 2.1, (2.2) is equivalent to

$$\int_{\Omega^\varepsilon} \nabla\varphi_\varepsilon \nabla\varphi = \int_{\Omega_1} \{\varepsilon^{\gamma-2} f\varphi - \varepsilon^{\gamma-2} \nabla v \nabla \varphi - \nabla\{C_0 u$$
$$+ f(\omega^\varepsilon - C_0)\}\nabla\varphi\} + \int_{\Omega_2^\varepsilon} \{\varepsilon^{-2} f\varphi - \nabla(w^\varepsilon f + \omega^\varepsilon f)\nabla\varphi\}.$$

To proceed further we use integration by parts and get

$$\int_{\Omega^\varepsilon} \nabla\varphi_\varepsilon \nabla\varphi = \int_{\Omega_1} \varphi\{\varepsilon^{\gamma-2}(f + \delta v) - \delta f(\omega^\varepsilon - C_0)$$
$$+ 2\mathrm{div}\{\nabla f((\omega^\varepsilon - C_0)\}\} + \int_{\Omega_2^\varepsilon} \varphi\{2\mathrm{div}\{\nabla f(\omega^\varepsilon + w^\varepsilon)$$
$$- \delta f(\omega^\varepsilon + w^\varepsilon)\} + \int_\Sigma \varphi\{C_0 \tfrac{\partial u}{\partial x_n}(\tilde{x}, +0) + \varepsilon^{\gamma-2} \tfrac{\partial v}{\partial x_n}(\tilde{x}, +0)$$
$$\tfrac{\partial f}{\partial x_n}(\tilde{x}, 0)\{\omega^\varepsilon(\tilde{x}, +0) - \omega^\varepsilon(\tilde{x}, -0) - \omega^\varepsilon(\tilde{x}, 0) - C_0\}$$
$$+ f(\tilde{x}, 0)\{\partial\omega^\varepsilon \partial x_n(\tilde{x}, +0) - \partial\omega^\varepsilon \partial x_n(\tilde{x}, -0) - \partial\omega^\varepsilon \partial x_n(\tilde{x}, 0)\}\}. \ (2.3)$$

After recalling equations (1.30), (1.11) and (1.12) and transforming terms in divergence form we find simplified form of (2.3):

$$\int_{\Omega^\varepsilon} \nabla\varphi_\varepsilon \nabla\varphi = \int_\Sigma \varphi\{C_0(\tfrac{\partial u}{\partial x_n}(\tilde{x}, +0)) - (\tfrac{\partial f}{\partial x_n}(\tilde{x}, 0)) + \varepsilon^{\gamma-2}(\tfrac{\partial v}{\partial x_n}(\tilde{x}, +0))\}$$
$$- \int_{\Omega_1} \varphi\delta f(\omega^\varepsilon - C_0) - - \int_{\Omega_2^\varepsilon} \varphi\delta f(\omega^\varepsilon + w^\varepsilon) - 2 - \int_{\Omega_1} \varphi\nabla f(\omega^\varepsilon - C_0)\nabla\varphi$$
$$-2 \int_{\Omega_2^\varepsilon} \nabla\varphi\nabla f(\omega^\varepsilon + w^\varepsilon) \quad \forall\varphi \in H_0^1(\Omega^\varepsilon). \qquad (2.4)$$

Now we insert $\varphi = \varphi_\varepsilon$ as a test function in (2.4). Using Proposition 1.1. we get

$$\begin{aligned}\int\limits_{\Omega^\varepsilon} |\nabla \varphi_\varepsilon|^2 \leq C\{\varepsilon^{1/2} + \varepsilon^{\gamma-3/2}\}\|\varphi_\varepsilon\|_{L^2(\Omega_2^\varepsilon)^n}\\ +C\|\varphi_\varepsilon\|_{L^2(\Omega_1)}\varepsilon^{1/2} + +C\varepsilon\|\varphi_\varepsilon\|_{L^2(\Omega_2^\varepsilon)^n}(1+\varepsilon^{1/2})\\ +C\|\varphi_\varepsilon\|_{L^2(\Omega_1)^n}\varepsilon(1/2) + C\|\varphi_\varepsilon\|_{L^2(\Omega_2^\varepsilon)^n}(1+\varepsilon^{1/2}).\end{aligned} \tag{2.5}$$

Consequently, for $\gamma \geq 3/2$ we have $\|\varphi_\varepsilon\|_{L^2(\Omega^\varepsilon)^n} \leq C$.
Using once more Proposition 1.1, from (2.5) we obtain from (2.5)

$$\|\varphi_\varepsilon\|_{L^2(\Omega_2^\varepsilon)} \leq C\varepsilon; \tag{2.6}$$

$$\|\varphi_\varepsilon\|_{L^2(\Sigma)} \leq C\varepsilon^{1/2}. \tag{2.7}$$

Obviously, (2.6) is equivalent to (1.46) and it remains to prove (1.45).

By the above calculations $\varphi_\varepsilon$ satisfies

$$\begin{cases} -\Delta\varphi_\varepsilon = -\Delta f(\omega^\varepsilon - C_0) + 2div\{\nabla f(\omega^\varepsilon - C_0)\} & \text{in } \Omega_1;\\ \|\varphi_\varepsilon\|_{L^2(\partial\Omega_1)} \leq C\varepsilon^{1/2}. \end{cases} \tag{2.8}$$

Furthermore, we have

$$\|\text{div}\{\nabla f(\omega^\varepsilon - C_0)\}\|_{H^{-1}(\Omega_1)} = \|\nabla f(\omega^\varepsilon - C_0)\}\|_{L^2(\Omega_1)^n} \leq C\varepsilon^{1/2} \tag{2.9}$$

$$\|\Delta f(\omega^\varepsilon - C_0)\|_{L^2(\Omega_1)} \leq C\varepsilon^{1/2}. \tag{2.10}$$

Therefore (2.8), (2.9), (2.10) and Proposition 1.2 imply

$$\|\varphi_\varepsilon\|_{L^2(\Omega_1^\varepsilon)} \leq C\varepsilon^{1/2}, \tag{2.11}$$

and (1.45) is proved.

Concerning the corollaries, obviously it suffices to prove Corollary 1.1; proofs of the others are identical.

**Proof of Corollary 1.1.**

Let us note that (1.47) is an obvious consequence of (1.45)–(1.46) and (1.9). Proofs of (1.48) and (1.49) are almost identical; we give a proof for (1.49):

Let $\psi \in C_0^\infty(\Omega)^n$. Then

$$\int_\Omega \nabla\{\tfrac{\partial \tilde{u}^\varepsilon}{\partial \varepsilon^2} - \tilde{w}^\varepsilon f(1/H(x_n))\}\psi =$$
$$\int_\Omega \nabla\varphi_\varepsilon\psi + \int_{\Omega_1}\{C_0\nabla u + \varepsilon^{-2}\nabla v + (\omega^\varepsilon - C_0)\nabla f\}\psi + \int_{\Omega_2^\varepsilon} \omega^\varepsilon \nabla f\psi$$
$$+ \int_{\Omega^\varepsilon} \nabla\{\omega^\varepsilon - C_0H(x_n)\}f\psi = \int_\Omega \nabla\varphi_\varepsilon\psi + \int_\Sigma\{C_0 - w^\varepsilon(\tilde{x},0)\}f\psi_n$$
$$+ \int_{\Omega_1}\{C_0\nabla u + \varepsilon^{\gamma-2}\nabla v + (\omega^\varepsilon - C_0)\nabla f\}\psi + \int_{\Omega_2^\varepsilon}\omega^\varepsilon\nabla f\psi$$
$$- \int_{\Omega^\varepsilon}\{\omega^\varepsilon - C_0H(x_n)\}\mathrm{div}\{f\psi\}$$
$$\longrightarrow \int_{\Omega_1} C_0\nabla u\psi + \int_\Sigma (C_0 - k_\Sigma)f(\tilde{x},0)\psi_n(\tilde{x},0), \tag{2.12}$$

where $k_\Sigma = \int_{[0,1]^{n-1}} w(\tilde{y},0)d\tilde{y}$.

Since $C_0^\infty(\Omega)^n$ is dense in $C_0(\Omega)^n$ and $\nabla\{\omega^\varepsilon - C_0H(x_n)\}$ is uniformly bounded in $L^1(\Omega)^n$, we conclude that (1.49) holds true.

## *2.2 Proof of Theorem 1.2*

We introduce function $\psi_\varepsilon$ by

$$\begin{aligned}\psi_\varepsilon(x) &= \tfrac{u^\varepsilon}{\varepsilon^2} - C_0uH(x_n) - \varepsilon^{\gamma-2}vH(x_n) - \{\omega^\varepsilon - C_0H(x_n)\}f\\ &-\{1-H(x_n)\}w^\varepsilon f - \varepsilon^{\gamma-2}V(\beta^\varepsilon - \varepsilon C_\beta H(x_n)) - \varepsilon^{\gamma-1}C_\beta\bar{V}H(x_n)\\ &= \varphi_\varepsilon(x) - \varepsilon^{\gamma-2}V(\beta^\varepsilon - \varepsilon C_\beta H(x_n)) - \varepsilon^{\gamma-1}C_\beta\bar{V}H(x_n).\end{aligned} \tag{2.13}$$

We have the following auxiliary result:

**Lemma 2.2.** $\psi_\varepsilon \in H^1(\Omega^\varepsilon)$.

However, $\psi_\varepsilon$ is not zero at $\partial\Omega_1$, since $V$ takes nonzero values at $x_n = L$ and $x_n = -L$. We should correct values of $\psi_\varepsilon$ at those parts of $\partial\Omega$.

Using partition of unity we construct a function $\alpha_\varepsilon$ such that

$$\begin{cases}\alpha_\varepsilon \in C^\infty(\bar{\Omega}); \quad \alpha_\varepsilon(x) = 1 \text{ for } \mathrm{dist}(x,\partial\Omega) \le d_z\varepsilon;\\ 0 \le \alpha_\varepsilon \le 1;\ |\nabla\alpha_\varepsilon| \le C/\varepsilon,\ \alpha_\varepsilon(x) = 0 \text{ for } \mathrm{dist}(x,\partial\Omega) \ge 2d_z\varepsilon,\end{cases} \tag{2.14}$$

where $d_z$ is such that $\alpha_\varepsilon$ is identically zero around the perforations.

**Proposition 2.1.** *Let $A_\varepsilon$ be defined by*

$$A_\varepsilon(x) = -\alpha_\varepsilon(x)V(\tilde{x})\{\beta^\varepsilon(x) - \varepsilon C_\beta H(x_n)\}, \quad \forall x \in \Omega^\varepsilon \tag{2.15}$$

*Then* $\psi_\varepsilon = A_\varepsilon \varepsilon^{\gamma-2}$ *on* $\partial\Omega_1$. *Furthermore,* $A_\varepsilon$ *satisfies*

$$\|\nabla A_\varepsilon\|_{L^2(\Omega^\varepsilon)^n} \leq C\varepsilon^{1/2} e^{-C/\varepsilon}. \tag{2.16}$$

**Proof.**

It is easy to see that

$$\|\nabla A_\varepsilon\|^2_{L^2(\Omega^\varepsilon)^n} \leq C \int_{[0,L]^{n-1}} \int_{L-2d_z\varepsilon}^{L} \{|\nabla \alpha_\varepsilon|^2 |\beta^\varepsilon - \varepsilon C_\beta|^2,$$
$$|\alpha_\varepsilon|^2 |\nabla \beta^\varepsilon|^2\} dx \leq C\varepsilon e^{-C/\varepsilon},$$

by (1.21), (1.23), and (2.14).

Now we are ready to prove Theorem 1.1.

**Proof of Theorem 1.2.**

By definition of $\psi_\varepsilon$ (1.1)–(1.2) is equivalent to

$$\int_{\Omega^\varepsilon} \nabla\psi_\varepsilon \nabla\varphi = \int_{\Omega^\varepsilon} \nabla\varphi_\varepsilon \nabla\varphi - \varepsilon^{\gamma-2} \int_{\Omega_1} \nabla\{V(\beta^\varepsilon - \varepsilon C_\beta)\}\nabla\varphi$$
$$-\varepsilon^{\gamma-1} C_\beta \int_{\Omega_1} \nabla\bar{V}\nabla\varphi - \varepsilon^{\gamma-2} \int_{\Omega_2^\varepsilon} \nabla\{V\beta_\varepsilon\}\nabla\varphi, \quad \varphi \in H^1_0(\Omega^\varepsilon). \tag{2.17}$$

Using definitions of $V, \bar{V}$ and $\beta^\varepsilon$ we transform (2.17) to

$$\int_{\Omega^\varepsilon} \nabla\psi_\varepsilon \nabla\varphi = \int_{\Omega^\varepsilon} \nabla\varphi_\varepsilon \nabla\varphi + \varepsilon^{\gamma-1} C_\beta \int_\Sigma \varphi \frac{\partial \bar{V}}{\partial x_n}(\tilde{x}, +0) - \varepsilon^{\gamma-2} \int_\Sigma V\varphi$$
$$+\varepsilon^{\gamma-2} \int_{\Omega_1} \varphi\{2\text{div}_{\tilde{x}}(\nabla_{\tilde{x}} V(\beta\varepsilon - \varepsilon C_\beta)) - \Delta_{\tilde{x}} V(\beta^\varepsilon - \varepsilon C_\beta)\}$$
$$+\varepsilon^{\gamma-2} \int_{\Omega_2^\varepsilon} \varphi\{2\text{div}_{\tilde{x}}(\nabla_{\tilde{x}} V\beta^\varepsilon) - \Delta_{\tilde{x}} V\beta^\varepsilon\} =$$
$$= \int_\Sigma \varphi \left\{C_0 \left(\frac{\partial u}{\partial x_n}(\tilde{x}, +0) - \frac{\partial f}{\partial x_n}\right)(\tilde{x}, 0) + \varepsilon^{\gamma-1} C_\beta \frac{\partial \bar{V}}{\partial x_n}(\tilde{x}, +0)\right\}$$
$$- \int_{\Omega_1} \varphi\{\Delta f(\omega^\varepsilon - C_0) + \varepsilon^{\gamma-2}\Delta_{\tilde{x}} V(\beta^\varepsilon - \varepsilon C_\beta)\} - \int_{\Omega_2^\varepsilon} \varphi\{\Delta f(\omega^\varepsilon + w^\varepsilon)$$
$$+\varepsilon^{\gamma-2}\Delta_{\tilde{x}} V\beta^\varepsilon\} - \int_{\Omega_2^\varepsilon} 2\nabla\varphi\nabla f w^\varepsilon - \int_{\Omega^\varepsilon} 2\nabla\varphi\{\varepsilon^{\gamma-2}\nabla_{\tilde{x}} V(\beta^\varepsilon - \varepsilon C_\beta H(x_n))$$
$$+\nabla f(\omega^\varepsilon - C_0 H(x_n))\}, \quad \forall \varphi \in H^1_0(\Omega^\varepsilon). \tag{2.18}$$

Now estimates (1.23)–(1.24) for $\beta^\varepsilon$, Proposition 1.1, regularity properties of $V$ and $\bar{V}$, together with (2.5), applied to (2.18) yield

$$\int_{\Omega^\varepsilon} \nabla\psi_\varepsilon \nabla\varphi \leq C\{|C_0|\varepsilon^{1/2} + |C_\beta|\varepsilon^{\gamma-1/2}\}\|\nabla\varphi_\varepsilon\|_{L^2(\Omega_2^\varepsilon)^n}$$
$$+C\|\varphi_\varepsilon\|_{L^2(\Omega_1)}\{\varepsilon^{1/2} + \varepsilon^{\gamma-1/2}\} + C\varepsilon\|\nabla\varphi_\varepsilon\|_{L^2(\Omega_2^\varepsilon)^n}(1 + \varepsilon^{1/2} + \varepsilon^{\gamma-1/2})$$
$$+C\|\nabla\varphi_\varepsilon\|_{L^2(\Omega_2^\varepsilon)^n} + C\|\nabla\varphi_\varepsilon\|_{L^2(\Omega^\varepsilon)^n}(\varepsilon^{\gamma-1/2+\varepsilon^{1/2}}). \tag{2.19}$$

Therefore, inserting $\varphi = \psi_\varepsilon - \varepsilon^{\gamma-2}A_\varepsilon \in H_0^1(\Omega^\varepsilon)$ in (2.19) and using (2.16) we get

$$\|\nabla\varphi_\varepsilon\|_{L^2(\Omega^\varepsilon)^n} \leq C$$

for $\gamma \geq \frac{1}{2}$, and consequently

$$\|\varphi_\varepsilon\|_{L^2(\Omega_2^\varepsilon)^n} \leq C\varepsilon; \tag{2.20}$$

$$\|\varphi_\varepsilon\|_{L^2(\Sigma)} \leq C\varepsilon^{1/2}. \tag{2.21}$$

Now (1.57) is proved. Using the same argument as at the end of the proof of Theorem 1.1 we obtain (1.56), but only for $\gamma \geq 1$. For $1 > \gamma \geq 1/2$ we get

$$\|\psi_\varepsilon\|_{L^2(\Omega_1)} \leq C\varepsilon^{\gamma-1/2}.$$

This completes the proof of Theorem 1.2.

## *2.3 Proof of Theorem 1.3*

Situation here is more complicated than in the two preceding proofs, since it is necessary to correct one volume term.

As above, we introduce the function to be estimated

$$\begin{aligned}
\xi_\varepsilon(x) = \tfrac{u^\varepsilon}{\varepsilon^2} - C_0uH(x_n) - \varepsilon^{\gamma-2}vH(x_n) - \{\omega^\varepsilon - C_0H(x_n)\}f \\
-\{1 - H(x_n)\}w^\varepsilon f - \varepsilon^{\gamma-2}V(\beta^\varepsilon - \varepsilon C_\beta H(x_n)) - \varepsilon^{\gamma-1}C_\beta VH(x_n) \\
-2\varepsilon^{\gamma-2}\sum_{i=1}^{n-1}\tfrac{\partial V}{\partial x_i}(\tilde{x})(\zeta_i^\varepsilon - \varepsilon^2C_{i,\zeta}H(x_n)) - 2\varepsilon^\gamma\sum_{i=1}^{n-1}C_{i,\zeta}U_iH(x_n) \\
-\varepsilon^\gamma C_\beta^2\bar{U}H(x_n) - \varepsilon^{\gamma-1}C_\beta\tilde{V}(\tilde{x})(\beta^\varepsilon - \varepsilon C_\beta H(x_n)) = \psi_\varepsilon(x) \\
-2\varepsilon^{\gamma-2}\sum_{i=1}^{n-1}\tfrac{\partial V}{\partial x_i}(\tilde{x})(\zeta_i^\varepsilon - \varepsilon^2C_{i,\zeta}H(x_n)) - 2\varepsilon^\gamma\sum_{i=1}^{n-1}C_{i,\zeta}U_iH(x_n) \\
-\varepsilon^\gamma C_\beta^2\bar{U}H(x_n) - \varepsilon^{\gamma-1}C_\beta\tilde{V}(\tilde{x})(\beta^\varepsilon - \varepsilon C_\beta H(x_n)).
\end{aligned} \tag{2.22}$$

We have the following result:

**Lemma 2.3.** $\xi \in H^1(\Omega^\varepsilon)$.

However, $\xi_\varepsilon$ is not zero on $\partial\Omega_1$ since $V, \nabla_{\tilde{x}}V$ and $\bar{V}$ take nonzero values at $x_n = L$ and $x_n = -L$. We should correct values of $\xi_\varepsilon$ at those parts of $\partial\Omega$. As we have shown in Section 2.2 the correction is, in fact, exponentially small.

**Proposition 2.2** *Let $B_\varepsilon$ be defined by*

$$B_\varepsilon = -\alpha_\varepsilon\{V(\beta^\varepsilon - \varepsilon C X_\beta H(x_n)) + 2\sum_{i=1}^{n-1} \frac{\partial V}{\partial x_i}(\tilde{x})(\zeta_i^\varepsilon - \varepsilon^2 C_{i,\zeta} H(x_n))+ \\ +\varepsilon C_\beta \tilde{V}(\beta^\varepsilon - \varepsilon C_\beta H(x_n))\}. \tag{2.23}$$

*Then $\xi_\varepsilon = B_\varepsilon \varepsilon^{\gamma-2}$ on $\partial\Omega_1$. Furthermore, $B_\varepsilon$ satisfies*

$$\|\nabla B_\varepsilon\|_{L^2(\Omega^\varepsilon)^n} \leq C\varepsilon^{1/2} e^{-C/\varepsilon}. \tag{2.24}$$

**Proof.**
It is almost identical to the proof of Proposition 2.1.
In addition we also need the following auxiliary result:

**Proposition 2.3** *Let $\varphi \in H_0^1(\Omega^\varepsilon)$.*
*Then*

$$\varepsilon^{\gamma-2}|\int_{\Omega^\varepsilon} \Delta_{\tilde{x}} V\varphi(\beta^\varepsilon - \varepsilon C_\beta H(x_n))| \leq C\varepsilon^{\gamma+1/2}\|\nabla\varphi\|_{L^2(\Omega_2^\varepsilon)^n}. \tag{2.25}$$

**Proof.**
Let $\Xi$ be defined by (A.16)–(A.17). We define $\Xi^\varepsilon$ as $\Xi^\varepsilon(x) = \varepsilon^3\Xi(x/\varepsilon)$. Then $\Xi^\varepsilon = 0$ on $\partial\Omega^\varepsilon \setminus \partial\Omega$ and it satisfies

$$-\Delta\Xi^\varepsilon = \beta^\varepsilon - \varepsilon C_\beta H(x_n)$$

in $\Omega^\varepsilon$. Therefore

$$|\int_{\Omega^\varepsilon} \Delta_{\tilde{x}} V\varphi(\beta^\varepsilon - \varepsilon C_\beta H(x_n))| = |\int_{\Omega^\varepsilon} \nabla(\Delta_{\tilde{x}} V\varphi)\nabla\Xi^\varepsilon| \leq \\ C\|\nabla\Xi^\varepsilon\|_{L^2(\Omega^\varepsilon)^n}\|\nabla\varphi\|_{L^2(\Omega^\varepsilon)^n} \leq C\varepsilon^{5/2}\|\nabla\varphi\|_{L^2(\Omega^\varepsilon)^n}$$

and (2.25) is proved.
Now we are ready for the proof of Theorem 1.3:

**Proof of Theorem 1.3.**
By definition of $\xi_\varepsilon$ (1.1)–(1.2) is equivalent to

$$\int_{\Omega^\varepsilon} \nabla\xi_\varepsilon\nabla\varphi = \int_{\Omega^\varepsilon} \nabla\psi_\varepsilon\nabla\varphi - 2\varepsilon^{\gamma-2}\int_{\Omega_1} \nabla\left\{\sum_{i=1}^{n-1} \frac{\partial V}{\partial x_i}(\tilde{x})(\zeta_i^2 - \varepsilon^2 C_{i,\zeta})\right. \\ \left. +\varepsilon^2\sum_{i=1}^{n-1} C_{i,\zeta} U_i\right\}\nabla\varphi - 2\int_{\Omega_2^\varepsilon} \varepsilon^{\gamma-2}\nabla\varphi\nabla\left\{\sum_{i=1}^{n-1} \frac{\partial V}{\partial x_i}(\tilde{x})\zeta_i^3\right\}$$

$$-\varepsilon^{\gamma-1}C_\beta \int_{\Omega_1} \nabla\{\tilde{V}(\tilde{x})(\beta^\varepsilon - \varepsilon C_\beta) + \varepsilon C_\beta \bar{U}\}\nabla\varphi - \int_{\Omega_2^\varepsilon} \varepsilon^{\gamma-1}C_\beta \nabla\left\{\tilde{V}\beta^\varepsilon\right\}\nabla\varphi =$$

$$= \int_{\Omega^\varepsilon} \nabla\psi_\varepsilon\nabla\varphi + 4\varepsilon^{\gamma-2}\int_{\Omega_1\cup\Omega_2^\varepsilon} \varphi \mathrm{div}_{\tilde{x}}\left\{\sum_{i=1}^{n-1}\frac{\partial V}{\partial x_i}(\zeta_i^\varepsilon - \varepsilon^2 C_{i,\zeta}H(x_n))\right\}$$

$$-2\varepsilon^{\gamma-2}\int_{\Omega_1\cup\Omega_2^\varepsilon}\varphi\nabla_{\tilde{x}}V\nabla_{\tilde{x}}(\beta^\varepsilon - \varepsilon C_\beta H(x_n))$$

$$+2\varepsilon^\gamma\int_\Sigma \varphi\sum_{i=1}^{n-1}C_{i,\zeta}\frac{\partial}{\partial x_n}U_i(\tilde{x},+0) - C_\beta\varepsilon^{\gamma-1}\int_\Sigma\varphi\tilde{V}$$

$$+2C_\beta\varepsilon^{\gamma-1}\int_{\Omega_1\cup\Omega_2^\varepsilon}\varphi\mathrm{div}_{\tilde{x}}\left\{\nabla_{\tilde{x}}\left\{\tilde{V}'x(\beta^\varepsilon - \varepsilon C_\beta H(x_n))\right\}\right\}$$

$$+C_\beta^2\varepsilon^\gamma\int_\Sigma\varphi\frac{\partial}{\partial x_n}\bar{U}(\tilde{x},+0) - 2\varepsilon^{\gamma-2}\int_{\Omega^\varepsilon}\varphi\sum_{i=1}^{n-1}\Delta_{\tilde{x}}\frac{\partial V}{\partial x_i}(\zeta_i^3 - \varepsilon^2 C_{i,\zeta}H(x_n))$$

$$-2C_\beta\varepsilon^{\gamma-1}\int_{\Omega^\varepsilon}\varphi\Delta_{\tilde{x}}\tilde{V}(\beta^\varepsilon - \varepsilon C_\beta H(x_n)) =$$

$$\int_{\Omega^\varepsilon}\nabla\psi_\varepsilon\nabla\varphi + 2\varepsilon^\gamma\int_\Sigma\varphi\sum_{i=1}^{n-1}C_{i,\zeta}\frac{\partial}{\partial x_n}U_i(\tilde{x},+0) + C_\beta^2\varepsilon^\gamma\int_\Sigma\varphi\frac{\partial}{\partial x_n}\bar{U}(\tilde{x},+0)$$

$$-2\varepsilon^{\gamma-2}\int_{\Omega_1\cup\Omega_2^\varepsilon}\varphi\nabla_{\tilde{x}}V\nabla_{\tilde{x}}(\beta^\varepsilon - \varepsilon C_\beta H(x_n))$$

$$-4\varepsilon^{\gamma-2}\int_{\Omega^\varepsilon}\nabla_{\tilde{x}}\varphi\{\sum_{i=1}^{n-1}\nabla_{\tilde{x}}\frac{\partial V}{\partial x_i}(\zeta_i^3 - \varepsilon^2 C_{i,\zeta}H(x_n))\} - C_\beta\varepsilon^{\gamma-1}\int_\Sigma\varphi\tilde{V}$$

$$-2C_\beta\varepsilon^(\gamma-1)\int_{\Omega^\varepsilon}\nabla_{\tilde{x}}\varphi\nabla_{\tilde{x}}\tilde{V}(\beta^\varepsilon - \varepsilon C_\beta H(x_n))$$

$$-2\varepsilon^{\gamma-2}\int_{\Omega^\varepsilon}\varphi\sum_{i=1}^{n-1}\Delta_{\tilde{x}}\frac{\partial V}{\partial x_i}(\zeta_i^3 - \varepsilon^2 C_{i,\zeta}H(x_n))$$

$$-2C_\beta\varepsilon^{\gamma-1}\int_{\Omega^\varepsilon}\varphi\Delta_{\tilde{x}}\tilde{V}(\beta^\varepsilon - \varepsilon C_\beta H(x_n)) = \int_\Sigma\varphi\{C_0(\frac{\partial u}{\partial x_n}(\tilde{z},+0) + \frac{\partial f}{\partial x_n}(\tilde{x},0))$$

$$+\varepsilon^\gamma(2\sum_{i=1}^{n-1}C_{i,\zeta}\frac{\partial}{\partial x_n}U_i(\tilde{x},0) + C_\beta^2\varphi\frac{\partial}{\partial x_n}\bar{U}(\tilde{x},+0))\} - \int_{\Omega_1}\varphi\{\Delta f(\omega^\varepsilon - C_0)$$

$$+2\varepsilon^{\gamma-2}\sum_{i=1}^{n-1}\Delta_{\tilde{x}}\frac{\partial V}{\partial x_i}(\zeta_i^\varepsilon - \varepsilon^2 C_{i,\zeta}H(x_n)) + 2C_\beta\varepsilon^{\gamma-1}\Delta_{\tilde{x}}\tilde{V}(\beta^\varepsilon - \varepsilon C_\beta)\}$$

$$-\int_{\Omega_2^\varepsilon}\varphi\{\Delta f(\omega^\varepsilon + w^\varepsilon) + 2\varepsilon^{\gamma-2}\sum_{i=1}^{n-1}\Delta_{\tilde{x}}\frac{\partial V}{\partial x_i}\zeta_i^\varepsilon + 2C_\beta\varepsilon^{\gamma-1}\Delta_{\tilde{x}}\tilde{V}\beta^\varepsilon\}$$

$$-\int_{\Omega_2^\varepsilon}2\nabla\varphi\nabla f w^\varepsilon - 2\varepsilon^{\gamma-2}\int_{\Omega_1\cup\Omega_2^\varepsilon}\nabla_{\tilde{x}}V\nabla_{\tilde{x}}\{\varphi(\beta^\varepsilon - \varepsilon C_\beta H(x_n))\}$$

$$-2C_\beta\varepsilon^{\gamma-1}\int_{\Omega^\varepsilon}\nabla_{\tilde{x}}\varphi\nabla_{\tilde{x}}\tilde{V}(\beta^\varepsilon - \varepsilon C_\beta H(x_n))$$

$$-4\varepsilon^{\gamma-2}\int_{\Omega^\varepsilon}\nabla_{\tilde{x}}\varphi\{\sum_{i=1}^{n-1}\nabla_{\tilde{x}}\frac{\partial V}{\partial x_i}(\zeta_i^\varepsilon - \varepsilon^2 C_{i,\zeta}H(x_n))\} - \int_{\Omega^\varepsilon}3\nabla\varphi\nabla f(\omega^\varepsilon - C_0 H(x_n))$$

$$-\varepsilon^{\gamma-2} \int\limits_{\Omega^\varepsilon} \Delta_{\tilde{x}} V \varphi(\beta^\varepsilon - \varepsilon C_\beta H(x_n)). \tag{2.26}$$

Now we use the identity

$$-2\varepsilon^{\gamma-2} \int\limits_{\Omega_1 \cup \Omega_2^\varepsilon} \nabla + \tilde{x} V \nabla_{\tilde{x}} \{\varphi(\beta^\varepsilon - \varepsilon C_\beta H(x_n))\}$$
$$= 2\varepsilon^{\gamma-2} \int\limits_{\Omega^\varepsilon} \Delta_{\tilde{x}} V \varphi(\beta^\varepsilon - \varepsilon C_\beta H(x_n)).$$

Proposition 2.3, definitions of $\beta^\varepsilon$ and $\xi^\varepsilon$ and regularity of $V, \bar{V}, \bar{U}, U_i$ and $\tilde{V}$ to obtain

$$\int\limits_{\Omega^\varepsilon} \nabla \xi_\varepsilon \nabla \varphi \leq C \|\nabla \varphi_\varepsilon\|_{L^2(\Omega^\varepsilon)^n}, \qquad \forall \varphi \in H_0^1(\Omega^\varepsilon).$$

Now we proceed as at the end of the proof of Theorem 1.2 and using Propositions 2.2 and 1.1 we obtain (1.61)–(1.62).

## Appendix A

In this section the solubility and asymptotic behavior as $|y_n| \to \infty$ of solutions of the auxiliary problems are considered. We shall first formulate a general problem, establish its solubility and asymptotic behavior and then derive the results for the auxiliary problems from Section 1.2 as corollaries.

Most of these results can be found in [4]; we collect them for the comfort of the reader. Results in this direction, but more complete and for more complicated elliptic system of linear elasticity can be found in recent book Oleinik et al [5], which came to our attention after this work had been completed.

In order to formulate a general problem we require a precise definition of geometry.

Let $Z^*$ be a $C^{0,1}$-open set strictly included in $Z = ]0,1[^n$ and let $Y^* = Z \setminus \bar{Z}^*$. We suppose that $Y^*$ (the fluid part) is connected and that $Z^*$ consists of a finite number of the component homeomorphic to a unit ball. We set $Z^- = \cup_{k=1}^\infty \{Y^* - k\vec{e}_n\}$, $S = ]0,1[^{n-1} \times \{0\}$, $S_b = ]0,1[^{n-1} \times \{b\}$, $Z^+ = ]0,1[^{n-1} \times ]0,+\infty[$ and $Z_{BL} = Z^- \cup S \cup Z^+$ (the flow region). Let $\Pi = \cup_{k=1}^\infty \{\partial Z^* - k\vec{e}_n\}$.

We define a vector space $V$ by

$$V = \{z \in L^2_{loc}(Z_{BL}) : \nabla_y z \in L^2(Z_{BL})^n;\ z \in L^2(Z^-);$$
$$z = 0 \text{ on } \Pi;\ z \text{ is } \tilde{y} - \text{periodic}\} \tag{A.1}$$

and functional space $W$ as a completion of $V$ in the norm

$$\|z\|_W = \|\nabla_y z\|_{L^2(Z_{BL})^n}. \tag{A.2}$$

It should be noted that for all $z$ in $W$ we have the Poincaré inequality

$$\|z\|_{L^2(Z^-)} \leq C_p \|\nabla_y z\|_{L^2(Z^-)^n}. \tag{A.3}$$

Having defined geometry and functional space we are ready to formulate our basic variational problem:

**Problem (AUX)** *For a given $\rho \in L^2(Z_{BL})$, such that $e^{\gamma_1|y_n|}\rho \in L^2(Z_{BL})$, and for a given $\sigma \in L^2(S)$ we seek $\kappa \in W$ satisfying*

$$\int \nabla_y \kappa \nabla_y \varphi = \int \varrho\varphi + \int_S \varphi\sigma, \quad \forall \varphi \in W. \tag{A.4}$$

The existence and uniqueness of solution of the problem (A.4) is an immediate consequence of Lax-Milgram's lemma.

**Lemma A1.** *Problem (AUX) is uniquely solvable.*

The next result concerns regularity. It is a consequence of the classical elliptic regularity for the Laplace's equation.

**Lemma A2.** *Let $\kappa \in W$ be a solution of (AUX).*

*Then $\kappa \in H^2_{loc}(Z^+ \cup Z^-)$.*

Finally we turn our attention to asymptotic behavior of $\kappa$ as $\|y_n\| \to \infty$. We have

**Proposition A1** *Let $\kappa \in W$ be a solution of* **(AUX)**.

*Then there exists $\gamma \in ]0, \gamma_1]$ such that $e^{\gamma_0\|y_n\|}\kappa \in H^1(Z^-)$.*

**Proof.**

It is given in [4], pp. 76-80.

**Proposition A2** *Let $\kappa \in W$ be a solution of* **(AUX)**.

*Then there exist constants $\gamma_0 \in ]0, \gamma_1]$ and $C_\kappa$ such that $e^{\gamma_0\|y_n\|}(\kappa - C_\kappa) \in L^2(Z^+)$ and $e^{\gamma_0\|y_n\|}\nabla\kappa \in L^2(Z^+)^n$.*

**Proof.**

It relies on Tartar's lemma (see [4], pp. 51-53) and can be found in [4], pp. 54-58.

**Lemma A3.** *Let $\kappa \in W$ be a solution of (AUX) with $\rho \in H^1(Z^+)$ and let $\gamma_0$ and $C_\kappa$ be as in Proposition A2.*

*Then we have*

$$\begin{cases} C_\kappa = \lim\limits_{a\to\infty} \int_{[0,1]^{n-1}} \kappa(\xi, \alpha)d\xi; \\ \|\kappa(\cdot, y_n) - C_k\|_{L^2(]0,1[^{n-1})} \leq Ce^{-\gamma_0|y_n|/2} \end{cases} . \tag{A.5}$$

**Proof.**

We start with a simple inequality

$$|\kappa(\tilde{y}, y_n) - \int_{[0,1]^{n-1}} \kappa(\xi, a)d\xi| \le C\|e^{\gamma_0|\cdot|}\frac{\partial\kappa}{\partial y_n}\|_{L^2(Z^+)}|e^{-\gamma_0|y_n|} - e^{-\gamma_0 a}|$$
$$+e^{-\gamma_0|y_n|}e^{\gamma_0|y_n|}|\kappa(\tilde{y}, y_n) - \int_{[0,1]^{n-1}} \kappa(\xi, y_n)d\xi|, \quad \forall\tilde{y} \in ]0,1[^{n-1},$$
$$\forall a > y_n > 0. \tag{A.6}$$

Applying the Poincaré inequality

$$e^{\gamma_0|y_n|}\left\|\kappa(\cdot, y_n) - \int_{[0,1]^{n-1}} \kappa(\xi, y_n)d\xi\right\|_{L^2(]0,1[^{n-1})} \le$$
$$\le C_p e^{\gamma_0|y_n|}\left\|\frac{\partial\kappa}{\partial y_n}(\cdot, y_n)\right\|_{L^2(]0,1[^{n-1})}$$

and Proposition A2 to (A.6) we get

$$|\kappa(\tilde{y}, y_n) - \int_{[0,1]^{n-1}\kappa(\xi,a)d\xi}| \le C\{|e^{-\gamma_0|y_n|/2} - e^{-\gamma_0 a/2}| \tag{A.7}$$
$$+e^{-\gamma_0|y_n/2|}\|e^{\gamma_0|y_n|}\nabla_y\kappa\|_{L^2(]0,1[^{n-1})} \quad \forall y_n \in R_+. \tag{A.8}$$

Using regularity theory we obtain

$$e^{\gamma_0\|y_n\|}\frac{\partial^2\kappa}{\partial y_i\partial y_j} \in L^2(Z^+)$$

and, consequently, the pointwise convergence of $e^{\gamma_0\|y_n\|}\nabla_y\kappa$ as $y_n \to \infty$. Hence in the limit $a \to \infty$ we get (A.5).

It remains to apply our results to the auxiliary problems introduced in Section 1.2.

The first problem under consideration is the problem (1.10)–(1.13).

In order to write variational formulation corresponding to (1.10)–(1.13) we have to eliminate the jump condition (1.12).

Let $w_1 = w_1(y) \in H^3(Z^+)$ be a periodic function in $\tilde{y}$ such that $w_1(\tilde{y}, +0) = w(\tilde{y}, 0)$ and $\frac{\partial}{\partial y_n}w_1(\tilde{y}, +0) = \frac{\partial}{\partial y_n}w(\tilde{y}, 0)$. Furthermore, let $k = k(y_n)$ be such that $k \in C^\infty[0, \infty[$, $k = 0$ for $y_n \ge 1$ and $k'(0) = 0$. Then we introduce $w_2(y) = k(y_n)\, w_1(y)$ for $y_n > 0$. Obviously, $w_2$ has a compact support. We proceed with a new unknown function

$$\psi = \begin{cases} \omega - \omega_2 & \text{for } y_n > 0; \\ \omega & \text{for } y_n < 0. \end{cases} \tag{A.9}$$

Then the variational problem corresponding to (1.10)–(1.13) is the following:

Find $\psi \in W$ satisfying

$$\int_{Z_{BL}} \nabla_y \psi \nabla_y \varphi = \int_{Z_{BL}} \Delta w_2 \varphi, \quad \forall \varphi \in W. \tag{A.10}$$

As a direct consequence of Lemmas A1-A3 and Propositions A1-A2, we have

**Corollary A1.** *The problem (A.10) is uniquely solvable and its solution* $\psi \in H^2_{loc}(Z_{BL})$. *Furthermore, the problem (1.10)–(1.13) is uniquely solvable,* $\omega \in H^2_{loc}(Z^+ \cup Z^-)$, *and there exists a constant* $\gamma_0$ *such that* $e^{\gamma_0 \|y_n\|} \nabla_y \in L^2(Z^+ \cup Z^-)^n$, $e^{\gamma_0\|y_n\|}\omega \in L^2(Z^-)$ *and* $e^{\gamma_0\|y_n\|}(\omega - C_0) \in L^2(Z^+)$, *where*

$$C_0 = \lim_{a \to \infty} \int_{]0,1[^{n-1}} \omega(\tilde{y}, a) d\tilde{y}. \tag{A.11}$$

**Lemma A4.** *Let* $\omega \in W$ *be a solution of (1.10)–(1.13) and let* $C_0$ *be defined by (A.11).*

*Then we have*

$$C_0 = \int_{]0,1[^{n-1}} \omega(\tilde{y}, +0) d\tilde{y}. \tag{A.12}$$

**Proof.**

We have

$$0 = \int_{]0,1[^{n-1}} \int_0^\infty \frac{\partial \omega}{\partial y_n} = -C_0 + \omega \int_{]0,1[^{n-1}} \omega(\tilde{y}, +0) d\tilde{y}$$

and the lemma is proved.

The next problem under consideration is the system (1.17)–(1.20). The corresponding variational formulation is:

Find $\beta \in W$ satisfying

$$\int_{Z_{BL}} \nabla_y \beta \nabla_y \varphi = \int_S \varphi, \quad \forall \varphi \in W. \tag{A.13}$$

Solubility and asymptotic behavior for (A.13) follow immediately from Lemmas A1-A3 and Propositions A1–A2.

**Corollary A2.** *The problem (A.13) is uniquely solvable and its solution* $\beta \in H^2_{loc}(Z_{BL})$. *Furthermore, there exists a constant* $\gamma_0$ *such that* $e^{\gamma_0\|y_n\|}\nabla_y\beta \in L^2(Z_{BL})^n$, $e^{\gamma_0\|y_n\|}\beta \in L^2(Z^-)$ *and* $e^{\gamma_0\|y_n\|}(\beta - C_\beta) \in L^2(Z^+)$, *where*

$$C_\beta = \lim_{a\to\infty} \int\limits_{[0,1]^{n-1}} \beta(\tilde{y},a)d\tilde{y} = \int\limits_{[0,1]^{n-1}} \beta(\tilde{y},0)d\tilde{y}. \tag{A.14}$$

We proceed with problem (1.25)–(1.26). The corresponding variational formulation is: Find $\zeta_i \in W$ satisfying

$$\int\limits_{Z_{BL}} \nabla_y\zeta - i\nabla_y\varphi = \int\limits_{Z_{BL}} \{\beta - C_\beta H(x_n)\}\frac{\partial\varphi}{\partial y_i} + C_\beta\delta_{in}\int\limits_S \varphi, \quad \forall\varphi \in W \tag{A.15}$$

As above, we get the following result:

**Corollary A3.** *The problem (A.15) is uniquely solvable and its solution* $\zeta_i \in H^2_{loc}(Z^+ \cup Z^-)$. *Furthermore, there exists a constant* $\gamma_0$ *such that* $e^{\gamma_0\|y_n\|}\nabla_y\zeta_i \in L^2(Z_{BL})^n$, $e^{\gamma_0\|y_n\|}\zeta_i \in L^2(Z^-)$ *and* $e^{\gamma_0\|y_n\|}(\zeta_i - C_{i,\zeta}) \in L^2(Z^+)$, *where*

$$C_{i,\zeta} = \lim_{a\to\infty} \int_{[0,1]^{n-1}} \zeta_i(\tilde{y},a)d\tilde{y} = \int_{[0,1]^{n-1}} \zeta_i(\tilde{y},0)d\tilde{y}, \quad 1 \le i \le n-1,$$

$$C_{n,\zeta} = \lim_{a\to\infty} \int_{[0,1]^{n-1}} \zeta_n(\tilde{y},a)d\tilde{y} = \int_{[0,1]^{n-1}} \zeta_n(\tilde{y},0)d\tilde{y} - \int_{[0,1]^{n-1}} \int_0^\infty (\beta - C_\beta)dy.$$

Finally we consider a problem

$$-\Delta\Xi = \beta - C_\beta H(x_n) \quad \text{in } Z_{BL} \tag{A.16}$$

$$\Xi = 0 \quad \text{on } \Pi, \quad \Xi \text{ is } \tilde{y}\text{–periodic.} \tag{A.17}$$

The corresponding variational formulation is:
Find $\Xi \in W$ satisfying

$$\int\limits_{Z_{BL}} \nabla_y\Xi\nabla_y\varphi = \int\limits_{Z_{BL}} \}\beta - C_\beta H(x_n)\}\varphi, \quad \forall\varphi \in W. \tag{A.18}$$

In analogy with the above results we have

**Corollary A4.** *The problem (A.18) is uniquely solvable and its solution* $\Xi \in H^2_{loc}(Z_{BL})$. *Furthermore, there exists a constant* $\gamma_0$ *such that* $e^{\gamma_0\|y_n\|}\nabla_y\Xi \in L^2(Z_{BL})^n$, $e^{\gamma_0\|y_n\|}\Xi \in L^2(Z^-)$ *and* $e^{\gamma_0\|y_n\|}(\Xi - C_\Xi) \in L^2(Z^+)$, *where*

$$C_\Xi = \lim_{a\to\infty} \int\limits_{[0,1]^{n-1}} \Xi(\tilde{y},a)d\tilde{y}.$$

## References


1. N. Bakhvalov and G. Panasenko, Homogenisation: Averaging Processes, *Periodic Media*, Kluwer, Dordrecht, 1989.
2. W. Jäger, O.A. Oleinik and A.S. Shamaev, On a homogenization problem for the Laplace operator in a partially perforated domain with the Neumann condition on holes, preprint ICTP no. SMR.719/3, Second Workshop on Composite Media and Homogenization, Trieste, September 1993.
3. J.L. Lions and E. Magenes, Problèmes aux limites non homogènes et applications, Vol. 1, Dunod, Paris, 1968.
4. J.L. Lions, Some Methods in the Mathematical Analysis of Systems and Their Control, Gordon and Breach, New York, 1981.
5. O.A. Oleinik, A.S. Shamaev and G.A. Yosifian, Mathematical Problems, *Elasticity and Homogenization*, Elsevier, Amsterdam, 1992.
6. E. Sanchez-Palencia, Non Homogeneous Media and Vibration Theory, Lecture Notes in Physics, 127, Springer, Berlin, 1980.

# CONTROL IN THE COEFFICIENTS OF LINEAR HYPERBOLIC EQUATIONS VIA SPACIO-TEMPORAL COMPONENTS

K.A. LURIE

*Department of Mathematical Sciences, Worcester Polytechnic Institute, 100 Institute Road, Worcester, MA 01609*

In this paper, we give illustrations of some specific features of control over the solutions of linear hyperbolic equations produced by the variation of their senior coefficients.

## 1 Introduction

Control in the coefficients is known to be very effective in *elliptic* case where the idea of artificially assembled composites has appeared to be the key issue. Much less is known, however, about the effect produced by variable, particularly chattering coefficients upon the solutions of *hyperbolic* equations. The conventional technique of homogenization reveals some dramatic consequences of chattering, among them the possibility of a total screening of some extended parts of a system from the invasion of long wave dynamic disturbances. The transformation of energy that accompanies oscillations in the coefficients may create instabilities and self excitation of a parametric resonance type. This effect is considered desirable and is, therefore, stimulated in some cases, in others it is suppressed. The position/time dependence of the coefficients may be materialized through appropriate external influences that may be specific for each individual problem.

The chattering coefficients mathematically reflect the material mixing operation. In the hyperbolic case, the chattering and, consequently, the physical mixing, may occur in space-time, and the relevant microstructures appear to be spatial-temporal composites. With regard to them, there arises the G-closure problem just as it arose for the relevant spatial mixtures in the elliptic case. The problem of hyperbolic (spatial-temporal) G-closure obtains a covariant formulation in this paper, and some features of a G-closure are specified in the case of one-dimensional wave propagation.

## 2 Possible applications

Distributed systems being able to transport waves and generate them through the parametric excitation are widely used in electrical engineering [1]. Such devices as slow wave circuits and the traveling wave tubes form a basis for the effective coupling and transport of energy between the electron beams and transmission lines. One of the important implications of coupling is the frequency conversion desirable in many applications. Also, the parametric amplification of space-charge waves can be made practically noiseless.

These effects are produced due to an appropriate behavior of distributed material parameters of a system as functions of position and time. The specific electrical characteristics of the line, i.e. its linear inductance and capacitance, are subjected to an external activation in the form of a "pump wave" traveling along the line. This activation provides the energy needed for an effective coupling. The capacitance is activated through the use of P-N junction diodes distributed along the line and thus serving as variable capacity elements. The activation of inductance is accomplished through the use of a ferromagnetic medium into which the line is embedded. Both types of activation are maintained electronically at a desired speed.

In acoustics, a similar type of control may be implemented in the propagation of sound waves through ducts with variable cross section. If a duct is filled by a mixture of water and air (or vapor) bubbles, then under a suitable value of bubbles' concentration, the effective velocity of sound in the mixture may be made fairly low, of the order of meters per/sec, and the cross-section of the duct may be subjected to a mechanical change in the form of a periodic wave traveling along the duct at a comparable speed. The duct will then be acting like a transmission line with variable parameters [2].

In structural dynamics, there are numerous examples of catastrophic consequences of dynamic instability caused by parametric resonance (the famous Tacoma bridge collapse in 1940 is believed to be caused by this phenomenon [3]). A suitable control of the coefficients, i.e. the spatial-temporal distribution of material characteristics, may provide an effective protection against the development of unstable and excessively powerful vibrations.

This list of possible applications is far from complete; it reveals the desirability of optimization produced by the variable coefficients.

## 3 The structure of the paper

The paper consists of two parts. In Part I, we explore the *spatial-temporal laminates* - examples of chattering microstructures in space-time. These laminates are introduced in connection with the simplest hyperbolic equation

$$(\rho v_t)_t - (k v_z)_z = 0 \tag{3.1}$$

with variable coefficients $\rho = \rho(t,z), k = k(t,z)$ taking values from an admissible set $U$. An elliptic counterpart of this situation is given by the equation

$$(\lambda w_x)_x + (\lambda w_y)_y = 0 \tag{3.2}$$

with $\lambda = \lambda(x,y)$; the spatial variability of coefficients is then particularly implemented in the purely spatial laminates.

Homogenization procedure applied to (1) generates formulas for the *effective material parameters* of a hyperbolic system. These parameters participate in the characterization of a hyperbolic G-closure (cf. [4] where this notion is discussed for an elliptic case) of the original set $U$ formed by the admissible values of $\rho$ and $k$. The laminates constitute a part of the G-closure, but, in contrast to an elliptic case [4], it is still unknown whether this entire set may ever be covered by laminates alone. The description of a hyperbolic G-closure is significant for optimization purposes. Unfortunately, however, for Eq. (1) such a description is still inaccessible, and both Part I and Part II of the paper contain information that may help to build it.

G-closure is an invariant object, and its appropriate definition requires a covariant formulation of the basic equations, particularly, of Eq. (1). In Part II we show that to obtain such a formulation, we should perceive Eq. (1) *as a part of the Maxwell's system* formulated for the relevant material continuum. In this interpretation, $\rho$ may be treated as the magnetic permeability $\mu$ of the medium whereas $k$ is interpreted as its inverse dielectric permittivity $1/\epsilon$.

The Maxwell's system incorporates the fundamental relativistic invariant properties of space-time and, therefore, provides the required covariant description. Particularly, the material tensor $s$ of the medium is defined by parameters $\mu$ and $\epsilon$.

The Maxwell's system is of course formulated for the most general three dimensional case. In the case of one spatial dimension illustrated by Eq. (1), the G-closure possesses a special property: if the elements of $U$ have the same value of $\det s$, then $\det s_0 = \det s$, where $s_0$ is an effective tensor of the medium after homogenization. This is a hyperbolic analog of the conservation law that holds for a two-dimensional polycrystal appearing in the relevant problems for Eq. (2)[4].

## Part I. Spatial-temporal laminar microstructures

### 4 Second order hyperbolic equation with chattering coefficients

The effect of chattering coefficients will be demonstrated here for the model equation (1). We will assume that the coefficients $\rho, k$ of this equation are fast periodic, i.e., that they are periodic functions of the argument $(z - Vt)/\epsilon$ where $\epsilon$ is a small parameter and $V$ is the velocity of the "properties' wave"

$$\rho = \rho\left(\frac{z - Vt}{\epsilon}\right), \quad k = k\left(\frac{z - Vt}{\epsilon}\right). \tag{4.1}$$

At $t = 0$, this wave becomes a periodic pattern

$$\rho = \rho(z/\epsilon), \quad k = k(z/\epsilon),$$

and (3) may be perceived as this pattern moving with velocity $V$ along the $z$-axis.

To make this description more specific, we will assume that:

(a) at each point $(z, t)$ the pair of controls $\rho$ and $k$ can take either the set of values $(\rho_1, k_1)$ or the set $(\rho_2, k_2)$ ( we will refer to "material 1" and "material 2", respectively; the set $U$ becomes defined as $(\rho_1, k_1) \cup (\rho_2, k_2)$);

(b) these materials are placed within alternating layers having the slope $dz/dt = V$ in the $(z, t)$ - plane;

(c) the period of the control pattern is combined of two successive layers filled, respectively, by materials 1 and 2, the volume fractions of these layers being $m_1$ and $m_2$ $(m_1,\ m_2 \geq 0,\ m_1 + m_2 = 1)$.

We will consider solutions of Eq. (1) defined in a spatial-temporal region $\Delta$ : $((0, \ell) \times (0, T))$ and satisfying the boundary and initial conditions

$$u(0,t) = g_1(t), \quad u(\ell, t) = g_2(t),$$

$$u(z,0) = v(z,0) = 0;$$

the functions $g_i$ will be assumed differentiable infinitely many times, and the compatibility conditions

$$\frac{d^j g_i(0)}{dt^j} = 0$$

satisfied for all $j \geq 0$.

The relevant solutions will be assumed smooth, i.e. belonging to $W_2^1((0, \ell) \times (0, T))$.

The smoothness property implies certain restrictions on the shape of the interfaces $L$ between the neighboring subregions of $\Delta$ occupied by different materials. Each interface will be assumed piecewise smooth, i.e. possessing no more than finite number of corner points. Between those points, the slope $V = dz/dt$ of the interface should satisfy the inequality

$$\frac{V^2 - a_1^2}{V^2 - a_2^2} \geq 0 \tag{4.2}$$

where $a_i = \sqrt{k_i/\rho_i}$, $i = 1, 2$ stands for the squares of the phase velocity in materials 1 and 2.

Inequality (4.2) guarantees the regular transport of the field vectors across the interface $L$ with the observance of the compatibility conditions

$$[v_t + V v_z]_1^2 = [k v_z + V \rho v_t]_1^2 = 0.$$

To satisfy these conditions, we ultimately need as many as two outgoing characteristics on the interface. Assume that $a_2 > a_1$, then Ineq. (4.2) will be satisfied either for $V^2 \leq a_1^2$ or for $V^2 \geq a_2^2$. In the first case, there will be one outgoing characteristic pointing to material 1 and another one pointing to material 2 (Fig. 1). In the second case, both characteristics will point into one of those materials (material 2 in Fig. 2). When $a_1 = a_2$, the banned interval $(a_1^2, a_2^2)$ for $V^2$ shrinks to a point. Ineq. (4.2) is obviously satisfied if either $V = 0$ or $V = \infty$.

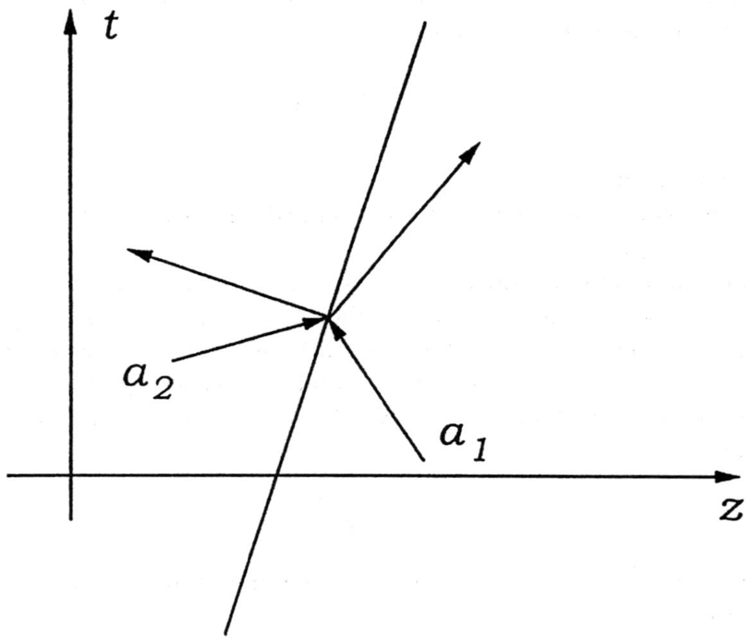


Figure 1: Case $V^2 < a_1^2$.

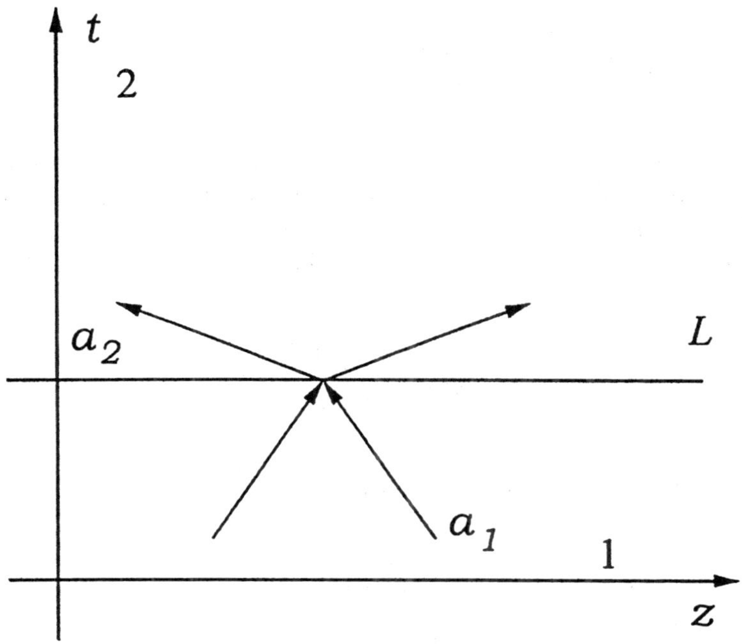


Figure 2: Case $V^2 > a_2^2$.

Applying the standard procedure of homogenization [5], we arrive at the following differential equation for the function $\langle v \rangle$ - the value of $v$ averaged over the period of the array [6]:

$$\frac{1}{a_1^2 a_2^2}\left[V^2 - \bar{k}\left(\overline{\frac{1}{\rho}}\right)\right]\langle v\rangle_{tt} + 2V\left[\bar{\rho}\left(\overline{\frac{1}{k}}\right) - \left(\overline{\frac{1}{a^2}}\right)\right]\langle v\rangle_{tz}$$
$$- \bar{\rho}\left(\overline{\frac{1}{k}}\right)\left[V^2 - \frac{1}{\bar{\rho}\left(\overline{\frac{1}{k}}\right)}\right]\langle v\rangle_{zz} = 0. \qquad (4.3)$$

Here, $\langle v\rangle = m_1 v_1 + m_2 v_2$, $\bar{k} = m_1 k_2 + m_2 k_1$, etc.; the subscripts 1 and 2 are related to materials 1 and 2, respectively. Eq. (5) is hyperbolic once Ineq. (4) is satisfied. It has no dispersion, i.e. it possesses wave solutions $\langle v\rangle(z - \lambda_1 t)$, $\langle v\rangle(z - \lambda_2 t)$. The characteristic velocities $\lambda_1$ and $\lambda_2$ have the product

$$\lambda_1\lambda_2 = -a_1^2 a_2^2 \bar{\rho}\left(\overline{\frac{1}{k}}\right)\frac{V^2 - \frac{1}{\bar{\rho}\left(\overline{\frac{1}{k}}\right)}}{V^2 - \bar{k}\left(\overline{\frac{1}{\rho}}\right)}. \qquad (4.4)$$

Assuming that $a_2 = \sqrt{k_2/\rho_2} \geq a_1 = \sqrt{k_1/\rho_1}$, we shall consider two cases:

(a) $k_2 \geq k_1$, $\rho_2 \leq \rho_1$ (regular mode),

(b) $k_2 \geq k_1$, $\rho_2 \geq \rho_1$ (irregular mode).

Inequality $a_2 \geq a_1$ that apparently holds for case (a), will be assumed to hold for case (b) as well.

Equations (5), (6) show that

$$\lambda_1 = -\lambda_2 = \sqrt{\langle k^{-1}\rangle^{-1}\langle \rho\rangle^{-1}} \text{ for } V = 0 \text{ ( static spatial pattern)}, \qquad (4.5)$$
$$\lambda_1 = -\lambda_2 = \sqrt{\langle k\rangle\langle \rho^{-1}\rangle} \text{ for } V = \infty \text{ ( dynamic temporal pattern)}.$$

It can be seen from these formulas that for the regular mode the long wave disturbances may propagate at the speed that always belongs to the interval $[a_1, a_2]$. For the irregular mode, however, this speed *may lie beyond* this interval. In particular, if $V = 0$ then always $\lambda_1^2 = \lambda_2^2 \leq a_2^2$;

however, both of the inequalities $\lambda_1^2 = \lambda_2^2 \geq a_1^2$ and $\lambda_1^2 = \lambda_2^2 \leq a_1^2$ may occur. Also, if $V = \infty$, then always $\lambda_1^2 = \lambda_2^2 \geq a_1^2$, but there may either be $\lambda_1^2 = \lambda_2^2 \leq a_2^2$ or $\lambda_1^2 = \lambda_2^2 \geq a_2^2$. Long wave disturbances are slowed down by a static pattern and may be speeded up by a dynamic pattern.

In the first case, high frequency harmonics are suppressed because of redistribution of energy through the multiple reflections and refractions on the interfaces that do not move. In the second case ($V = \infty$), the situation is different. Energy is then *pumped* into the system each time the value of $\rho$ decreases, and it is *withdrawn* from it as $\rho$ increases. The phasing of these events, by a suitable choice of $m_1$, affects the phase velocity of the long wave disturbances.

Another phenomenon associated with the irregular mode occurs when $V \neq 0(\infty)$. Eq. (6) shows that $\lambda_1\lambda_2$ may become positive. This happens when $V^2$ belongs to one of the following two ranges:

$$\frac{1}{\bar{\rho}\left(\overline{\frac{1}{k}}\right)} < V^2 < a_1^2 < a_2^2 (\text{ slow range}).$$

$$a_1^2 < a_2^2 < V^2 < \bar{k}\left(\overline{\frac{1}{\rho}}\right) (\text{fast range}).$$

Both ranges are possible for the irregular mode and do not apply for the regular mode.

With this choice of $V^2$, one obtains spatial-temporal arrays (laminates) for which both effective phase velocities $\lambda_1, \lambda_2$ become of the same sign. This sign may be switched to opposite as we go from $V$ to $-V$.

The relevant waves will become *coordinated*: both of them will propagate in the same direction relative to a laboratory frame. We will then refer to the right (left) laminate if both waves travel to the right (left).

The use of such laminates may implement the effect of screening mentioned in the introduction. To this end we may place the right laminate in the first quadrant $t \geq 0, x > 0$ of the $(t, x)$-plane, and the left laminate in the second quadrant. Within a sector bounded by the characteristic rays starting at the origin and having slopes $-\lambda_2, \lambda_2(|\lambda_1| > |\lambda_2|)$, there appears to be a shadow zone protected from the invasion of long wave disturbances initiated at $t = 0$. Such disturbances will propagate into the rest of the half plane $t > 0$.

The shadow zone was observed by the present author [7] in a control problem for the first order equation

$$v_t + kv_z = 0$$

with $k(t, z)$ treated as a control. This zone was then totally screened from the invasion of dynamic disturbances of all frequencies.

In contrast to this, the screening now affects the long-wave disturbances alone. If the system occupies a segment $[a, b]$ of the $x$-axis, then the boundary layers will arise close to its endpoints. High frequency waves will be reflected from the endpoints and travel back to the interior of $[a, b]$, so the coordinated wave motion will not apply for such waves.

Coordinated wave motion will not apply also for long-wave disturbances if materials 1 and 2 differ either *only in* $\rho$ or *only in* $k$. Both parameters should be different for these materials to make coordinated wave motion possible.

## 5 Application of the Floquet theory

Along with the standard homogenization technique [5], the analysis of Eq. (1) with the coefficients periodic in the argument $z - Vt$ can be carried out with the means of Floquet theory. Eq. (5) then appears to be a long wave Floquet approximation related to a low frequency passing band.

To implement this analysis, we start from the system

$$\rho v_t = u_z, \quad kv_z = u_t \tag{5.1}$$

that replaces Eq. (1).

Introducing new variables

$$\zeta = z - Vt, \quad \tau = t, \tag{5.2}$$

we replace (8) by an equivalent system ($a^2 = k/\rho$)

$$\begin{aligned} v_\zeta &= \frac{Vv_\tau - \frac{1}{\rho}u_\tau}{V^2 - a^2}, \\ u_\zeta &= -\frac{kv_\tau - Vu_\tau}{V^2 - a^2}. \end{aligned} \tag{5.3}$$

Parameters $k, \rho, a$ in these equations are assumed to be periodic functions of the argument $\zeta$. Applying the Laplace transform with respect to $\tau$ and preserving the symbols $v$ and $u$ for transforms, we reduce (10) to the system

$$v_\zeta = p\frac{Vv - \frac{1}{\rho}u}{V^2 - a^2},$$
$$u_\zeta = -p\frac{kv - Vu}{V^2 - a^2} \tag{5.4}$$

with periodic coefficients. Let $d = \ell_1 + \ell_2$ be the period; the part $(-\ell, 0)$ of the periodic cell $(-\ell_1, \ell_2)$ will be occupied by material 1, and the part $(0, \ell_2)$ by material 2. In accordance with Floquet theory, we shall seek for a solution $v, u$ of (11) possessing the property

$$v(\zeta) = e^{\mu d} v(\zeta - d), \quad u(\zeta) = e^{\mu d} u(\zeta - d) \tag{5.5}$$

with parameter $\mu$ to be determined.

We define such a solution by the formulas

$$\left.\begin{aligned} v &= Ae^{\frac{p\zeta}{V-a_1}} + Be^{\frac{p\zeta}{V+a_1}} \\ u &= -\rho_1 a_1 Ae^{\frac{p\zeta}{V-a_1}} + B\tfrac{k_1}{a_1}e^{\frac{p\zeta}{V+a_1}} \end{aligned}\right\} - \ell_1 \le \zeta \le 0,$$

$$\left.\begin{aligned} v &= Ce^{\frac{p\zeta}{V-a_2}} + De^{\frac{p\zeta}{V+a_2}} \\ u &= -\rho_2 a_2 Ce^{\frac{p\zeta}{V-a_2}} + D\tfrac{k_2}{a_2}e^{\frac{p\zeta}{V+a_2}} \end{aligned}\right\} 0 \le \zeta \le \ell_2,$$

$$\left.\begin{aligned} v &= Ae^{\mu d}e^{p\frac{\zeta-d}{V-a_1}} + Be^{\mu d}e^{p\frac{\zeta-d}{V+a_1}} \\ u &= -\rho_1 a_1 Ae^{\mu d}e^{p\frac{\zeta-d}{V-a_1}} + B\tfrac{k_1}{a_1}e^{\mu d}e^{p\frac{\zeta-d}{V+a_1}} \end{aligned}\right\} \ell_2 \le \zeta \le \ell_1 + \ell_2.$$

We demand that $v(0^-) = v(0^+)$, $u(0^-) = u(0^+)$, $v(\ell_2 - 0) = v(\ell_2 + 0)$, $u(\ell_2 - 0) = u(\ell_2 + 0)$. In view of (12), these conditions yield, respectively,

$$\begin{cases} A + B = C + D, \\ -A\rho_1 a_1 + B\frac{k_1}{a_1} = -C\rho_2 a_2 + D\frac{k_2}{a_2}, \\ Ae^{\mu d - \frac{p\ell_1}{V-a_1}} + Be^{\mu d - \frac{p\ell_1}{V+a_1}} = Ce^{\frac{p\ell_2}{V-a_2}} + De^{\frac{p\ell_2}{V+a_2}}, \\ -A\rho_1 a_1 e^{\mu d - \frac{p\ell_1}{V-a_1}} + B\frac{k_1}{a_1}e^{\mu d - \frac{p\ell_1}{V+a_1}} = -C\rho_2 a_2 e^{\frac{p\ell_2}{V-c_2}} + D\frac{k_2}{a_2}e^{\frac{p\ell_2}{V+a_2}}. \end{cases} \tag{5.6}$$

We set the determinant of this system equal to zero; introducing the volume fractions $m_1 = \ell_1/d_1$, $m_2 = \ell_2/d$, we arrive at the equation

$$e^{2\mu d} e^{-2pd\frac{m_1 V}{V^2-a_1^2}} - 2e^{\mu d}\left[C_1C_2 + \frac{1}{2}\left(\frac{\gamma_1}{\gamma_2} + \frac{\gamma_2}{\gamma_1}\right) S_1S_2\right] + e^{2pd\frac{m_2 V}{V^2-a_2^2}} = 0 \tag{5.7}$$

that defines unknown parameter $\mu$. In (14), the symbols $C_1, \ldots, \gamma_2$ are defined by the formulas

$$C_1 = \frac{1}{2}\left[e^{-pd\frac{m_1}{V-a_1}} + e^{-pd\frac{m_1}{V+a_1}}\right],$$

$$C_2 = \frac{1}{2}\left[e^{pd\frac{m_2}{V-a_2}} + e^{pd\frac{m_2}{V+a_2}}\right],$$

$$S_1 = \frac{1}{2}\left[e^{-pd\frac{m_1}{V-a_1}} - e^{-pd\frac{m_1}{V+a_1}}\right],$$

$$S_2 = -\frac{1}{2}\left[e^{pd\frac{m_2}{V-a_2}} - e^{pd\frac{m_2}{V+a_2}}\right],$$

$$\gamma_1 = \frac{k_1}{a_1} = \rho_1 a_1 = \sqrt{k_1\rho_1}, \quad \gamma_2 = \sqrt{k_2\rho_2}.$$

The coefficients in (14) depend on $s = pd$; this parameter will be assumed small (compared with $a_1$), and the roots of (14) evaluated accordingly.

After calculations (requiring power expansions of the coefficients up to terms quadratic in $s$) we arrive at the following formula for the roots;

$$(e^{\mu d})_{1,2} = 1 + \frac{s}{(V^2 - a_1^2)(V^2 - a_2^2)}$$
$$\left[V(V^2 - \bar{a^2}) \pm a_1a_2\sqrt{\bar{\rho}\left(\bar{\frac{1}{k}}\right)\left(V^2 - \frac{\bar{k}}{\bar{\rho}}\right)\left(V^2 - \frac{\left(\bar{\frac{1}{\rho}}\right)}{\left(\bar{\frac{1}{k}}\right)}\right)}\right].$$

Applying the linear approximation

$$e^{\mu d} \sim 1 + \mu d,$$

we obtain

$$\mu_{1,2} = \frac{p}{(V^2 - a_1^2)(V^2 - a_2^2)}$$
$$\left[V(V^2 - \bar{a^2}) \pm a_1a_2\sqrt{\bar{\rho}\left(\bar{\frac{1}{k}}\right)\left(V^2 - \frac{\bar{k}}{\bar{\rho}}\right)\left(V^2 - \frac{\left(\bar{\frac{1}{\rho}}\right)}{\left(\bar{\frac{1}{k}}\right)}\right)}\right] \tag{5.8}$$

If the materials 1 and 2 are identical, then

$$\mu_{1,2} = \frac{p}{V \pm a}$$

where $a$ is the common phase velocity. Plus and minus signs are related to the waves traveling in opposite directions with respect to a laboratory frame.

In the case of different materials, the waves propagate with velocities $\lambda_1, \lambda_2$ relative to the same frame, and we get

$$\mu_{1,2} = \frac{p}{V + \lambda_{1,2}}.$$

Referring to (15) we obtain for $\lambda_{1,2}$ the expressions

$$\lambda_{1,2} = \frac{V a_1^2 a_2^2 \left[\bar{\rho}\left(\frac{\bar{1}}{k}\right) - \left(\frac{\bar{1}}{a_2}\right)\right] \mp a_1 a_2 \sqrt{\bar{\rho}\left(\frac{\bar{1}}{k}\right)\left(V^2 - \frac{\bar{k}}{\bar{\rho}}\right)\left(V^2 - \frac{\left(\frac{\bar{1}}{\rho}\right)}{\left(\frac{\bar{1}}{k}\right)}\right)}}{V^2 - \bar{k}\left(\frac{\bar{1}}{\rho}\right)}.$$

It can be verified by a direct inspection that these expressions represent the phase velocities related to Eq. (5).

## Part II. Covariant formulation of the problem of spatial – temporal material mixing

The problem of material mixing has been intensively explored in statics with the original materials distributed in space alone (see [4] and bibliography cited therein). In this part of the paper, we investigate a similar problem for a *spatial-temporal* distribution of original materials. Specifically, we consider a dielectric continuum being able to take either the values $(\epsilon_1, \mu_1)$ or the values $(\epsilon_2, \mu_2)$ of its dielectric permittivity and magnetic permeability at each point $(x, y, z)$ at every instant $t$ of time. This scheme does not necessarily require the physical motion of a material medium to become feasible. The space-time dependent properties can be created through the appropriate "switching" of material constants generated by a suitable electronic scheme. For example, the transmission line may have its linear inductance $L$ and capacitance $C$ take one of the two admissible pairs of values $(L_1, C_1)$ and $(L_2, C_2)$ at each point $(z, t)$ of space-time, the necessary adjustment being implemented due to the electronic (or even mechanical) switching. Note that this switching may be conducted at *any* speed; in particular, it may be instantaneous.

In what follows, we give a mathematical formulation of a problem of spatial-temporal mixing of two originally given dielectrics in the framework of Maxwell's theory. Our reference to Maxwell's theory is more than a reference to a special physical example.

The formalism of this theory is *universal* in a sense that it can be directly applied to an arbitrary physical embodiment, e.g. to problems of structural mechanical vibrations of non-homogeneous elastic bodies. In all cases, the material medium will be assumed immovable, and the required spatial-temporal dependence of its properties implemented by the artificial means mentioned above.

## 6 The one-dimensional wave propagation: a special use of Maxwell's system

To simplify technicalities, we begin with the case of a plane electromagnetic wave propagating along the $z$-axis through the medium characterized by isotropic dielectric and magnetic permeabilities $\epsilon(z,t)$ and $\mu(z,t)$. The pair $(\epsilon(z,t),\mu(z,t))$ is assumed to take at each point $(z,t)$ of space-time only one of two admissible pairs of values:

$$(\epsilon(z,t),\mu(z,t)) = \begin{cases} (\epsilon_1,\mu_1) \ (\text{"material 1"}), \\ (\epsilon_2,\mu_2) \ (\text{"material 2"}). \end{cases} \tag{6.1}$$

Any relevant region in the $(z,t)$-plane is thus divided into subregions occupied either by material 1 or by material 2. In the absence of currents and charges, the Maxwell's system includes the equations

$$\begin{aligned} \text{curl}\mathbf{E} &= -\mathbf{B}_t, \\ \text{div}\mathbf{B} &= 0, \\ \text{curl}\mathbf{H} &= \mathbf{D}_t, \\ \text{div}\mathbf{D} &= 0. \end{aligned} \tag{6.2}$$

For a plane electromagnetic wave, this system reduces to

$$\begin{aligned} E_z &= B_t, \\ H_z &= D_t \end{aligned} \tag{6.3}$$

for the relevant components $E, B, H, D$ of the basic vectors $\mathbf{E}(z,t)$, $\mathbf{B}(z,t)$, $\mathbf{H}(z,t), \mathbf{D}(z,t)$ :

$$\mathbf{E} = E\mathbf{j}, \quad \mathbf{B} = B\mathbf{i}, \quad \mathbf{H} = H\mathbf{i}, \quad \mathbf{D} = D\mathbf{j}. \tag{6.4}$$

Eqs. (6.3) correspond to the first and the third of Eqs. (6.2), respectively; the second and the fourth equations of this system are satisfied identically.

By introducing the scalar potentials $\varphi, \varphi^*$ and the vector potentials $\mathbf{A}, \mathbf{A}^*$ through the formulas

$$\begin{aligned} \mathbf{E} &= -\text{grad}\varphi - \mathbf{A}_t, \quad \mathbf{B} = \text{curl}\mathbf{A}, \\ \mathbf{H} &= -\text{grad}\varphi^* + \mathbf{A}^*_t, \quad \mathbf{D} = \text{curl}\mathbf{A}^*, \end{aligned} \tag{6.5}$$

and by taking

$$\mathbf{A} = -u\mathbf{j}, \quad \mathbf{A}^* = v\mathbf{i}, \quad \varphi = \varphi^* = 0, \tag{6.6}$$

we refer to (6.4) and obtain

$$E = u_t, \quad B = u_z, \quad H = v_t, \quad D = v_z. \tag{6.7}$$

Eqs. (6.3) are thereby satisfied. We now have to apply the material equations to make the Maxwell's system complete. In a laboratory frame (with respect to which a medium is at rest), these equations are given by

$$\mathbf{D} = \epsilon\mathbf{E}, \quad \mathbf{B} = \mu\mathbf{H}. \tag{6.8}$$

We arrive, finally, at the system

$$v_z = \epsilon u_t, \quad u_z = \mu v_t \tag{6.9}$$

governing the wave propagation in one spatial dimension. Clearly, this system will become identical to (8) if we replace $\mu$ by $\rho$ and $\epsilon$ by $1/k$. The phase velocity $c = 1/\sqrt{\epsilon\mu}$ will then be replaced by $a = \sqrt{k/\rho}$.

## 7 The Maxwell's system for a moving dielectric medium

If a dielectric medium is at rest with respect to the laboratory frame $(x, y, z)$, then in the absence of currents and charges the electromagnetic field in it is described by Eqs. (6.2) and (6.8).

The pairs of vectors $\mathbf{B}, \mathbf{E}$ and $\mathbf{H}, \mathbf{D}$ generate, respectively, two skew-symmetric tensors in 4-space ($x_1 = x, \quad x_2 = y, \quad x_3 = z, \quad x_4 = ict$)

$$F = (c\mathbf{B}, -i\mathbf{E}) = \begin{pmatrix} 0 & cB_3 & -cB_2 & -iE_1 \\ -cB_3 & 0 & cB_1 & -iE_2 \\ cB_2 & -cB_1 & 0 & -iE_3 \\ iE_1 & iE_2 & iE_3 & 0 \end{pmatrix}, \tag{7.1}$$

$$f = (\mathbf{H}, -ic\mathbf{D}) = \begin{pmatrix} 0 & H_3 & -H_2 & -icD_1 \\ -H_3 & 0 & H_1 & -icD_2 \\ H_2 & -H_1 & 0 & -icD_3 \\ icD_1 & icD_2 & icD_3 & 0 \end{pmatrix}. \tag{7.2}$$

Here, $c$ stands for the velocity $1/\sqrt{\epsilon_0\mu_0}$ of light in a vacuum, and $B_1, \ldots, D_3$ are the relevant vector components.

The Maxwell's system now takes a tensor form incorporated in the equations

$$\frac{\partial F^*_{ik}}{\partial x_k} = 0, \quad \frac{\partial f_{ik}}{\partial x_k} = 0,$$

where $F^*_{ik}$ is a tensor dual to $F_{ik}$, i.e.

$$F^*_{ik} = \frac{1}{2} e_{iklm} F_{lm}.$$

Here, $e_{iklm}$ is a completely antisymmetric tensor of the forth rank.

*Remark*

In a special case (6.4), (6.7), we put $F_{1i} = -F_{i1} = F_{34} = -F_{43} = f_{1i} = -f_{i1} = f_{34} = -f_{43} = 0$ for $i = 1, \ldots, 4$, and the tensors $F$ and $f$ reduce to

$$F = \begin{pmatrix} 0 & cB & -iE \\ -cB & 0 & 0 \\ iE & 0 & 0 \end{pmatrix} = \begin{pmatrix} 0 & cu_z & -iu_t \\ -cu_z & 0 & 0 \\ iu_t & 0 & 0 \end{pmatrix} = c \begin{pmatrix} 0 & u_{x3} & u_{x4} \\ -u_{x3} & 0 & 0 \\ -u_{x4} & 0 & 0 \end{pmatrix} \tag{7.3}$$

$$f = \begin{pmatrix} 0 & H & -icD \\ -H & 0 & 0 \\ icD & 0 & 0 \end{pmatrix} = \begin{pmatrix} 0 & v_t & -icv_z \\ -v_t & 0 & 0 \\ icv_z & 0 & 0 \end{pmatrix} = ic \begin{pmatrix} 0 & v_{x4} & -v_{x3} \\ -v_{x4} & 0 & 0 \\ v_{x3} & 0 & 0 \end{pmatrix} \tag{7.4}$$

Because these matrices appear to be the $3 \times 3-$ submatrices of (7.1) and (7.2) standing in their right low corners, the numbers $2, 3, 4$ generate the relevant indexing.

The material equations (6.8) are combined in a single tensor equation — a linear relationship between tensors $f$ and $F$:

$$f_{ik} = s_{iklm} F_{lm}. \tag{7.5}$$

This equation introduces the fourth rank tensor $s = (s_{iklm})$. This tensor is symmetric with respect to a simultaneous permutation of indices in the pairs $(i, l)$ and $(k, m)$ as well as to the permutation of pairs $(il)$ and $(km)$:

$$s_{iklm} = s_{lmik} = s_{kiml}. \tag{7.6}$$

In a laboratory frame, the non-zero components of $s$ are defined for an immovable isotropic dielectric by the formulas:

$$\begin{aligned} &s_{1212} = s_{2121} = -1/\mu c, \quad s_{1313} = s_{3131} = -1/\mu c, \quad s_{2323} = s_{3232} = -1/\mu c, \\ &s_{1414} = s_{4141} = -\epsilon c, \quad s_{2424} = s_{4242} = -\epsilon c, \quad s_{3434} = s_{4343} = -\epsilon c. \end{aligned} \tag{7.7}$$

In the space of skew-symmetric tensors of the second rank in four dimensions we introduce an orthogonal basis defined as a set of skew-symmetric second rank tensors $a_{ik}(i, k = 1, 2, 3)$ specified by the formulas

$$\begin{aligned} a_{12} = \frac{1}{\sqrt{2}}(i_1 i_2 - i_2 i_1), \ a_{13} &= \frac{1}{\sqrt{2}}(i_1 i_3 - i_3 i_1), \ a_{14} = \frac{1}{\sqrt{2}}(i_1 i_4 - i_4 i_1), \\ a_{23} &= \frac{1}{\sqrt{2}}(i_2 i_3 - i_3 i_2), \ a_{24} = \frac{1}{\sqrt{2}}(i_2 i_4 - i_4 i_2), \\ & \qquad a_{34} = \frac{1}{\sqrt{2}}(i_3 i_4 - i_4 i_3). \end{aligned} \tag{7.8}$$

Here, $i_1, \ldots, i_4$ denote the unit vectors of Cartesian axes $x_1, \ldots, x_4$ in a four dimensional Euclidean space. The formulas

$$a_{ik} : a_{lm}^T = \begin{cases} 1, \ i = l, k = m, \\ 0 \quad \text{otherwise} \end{cases}$$

allow to represent every skew-symmetric tensor

$$q = \begin{pmatrix} 0 & q_{12} & q_{13} & q_{14} \\ -q_{12} & 0 & q_{23} & q_{24} \\ -q_{13} & -q_{23} & 0 & q_{34} \\ -q_{14} & -q_{24} & -q_{34} & 0 \end{pmatrix}$$

in the form

$$q = \bar{q}_{12}a_{12} + \bar{q}_{13}a_{13} + \bar{q}_{14}a_{14} + \bar{q}_{23}a_{23} + \bar{q}_{24}a_{24} + \bar{q}_{34}\bar{a}_{34}$$

where $\bar{q}_{ik} = \sqrt{2}q_{ik} = q \ : \ a_{ik}^T$.

In particular, every skew-symmetric tensor in a 3-subspace $(x_2, x_3, x_4)$

$$q = \begin{pmatrix} 0 & q_{23} & q_{24} \\ -q_{23} & 0 & q_{34} \\ -q_{24} & -q_{34} & 0 \end{pmatrix}$$

is represented in the form

$$q = \bar{q}_{23}a_{23} + \bar{q}_{24}a_{24} + \bar{q}_{34}\bar{a}_{34}$$

with tensors $a_{23}, a_{24}, a_{34}$ constituting the relevant basis .

In a special case (6.4), (6.7), the $F$ - and $f$ - tensors are given by the formulas (see(7.3), (7.4) and (6.7))

$$\begin{aligned} F &= \sqrt{2}(cBa_{23} - iEa_{24}) = \sqrt{2}c(u_{x_3}a_{23} + u_{x_4}a_{24}), \\ f &= \sqrt{2}(Ha_{23} - icDa_{24}) = \sqrt{2}ic(v_{x_4}a_{23} - v_{x_3}a_{24}). \end{aligned} \tag{7.9}$$

Getting back to a general case, let us consider tensors $a_{ik}$ of the set (7.8) as the primary entities and introduce the elementary symmetric functions of the second degree of these tensors. We obtain as many as 21 such functions given by the following table:

$$a_{12}a_{12}, a_{12}a_{13} + a_{13}a_{12}, a_{12}a_{14} + a_{14}a_{12}, a_{12}a_{23} + a_{23}a_{12},$$
$$a_{12}a_{24} + a_{24}a_{12}, a_{12}a_{34} + a_{34}a_{12},$$

$$a_{13}a_{13}, a_{13}a_{14} + a_{14}a_{13}, a_{13}a_{23} + a_{23}a_{13}, a_{13}a_{24} + a_{24}a_{13},$$
$$a_{13}a_{34} + a_{34}a_{13},$$

$$a_{14}a_{14}, a_{14}a_{23} + a_{23}a_{14}, a_{14}a_{24} + a_{24}a_{14},$$
$$a_{14}a_{34} + a_{34}a_{14}, \quad (7.10)$$

$$a_{23}a_{23}, a_{23}a_{24} + a_{24}a_{23},$$
$$a_{23}a_{34} + a_{34}a_{23},$$

$$a_{24}a_{24},$$
$$a_{24}a_{34} + a_{34}a_{24},$$

$$a_{34}a_{34}.$$

The most general linear form of these functions represents the second rank symmetric tensor in the space of $a_{ik}$, i.e. the fourth rank tensor in the original space $(x_1, x_2, x_3, x_4)$. This latter tensor possesses a special symmetry of indices given by formulas (7.6). For example, we have

$$a_{12}a_{23} + a_{23}a_{12} = \frac{1}{2}[i_1i_2i_2i_3 + i_2i_1i_3i_2 - i_1i_2i_3i_2 - i_2i_1i_2i_3 + i_2i_3i_1i_2 + i_3i_2i_2i_1 - i_2i_3i_2i_1 - i_3i_2i_1i_2],$$

and the required symmetry is directly observed. A unit tensor $e$ in the space of skew-symmetric second rank tensors $a_{ik}$ is given by

$$e = -a_{12}a_{12} - a_{13}a_{13} - a_{14}a_{14} - a_{23}a_{23} - a_{24}a_{24} - a_{34}a_{34}. \quad (7.11)$$

The $s$-tensor for an isotropic dielectric medium that does not move with respect to a laboratory frame is given in this frame by the formula

$$s = -\frac{1}{\mu c}(a_{12}a_{12} + a_{13}a_{13} + a_{23}a_{23}) - \epsilon c(a_{14}a_{14} + a_{24}a_{24} + a_{34}a_{34}). \quad (7.12)$$

When $1/(\mu c) = \epsilon c$, i.e. $c = 1/\sqrt{\epsilon\mu}$, then $s$ becomes proportional to a unit tensor $e$. In particular, in the 3-dimensional space $(x_2, x_3, x_4)$, the relevant set of symmetric functions is given by the last three lines of (7.10), i.e. by the functions

$$a_{23}a_{23}, a_{23}a_{24} + a_{24}a_{23}, a_{23}a_{34} + a_{34}a_{23},$$
$$a_{24}a_{24}, a_{24}a_{34} + a_{34}a_{24}, \quad (7.13)$$
$$a_{34}a_{34}.$$

We thus obtain 6 such functions, and the most general linear form of them defines the forth rank tensor in $(x_2, x_3, x_4)$ possessing the symmetry of indices specified by (7.6).

A unit tensor in the subspace of tensors $a_{23}, a_{24}, a_{34}$ is given by

$$e = -a_{23}a_{23} - a_{24}a_{24} - a_{34}a_{34}; \tag{7.14}$$

the $s$-tensor in the same space is defined as

$$s = -\frac{1}{\mu c}a_{23}a_{23} - \epsilon c(a_{24}a_{24} + a_{34}a_{34}). \tag{7.15}$$

Eq. (7.5) is a tensor equation between two tensors $F$ and $f$ (see (7.1), (7.2)) in a Euclidean 4-space $(x_1, x_2, x_3, x_4)$. The proportionality factor $s$ in this equation is characterized by a clear geometric structure in the case of isotropic dielectric. This structure is incorporated in Eqs. (7.12) and (7.8). We observe that a purely spatial rotation does not affect either of the tensors $a_{12}a_{12} + a_{13}a_{13} + a_{23}a_{23}$ and $a_{14}a_{14} + a_{24}a_{24} + a_{34}a_{34}$ appearing in parentheses at the right hand side of (7.12). A uniform motion of the coordinate frame with the velocity $V$ along the $x_3$-axis is equivalent to its rotation about the $(x_1, x_2)$-plane by an imaginary angle $i\psi$ in the 4-space, with the value of $\psi$ being completely determined by $V$. In a moving ("primed") frame, the components of $F$ and $f$ may be expressed through their values in a laboratory frame by virtue of the formulas expressing the effect of rotation (Lorentz transformation). This procedure will affect all of the tensor dyads in (7.12) except $a_{12}a_{12}$ and $a_{34}a_{34}$ generating the well-known Minkowski's material relations for a moving medium. The $s$-tensor (7.12) is thus isotropic with respect to purely spatial rotations and anisotropic with respect to the motion of the coordinate frame. It will become *completely isotropic*, i.e. isotropic with respect to *all* rotations in 4-space, provided that $1/(\mu c) = \epsilon c$, i.e. $c^2 = 1/(\epsilon\mu)$. This case is exceptional: it holds for a vacuum where $\epsilon = \epsilon_0,\ \ \mu = \mu_0,\ \ c^2 = 1/(\epsilon_0\mu_0)$, and the $s$-tensor becomes proportional to a unit tensor (7.11)

$$s = \sqrt{\frac{\epsilon_0}{\mu_0}}e.$$

For isotropic, but not completely isotropic materials, we shall assume that their refraction index

$$n = c\sqrt{\epsilon\mu}$$

is never less than unity:

$$\epsilon\mu \geq 1/c^2, \quad i.e. \ \ \epsilon c \geq 1/(\mu c). \tag{7.16}$$

Since the eigenvalues of the isotropic tensor (7.12) are defined as

$$d_1 = \frac{1}{\mu c}, \quad d_2 = \epsilon c, \tag{7.17}$$

the assumption (7.16) means that

$$d_2 \geq d_1. \tag{7.18}$$

For different original materials, there arise various problems of mixing them on a fine scale in space-time. We may consider the spatial-temporal microstructures generated by two different isotropic materials occupying periodic cells in space-time but *motionless in a laboratory frame.* This means that the relevant tensors $s$ differ only in their eigenvalues, their eigentensors $a_{ik}$ being preserved identical. This problem belongs to the type mentioned in the beginning of Part II: the temporal variability of material properties is then achieved through switching.

Another thinkable problem may be that of a "spatial-temporal polycrystal" in which there appear microstructures assembled from fragments of one and the same isotropic material brought each to *its own motion.* This pattern represents a direct analog of a polycrystal assembled from fragments that differ only by their orientation with regard to laboratory axes. In space-time, the difference in orientation is due to a relative motion.

This type of microstructure is conceptually possible though its practical implementation is yet to come.

Returning to the first of the mixing problems, we show in the next section that the effective properties of the relevant microstructures allow for an important detailing in a special case of one-dimensional wave propagation when the $s$-tensors of participating materials have the same determinant, i.e. the product of their eigenvalues $d_1 d_2$, or the same ratio $\epsilon/\mu$.

## 8 One-dimensional wave propagation: conservation of the determinant of the material tensor

Because of (7.9), we will be interested in rotation of a coordinate frame $(x_1, x_2, x_3, x_4)$ that leaves intact the $x_1$ and $x_2$-axes. This rotation introduces a primed coordinate frame; it affects the vectors $i_3, i_4$ and, consequently, all of the tensors $a_{ik}$ listed in (7.8) except $a_{12}$. The vectors $i_3'$ and $i_4'$ will become linear functions of $i_3$ and $i_4$, and tensors $a_{13}'$ and $a_{14}'$ will consequently become linear functions of $a_{13}$ and $a_{14}$, and tensors $a_{23}', a_{24}'$ will depend on $a_{23}$ and $a_{24}$. As to $a_{34}$, this tensor will not be affected by this rotation: $a_{34}' = a_{34}$.

The described rotation occurs in a subspace of tensors $a_{23}, a_{24}$; the unit tensor in this subspace becomes

$$e = -a_{23}a_{23} - a_{24}a_{24}, \tag{8.1}$$

whereas the $s$-tensor in it may be specified as

$$s = -\frac{1}{\mu c}a_{23}a_{23} - \epsilon c a_{24}a_{24}. \tag{8.2}$$

Consider the tensor

$$O = a_{23}a_{24} - a_{24}a_{23} \tag{8.3}$$

and introduce two linearly independent tensors $F(1)$ and $F(2)$ (see (7.9)) generated, respectively, by two linearly independent test fields $u(1)$ and $u(2)$. The expression

$$\begin{aligned} F(2) : O : F(1) = 2c^2(u_{x_3}(1)u_{x_4}(2) - u_{x_4}(1)u_{x_3}(2)) = \\ = 2c^2 \det(\nabla u(1), \nabla u(2)) \end{aligned} \tag{8.4}$$

is quasiaffine in $W_2^1(\Delta)$, i.e.

$$\lim \mathrm{wk}_{L_1(\Delta)} 2c^2 \det(\nabla u^{(r)}(1), \nabla u^{(r)}(2)) = 2c^2 \det(\nabla u^{(0)}(1), \nabla u^{(0)}(2)), \tag{8.5}$$

where $r$ is the index of partition of $\Delta$ into subdomains occupied by materials 1 or 2, and

$$\lim \mathrm{wk}_{W_2^1(\Delta)} u^{(r)}(i) = u^{(0)}(i), \ \ i = 1, 2.$$

A similar behavior is demonstrated by the expression

$$\begin{aligned} f(2) : O : f(1) = 2c^2(v_{x_3}(1)v_{x_4}(2) - v_{x_4}(1)v_{x_3}(2)) = \\ 2c^2 \det(\nabla v(1), \nabla v(2)). \end{aligned}$$

This is also quasiaffine, i.e.

$$\lim \mathrm{wk}_{L_1(\Delta)} 2c^2 \det(\nabla v^{(r)}(1), \nabla v^{(r)}(2)) = 2c^2 \det(\nabla v^{(0)}(1), \nabla v^{(0)}(2)) \tag{8.6}$$

where

$$\lim \mathrm{wk}_{W_2^1(\Delta)} v^{(r)}(i) = v^{(0)}(i), \ i = 1, 2.$$

Because $f = s : F$, we rewrite (8.6) as

$$\begin{aligned} &\lim \mathrm{wk}_{L_1(\Delta)} 2c^2 \det(s^{(r)} : \nabla u^{(r)}(1), s^{(r)} : \nabla u^{(r)}(2)) \\ &= 2c^2 \det(s_0 : \nabla u^{(0)}(1), s_0 : \nabla u^{(0)}(2)). \end{aligned} \tag{8.7}$$

Here we introduced $s_0$ - an effective material tensor, i.e. the symmetric tensor that appears in the homogenized material equation

$$f^{(0)} = s_0 : F^{(0)}. \tag{8.8}$$

The symbols $F^{(0)}$ and $f^{(0)}$ are defined as weak limits of tensors $F^{(r)}$ and $f^{(r)}$, respectively. Bearing in mind that for $(2 \times 2)$-matrices

$$\det(s : \nabla u(1), s : \nabla u(2)) = (\det s) \det(\nabla u(1), \nabla u(2)),$$

we transform (8.7) to

$$\begin{aligned} &\lim \mathrm{wk}_{L_1(\Delta)} (\det s^{(r)}) \det(\nabla u^{(r)}(1), \nabla u^{(r)}(2)) \\ &\qquad = (\det s_0) \det(\nabla u^{(0)}(1), \nabla u^{(0)}(2)). \end{aligned} \tag{8.9}$$

Assume now that

$$\det s^{(r)} = \epsilon/\mu = \mathrm{const}(r);$$

then, comparing (8) and (8.5), we conclude that

$$\det s_0 = \epsilon/\mu. \tag{8.10}$$

In other words, if the determinant of the $s$-tensor (8.2) is the same for both of the original materials (16), then the effective tensor $s_0$ possesses the same determinant. This is a hyperbolic analog of the conservation law that holds for a two-dimensional polycrystal appearing in planar elliptic problems of the second order as a two-dimensional composite microstructure (cf. [4]). We see that a similar law also holds for spatial-temporal composites in one spatial dimension; the microgeometry of

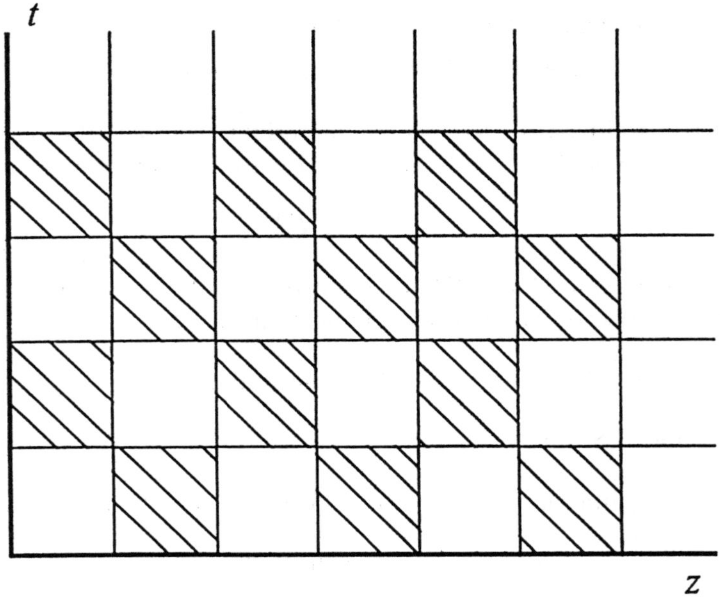


Figure 3: An assemblage for which (52) holds.

such composites in space-time should, however, be consistent with Ineq. (4.2). An example of such a microgeometry is given by a chess-board pattern shown in Fig. 3.

The conservation law (8.10) can be directly checked for spatial-temporal laminates of the first rank considered in Part I. The relevant formulas are reproduced in Appendix where we also verify the law for laminates of the second rank.

**Acknowledgments**

The author gratefully acknowledges the National Science Foundation, through grant DMS–9803476.

## Appendix A. Spatial-Temporal Laminates of Rank 1

We will reproduce in this Appendix some details of the homogenization procedure carried out in [6] for rank 1 spatial-temporal laminate, this procedure leading, in particular, to Eq. (5). We will refer to the system (24) with the coefficients $\epsilon, \mu$ characterized by properties (a)-(c), Section 4, where we bear in mind the correspondence $\mu \to \rho, \quad \epsilon \to 1/k$ mentioned at the end of Section 6.

After homogenization, Eqs. (24) are replaced by

$$\begin{aligned} \alpha u_z + \beta V u_t &= V v_z + v_t, \\ V u_z + u_t &= \theta(\alpha v_z + \beta V v_t), \end{aligned} \tag{A.1}$$

with parameters $\alpha, \beta, \theta$ defined as [6]

$$\begin{aligned} \alpha &= \frac{1}{\mu_1\mu_2}\frac{\overline{\left(\frac{1}{c^2}\right)V^2 - 1}}{\bar{\epsilon}V^2 - \overline{\left(\frac{1}{\mu}\right)}}, \\ \beta &= \epsilon_1\epsilon_2\frac{V^2 - \bar{c^2}}{\bar{\epsilon}V^2 - \overline{\left(\frac{1}{\mu}\right)}}, \\ \theta &= \frac{\bar{\epsilon}V^2 - \overline{\left(\frac{1}{\mu}\right)}}{\epsilon_1\epsilon_2\langle\frac{1}{\mu}\rangle V^2 - \frac{1}{\mu_1\mu_2}\langle\epsilon\rangle}, \end{aligned} \tag{A.2}$$

$c = 1/\sqrt{\epsilon\mu}$, $\bar{\epsilon} = m\epsilon_2 + (1-m)\epsilon_1$, $\langle\epsilon\rangle = m\epsilon_1 + (1-m)\epsilon_2$, etc. In (A.1), we preserve the original symbols $u, v$ to denote the weak limits of these quantities, i.e. their values averaged over the cell of periodicity.

An equivalent systems of equations is given by the formulas

$$\begin{aligned} pu_z - qu_t &= v_t, \\ qu_z + ru_t &= v_z, \end{aligned} \tag{A.3}$$

with parameters $p, q, r$ defined as

$$p = \frac{V^2 - \theta\alpha^2}{\theta(V^2\beta - \alpha)}, \quad q = \frac{V(1-\theta\alpha\beta)}{\theta(V^2\beta - \alpha)}, \quad r = -\frac{1 - V^2\theta\beta^2}{\theta(V^2\beta - \alpha)}. \tag{A.4}$$

The system (A.3) can be simplified if we introduce a moving coordinate frame $z', t'$ by the formulas

$$z' = z - (q/r)t, \quad t' = t. \tag{A.5}$$

The frame $(z', t')$ moves relative to that of $(z, t)$ with a transfer velocity $v_T = q/r$. In the primed frame, Eqs. (A.3) become

$$\begin{cases} \frac{pr+q^2}{r} u_{z'} = v_{t'}, \\ ru_{t'} = v_{z'}. \end{cases} \tag{A.6}$$

If we compare this system with Eqs. (24), then the complex $(pr + q^2)/r$ will be treated as an inverse effective magnetic permeability $1/M$, and the factor $r$ will appear to be an effective dielectric permeability $\mathcal{E}$.

We apply Eqs. (A.1) to show that

$$pr + q^2 = 1/\theta, \tag{A.7}$$

and Eqs. (A.2) to obtain the formulas

$$\begin{aligned} \frac{pr+q^2}{r} &= \frac{1}{\mu_1\mu_2}\frac{\bar{\mu}V^2 - \overline{\left(\frac{1}{\epsilon}\right)}}{V^2 - \overline{\left(\frac{1}{\mu}\right)\left(\frac{1}{\epsilon}\right)}} = \frac{1}{M}, \\ r &= \epsilon_1\epsilon_2 \frac{V^2 - \overline{\left(\frac{1}{\mu}\right)\left(\frac{1}{\epsilon}\right)}}{\bar{\epsilon}V^2 - \overline{\left(\frac{1}{\mu}\right)}} = \mathcal{E}. \end{aligned} \tag{A.8}$$

The last two formulas are equivalent to the relationship

$$\frac{1}{\langle\mu\rangle\langle\frac{1}{\mu}\rangle}\frac{M-\langle\mu\rangle}{M-\frac{1}{\langle\frac{1}{\mu}\rangle}}=\frac{1}{\langle\epsilon\rangle\langle\frac{1}{\epsilon}\rangle}\frac{\mathcal{E}-\langle\epsilon\rangle}{\mathcal{E}-\frac{1}{\langle\frac{1}{\epsilon}\rangle}}=\frac{V^2}{\langle\frac{1}{\mu}\rangle\overline{\left(\frac{1}{\epsilon}\right)}}. \tag{A.9}$$

If $\epsilon_1/\mu_1=\epsilon_2/\mu_2$, then the first Eq. (A.9) shows that

$$\epsilon_1/\mu_1=\epsilon_2/\mu_2=\mathcal{E}/M, \tag{A.10}$$

which confirms Eq. (52).

The transfer velocity $q/r$, by virtue of Eqs. (A.2), (A.4), is computed as

$$\frac{q}{r}=-c_1^2c_2^2V\frac{\bar{\epsilon}\bar{\mu}-\overline{\left(\frac{1}{c^2}\right)}}{V^2-\overline{\left(\frac{1}{\mu}\right)\left(\frac{1}{\epsilon}\right)}}. \tag{A.11}$$

In a laboratory frame, the effective phase velocities $\wedge_1, \wedge_2$ of disturbances defined by Eqs. (A.1) or (A.3) are equal to

$$\begin{aligned}\wedge_{1,2}&=v_T\pm 1/\sqrt{\mathcal{E}M}=\frac{1}{r}\left(q\pm\sqrt{pr+q^2}\right)\\&=c_1c_2\frac{-Vc_1c_2\left[\bar{\epsilon}\bar{\mu}-\overline{\left(\frac{1}{c^2}\right)}\right]\pm\sqrt{\left[\bar{\epsilon}V^2-\overline{\left(\frac{1}{\mu}\right)}\right]\left[\bar{\mu}V^2-\overline{\left(\frac{1}{\epsilon}\right)}\right]}}{V^2-\overline{\left(\frac{1}{\mu}\right)\left(\frac{1}{\epsilon}\right)}}.\end{aligned}$$

## Appendix B. Modulation: Spatial – Temporal Laminates of Rank 2

We consider a rank two spatial-temporal laminate assembled from rank one composites $C1$ and $C2$ used as original materials instead of materials 1 and 2 above. These composites are in their turn laminates constructed from materials 1 and 2 applied on a smaller scale.

Composites $C1$ and $C2$ differ from each other by the values of $m, V$, and, consequently, $\wedge_1$ and $\wedge_2$. The relevant values for $C1$ will be denoted as $m_{11}, V_1, \wedge_{11}, \wedge_{12}$, and for $C2$ as $m_{21}, V_2, \wedge_{21}, \wedge_{22}$.

The rank two composite will be characterized by the volume fraction $m_1$ of $C1$ in it, as well as by the slope $U$ of alternating layers filled by

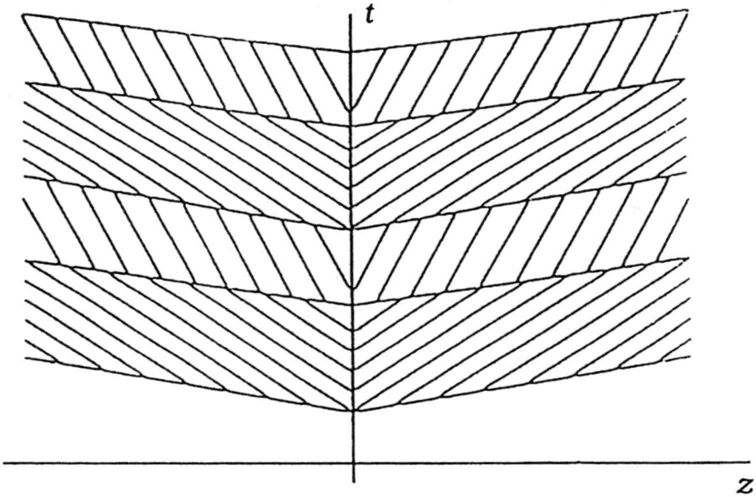


Figure 4: Rank two composite in space-time.

$C1$ and $C2$ (Fig. 4). The overall volume fraction $m$ of material 1 in the rank two composite will be equal to

$$m = m_1 m_{11} + (1 - m_1) m_{21}, \tag{B.1}$$

and material 2 in this composite will be represented in a volume fraction

$$1 - m = m_1(1 - m_{11}) + (1 - m_1)(1 - m_{21}). \tag{B.2}$$

The slopes $V_1, V_2$ should both satisfy Ineq. (4), and the slope $U$ must obey the inequality

$$\frac{(U - \wedge_{11})(U - \wedge_{12})}{(U - \wedge_{21})(U - \wedge_{22})} \geq 0 \tag{B.3}$$

in order to observe compatibility conditions on the interface between the layers occupied by $C1$ and $C2$. The effective behavior of the rank two composite will be given by the system

$$\begin{aligned} Au_z + BUu_t &= Uv_z + v_t, \\ Uu_z + u_t &= \Theta(Av_z + BUv_t), \end{aligned} \tag{B.4}$$

similar to (A.1).

As before, we preserve symbols $u$ and $v$ to designate the weak limits of these quantities, the symbols $A, B, \Theta$ are defined by

$$A = \frac{\langle \frac{\pi}{D} \rangle}{\langle \frac{1}{D} \rangle}, \quad BU = \frac{\langle \frac{\rho}{D} \rangle}{\langle \frac{1}{D} \rangle}, \quad \Theta = \frac{\langle \frac{1}{D} \rangle}{\langle \frac{pr+q^2}{D} \rangle}, \tag{B.5}$$

where

$$\pi = p + qU, \quad \rho = rU - q, \quad D = \rho U - \pi. \tag{B.6}$$

Parameters $p, q, r$ in these formulas are defined for each of the components $C1$ and $C2$ by the relevant Eqs. (A.4); we apply notation $p_1, q_1, r_1$ and $p_2, q_2, r_2$ for these composites, as well as the concurrent notation $\alpha_1, \beta_1, \theta_1, V_1$ and $\alpha_2, \beta_2, \theta_2, V_2$. The operation $\langle \cdot \rangle$ in (B.5) is defined as

$$\langle \cdot \rangle = m_1 (\cdot)_1 + (1 - m_1)(\cdot)_2. \tag{B.7}$$

After some calculation we obtain the formulas (cf (A.2))

$$A = \frac{\pi_1 \pi_2 \left[ U \overline{\left( \frac{\rho}{\pi} \right)} - 1 \right]}{U \bar{\rho} - \bar{\pi}},$$

$$BU = \frac{\rho_1\rho_2\left[U - \overline{\left(\frac{\pi}{\rho}\right)}\right]}{U\bar{\rho} - \bar{\pi}}, \tag{B.8}$$
$$\Theta = \frac{U\bar{\rho} - \bar{\pi}}{\kappa},$$

$$\kappa = \rho_1\rho_2\langle p\rangle - \pi_1\pi_2\langle r\rangle + \langle q\rangle(\rho_1\pi_2 + \rho_2\pi_1).$$

Here, again, $\pi_1 = p_1 + q_1 U$, $\langle p\rangle = m_1 p_1 + (1 - m_1)p_2$, $\bar{\pi} = m_1\pi_2 + (1 - m_1)\pi_1$, etc. We now introduce the quantities $P, Q, R$ linked with $A, B, \Theta, U$ by the formulas (A.4) in which $P, Q, R$ replace $p, q, r$, and $A, B, \Theta, U$ replace $\alpha, \beta, \theta, V$. The system (B.4) is equivalent to

$$\begin{aligned} Pu_z - Qu_t &= v_t, \\ Qu_z + Ru_t &= v_z, \end{aligned} \tag{B.9}$$

(an analog of (A.3)). With the aid of new coordinates

$$z' = z - (Q/R)t, \quad t' = t,$$

we replace (B.9) by

$$\begin{aligned} \frac{PR + Q^2}{R} u_{z'} &= v_{t'}, \\ Ru_{t'} &= v_{z'}, \end{aligned}$$

and qualify the complex $(PR + Q^2)/R$ as an inverse effective magnetic permeability $1/M$, and the factor $R$ as an effective dielectric permeability $\mathcal{E}$. As before, we prove that

$$PR + Q^2 = 1/\Theta,$$

and find that

$$\frac{PR + Q^2}{R} = \frac{1}{M} = \frac{1}{\Theta\mathcal{E}},$$

i.e.

$$\frac{\mathcal{E}}{\mathcal{M}} = \frac{1}{\Theta}. \tag{B.10}$$

Assuming now that $\epsilon_1/\mu_1 = \epsilon_2/\mu_2$, we will show that

$$\Theta = \frac{\mu_1}{\epsilon_1} = \frac{\mu_2}{\epsilon_2}, \tag{B.11}$$

and, consequently,

$$\frac{\mathcal{E}}{\mathcal{M}} = \frac{\epsilon_1}{\mu_1} = \frac{\epsilon_2}{\mu_2}.$$

To demonstrate (B.11), apply direct calculation to show that

$$U\bar{\rho}-\bar{\pi} = m_1(U^2r_2-2Uq_2-p_2)+m_2(U^2r_1-2Uq_1-p_1) = m_1S_2+m_2S_1 = \bar{S},$$

and

$$\kappa = m_1(p_1r_1 + q_1^2)S_2 + m_2(p_2r_2 + q_2^2)S_1.$$

Once $\epsilon_1/\mu_1 = \epsilon_2/\mu_2$, we have by (A.7)-(A.10)

$$p_1r_1 + q_1^2 = \frac{1}{\theta_1} = p_2r_2 + q_2^2 = \frac{1}{\theta_2} = \frac{\epsilon_1}{\mu_1} = \frac{\epsilon_2}{\mu_2},$$

and, consequently,

$$\frac{1}{\Theta} = \frac{\kappa}{U\bar{\rho} - \bar{\pi}} = \frac{(p_1r_1 + q_1^2)\bar{S}}{\bar{S}} = p_1r_1 + q_1^2 = \frac{\epsilon_1}{\mu_1} = \frac{\epsilon_2}{\mu_2}, \quad QED.$$

## References


1. W.H. Louisell, Coupled Mode and Parametric Electronics, Wiley, 1960.
2. P.M. Morse, and K. Uno Ingard, Theoretical Acoustics, McGraw Hill, 1968.
3. V.V. Bolotin, The Dynamic Stability of Elastic Systems, Holden Day, San Francisco, 1964.
4. K.A. Lurie, and A.V. Cherkaev,Effective Characteristics of Composite Materials and the Optimal Design of Structural Elements, in Topics in Mathematical Modeling, ed. by A.V. Cherkaev and R.V. Kohn, Birkhauser, 1997.
5. N.S. Bakhvalov, and G.P. Panasenko, Homogenisation: Averaging Processes in Periodic Media, Kluwer, Dordrecht, 1989.
6. K.A. Lurie, Effective Properties of Smart Elastic Laminates and the Screening Phenomenon, *International Journal of Solids and Structures*, **34** (13), 1633–1643 (1997).
7. K.A. Lurie, Discontinuities in Hyperbolic Optimum Problems, *Journal of Optimization Theory and Applications*, **7** (5), (1971).

# MULTISCALE AVERAGING FOR ODINARY DIFFERENTIAL EQUATIONS. APPLICATIONS TO THE SPECTRAL THEORY OF ONE-DIMENSIONAL SCHRÖDINGER OPERATOR WITH SPARSE POTENTIALS

S.A. MOLCHANOV

*Department of Mathematics, University of North Carolina, Charlotte, NC–28223, USA*

The paper contains the spectral theory of one-dimensional Schrödinger operator with deterministic sparse potential. It generalizes and extends the classical results by D. Pearson and the recent results by B. Simon and his group (Caltech). The technical tools are: phase formalism, martingale methods, multiscaling averaging.

## 1 Introduction

The simplest forms of the averaging principle are the statements about the integrals with fast oscillation. A typical example:

$$\text{if} \quad \varphi(\cdot) \in C([0,1]) \text{ then for } \quad n \to \infty$$
$$\lim \int_0^1 \varphi(x) |\cos nx| \, dx = \frac{2}{\pi} \int_0^1 \varphi(x) \, dx.$$

Slightly more complex (but still elementary ones) are the results in the spirit of Krilov-Bogoljubov theory of the averaging of the ODE's containing "fast" and "slow" variables. Let us consider, for instance, the following equation in $R^d$ :

$$\dot{X}_t^\varepsilon = b\left(X_t^\varepsilon, \xi_{t/\varepsilon}\right) \ , \ X_0^\varepsilon = x_0. \tag{1.1}$$

If the function $b(\cdot,\cdot)$ is regular enough and

$$\lim_{T\to\infty} \frac{1}{T} \int_0^T b(x, \xi_s) \, ds = \bar{b}(x) \tag{1.2}$$

(averaging is made with respect to the second fast oscillating variable $\xi$; the relation (1.2) is fulfilled in many cases: process $\xi_s$ is periodic, almost periodic, random, ergodic etc.), then, under natural technical conditions, for fixed $T$

$$\sup_{t\in[0,T]} \left|X_t^\varepsilon - \bar{X}_t\right| \to 0\ ,\ \varepsilon \to 0$$

where $\bar{X}_t$ is the solution of the problem

$$\dot{\bar{X}}_t = \bar{b}\,(X_t)\ ,\ \bar{x}_0 = x_0.$$

Unfortunately, this and similar results generally speaking are valid only for the finite time intervals. Results for infinite time interval, especially the stability or instability criteria for (1.1), are extremely important in the physical applications. The same is true in the cases when the system of ODE's contains not only one scaling parameter $\varepsilon$ but many (maybe infinite number) of such parameters.

We will discuss in this paper the idea of the multiscale averaging in the special context: scattering on the system of very distant (deterministic) bumps.

It is interesting that a pure probabilistic tool is used in this deterninstic problem.

I want to note that my scientific contacts with S. Kozlov started from discussing a similar situation. During one of our conversations I remarked that his idea of the harmonic coordinate in the homogenization of the (symmetric) elliptical or parabolic equation is similar to the martingale representation of the Markov processes and that the CLT for the martingales is well developed. With typical to Serguei rapid reaction he absorbed in one week all the circle of ideas around Billingsley-Brown form of the CLT for the martingale-differences. We published very soon the small paper [1]. Later [2] he applied the general martingale ideology of [1] to the homogenization of the lattice systems, more precisely to the random walks in the random environments with additional symmetry conditions (see also [9] for additional information).

In this paper I will use mainly not CLT for the martingales but more elementary result on the convergence of the series of the martingale differences (see Section 5), however, there exists a clear similarity between homogenization (in the spirit of [1]) and the multiscale averaging.

Let us return to the specific situation of the sparse potentials and give a brief review of the principle results in this area.

Many years ago D. Pearson [3] proved the following, now classical result. Let $\varphi(x)$, $x \in \mathbf{R}^1$ be a fixed continuous function with compact support (bumps shape), $\{a_n\}$, $a_n \to 0$ be a decreasing sequence of real numbers (bumps amplitudes), $\{x_n\}$ be a very fast increasing (roughly like $\exp(\exp cn)$, $c > 0$ sequence of the bumps centers. By definition, sparse potential has the form

$$q(x) = \sum_{n=1}^{\infty} a_n \varphi(x - x_n) \tag{1.3}$$

and one can analyze the spectral properties of the Schrödinger operator (Hamiltonian) $H^{\theta_0} = H$ given by

$$\begin{gathered} H\psi(x) = -\psi'' + q(x)\psi \\ \psi(0)\cos\theta_0 - \psi'(0)\sin\theta_0 = 0, \quad \theta_0 \in [0, \pi) \end{gathered} \tag{1.4}$$

on the space $L^2(\mathbf{R}^1_+)$. For $\lambda < 0$ the spectral problem $H\psi = \lambda\psi$ has only discrete spectrum (if any) with possible accumulation point $\lambda_0 = 0$. For $\lambda > 0$ the complete answer is given by
**Theorem A** (Pearson [3]). *If* $\sum_n a_n^2 < \infty$ *then* [1]

$$\begin{aligned} \mu\,(H) &= \mu_{\text{ac}}\,(H) \quad \text{on } [0, \infty) \\ \mu\,(H) &= \mu_{\text{d}}\,(H) \quad \text{on } (-\infty, 0] \end{aligned}$$

*if* $\sum_n a_n^2 = \infty$, $a_n \to 0$ *then*

$$\begin{aligned} \mu\,(H) &= \mu_{\text{sc}}\,(H) \quad \text{on } [0, \infty) \\ \mu\,(H) &= \mu_{\text{d}}\,(H) \quad \text{on } [0, \infty) \end{aligned}$$

It was the first general result on $\Sigma_{\text{sc}}$, the subject which is popular now due to recent activity of B. Simon and his group (see references).

Recent progress in the theory of 1-D Schrödinger operator was presented in the talk by B. Simon [4] at the ICDEMP (Atlanta, 97) and in the preprints [5], [6]. For a different approach to the same problem, see below, Section 7. Overlap of our paper with [4]-[5] is minimal.

[1]The reader who is not familiar with the subject may find explanations of the notation at the end of this section.

The main goal of this paper is to extend the Pearson's results in another direction. To stress the important role of the reflection properties of the bumps we eliminate the condition $a_n \to 0$, the bumps will be only bounded [2]. Condition for the rate of increasing of the bump centers $\{x_n\}$ will have a form

$$\sum \frac{x_n}{x_{n+1}} < \infty \qquad (\mathcal{SL}) \tag{1.5}$$

(superlacunarity, roughly $x_n \sim \exp(cn \ln n)$, $c > 1$), this condition is final for many of our results. Most important, that bumps will not have compact support, corresponding reflection coefficients will be only continuous (but not analytic as in [3]). Formally, we will work with the sparse potentials of the form:

$$q(x) = \sum_{n=1}^{\infty} g_n(x - x_n) \tag{1.6}$$

a) for some fixed constant $C > 0$, $\varepsilon > 0$

$$|g_n(x)| \leq \frac{C}{1 + |x|^{1+\varepsilon}}, \quad g_n(\cdot) \in C(\mathbf{R}^1) \tag{1.7}$$

b) $\sum_n \frac{x_n}{x_{n+1}} < \infty \quad (\mathcal{SL})$.

Let us formulate the main results of the paper. The statements below do not correspond to the specific theorems (which will have more general form): in the introduction we give everything in the simplest form and under the most understandable conditions.

**Theorem B.** *Let $\{\lambda_{n,i}\}$ be the set of negative eigenvalues for the individual scattering problem on $L^2(\mathbf{R}^1)$ :*

$$H_n\psi = -\psi'' + g_n(x)\psi = \lambda\psi, \ \lambda < 0, n = 1, 2...$$

*Then*

*a) $\sum_{\text{ess}} \left(H^{\theta_0}\right) \cap [-\infty, 0)$ is the set of limiting points for $\cup_n \{\lambda_{n,i}\}$. It means that*

$$\lambda \in \sum\nolimits_{\text{ess}} \left(H^{\theta_0}\right) \cap (-\infty, 0) \Rightarrow \lambda = \lim_{n' \to \infty} \lambda_{n', i(n')}$$

[2]The theory of increasing bumps is well understood, starting from the pioneering work by B. Simon, T. Spencer [7]. For additional information see [8]-[10]. In this case the spectrum as a rule is $\Sigma_{\text{sing}}$ and very often $\Sigma_{\text{pp}}$.

*for suitable sequence* $(n', i(n'))$.

*b) For a.e.* $\theta_0 \in [0, \pi]$

$$\mu\left(H^{\theta_0}\right) = \mu_{\text{pp}}\left(H^{\theta_0}\right) \text{ on } (-\infty, 0)$$

This p.p. spectrum is generally speaking nontrivial and contains two parts: $\Sigma_{\text{d}}(H^{\theta_0})$ and dense $\Sigma_{\text{pp}}$ on Int $(\Sigma_{\text{ess}}(H))$ (if the latter set is not empty).

Theorem B gives complete description of the spectrum for negative energies. Let now $\lambda = k^2 > 0$. Individual equations $H_n\psi = k^2\psi$ have unique Jost solutions for each $k = \sqrt{\lambda} > 0$ and $n = 1, 2, \cdots$

$$\begin{aligned} \psi_k^{(n)}(x) &= \exp(-ikx) + o(1), \quad x \to -\infty \qquad (1.8) \\ \psi_k^{(n)}(x) &= A_n(k)\exp(-ikx) + B_n(k)\exp(ikx) + o(1), \quad x \to -\infty \end{aligned}$$

Coefficients $A_n(k)$, $B_n(k)$ have well-known analytical properties, among them identity $|A|^2 = 1 + |B|^2$ and are related to inverse scattering problem (see [6] [12], [13]). Under our condition (1.7) they are continuous and even from the Hölder class (see later) but not analytic as in [3].

Expressions

$$a_n(k) = \frac{1}{|A_n(k)|}, \quad b_n(k) = \frac{|B_n(k)|}{|A_n(k)|}, \quad a_n^2 + b_n^2 = 1$$

are known as transmission and reflection coefficients on frequency $k > 0$ (and bump $g_n$).

The spectral properties of the Hamiltonian $H^{\theta_0}$ on the positive axis $\lambda = k^2 > 0$ depend on the convergence of the series $\sum_n |B_n(k|^2$. Let us put

$$\begin{aligned} \mathcal{A}_0 &= \{k : \sum_n |B_n(k)|^2 < \infty\} \qquad (1.9) \\ \mathcal{A}_1 &= \{k : \sum_n |B_n(k)|^2 = \infty\} \end{aligned}$$

$\widetilde{\mathcal{A}_0^c} \subset \mathcal{A}_0$ is the essential closure of $\mathcal{A}_0$, i.e., $k_0 \in \widetilde{\mathcal{A}_0^c}$ if for any $\varepsilon > 0$

$$|\mathcal{A}_0 \cap [k_0 - \varepsilon, k_+\varepsilon]| > 0.$$

Here and in what follows $|A|$ is the Lebesgue measure of the Borel set $A$.

**Theorem C.** For any $\theta_0$ and potential (1.5)

$$\sum{}_{\text{ac}}\left(H^{\theta}\right) = \widetilde{\mathcal{A}}_0^c\,,\ \sum{}_p\,(H) \cap [0,\infty) = \emptyset. \tag{1.10}$$

Let us remark that in many cases $\widetilde{\mathcal{A}_0}^{\mathrm{c}} = [0,\infty)$. We can not guarantee the absence of $\Sigma_{\text{sing}}$ on $\widetilde{\mathcal{A}}_0^c$. Sufficient condition for the pure absolutely continuous spectrum has a form

**Theorem D.** *If in some closed interval* $\triangle \subset (0,\infty]$ *the series* $\sum_n |B_n(k|^2$ *has dominated convergence, i.e., for appropriate* $\{\beta_n\}$

$$|B_n(k)|^2 \le \beta_n,\ k \in \triangle;\ \sum \beta_n < \infty$$

*then*

$$\mu\left(H^{\theta_0}\right) = \mu_{\text{ac}}\left(H^{\theta_0}\right) \text{ on } \triangle \tag{1.11}$$

There are different conditions for pure $\Sigma_{\text{ac}}$, for example, convergence $\sum_n |B_n^2(k)|$ in the suitable Hölder's norm.

Situation with $\Sigma_{\text{sing}}$ is similar. It must be concentrated on $\mathcal{A}_1$ or on the closure of some subset of $\mathcal{A}_1$. The problem is that both sets $\mathcal{A}_0, \mathcal{A}_1$ generally speaking are dense on $[0,\infty)$, spectra $\Sigma_{\text{ac}}$ and $\Sigma_{\text{sing}}$ ($\Sigma_{\text{pp}}$ is empty on $[0,\infty)$) can overlap. It is difficult to prove something about $\mu_{\text{sing}}$ on $\sum_{\text{ac}}$, there is no direct methods for $\Sigma_{\text{sing}}$ (the latter fact has been emphasized many times by B. Simon). The best result in this direction is the following one

**Theorem E.** *Assume that for a fixed open interval* $\triangle \subset (0,\infty)$

$$\sum_n |B_n^2(k)| = \infty \text{ a.e. on } \triangle\,. \tag{1.12}$$

*Then for all* $\theta_0 \in [0,\pi]$

$$\mu(H) = \mu_{\text{sc}}(H) \text{ on } \triangle\,.$$

From the Theorems D,E it follows

**Theorem F.** *Assume that for a given sparse potential* $q(\cdot)$ *once can construct partition* $(0,\infty) = A_0 \cup B \cup A_1$, *where* $A_0, A_1$ *are open,* $B$ *is closed,* $|B| = 0$. *Let* $A_0 = \cup\triangle_j^{(0)}$, $A_1 = \cup\triangle_j^{(1)}$, *are the representation of the corresponding open sets in terms of connected components (nonintersected open subintervals). If on each* $\triangle_i^{(0)}$ *condition of dominated convergence from Theorem D holds for any closed subinterval* $\widetilde{\triangle}_i^{(0)} \subset \triangle_i^{(0)}$, *on each* $\triangle_j^{(1)}$ *condition of Theorem E holds* $\left(\sum_n |B_n(k|^2 = \infty \text{ a.e. on } \triangle_j^{(1)}\right)$, *then*

$$\sum{}_{\mathrm{ac}}\left(H^{\theta_0}\right) = A_0{}^{\mathrm{c}}, \sum{}_{\mathrm{sc}}\left(H^{\theta_0}\right) = A_1{}^{\mathrm{c}}. \tag{1.13}$$

If the bumps have decaying amplitudes $\{a_n\}$, $a_n \to 0$ then the reflection coefficients $B_n(k)$ can be evaluated by the series of perturbation theory (like in [3] where D. Pearson used the first order perturbations).

**Theorem G.** *If*

$$q(x) = \sum_{n=1}^{\infty} a_n g_n (x - x_n) \tag{1.14}$$

*with conditions (1.4, 1.7) ) on* $\{x_n\}, \{g_n\}$ *then*

*a)* $\Sigma\left(H^{\theta_0}\right) \cap (-\infty, 0) = \Sigma_d\left(H^{\theta_0}\right), \Sigma_{pp}\left(H^{\theta_0}\right) \cap (0, \infty) = \emptyset.$

*b) Convergence of the series* $\sum_n a_n^2$ *implies*

$$\mu\left(H^{\theta_0}\right) = \mu_{ac}\left(H^{\theta_0}\right) \ \ on\ [0, \infty).$$

*c) If* $\sum_n |a_n|^3 < \infty$ *then*

$$\sum{}_{ac}\left(H^{\theta}\right) = \widetilde{\mathcal{A}_0^c}$$

*where* $\mathcal{A}_0 = \{k : \sum_n a_n^2 |\widehat{g}_n(2k)|^2 < \infty\}, \widetilde{\mathcal{A}_0^c}$ *is the essential closure of* $\mathcal{A}_0$.

*d) If for closed interval* $\triangle \subset (0, \infty)$

$$a_n^2 |\widehat{g}_n(2k)|^2 \leq \beta_n, \ \sum_n \beta_n < \infty, \ k \in \triangle \tag{1.15}$$

*then*

$$\mu\left(H^{\theta_0}\right) = \mu_{ac}\left(H^{\theta_0}\right) \ \ on\ \triangle.$$

*e) If* $\sum_n a_n^2 |\widehat{g}_n(2k)|^2 = \infty$ *a.e. on* $\triangle$ *then*

$$\mu\left(H^{\theta_0}\right) = \mu_{sc}\left(H^{\theta_0}\right) \ \ on\ \triangle. \tag{1.16}$$

**Remark.** In this paper we will not try to present the results in the most general form. In fact, instead of condition (1.7) one can use boundedness and tails estimations in terms of local $L^2$-norms. The same method works for Hamiltonian $H$ on $L^2(\mathbf{R}^1)$ with the potential $q(\cdot)$ sparse in both directions $x \to \pm\infty$, for the lattice models on $l^2\left(\mathbf{Z}_+^1\right), l^2(\mathbf{Z}_1)$ etc. The central idea here is to show how one can use new method based on the multistep homogenization and martingale theory in the spectral analysis. More complete text containing in addition many specific examples and counter examples will be published later.

The structure of the paper:

1. Introduction
2. Technical tools
3. Scattering theory
4. Tests for the different types of spectra.
5. Martingales and homogenization lemmas
6. Theorems for nondecreasing sparse potentials
7. Perturbation theory and decaying sparse potentials.
8. References.

*Notations.* Majority of notations are standard. We will mention only some of them, specific for the theory of one-dimensional Schrödinger operators

1. $H^\theta \psi = -\psi'' + q\psi\ ,\ \psi(0)\cos\theta - \psi'(0)\sin\theta = 0$ :

Hamiltonian, i.e., Schrödinger operator on $L^2(R_+)$ with boundary condition, depending on the boundary phase $\theta$.

2. $\mu(H)$ : spectral measure of $H = H^\theta$ on $L^2(R_+)$, $\mu_{\rm ac}(H)$, $\mu_{\rm sc}(H)$, $\mu_{\rm p}(H)$, $\mu_{\rm pp}(H)$, $\mu_{\rm d}(H)$ are different components of $\mu(H)$, absolutely continuous, singular continuous, point, pure point, discrete

3. $\sum_\mu = \sum_{\mu(H)} = \sum(H)$ the closed support of the spectral measure $\mu(H)$, i.e., spectrum of $H$, similar sense have $\sum_{\rm ac}(H), \sum_{\rm sc}(H)$ etc.

4. $\rho_\lambda(x)$, $\theta_\lambda(x)$ are the amplitude and the phase for the solution of initial-value problem $H\psi = \lambda\psi$, $x \geq 0$, $\lambda \in R^1$ given by polar substitution $\psi_\lambda = \rho(x)\sin\theta_\lambda(x)$ , $\psi'_\lambda = \rho_\lambda\cos\theta_\lambda$; modified amplitude and phase $r_k(x)$, $t_k(x)$ given by similar formulas: for $\lambda = k^2 > 0$ one has $\psi_k(x) = r_k(x)\sin t_k(x)$, $\psi'_k(x) = k\, r_k(x)\cos t_k(x)$.

5. If $q(x) = \sum\limits_{n=1}^{\infty} g_n(x - x_n)$ is a sparse potential then for individual (integrable) bump $g_n(\cdot)$ the scattering data have a following notations: $\{\lambda_{n,i} < 0\}$ is the set of negative eigenvalues for $H_n\psi = -\psi'' + g_n(\cdot)\psi$ on $L^2(R)$, this set can be empty or countable, $A_n(k)$, $B_n(k)$, $k \geq 0$ are transmission and reflection coefficients for the problem $H\psi = k^2\psi$ given by condition that the solution $\psi = \psi_k(x)$ has the following asymptotics: $\psi_k(x) = \exp(+ikx) + o(1), x \to -\infty$, $\psi_k(x) \sim A(k)\exp(ikx) + B(k)\exp(-ikx) + o(1)$, $x \to +\infty$.

6. $M_\lambda([a,b])$ are monodromy operators for the problem $H\psi = \lambda\psi$ on $[a,b]$, $||M_\lambda(\cdot)||$ is Hilbert-Schmidt norm, $||M_\lambda(\cdot)||_0$ is corresponding operator norm.

## 2 Technical tools

### *2.1 Spectral measure*

In this paper we will use only direct methods based on the phase-amplitude formalism. It is convenient in this context to introduce spectral measure $\mu^{\theta_0}(d\lambda)$ of the Hamiltonian (1.2) using Levitan's method [14]. Let us remark, that potential (1.7) under condition (1.5), (1.6) is continuous and bounded, $\|q(\cdot)\|_\infty = C_0 < \infty$. As a result the spectral measure $\mu(d\lambda)$ is well-defined.

For each $\lambda \in \mathbf{R}^1$ the boundary-value problem

$$\begin{aligned} H\psi_\lambda(x) &= -\psi'' + q(x)\psi = \lambda\psi \\ \psi_\lambda(0) &= \sin\theta_0, \quad \psi'_\lambda(0) = \cos\theta_0, \end{aligned} \tag{2.1}$$

has unique solution $\psi_\lambda(x) \in C^2_{\text{loc}}(\mathbf{R}^1)$.

If $f \in C_0(\mathbf{R}^1_+)$ one can define the generalized Fourier transform by the formula

$$\widehat{f}(\lambda) = \int_0^\infty \psi_\lambda(x) f(x) dx. \tag{2.2}$$

Spectral measure $\mu(d\lambda) = \mu^{\theta_0}(d\lambda)$ is the principal element in the inverse transform

$$f(x) = \int_{\mathbf{R}^1} \widehat{f}(\lambda)\psi_\lambda(x)\mu(d\lambda) \tag{2.3}$$

and Parseval isomorphism

$$\begin{aligned} (f_1, f_2) &= \int_0^\infty f_1(x) f_2(x) dx \\ &= \left(\widehat{f_1}, \widehat{f_2}\right)_\mu = \int_{\mathbf{R}^1} \widehat{f_1}(\lambda)\widehat{f_2}(\lambda)\mu(d\lambda). \end{aligned} \tag{2.4}$$

It is a weak limit of the discrete spectral measures $\mu_L(d\lambda)$ constructed in terms of the spectrum of the following operator $H_L^{\theta_0,\theta_L}$ on $L^2([0,L])$:

$$\begin{aligned} H\psi &= -\psi'' + q(x)\psi, \qquad x \in [0,L] \\ \psi(0)\cos\theta_0 - \psi'(0)\sin\theta_0 &= 0, \\ \psi(L)\cos\theta_L - \psi'(L)\sin\theta_L &= 0. \end{aligned} \tag{2.5}$$

In obvious notation

$$\mu_L(d\lambda) = d\lambda \cdot \left( \sum_{i=1}^{\infty} \frac{\delta(\lambda - \lambda_i(L))}{\int_0^L \psi_{\lambda_i}^2(x)dx} \right) \tag{2.6}$$

Here we used the normalization $\psi_{\lambda_i}^2(0) + (\psi'_{\lambda_i})^2(0) = 1$.

If $L \to \infty$ for arbitrary choice of $\{\theta_L\}$

$$\mu_L(d\lambda) \Rightarrow \mu(d\lambda).$$

Decomposition of $\mu(d\lambda)$ into three parts

$$\mu(d\lambda) = \mu_{\text{ ac}}(d\lambda) + \mu_{\text{ sc}}(d\lambda) + \mu_{\text{ p}}(d\lambda)$$

and analysis of the supports of the corresponding components, i.e., sets $\Sigma_{\text{ ac}}$, $\Sigma_{\text{ sc}}$, $\Sigma_{\text{ p}}$ is the main problem of the qualitative spectral analysis.

In our case $\Sigma_{\text{ p}} \cap [0, \infty) = \emptyset$. To study the continuous components of $\mu(d\lambda)$ it is convenient to represent $\mu(d\lambda)$ as a weak limit of the a.c. measures

$$\bar{\mu_L}(d\lambda) = \frac{1}{\pi} \int_0^{\pi} \mu_L^{\theta_0, \theta_L}(d\lambda) d\theta_L.$$

Measure $\bar{\mu_L}(d\lambda)$ has a very simple density, which can be expressed in terms of the amplitude of the solution $\psi_\lambda(x)$ (see later).

### *2.2 Phase and amplitude*

For the solution $\psi_\lambda(x)$ of boundary-value problem (2.1) phase and amplitude are given by the standard formulas (Atkinson [15])

$$\psi_\lambda(x) = \rho_\lambda(x) \sin \theta_\lambda(x), \quad \psi'_\lambda(x) = \rho_\lambda(x) \cos \theta_\lambda(x). \tag{2.7}$$

Then

$$\begin{aligned} \theta'_\lambda &= \cos^2 \theta_\lambda + (\lambda - q(x)) \sin^2 \theta_\lambda, \quad \theta_\lambda(0) = \theta_0 \\ \rho'_\lambda &= \frac{1}{2} \rho_\lambda(x)(1 + q(x) - \lambda) \sin 2\theta_\lambda, \quad \rho_\lambda(0) = 1 \\ \rho_\lambda &= \exp\left( \frac{1}{2} \int_0^x (1 + q(z) - \lambda) \sin 2\theta_\lambda(z) dz \right) \end{aligned} \tag{2.8}$$

The following result is fundamental [15]:

**Lemma 1.** If $q(\cdot) \geq q_0 > -\infty$ then

$$\bar{\mu_L}(\lambda) = \frac{d\lambda}{\rho_\lambda^2(L)}, \ \bar{\mu_L}(\lambda) \Rightarrow \mu(d\lambda) = \mu^{(\theta_0)}(d\lambda) \quad (2.9)$$

$$\frac{1}{\pi}\int_0^\pi \mu^{\theta_0}(d\lambda)\, d\theta_0 = d\lambda.$$

Spectral properties for $H^{\theta_0}$ depend on the behavior of $\rho_\lambda(L), L \to \infty$ (boundedness, growth, Lyapunov exponents, etc.). The results on the negative part of the spectrum $\Sigma(H^{\theta_0})$ $(\lambda < 0)$ usually are simpler. For instance, if $q(x) \to 0,\ x \to \infty$, then $\Sigma(H^{\theta_0}) \cap (-\infty, 0) = \Sigma_{\text{d}}(H^{\theta_0}) \cap (-\infty, 0)$. For positive energies $\lambda$ it is useful to work with frequency $k = \sqrt{\lambda} > 0$. We will use notations $\psi_\lambda(x)$, $\psi_{k^2}(x)$, $\psi_k(x)$ as synonyms. Similarity with WKB approach suggests slightly different definition of the phase and amplitude (the so-called Prüfer transformation):

$$\begin{aligned} \psi_k(x) &= r_k(x)\sin t_k(x) \\ \psi_k'(x) &= k r_k(x)\cos t_k(x). \end{aligned} \quad (2.10)$$

Then

$$\begin{aligned} t_k'(x) &= k - \frac{q(x)\sin^2 t_k(x)}{k} \\ r_k'(x) &= \frac{q(x)\sin 2t_k(x)}{2k} r_k \end{aligned} \quad (2.11)$$

with the initial conditions

$$\begin{aligned} \operatorname{cotan} t_k(0) &= \frac{1}{k}\operatorname{cotan}\theta_0 \\ r_k(0) &= \sqrt{\sin^2\theta_0 + \frac{1}{k^2}\cos^2\theta_0} \end{aligned} \quad (2.12)$$

In particular, if $\theta_0 = \pi/2$

$$t_k(0) = 0, \quad r_k(0) = 1. \quad (2.13)$$

Prespectral measure $\bar{\mu_L}(d\lambda)$ has in a new variable the representation

$$\bar{\mu_L}(d\lambda) = \frac{2k\ dk}{\rho_{k^2}^2(L)}$$

If $\Delta = [\alpha, \beta] \subset (0, \infty)$ is a fixed interval on the positive energy axis then on the frequency axis it transforms to $\widetilde{\Delta} = [\sqrt{\alpha}, \sqrt{\beta}]$. As easy to see for appropriate constants $C^{\pm}(\Delta))$ and arbitrary $L > 0$

$$C^{-}(\Delta)\frac{1}{r_k^2(L)} \leq \frac{2k}{\rho_{k^2}^2(L)} \leq C^{+}(\Delta)\frac{1}{r_k^2(L)} \tag{2.14}$$

i.e., the spectral properties of the measure on $\widetilde{\Delta}$

$$\widetilde{\mu_L}(dk) = \frac{dk}{r_k^2(L)}$$

in the limit $L \to \infty$ are exactly the same as the properties of $\mu(d\lambda)$ on the corresponding interval $\Delta$ of the energy axis.

The key moment in the classification of the spectra is the Schnoll's lemma [16], details in [17]:

**Lemma 2.** *Let* $q(x) \geq q_0 > -\infty$. *For any bounded interval* $\Delta$ *on the energy (frequency) axis one can find positive constants* $x_0 = x_0(q_0, \Delta), C_0 = C_0(q_0, \Delta), \delta_0 = \delta_0(q_0, \Delta)$ *such that for all* $x \geq x_0$ *and* $\lambda \in \Delta$

a)

$$\int_{\Delta} \psi_\lambda^2(x)\, \mu(d\lambda) \leq c_0 \tag{2.15}$$

b) $|\psi_\lambda(z)| \geq \frac{1}{2}|\psi_\lambda(x)|$ *either for* $z \in [x_0, x_0 - \delta_0]$ *or for* $z \in [x_0, x_0 + \delta_0]$ *(absence of the sharp peaks!).*

If the potential $q(\cdot)$ is uniformly bounded, $\|q(\cdot)\|_\infty \leq q_0 < \infty$ then one can estimate (see, for instance [18]) the derivative $\psi'_\lambda(x)$ of the arbitrary solution in terms of local $L^2$-norm of the solution $\psi_\lambda(x)$:

$$|\psi'_\lambda(x)|^2 \leq \frac{c_0}{2\delta}\int_{x-\delta}^{x+\delta} \psi_\lambda^2(z)dz \tag{2.16}$$

(in the spirit of Sobolev's embedding theorems). It gives immediately the following extension of the Schnoll's lemma:

**Lemma 2 $'$.** If $\|q(\cdot)\|_\infty \leq q_0 < \infty$ and $\Delta$ is a fixed interval on the energy axis then for appropriate const $x_0, c_0$,(depending only on $q_0, \Delta$) and all $x \geq x_0$

$$\int_{\Delta} \rho_\lambda^2(x)\, \mu(d\lambda) \leq c_0. \tag{2.17}$$

**Corollaries of Lemmas 2, 2 $'$.** Under conditions of the Lemma 2$'$ for arbitrary fixed sequence $\{x_n\}, x_n \to \infty$ and $\mu$-a.e. $\lambda \in \Delta, \varepsilon > 0$

a)

$$\rho_\lambda(x_n) \le C(\lambda,\varepsilon)\, n^{\frac{1}{2}+\varepsilon} \tag{2.18}$$

b) For all $x \ge x_0$

$$\rho_\lambda(x) \le C(\lambda,\varepsilon)\, x^{\frac{1}{2}+\varepsilon}. \tag{2.19}$$

## 3 Scattering Theory

Let us recall now a few definitions and formulas from the scattering theory. Monodromy operator (fundamental matrix) $M_\lambda(x_1, x_2)$ for the problem $H\psi = \lambda\psi$ is the unimodular $2 \times 2$ matrix function, such that for $x_2 > x_1$,

$$-\frac{\partial^2 M_\lambda}{\partial x_2^2} + q(x_2)M_\lambda = \lambda M_\lambda, \qquad M_\lambda(x_1, x_1) = I. \tag{3.1}$$

Of course, $M_\lambda(x_1,x_2)M_\lambda(x_2,x_3) = M_\lambda(x_1,x_3)$, $x_1 < x_2 < x_3$. In terms of $\rho_\lambda$, $\theta_\lambda$ monodromy operator has the following representation: if $\theta_\lambda^{(i)}$, $\theta_\lambda^{(i)}$, $i = 1,2$ are the solutions of the equations (18′) with initial data

$$\theta_\lambda^{(1)}(x_1) = \frac{\pi}{2}, \quad \theta_\lambda^{(2)}(x_1) = 0, \quad \rho_\lambda^{(1)}(x_1) = \rho_\lambda^{(2)}(x_1) = 1, \tag{3.2}$$

then

$$M_\lambda(x_1,x_2) = \begin{bmatrix} \rho_\lambda^{(1)}(x_2)\ \sin\theta_\lambda^{(1)}(x_2) & \rho_\lambda^{(1)}(x_2)\ \cos\theta_\lambda^{(1)}(x_2) \\ \rho_\lambda^{(2)}(x_2)\ \sin\theta_\lambda^{(2)}(x_2) & \rho_\lambda^{(2)}(x_2)\ \cos\theta_\lambda^{(2)}(x_2) \end{bmatrix} \tag{3.3}$$

For $\lambda = k^2 > 0$ we will use monodromy operator in "$k$-phase" representation based on the formulas (6.34). In obvious notations

$$M_k(x_1,x_2) = \begin{bmatrix} r_k^{(1)}\ \sin t_k^{(1)}(x_2) & r_k^{(1)}\ \cos t_k^{(1)}(x_2) \\ r_k^{(2)}\ \sin t_k^{(2)}(x_2) & r_k^{(2)}\ \cos t_k^{(2)}(x_2) \end{bmatrix} \tag{3.4}$$

Here $t_k^{(i)}(x)$, $r_k^{(i)}(x)$, $i = 1,2$ are the solutions of the equations (1.13) with the same boundary conditions (3.2). If $q(\,\cdot\,) \equiv 0$ on $[x_1, x_2]$ then

$$M_k(x_1,x_2) = \begin{bmatrix} \cos k(x_2 - x_1) & \text{-}\sin k(x_2 - x_1) \\ \sin k(x_2 - x_1) & \cos k(x_2 - x_1) \end{bmatrix} \tag{3.5}$$

i.e., orthogonal matrix. It is not true for $M_\lambda(x_1, x_2)$. Formula (3.4) is the main advantage of the "$k$-phase" representation.

The third expression for the same object can be done in terms of scattering (reflection and transmission) coefficients. Assume that potential $q(\cdot)$ is compactly supported and supp $q(\cdot) \subset [-l, l]$. Then the equation

$$-\psi'' + q(\cdot)\psi = k^2\psi \tag{3.6}$$

has two (the so-called Jost) solutions $\psi_k^{(i)}(x)$, $i = 1, 2$ such that

$$\begin{aligned} \psi^{(1)}(x) &= \exp(-ikx), \quad x \leq -l \\ \psi^{(1)}(x) &= A(k)\exp(-ikx) + B(k)\exp(ikx), \quad x \geq l \end{aligned} \tag{3.7}$$

and

$$\psi^{(2)}(x) = \bar{\psi^{(1)}}(x).$$

Functions $A(k), B(k)$ have many specific analytic properties (the subject of the scattering theory, see for instance monograph [13]) among them formula $|A(k)|^2 = 1 + |B(k)|^2$. If $A(k) = a_1(k) + ia_2(k)$, $B(k) = b_1(k) + ib_2(k)$ with real $a_i$, $b_i$, $i = 1, 2$ then simple calculations give the following expression

$$M_k(-l, l) =$$
$$\begin{bmatrix} a_1 \cos 2kl + a_2 \sin 2kl + b_1 - & a_1 \sin 2kl + a_2 \cos 2kl - b_2 \\ +a_1 \sin 2kl - a_2 \cos 2kl \; b_2 & a_1 \cos 2kl + a_2 \sin 2kl - b_1 \end{bmatrix} \tag{3.8}$$

Let us give the formulas for the transformation of the phase and amplitude of the solution, propagating through bump $q(\cdot)$ on $[-l, l]$.

If $\theta_k(-l) = \varphi$, $r_k(-l) = 1$ then to get $\theta_k(l)$, $r_k(l)$ we have to find vector $\vec{e}(l) = M_k\vec{e}(-l)$, where $\vec{e}(-l) = \begin{pmatrix} \sin\varphi \\ \cos\varphi \end{pmatrix}$. Calculations give

$$\vec{e}(l) = \begin{bmatrix} a_1 \sin\,(2kl + \varphi) - a_2 \cos(2kl + \varphi) + b_1 \sin\varphi - b_2 \cos\varphi \\ a_1 \cos\,(2kl + \varphi) - a_2 \sin(2kl + \varphi) + b_1 \cos\varphi + b_2 \sin\varphi \end{bmatrix}$$

and

$$r_k(l) = ||\vec{e}(l)|| = \left[|A|^2 + |B|^2 + 2|A| \cdot |B| \cos(2kl + 2\varphi + \theta)\right]^{1/2} \tag{3.9}$$
$$\cos\theta = \tfrac{a_2 b_2 - a_1 b_1}{|A| \cdot |B|}$$

If reflection coefficient $|B(k)| << 1$ then (due to $|A|^2 = 1 + |B|^2$)

$$\ln r_k(l) = \tfrac{1}{2}\ln\left(1 + 2|B|^2 + 2|A|\cdot|B|\cos(2kl + 2\varphi + \theta)\right)$$
$$= |A|\cdot|B|\cos(2kl + 2\varphi + \theta) + |B|^2\sin^2(2kl + 2\varphi + \theta) + O(|B|^3).$$

Formulas (3.6), (3.8), (3.9) will be the basis of the further analysis.

The norm of monodromy operators in different representations plays especially important role in our study. We will use one of the two standard definitions of the norm: Hilbert-Schmidt and operator norms.

If $M_\lambda(\cdot) = \begin{vmatrix} a & b \\ c & d \end{vmatrix}$ then

$$||M_\lambda(\cdot)||_{\text{HS}} = \sqrt{a^2 + b^2 + c^2 + d^2} = (Tr)\,(M_\lambda M_\lambda^*))^{1/2} \qquad (3.10)$$

and

$$||M_\lambda(\cdot)||_{\text{O}} = \left(S^\dagger(M_\lambda)\right)^{1/2}, \qquad (3.11)$$

where $S^\dagger$ is the greatest root of the characteristic equation for the matrix $M_\lambda \cdot M_\lambda^*$:

$$S^2 - Tr\,(M_\lambda \cdot M_\lambda^*))\,S + 1 = 0 \qquad (3.12)$$

(because $\det M_\lambda = \det M_k = 1$ identically).

Usually we will use HS-norm without indexes. As $M_\lambda$, $M_k \in SL\,(2, \mathbf{R})$,

$$||M_\lambda(\cdot)|| = ||M_\lambda^{-1}(\cdot)||, \quad ||M_k(\cdot)|| = ||M_k^{-1}(\cdot)||, \qquad (3.13)$$

(for both norms: $\|\ \|_{\text{HS}}$ and $\|\ \|_{\text{O}}$).

In particular

$$\vec{e_{\theta_0}} M_\lambda(\cdot) M_\lambda^{-1}(\cdot) = \vec{e_{\theta_0}}$$

i.e.,

$$1 \leq |\vec{e_{\theta_0}} M_\lambda(0,x)| \cdot ||M_\lambda^{-1}(0,x)|| = \rho_\lambda(x)||M_\lambda^{-1}(0,x)|| \qquad (3.14)$$
$$\frac{1}{\rho_\lambda^2(0,x)} \leq ||M_\lambda(0,x)||^2, \qquad \rho_\lambda^2(0,x) \geq \frac{1}{||M_\lambda(0,x)||^2}$$

and similar inequalities for $r_k(x)$, $\|M_k(0,x)\|$.

Trivial estimations (3.14) will be used a few times in Sec. 3 ("tests for the different types of spectra").

Remaining part of the Sec. 2 will be related with more specific analytical problem. We want to study the analytical properties of the reflection coefficient $B(k)$, $k \in \Delta \subset (0,\infty)$ for the scattering problem (3.6), (3.7).

**Lemma 3.** *Let us consider the problem*

$$-\psi'' + q_l(x)\,\psi = k^2\psi,\ k \in \Delta \subset (0,\infty)$$
$$q_l(x) = I_{|x|\le l}\, q(x)$$

*where for some fixed positive constants* $c_0, \delta > 0$

$$|q(x)| \le \frac{c_0}{1+|x|^{1+\delta}}$$

*Let also*

$$F_l(k,\varphi) = \ln|M_k(-l,l)\, e_{\overrightarrow{\varphi}}|,\ e_{\overrightarrow{\varphi}} = \begin{pmatrix} \cos\varphi \\ \sin\varphi \end{pmatrix}\ \varphi \in [0,2\pi].$$

*Then for appropriate constants* $c_i, i = 1,2$, *depending only on* $c_0, \delta, \Delta$ *(i.e., uniformly in* $l$*!)*

a)

$$|F_l(k,\varphi) - F_l(k',\varphi) \le c_1 |k-k'|^{\delta},\ k,k' \in \Delta \tag{3.15}$$

b) $\left\| \frac{\partial F_l}{\partial\varphi}(k,\varphi) \right\|_\infty \le c_2.$

To prove this lemma, first of all, we have to get the closed equation for the scattering coefficients $A_\ell(k)$, $B_\ell(k)$ and after to use the formulas (3.8), (3.9).

The Jost solution $\psi_k(x)$ given by the formulas (3.7) outside $[-\ell,\ell]$ satisfies the well-known integral equation for $x \in [-\ell,\ell]$

$$\psi_k(x) = e^{-ikx} + \int_{-\ell}^{x} q(z)\frac{\sin k(x-z)}{k}\psi_k(z)dz. \tag{3.16}$$

Simple transformations give

$$\begin{aligned}\psi_k(x) &= \exp(-ikx)\left[1 - \frac{1}{2ki}\int_{-\ell}^{x} q(z)\exp(ikz)\psi_k(z)dz\right] \\ &\quad + \exp(ikx)\cdot\frac{1}{2ki}\int_{-\ell}^{x} q(z)\exp(-ikz)\psi_k(z)dz \\ &= \exp(-ikx)\left(1 - \frac{\alpha_k(x)}{2ki}\right) + \exp(ikx)\frac{\beta_k(x)}{2ki}\end{aligned} \tag{3.17}$$

where

$$\alpha_k(x) = \int_{-\ell}^{x} q(z)\exp(ikz)\psi_k(z)dz \tag{3.18}$$
$$\beta_k(x) = \int_{-\ell}^{x} q(z)\exp(-ikz)\psi_k(z)dz.$$

Let us remark that

$$A_\ell(k) = 1 - \frac{\alpha_k(\ell)}{2ki} \tag{3.19}$$
$$B_\ell(k) = \frac{\beta_k(\ell)}{2ki}\,.$$

Substitution of the expressions (3.17) into initial problem (3.16) provides the closed system for $\alpha_k(\cdot)$, $\beta_k(\cdot)$:

$$\alpha_k(x) =$$
$$= \int_{-\ell}^{x} q(z)dz - \tfrac{1}{2ki}\int_{-\ell}^{x} q(z)\alpha_k(z)dz + \tfrac{1}{2ki}\int_{-\ell}^{x} q(z)e^{2ikz}\beta_k(z)dz \tag{3.20}$$
$$\beta_k(x) =$$
$$= \int_{-\ell}^{x} q(z)e^{-2ikz}dz - \tfrac{1}{2ki}\int_{-\ell}^{x} q(z)e^{-2ikz}\alpha_k(z)dz + \tfrac{1}{2ki}\int_{-\ell}^{x} q(z)\beta_k(z)dz$$

or in the vector form

$$\vec{\alpha_k}(x) = \begin{pmatrix} a_k \\ \beta_k \end{pmatrix} = \tag{3.21}$$
$$= \begin{pmatrix} \int_{-\ell}^{x} q\,(z)\,dz \\ \int_{-\ell}^{x} q\,(z)\,e^{-2ikz}dz \end{pmatrix}$$
$$+\frac{1}{2ki}\int_{-\ell}^{x} q(z)dz \begin{pmatrix} -1 & e^{2ikz}dz \\ -e^{-2ikz} & 1 \end{pmatrix} \begin{pmatrix} a_k \\ \beta_k \end{pmatrix}(z)dz$$
$$= \vec{h_k}(x) + \frac{1}{2ki}\int_{-\ell}^{x} q\cdot A_k(z)\vec{\alpha_k}(z)dz.$$

Vector function

$$\vec{h_k}(x) = \begin{pmatrix} \int_{-\ell}^{x} q\,(z)\,dz \\ \int_{-\ell}^{x} q\,(z)\,e^{-2ikz}dz \end{pmatrix} \tag{3.22}$$

and (degenerated !) matrix kernel

$$A_k(z) = \begin{pmatrix} -1 & e^{2ikz} \\ -e^{-2ikz} & 1 \end{pmatrix} \tag{3.23}$$

are known and we are looking for vector function $\vec{\alpha}_k(x), x \in [-\ell, \ell]$.

System (3.21) has Volterra's type and is solvable by usual sequential approximation. This fact is useful if potential $q(\cdot)$ has small parameter $\sigma : q \to \sigma q$, $\sigma << 1$. In such situation sequential approximations give asymptotical expansion of the scattering coefficients with respect to $\sigma$ (see later).

Let us get the upper estimation for the components of the vector $\alpha_k(\cdot)$. If follows from (3.21) that

$$\begin{aligned}|\vec{\alpha_k}(\cdot)| &\le 2||q_\ell||_1 + \tfrac{1}{2k}\textstyle\int_{-\ell}^{x}|q(z)|\cdot||A_k(z)||\cdot|\alpha_k(z)|dz\\ &\le 2||q_\ell(\cdot)||_1 + \tfrac{2}{k}\textstyle\int_{-\ell}^{x}|q(z)|\cdot|\alpha_k(z)|dz\end{aligned}$$

and the Bellman-Gronwall inequality gives

$$|\alpha_k(x)| \le 2||q_\ell(\cdot)||_1 \exp\left(\frac{2}{k}\int_{-\ell}^{x}|q(z)|dz\right)$$

i.e., uniformly in $\ell$

$$\begin{aligned}|A_\ell(k)| &= \left|1 - \frac{1}{2ik}\alpha_k(\ell)\right| \qquad (3.24)\\ &\le 1 + c_1(\Delta)\int_{-\infty}^{+\infty}\frac{c_0 dz}{1+|z|^{1+\delta}}\cdot\exp\left(c_2(\Delta)\int_{-\infty}^{+\infty}\frac{c_0 dz}{1+|z|^{1+\delta}}\right)\\ &\le c_2(c_0,\delta,\Delta)\\ |B_\ell(k)| &\le c_3(c_0,\delta,\Delta).\end{aligned}$$

As a result

$$c_4^{-1} \le ||M_k(-e,e)|| \le c_4(c_0,\delta,\triangle) \qquad (3.25)$$

and

$$c_4^{-1} \le |F_l(k,\varphi)| \le c_4(c_0,\delta,\triangle) \qquad (3.26)$$

(again uniformly in $l$, i.e., even for $l = +\infty$, in different terms we proved uniform in $n, k$ estimation of the $S_n(k)$-matrix norm for the family of Hamiltonians $H_n$ on $L^2(R)$

$$H_n\psi = -\psi'' + \varphi_n(x - x_n)\psi = k^2\psi\ ,\ S_n(k) = \begin{vmatrix} A_n(k) & B_n(k)\\ \bar{B}_n(k) & \bar{A}_n(k)\end{vmatrix}$$

if

$$|\varphi_n(z)| \le \frac{c_0}{1+|z|^{1+\delta}}.$$

It follows from (3.24) also that

$$0 \leq \frac{|B_n(k)|}{|A_n(k)|} \leq \rho_0(c_0, \delta, \triangle) < 1, \; n \geq 1.$$

This fact will be significant later.

Let us estimate the modules of continuity for the function $F(k, \varphi)$. First of all, if

$$\overrightarrow{e_l}(\varphi) = M_k(-l, l) \; \overrightarrow{e}(\varphi) \; , \; \overrightarrow{e}(\varphi) = \begin{pmatrix} \cos\varphi \\ \sin\varphi \end{pmatrix}$$

then

$$\frac{d\, e_l(\varphi)}{d\varphi} = M_k(-l, l) \begin{pmatrix} -\sin\varphi \\ \cos\varphi \end{pmatrix}$$

and

$$|F(k, \varphi)|^2 = \left(\overrightarrow{e_l}(\varphi), \overrightarrow{e_l}(\varphi)\right).$$

That is

$$\frac{\partial F(k, \varphi)}{\partial \varphi} = \frac{2\left(\overrightarrow{e_l}(\varphi), \frac{d}{d\varphi}\overrightarrow{e_l}(\varphi)\right)}{2F(k, \varphi)}$$

or (according to (3.25), (3.26))

$$\left|\frac{\partial F(k, \varphi)}{\partial \varphi}\right| \leq c_5(c_0, \delta, \triangle).$$

The proof of inequality (3.15) is not so trivial. It follows from equation (3.21) that

$$\overrightarrow{\alpha_k}(x) - \overrightarrow{\alpha_{k'}}(x) = \overrightarrow{h_k}(x) - \overrightarrow{h_{k'}}(x) + \frac{1}{2ki}\int_{-e}^{x} dz\, q(z) \cdot$$
$$\cdot A_k(z) \left(\overrightarrow{\alpha_k}(z) - \overrightarrow{\alpha_{k'}}(z)\right) dz - \frac{1}{2ki}\int_{-e}^{x} q(z)\,(A_k - A_{k'})(z)\, \overrightarrow{\alpha_{k'}}(z)\, dz$$

all the differences $\left(\overrightarrow{h_k} - \overrightarrow{h_k}\right)$, $(A_k - A_{k'})$ contain the expressions of the form $\exp(\pm 2ikz) - \exp(\pm 2ik'z)$. For fixed const $L > 0$

$$|e^{\pm 2ikz} - e^{\pm 2ik'z}| \leq \begin{cases} 2\pi|k - k'||z|, & |z| \leq L \\ 2 \;, & |z| > L \end{cases}$$

Let us put

$$\triangle H(x,L) = 2\pi\,|k-k'| \int_0^{L\wedge x} |z|\, q(z)\, dz \cdot I_{|x|\le L} +$$
$$+\left(2\pi\,|k-k'| \int_0^{L} |z|\,|q(z)|dz + \int_L^{L\wedge x} |q(z)|dz\right) I_{|x|>L}$$

It follows from the previous inequalities that

$$|\alpha_k(x) - \alpha_{k'}(x)| = |\triangle\alpha(x)| \le \triangle H(x,L) +$$
$$+c_1(\triangle) \textstyle\int_{-e}^{x} |q(z)|\,|\triangle\alpha(z)|dz + c_2(c_0,\delta,\triangle)\triangle H(x,L)$$

and the Bellman-Gronwall inequality gives again

$$|\triangle\alpha(x)| \le c_3(c_0,\delta,\triangle)\exp(c_1(\triangle)\,||q||_1)\,.$$

That is

$$|\triangle\alpha(x)| \le c_4(c_0,\delta,\triangle)\min_L\left[|k-k'|\int_{-L}^{L}|z|\,|q(z)|\right]dz +$$
$$+\int_{|z|>L} |q(z)|\,dz.$$

But

$$\int_{-L}^{L}|z|\,|q(z)|\,dz \le \int_{-L}^{L}\frac{|z|\,c_0\,dz}{1+|z|^{1+\delta}} \le c_5(c_0,\delta)\,L^{1-\delta}$$
$$\int_{|z|} |q(z)|dz \le c_6(c_0,\delta)\,|L|^{-\delta}$$

One can put $L = \frac{1}{|k-k'|}$. Than (uniformly in $x$ and $l$)

$$|\overrightarrow{\alpha_k}(x) - \overrightarrow{\alpha_{k'}}(x)| \le c_7(c_0,\delta,\triangle)\,|k-k'|^{\delta}$$

It gives the Hölder continuity for $|A_l(k)|, |B_l(k)|$ uniformly in $l$ and due to (3.9) the same is true for $F_l(k,\varphi)$.

Another application of the integral equations (3.20), (3.21) will be given in the Section 7. We will discuss the situation with potential $q(\cdot)$ contains small coupling parameter $\sigma : q(x) \to \sigma q(x), \sigma << 1$ and will give asymptotical formulas for $B_l(\sigma,k)$ of the form

$$|B_l(\sigma,k)|^2 = \sigma^2 a_0(k) + \sigma^3 a_1(k) + \sigma^4 a_2(k) + \underline{\underline{0}}\left(\sigma^5\right)$$

Coefficients $a_i(k)$, $i = 0,1,2$ will be expressed in terms of Fourier transfers for potential $q(\cdot)$ and associated functions.

## 4 Tests for the different types of spectra

Separating the different components of the spectral measure $\mu(dk)$, description of the corresponding supports $\Sigma$ ac, $\Sigma$ sc, $\Sigma$ pp, $\Sigma$ d, characterization of the (generalized) eigenfunctions are the main problem of the spectral analysis. Physical literature suggests a very simple scenario for this classification:

a) generalized eigenfunctions are bounded (if potential $q(\cdot)$ has no singular points),

b) on the set $\{k \mid ||M_k(x)|| \leq c(k) < \infty\}$ the spectrum is a.c. and corresponding eigenfunctions have a "wave structure",

c) on the set $\{k \mid ||M_k(x)|| \to \infty, \int_0^\infty ||M_k(x)||^{-2} < \infty\}$, i.e., set of the fast increasing of the monodromy operator, the spectrum is p.p.

d) on the set $\{k \mid ||M_k(x)|| \to \infty$, but $\int_0^\infty ||M_k(x)||^{-2} = \infty\}$ the spectrum is s.c.

This scenario as a general principle is correct, the mathematician has to specify the definitions of the boundedness, to understand what is going on in the sets of the 0-measure etc.

We will collect below a list of sufficient conditions for the different types of spectra. These tests are known to the specialists and already used, sometimes in a modified form. Some of simple observations are, probably, new at least in our specific context.

### *4.1 Tests for $\Sigma_{ac}$*

Majority of known results on $\Sigma_{\rm ac}$ for specific classes of one-dimensional operators were based on the following result by Hilbert and Pearson [19]. For a short and transparent proof see Simon [18].

**Lemma 4.** *Suppose that* $|q(\cdot)||_\infty < \infty$ *and* $S_0 = \{k| ||M_k(x)||$ *is bounded on* $[0, \infty]\}$. *Then the spectral measure* $\tilde{\mu}\,(dk)$ *is a.c. on* $S_0$ *in the sense that*

(i) *If* $T \subset S_0, |T| > 0$ *then* $\mu_{\rm ac}(T) > 0$.

*Here and in the future* $|A|$ *is the Lebesque measure of* $A$.

(ii) $\mu_{\rm sing}(S_0) = 0$.

Let us stress the absence of any uniform estimations on $||M_k(\cdot)||$. Non trivial part of the lemma is the second one, presence of the dense a.c. spectrum on $S_0$ is a simple fact, see below the more general criteria which guarantee $\Delta \subset \Sigma_{\rm ac}$.

Generalized eigenfunctions of $\Sigma_{\text{ac}}$ can be unbounded $\mu$-a.e., also it is easier to control solutions $\psi_k(\cdot)$ for a.e. $k \in \Delta$ but not for all $k \in \Delta$. The following lemma is very effective and will be used later.

**Lemma 5**. *If for some fixed sequence $X = \{x_n\}, x_n \to \infty$, fixed $\delta > 0$ and interval $\Delta \subset (0, \infty)$,*

$$\int_\Delta \frac{dk}{(r_k(x_n))^{2+2\delta}} \leq C(\Delta, X) < \infty \tag{4.1}$$

*then*

$$\mu(H) = \mu_{\text{ac}}(H) \text{ on } \Delta .$$

*Proof.* According to Lemma 1

$$\bar{\mu}_n(dk) = \frac{2kdk}{\rho^2_{k^2}(x_n)} \Rightarrow \mu(dk) .$$

Assume that $\mu_{\text{sing}}(\Delta) > 0$. For arbitrary small $\varepsilon > 0$ one can find continuous function $\varphi_\varepsilon(k)$ such that

$$\begin{array}{l} \text{(i) } 0 \leq \varphi_\varepsilon(k) \leq 1 \\ \text{(ii) } \mu_{\text{sing}}(\{k \in \Delta : \varphi_\varepsilon(k) = 1\} \geq \frac{1}{2}\mu_{\text{sing}}(\Delta) \\ \text{(iii) } |\{k : \varphi_\varepsilon(k) > 0\}| \leq \varepsilon. \end{array} \tag{4.2}$$

Then

$$\int_\Delta \varphi_\varepsilon(k)\bar{\mu}_n(dk) \leq c(\Delta) \int_\Delta \frac{dk}{(r_k^2(x_n))}$$
$$\leq \left(\int_\Delta \varphi_\varepsilon^{\frac{\delta}{1+\delta}} dk\right) \left(\int_\Delta \frac{dk}{r_k^{2+2\delta}(x_n)}\right)^{\frac{1}{1+\delta}} \leq c_1(\Delta, \delta)\,\varepsilon^{\frac{\delta}{1+\delta}}.$$

If $\varepsilon$ is small enough, it contradicts to (4.2) that proves lemma. □

**Lemma 6.** *Suppose that for fixed sequence $x = \{x_n\}, x_n \to \infty$, and for a.e. $k \in \Delta \subset (0, \infty)$,*

$$\limsup_{n\to\infty} r_k(x_n) < \infty, \tag{4.3}$$

*or*

$$\limsup_{n\to\infty} \left(\frac{1}{x_n}\int_0^{x_n} r_k^2(z)\,dz\right)^{\frac{1}{2}} < \infty, \tag{4.4}$$

*or*

$$\limsup_{n\to\infty}\left(\frac{1}{x_n}\int_0^{x_n}\ln r_k(z)\,dz\right)<\infty. \tag{4.5}$$

*Then* $\triangle\subset\Sigma_{\text{ac}}(H)$ *and for a.e.* $\theta_0$ *the measure* $\mu^{\theta_0}(d\lambda)$ *is pure a.e. on* $\triangle$.

*Proof.* Let us start from a trivial remark: if on $C(\triangle)$ for the measures $\mu_n,\mu,$

$$\mu_n(dk)\Rightarrow\mu(dk)$$

and for all $n\geq n_0$

$$\frac{d\mu_n}{dk}(k)\geq C_0>0,\ k\in A,\ |A|>0$$

then

$$\frac{d\mu}{dk}(k)\geq C_0\text{ on }A,\text{ i.e., }\bar{A}\subseteq\text{ supp }\mu_{\text{ ac}}.$$

If $\limsup_{n\to\infty}r_k(x_n)<\infty$ then for arbitrary $\varepsilon>0$ one can find a constant $C=C(\varepsilon)$ and set $A_\varepsilon,|\triangle-A_\varepsilon|\leq\varepsilon$ such that for all $n\geq n_0(\varepsilon)$

$$r_k(x_n)\leq C(\varepsilon),\text{ i.e., }\frac{2k}{\rho^2_{k^2}(x_n)}\geq\frac{C(\triangle)}{r_k^2(x_n)}\geq\frac{C(\triangle)}{C^2(\varepsilon)}$$

etc. It proves the lemma under condition (4.3). In the cases (4.4), (4.5) first one has to apply the Jensen inequalities and then the fact of the weak convergence in $C(\triangle)$ of the measures $\mu_n(dk)=p_n(k)\,dk, p_n(k)=\frac{2k}{x_n}\int_0^{x_n}\frac{dz}{\rho^2_{k^2}(z)}$, to $\mu(dk)$.

For instance

$$\begin{aligned}p_n(k)&=\frac{2k}{x_n}\int_0^{x_n}\frac{dz}{\rho^2_{k^2}(z)}\\&\geq C(\triangle)\frac{1}{x_n}\int_0^{x_n}\frac{dz}{r_k^2(z)}\\&\geq C(\triangle)\exp\left\{-\frac{2}{x_n}\int_0^{x_n}\ln r_k(z)\right\},\end{aligned}$$

etc. The last statement is the consequence of Lemma 4 and Lemma 1.

### *4.2 Tests for $\Sigma_{sc}$ and $\Sigma_{pp}$*

Simplest criterion for the absence $\Sigma_{sc}$ gives the Schnoll theory (see Lemma 2′).

**Lemma 7**. *If $||q||_\infty < \infty$ and for fixed sequence $x = \{x_n\}, x_n \to \infty$ a.e. on $\triangle$,*

$$\text{(i)} \lim r \to \infty r_k(x_n) = +\infty, \ k \in \triangle \tag{4.6}$$

*or*

$$\text{(ii)} \lim n \to \infty \left(\tfrac{1}{x_n} \int_0^{x_n} r_k^2(z)\, dz\right)^{1/2} = + \quad \infty \quad k \in \triangle \tag{4.7}$$

*then $\mu_{\text{ac}}(\triangle) = 0$.*

Different form of the same proposition (compare with Lemma 4):

**Lemma 7′**. *If $||q||_\infty < \infty$ and for fixed sequence $x = \{x_n\}, x_n \to \infty$ a.e. on $\triangle$,*

$$\lim x \to \infty ||M_k(0, x_n)|| = +\infty \tag{4.8}$$

*then $\mu_{\text{ac}}(\triangle) = 0$.*

*Proof.* Using the Borel-Cantelli argument one can select subsequence $\{x_{n_l}\}$ such that

$$\lim_{l\to\infty} \frac{||M_k(0.x_{n_l})||}{2^l} = +\infty \text{ a.e. on } \triangle\,. \tag{4.9}$$

If $A$ is a 2×2 unimodular matrix and $\overrightarrow{e_\varphi} = \begin{pmatrix} \sin\varphi \\ \cos\varphi \end{pmatrix}, \varphi \in [0, \pi)$ then

$$\left|\left\{\varphi : |A\overrightarrow{e_\varphi}| \le \sqrt{\|A\|}\right\}\right| \le \frac{C_0}{\sqrt{\|A\|}} \tag{4.10}$$

with a constant $C_0$ depending only on the choice of $\|\cdot\|$. In fact, for appropriate $\varphi_1 \in [0, \pi)$

$$\left|A\overrightarrow{e}_\varphi\right|^2 = \left|AA^T\overrightarrow{e_\varphi} \cdot \overrightarrow{e_\varphi}\right| = \lambda^2\cos^2(\varphi - \varphi_1) + \lambda^{-2}\sin^2(\varphi - \varphi_1)\,.$$

Here $\lambda^2 = \|A\|_0^2$ is the largest eigenvalue of the matrix $AA^T$. But

$$\left|\left\{\varphi : \left|A\overrightarrow{e_\varphi}\right| \le \sqrt{\|A\|_0}\right\}\right| = \left|\left\{\varphi : AA^T\overrightarrow{e_\varphi} \cdot \overrightarrow{e_\varphi} \le \|A\|_0\right\}\right| \le$$
$$\le \left|\left\{\varphi : \cos^2(\varphi - \varphi_1) \le \frac{1}{\|A\|_0}\right\}\right| \le \frac{C_0}{\sqrt{\|A\|_0}}.$$

Using (4.9), (4.10) and again the Borel-Cantelli lemma one can be convinced that for a.e. $\theta_0 \in [0, \pi]$ and a.e. $k \in \triangle$,

$$\lim_{l \to \infty} \frac{\left| M_k \left(0, x_{n_l}\right) \overrightarrow{e_{\theta_0}} \right|}{2^{l/2}} = +\infty \tag{4.11}$$

But $\Sigma_{\text{ac}}$ does not depend on $\theta_0$ and Lemma 7 complete the proof. □

Separation of $\Sigma_{\text{pp}}$ and $\Sigma_{\text{sc}}$ is a more delicate problem. The central criterion here is the Simon-Wolff theorem [20]. In the special case of operator (1.1) with boundary phase $\theta_0$ result goes to Kotani [21].

**Lemma 8** (Kotani-Simon-Wolff). *If for fixed* $x_0 \in \mathbf{R}^1_+$, *a.e.* $k \in \triangle \subset (0, \infty)$ *the resolvent kernel* $R^0_{k^2}(x_0, \cdot) \in L^2\left(\mathbf{R}^1_+\right)$ *then for a.e.* $\theta_0 \in [0, \pi)$ *the spectrum of* $H$ *is p.p. on* $\triangle$.

Here $R^0_k(x_0, x)$ is the resolvent kernel of the problem $H\psi - k^2\psi = \delta_{x_0}$ with Dirichlet boundary condition at $x = 0$. Square integrability can be understood in one of the two equivalent ways:

$$\lim_{\varepsilon \to 0} \int_0^\infty |R^0_{k^2+i\varepsilon}(x_0, x)|^2 dx < \infty \tag{4.12}$$

or

$$\lim_{L_n \to \infty} \int_0^{L_n} |R^{(0,L_n)}_{k^2}(x_0, x)|^2 dx < \infty. \tag{4.13}$$

Of course, $R^{(0,L)}_{k^2} = (H - k^2)^{-1}$ on $L^2(0, L)$ with Dirichlet boundary conditions. $R^{(0,L)}_{k^2}$ exists and is bounded except finite number of eigenvalues.

A slightly different form of square integrability condition is:
for a.e. $k \in \Delta$ the ODE

$$-\psi'' + q(x)\psi = k^2\psi$$

has solution $\widetilde{\psi_k}(x) \in L^2(\mathbf{R}^1_+)$.

Dense point spectrum (or even discrete spectrum on $\Sigma_{\text{ess}}$ in our case $\Sigma_{\text{ess}} = [0, \infty)$) is not stable: for the set $S \subset [0, \pi)$ of the second category and $\theta_0 \in S$ the $\Sigma_{\text{pp}}$ for the operator $H^{\theta_0}$ transforms to $\Sigma_{\text{sc}}$ (see [22], [23]). Usually $|S| = 0$.

The standard method to prove the square integrability of the resolvent $R^0_{k^2}(x_0, \cdot)$ is based on the notion of generalized Lyapunov exponent. The following lemma [24] is typical:

**Lemma 9.** *Let scale function $A(x) \uparrow$ and regular in Karamata sense. Operator H has generalized exponent $\gamma_A(k)$, $k \in \triangle$ if a.e. on $\triangle$*

$$\lim_{x\to\infty} \frac{\ln \|M_k(0,x)\|}{A(x)} = \gamma_A(k). \tag{4.14}$$

Assume that $\gamma_A(\lambda) > 0$ a.e. on $\Delta$ and $\lim_{x\to\infty} \frac{A(x)}{\ln x} = \infty$. Then $R_{k^2}(x_0, \cdot) \in L^2(\mathbf{R}^1_+)$ for a.e. $k \in \Delta$.

If $A(x) = \ln x$ the same is true with additional restriction $\gamma_A(k) > 3/2$. Situation $\gamma_A(k) < 3/2$ and especially the case $\gamma_A(k) < 1/2$ is not clear. In the case of sparse bumps with random distances we can get probably the coexistence of the dense $\Sigma$ pp and $\Sigma$ sc for appropriate small set of boundary phases $\theta_0$ and pure $\Sigma$ sc for complementary set. Such examples (important from the point of view of the understanding the transition $\Sigma$ pp $\to$ $\Sigma$ sc) and dual to the theory [22], [23] are not known.

To prove the absence of $\Sigma$ pp on some spectral interval we can use the following simple proposition (see [24], [11]).

**Lemma 10**. *If for any $k \in \triangle$*

$$\int_0^\infty \frac{dx}{\|M_k(0,x)\|^2} = +\infty \tag{4.15}$$

*then $\Sigma_{pp} \cap \triangle = \emptyset$. If in addition*

$$\lim_{x\to\infty} ||M_k(0,x)|| = \infty \ a.e. \ on \ \triangle \tag{4.16}$$

*then $\mu(H) = \mu_{SC}(H)$ on $\triangle$.*

The proof is based on the two obvious observations:

(i) $||M_k(0,x)|| = ||M_k^{-1}(0,x)||$ as $M_k(\cdot) \in \mathrm{SL}(2,\mathbf{R})$.

(ii) $1 = |\vec{e_\varphi}| = ||M_k^{-1}(x) \cdot M_k(x)\vec{e_\varphi}|| \le ||M_k^{-1}|| \cdot r_k(x)$, i.e.,

$$r_k^2(x) \ge \frac{1}{||M_k(0,x)||^2}. \tag{4.17}$$

See also Lemma 7.

## 5 Cutoff Lemma

This section contains result of a different nature. It is cutoff lemma which reduce problem of the spectral analysis for the sparse potential to the similar problem with compact supported bumps. It is the same idea as in the classical scattering theory: small perturbation of the potential cannot change the norm of the fundamental matrix dramatically though each particular solution can be perturbed strongly.

Next proposition will compare the norms of fundamental matrices for two equations $H\psi = -\psi'' + Q_0\psi = 0$ and $\widetilde{H}\psi = -\psi'' + (Q_0 + q)\psi = 0$ where function $q(\,\cdot\,)$ is fast decreasing near infinity.

**Lemma 11.** Suppose that $\{y_1(x),\ y_2(x)\}$ is a system of two solutions for

$$H\psi = -\psi'' + Q_0\psi = 0 \tag{5.1}$$

and

$$W_{y_1,y_2} = \det \begin{bmatrix} y_1 & y_1' \\ y_2 & y_2' \end{bmatrix} = \det M_{y_1,y_2}(0,x) = 1. \tag{5.2}$$

Assume also that

$$||M_{y_1,y_2}(0,x)|| \le A(x),\ A(x) \uparrow . \tag{5.3}$$

If $q(x)$ is the perturbation of $Q_0(\cdot)$ such that

$$\int_0^\infty |q(x)|\ A^2(x)dx < \infty \tag{5.4}$$

then the equation

$$\widetilde{H}\psi = -\psi'' + (Q_0 + q)(x)\psi = 0 \tag{5.5}$$

has two solutions $\widetilde{y}_1(x),\ \widetilde{y}_2(x)$ such that

$$W_{\widetilde{y}_1,\widetilde{y}_2} = 1,\ \frac{1}{4}||M_{y_1,y_2}(0,x)|| \le ||\widetilde{M}_{\widetilde{y}_1,\widetilde{y}_2}(0,x)|| \le 4||M_{y_1,y_2}(0,x)|| \tag{5.6}$$

for $x \ge x_0$. Point $x_0$ depends only on $q(\cdot), |A(\cdot)|$ but does not depend on specific boundary conditions for $y_1, y_2$.

*Proof.* The central idea is standard. We are looking for the solution $\widetilde{y}_1(x)$ in the form

$$\begin{aligned} \widetilde{y}_1(x) &= c_1(x)\ y_1(x) + c_2(x)\ y_2(x) \\ \widetilde{y}_1'(x) &= c_1(x)\ y_1'(x) + c_2(x)\ y_2'(x) \\ c_1(x) &\to 1,\ c_2(x) \to 0 \text{ if } x \to \infty. \end{aligned} \tag{5.7}$$

The second equality means that

$$c_1'\, y_1 + c_2'\, y_2 = 0. \tag{5.8}$$

Substitution of the formulas (5.7) into (5.5) gives additional equation

$$c_1'\, y_1 + c_2'\, y_2 = q(c_1y_1 + c_2y_2)\,. \tag{5.9}$$

As $W_{y_1,y_2} = 1$ it follows from (5.8), (5.9) that

$$\begin{aligned} c_1' &= q(c_1\, y_1\, y_2 + c_2\, y_2^2) \\ c_2' &= q_1(c_1\, y_1^2\, + c_2\, y_1\, y_2) \end{aligned} \tag{5.10}$$

and (at least formally)

$$\begin{aligned} c_1(x) &= 1 + \int_x^\infty q(z)\left[c_1(z)\, y_1\, y_2 + c_2\, y_2^2\right] dz \\ c_2(x) &= \int_x^\infty q(x)\left[c_1\, y_1^2\, + c_2\, y_1\, y_2\right](z)\, dz \end{aligned} \tag{5.11}$$

i.e., (in the vector form)

$$\vec{c} = \vec{e} + \int_x^\infty q(z)\, B(z)\, \vec{c}(z)\, dz \tag{5.12}$$

where

$$\vec{c}(z) = \begin{pmatrix} c_1(z) \\ c_2(z) \end{pmatrix},\ \vec{e} = \begin{pmatrix} 1 \\ 0 \end{pmatrix},\ B(z) = \begin{bmatrix} y_1y_2 & y_2^2 \\ y_1^2 & y_1y_2 \end{bmatrix}(z).$$

Simple calculations show that

$$||B(z)|| = (y_1^2 + y_2^2)(z) \le A^2(z)$$

and due to condition (5.5) for appropriate $x_0$

$$\int_{x_0}^\infty |q(z)|\, A^2(z) dz \le \frac{1}{2}.$$

Mapping in the right part of equation (5.12) is the contraction with coefficient $\le \frac{1}{2}$ in the space $C_{[x_0,\infty)}$, i.e., this equation has unique solution at $[0,\infty)$ and $||\vec{c}(x)||^2 \le 2$ for $x \ge x_0$.

Similar analysis shows the existence of the second solution

$$\begin{aligned} \widetilde{y}_2(x) &= d_1(x)y_1 + d_2(x)\, y_2 \\ \widetilde{y}_2'(x) &= d_1(x)y_1' + d_2(x)\, y_2' \\ d_1(x) &\to 0,\ d_2(x) \to 0 \text{ if } x \to \infty \\ |\vec{d}(x)|^2 &\le 2\,,\ x \ge x_0. \end{aligned} \tag{5.13}$$

Then

$$\widetilde{M}_{\widetilde{y}_1,\widetilde{y}_2}(0,x) = \begin{bmatrix} \widetilde{y}_1 & \widetilde{y}_1' \\ \widetilde{y}_2 & \widetilde{y}_2' \end{bmatrix}(x) = \begin{bmatrix} c_1 & c_2 \\ d_1 & d_2 \end{bmatrix} \begin{bmatrix} y_1 & y_1' \\ y_2 & y_2' \end{bmatrix} \tag{5.14}$$

As $W_{\widetilde{y}_1,\widetilde{y}_2} = 1$ we have

$$\det \widetilde{M} = \det M = \det \begin{bmatrix} c_1 & c_2 \\ d_1 & d_2 \end{bmatrix} = 1.$$

According to our estimations of the coefficients $c_i, d_i, i = 1,2$ for $x \ge x_0$

$$\left|\left| \begin{matrix} c_1 & c_2 \\ d_1 & d_2 \end{matrix} \right|\right| \le 4\,,\quad \text{i.e., } ||\widetilde{M}|| \le 4||M||.$$

The second inequality in (5.7) follows from (5.14) and the trivial observation

$$\begin{bmatrix} c_1 & c_2 \\ d_1 & d_2 \end{bmatrix}^{-1} = \begin{bmatrix} d_2 & -d_1 \\ -c_2 & c_1 \end{bmatrix}.$$

**Corollary 1**. There exists const $\alpha^{\pm}$, depending only on $A(\cdot)$, $q(\cdot)$ such that for $x \ge x_0$

$$\alpha^- ||M(0,x)|| \le \left\| \widetilde{M}(0,x) \right\| \le \alpha^+ ||M(0,x)|| \tag{5.15}$$

where $M, \widetilde{M}$ are monodromy operators for $H, \widetilde{H}$ on [0,$x$] (i.e., fundamental matrices with initial data $M(0,0) = \widetilde{M}(0,0) = I$).

**Corollary 2**. Let $M_k(0,x)$ , $k \in \Delta \subset (0,\infty)$ be the monodromy operator for equation $H\psi = -\psi'' + Q_0\psi = k^2\psi$ and uniformly in $k \in \Delta$

$$||M_k(0,x)|| \le A(x) \uparrow \tag{5.16}$$

Assume that for perturbation $q(\cdot)$ of the potential $Q_0(\cdot)$ the inequality (4.1) is fulfilled. Then for appropriate const $\alpha^{\pm}$ and all $k \in \Delta$

$$\alpha^- ||M_k(0,x)|| \le ||\widetilde{M}_k(0,x)|| \le \alpha^+ ||M_k(0,x)||$$

The proof is based on the fact that $||M_k(0,x_0)|| \leq$ const depending on $A(\cdot)$ and $q(\cdot)$ but not $k \in \Delta$.

**Remark.** There is an essential difference between Lemma 11 and classical scattering theory. If $||M_k(0,x)|| \to \infty$ fast enough then (see Lemmas 7′, 9) one can construct for a.e. $k \in \Delta$ the decaying solution $\psi_k(x) \in L^2(R_+)$. Corresponding boundary phases $\widetilde{\theta} = \widetilde{\theta}(k)$ are known as a Weil's phases. It is well-known from localization theory (see review [9]) the following fact (goes to I. Goldshaidt [25]): in the localization regime Weil's phase has a.e. $k \in \Delta$ second type discontinuity, i.e., it is extremely irregular. In the scattering theory similar objects, Jost solutions, are regular (at least continuous functions) of the frequency $k$.

Norms of monodromy operator $||M_k(0,x)||$ or Lyapunov exponents $\gamma = \gamma(k)$ have essentially more regular structure (with respect to $k$) then Weil's solutions.

## *5.1 Martingales and homogenization lemmas*

The central part of the paper is related to the spectral analysis for the positive energies $\lambda = k^2 > 0$ (Sec. 6), corresponding results for $\lambda < 0$ (Sec. 6) are simpler. We will use the following strategy.

According to formulas (2.10), (6.34)

$$r_k(x) = \exp\left(\frac{1}{2k}\int_0^x q(z)\sin 2t_k(z)dz\right)$$

where

$$t_k'(x) = k - \frac{q(x)\sin^2 t_k(x)}{k}\ ,\ t_k(x) = t_k(0) + kx - \frac{1}{k}\int_0^x q(z)\sin^2 t_k(z)dz$$

with appropriate boundary conditions. Cutoff lemma (together with trivial estimations, see beginning of (6a)) reduces the problem to the case

$$q(x) = \sum_{n=1}^{\infty} g_n(x-x_n)\, I_{|x-x_n|\leq l_n}\ ,\ l_n = \exp(An). \tag{5.17}$$

If $x_n^{\pm} = x_n \pm l_n$ then

$$\begin{aligned} r_k\,(x_n^+) &= \sum_{i=1}^{n} \frac{1}{2k}\int_{x_i^-}^{x_i^+} q(z)\sin 2t_k(z)dz \\ &= \sum_{i=1}^{n} F_i(k,\ t_k(x_n^-)). \end{aligned} \tag{5.18}$$

Different expression for the same function $F_i(k,\varphi)$ give formulas (3.9), (3.15):

$$F_i(k,\varphi) = \ln |M_k(x_i^-, x_i^+)\overrightarrow{e_\varphi}|.$$

But

$$t_k(x_i^-) = L_i k + t_k(x_{i-1}^+)$$

where $L_i = x_i^- - x_{i-1}^+$ is the distance between bumps with the numbers $i-1$ and $i$.

Because of superlacunarity condition functions

$$f_n(k) = F_n(k, L_n k + t_k(x_{n-1}^+))$$

*have very fast oscillations with respect to the second argument.* The convergence (or divergence) of the functional series of the form

$$\sum_n F_n(k,\ L_n k + \tau_{n-1}(k)) \tag{5.19}$$

is the key moment of the theory. In the special form of decaying bumps:

$$g_n(x - x_n) = a_n\, \varphi_n(x - x_n)\ ,\ a_n \to 0$$

for (5.19) the perturbation theory gives the simplest "Fourier-like" series.

$$\sum_n a_n(k, \tau_{n-1}(k)) \cos L_n k + b_n(k, \tau_{n-1}(k)) \sin L_n k. \tag{5.20}$$

The central idea of our approach to the study of the series (5.19), (5.20) is going to the classical result by A. Kolmogorov on the lacunar trigonometric series.

Theorem $\mathcal{K}$. Let $S_n(x) = \sum_{i=1}^{n} (a_i \cos n_i x + b_i \sin n_i x)$ $x \in [0, 2\pi] = S^1$, $\frac{n_{i+1}}{n_i} \geq \gamma > 1$, $i \geq i_0$ is a lacunar Fourier series. Then

(i) Condition $\sum_i (a_i^2 + b_i^2) < \infty$ implies a.e. on $S^1$ convergence $S_n^i(x) \to S(x) = \lim_{n\to\infty} S_n(x)$

(ii) If $\sum_i (a_i^2 + b_i^2) = +\infty$, then

$$\limsup S_n(x) = +\infty, \text{ a.e. on } S^1$$
$$\liminf S_n(x) = -\infty \quad \text{a.e. on } S^1.$$

In fact, in the last case Kolmogorov gave even asymptotical formulas for the growth of $S_n(x),\ n \to \infty$.

There are different proofs of the Theorem $\mathcal{K}$ (see, for instance [26]). More important for us is not the theorem by itself but Kolmogorov's idea on the analogy between lacunar Fourier series and series of independent random variables.

We will use this idea to prove the generalization of the Theorem $\mathcal{K}$ to a wide class of the functional "Fourier-like" series. Preliminary result.

**Theorem 1**. Let us consider the following functional series

$$S(x) \sim \sum_{n=1}^{\infty} f_n(x)\cos 2\pi N_n x + g_n(x)\sin 2\pi N_n x,\ x \in [0,1]\,. \qquad (5.21)$$

Assume that $N_n = i_1 \cdot i_2 \ldots i_n, i_n > 1$ are integers and $\sum\limits_n \frac{N_{n-1}}{N_n} = \sum \frac{1}{i_n} < \infty$ (superlacunarity condition). Let us also $f_n(\cdot), g_n(\cdot) \in C^1([0,1])$ and $\|f_n\|_\infty \le 1, \|g_n(\cdot)\|_\infty < \infty, \|f_n'(\cdot)\|_\infty, \|g_n'(\cdot)\|_\infty \le N_{n-1}$ (i.e., coefficients are more regular then Fourier harmonics $\cos 2\pi N_n x, \sin 2\pi N_n x$). If $A_n(x) = \sum\limits_{i=1}^{n}(f_n^2(x) + g_n^2(x))$ and

$$\text{Œ}_0 = \left\{x : \lim_{n\to\infty} A_n(x) < \infty\right\},\ \bar{\text{Œ}}_0 = \left\{x : \lim_{n\to\infty} A_n(x) = +\infty\right\}$$

then

(i) a.e. on the set $\text{Œ}_0$ the partial sums

$$S_n(x) = \sum_{i=1}^{n} f_i(x)\cos 2\pi N_i x + g_i(x)\sin 2\pi N_i x \qquad (5.22)$$

converge to the limit

$$S(x) = \lim_{n\to\infty} S_n(x)$$

(ii) on the same set $l.i.m.S_n(x) = S(x)$, i.e.,

$$\int\limits_{\text{Œ}_0} |S_n(x) - S(x)|^2 dx < \infty$$

(iii) a.e. on the complimentary set $\bar{\text{Œ}}_0$

$$\limsup_{n\to\infty} \limsup S_n(x) = +\infty \qquad (5.23)$$

$$\limsup_{n\to\infty} S_n(x) = -\infty$$

and

$$S_n(x) = o(A_n^{1/2}(x) \ln^{1/2+\varepsilon} A_n(x)) \tag{5.24}$$

for arbitrary $\varepsilon > 0$, i.e.,

$$\lim_{n\to 0} \lim \frac{S_n(x)}{A_n^{1/2}(x) \ln^{1/2+\varepsilon} A_n(x)} = 0 \quad \text{a.e.on} \quad \bar{Œ}_0. \tag{5.25}$$

This theorem, of course, includes Kolmogorov's result although with a stronger conditions on $N_n$ which is simply lacunarity.

**Remark 5.1.1** *Slightly more complex analysis gives all the results of the Theorem 1 under a weaker condition* $\sum_n \left(\frac{N_n}{N_{n+1}}\right)^{1/2} < \infty\,(SL')$.

*Proof* of the theorem is based on the classical martingale theory (Doob-Kolmogorov), for the references see, for example, J. Neveu [27].

Let $\mathcal{F}_n$, $n = 1, 2, \ldots$ are the $\sigma-$algebras, generated by partitions of $\Omega = [0, 1]$ by the points $\left\{\frac{i}{N_n}\right\}, i = 1, 2, \ldots,\ N_n - 1$. Because $\frac{N_n}{N_{n-1}} \in Z_+^1$

$$\mathcal{F}_1 \subset \mathcal{F}_2 \subset \ldots \subset \mathcal{F}_n \subset \ldots$$

i.e., $\{\mathcal{F}_n\}$ is filtration. Functions $f_n(\cdot), g_n(\cdot), \sin(\cos)(2\pi N_n x)$ are not $\mathcal{F}_n$ measurable, but we can approximate them by the $\mathcal{F}_{\widetilde{n}}$ measurable functions (piecewise constant approximation) and this is the central moment

$$\text{Let } \varphi^{(n)}(x) = E\left[f_n(\cdot)|\mathcal{F}_n\right]\ ,\ \gamma^{(n)}(x) = E[g_n(\cdot)|\mathcal{F}_n], \tag{5.26}$$
$$s^{(n)}(x) = E[\sin 2\pi N_n \cdot |\mathcal{F}_{n+1}]\ , c^{(n)}(x) = E[\cos 2\pi N_n \cdot |\mathcal{F}_{n+1}].$$

One can recall that

$$\Phi^{(n)}(x) = E[F(\cdot)|\mathcal{F}_n] \equiv N_n \int_{\frac{i-1}{N_n}}^{\frac{i}{N_n}} F(z)dz\ ,\ x \in [\frac{i-1}{N_n}\ ,\ \frac{i}{N_n}], i = 1, 2, \ldots N_{n-1}.$$

It is clear that

$$||f_n(\cdot) - \varphi^{(n)}(x)||_\infty \le \frac{||f_n'(\cdot)||_\infty}{N_n} \le \frac{N_{n-1}}{N_n} = \frac{1}{i_n} \tag{5.27}$$
$$||\cos 2\pi N_n \cdot - c^{(n)}(\cdot)||_\infty \le \frac{N_n}{N_{n+1}}$$

with similar estimations for $g_n(\cdot), \sin 2\pi N_n \cdot$.

Because $\sum \frac{N_{n-1}}{N_n} < \infty$ and (5.27) the convergence (or divergence) of the initial functional series (5.21) is equivalent to the convergence (divergence) of the series

$$\sum_n \varphi^{(n)}(x)c^{(n)}(x) + \gamma^{(n)}(x)s^{(n)}(x) \tag{5.28}$$

(equiconvergence property).

Let us put

$$S^{(n)}(x) = \sum_{i=1}^{n}(\varphi^{(i)}c^{(i)} + \gamma^{(i)}s^{(i)}(x)).$$

The pair $(S^{(n)}(\cdot), \mathcal{F}_{n+1})$ is the martingale, i.e.,

$$\begin{array}{ll} (\text{ i}) & S^{(n)} \text{ is } \mathcal{F}_{n+1} - \text{ measurable} \\ (\text{ ii}) & E[S^{(n)}(\cdot)|\mathcal{F}_n] = S^{(n-1)}(\cdot) \end{array}$$

In fact,

$$\begin{aligned} E[S^{(n)}(\cdot)|\mathcal{F}_n] &= S^{(n-1)}(\cdot) + E[\varphi^{(n)}c^{(n)} + \gamma^{(n)}s^{(n)}|\mathcal{F}_n] \\ &= S^{(n-1)}(\cdot) + \varphi^{(n)}E[c^{(n)}|\mathcal{F}_n] + \gamma^{(n)}E(s^{(n)}|\mathcal{F}_n) = S^{(n-1)}(\cdot). \end{aligned}$$

Martingale $(S^{(n)}, \mathcal{F}_{n+1})$ is square integrable, i.e., $S^{(n)} \in L^2(\Omega)$. More (as shown by simple calculations)

$$\begin{aligned} V^{(n)}(x) = \int_0^1 [S^{(n)}(x)]^2 dx = E(S^{(n)})^2 = \sum_{i=1}^{n} \big\{ & E(\varphi^{(i)})^2 E(c^{(i)})^2 + \\ & + E(\gamma_{(\cdot)}^{(i)})^2 E(s^{(i)}(\cdot))^2 \big\}. \end{aligned} \tag{5.29}$$

But for arbitrary $F \in C^1(\Omega)$

$$E(F - E(F|\mathcal{F}_n))^2 = E(F^2) - E(E(F\ \mathcal{F}_n))^2$$

i.e.,

$$E(E(F|\mathcal{F}_n])^2 = EF^2 + O\left(\frac{||F'||_\infty}{N_n}\right)^2.$$

It gives that (for $A_n(x) = \sum_{i=1}^{n}(f_i^2(x) + g_i^2(x))$)

$$V^{(n)}(x) = \sum_{i=1}^{n} E\left\{\left[\varphi^{(i)}(x)c^{(i)}(x) + \gamma^{(i)}(x)s^{(i)}(x)\right]^2 |\mathcal{F}_n\right\} =$$

$$= \sum_{i=1}^{n} \frac{\left(\varphi^{(i)}_{(\cdot)}\right)^2 + \left(\gamma^{(i)}(\cdot)\right)^2}{2} + O\left(\sum_{i=1}^{n}\left(\frac{N_{i-1}}{N_i}\right)^2\right)$$

$$= \frac{1}{2}\sum_{i=1}^{n}(f_n^2(x) + g_n^2(x)) + O\left(\sum_{n}\frac{N_{n-1}}{N_n}\right) + O\left(\sum_{n=1}^{\infty}\left(\frac{N_{n-1}}{N_n}\right)^2\right) =$$

$$\frac{1}{2}A_n(x) + O(1).$$

In the last transition we used the boundedness of the functions $f_n, g_n$ :

$$||(f_n^2(\cdot))'||_\infty \leq 2||f_n||_\infty||f_n'(\cdot)||_\infty \leq 2N_{n-1} \text{ etc.}$$

Sequence of the functions $V^{(n)}(x)$ is the so-called quadratic characteristic of the martingale $(S^{(n)}, \mathcal{F}_{n+1})$. All statements of the theorem are direct consequences of the general martingale theory and equivalence of the series (5.21), (5.28) (see [27], Ch. IV, Sec 6).

All known "pure analytical proofs" of the Theorem 1 are the direct interpretations of the martingale proof. The central moment is the proof of Kolmogorov's inequality for the martingales.

Theorem 1 is intermediate between Theorem $\mathcal{K}$ and Theorem 2′ (see below) in which artificial condition $\frac{N_n}{N_{n-1}} \in Z_+^d$ will be removed and argument $k$ will run through the spectral interval $\triangle$ instead of the unite circle $S^1$. Theorem 2′ by itself is a consequence of essentially more general one.

**Theorem 2**. Let $F_n(\varphi, k)$ , $n = 1, 2, \ldots$ be the family of $C^1$- functions on $S^1 \times \triangle = [0, 2\pi] \times \triangle$, periodic in $\varphi : F_n(\varphi + 2\pi, \cdot) \equiv F_n(\varphi, \cdot)$. Assume that for appropriate const $A_i > 0, i = 0, 1, 2$, uniformly in $n$

$$||F_n(\cdot)||_\infty \leq A_0, || \bigtriangledown F_n(\cdot)||_\infty = ||\frac{\partial F_n}{\partial \varphi}||_\infty + ||\frac{\partial F_n}{\partial k}||_\infty \leq A_1 * . \quad (5.30)$$

Let also $\{L_n\}, n = 1, 2, \ldots$ be an increasing superlacunar sequence of the scaling parameters, $\sum_n \frac{L_{n-1}}{L_n} < \infty$ and $\{\tau_n(k)\}, n = 1, 2, \ldots$ be an additional family ("phase shifts") of the $C^1$ functions

$$||\frac{\partial \tau_{n-1}}{\partial k}(\cdot)||_\infty \leq A_2 L_{n-1}. \quad (5.31)$$

Let us put

$$S_n(k) = \sum_{i=1}^{n} F_i(\tau_{i-1}(k) + L_i k, k) \tag{5.32}$$

$$\bar{F}_n(k) = \frac{1}{2\pi}\int_0^{2\pi} F_n(\varphi, k)d\varphi \tag{5.33}$$

$$F_{n,c}(k) = F_n(\varphi, k) - \bar{F}_n(k) \tag{5.34}$$

$$S_{n,c}(k) = \sum_{i=1}^{n} \left[F_i(\tau_i(k) + L_i k, k) - \bar{F}_i(k)\right] \tag{5.35}$$

$$\bar{F}_{n,c}^{(2)} = \frac{1}{2\pi}\int_0^{2\pi} F_{n,c}^2(\varphi, k)d\varphi \tag{5.36}$$

$$D_n(k) = \sum_{i=1}^{n} \bar{F}_{n,c}^{(2)}(k). \tag{5.37}$$

Then the sequence $S_{n,c}(\cdot)$ converges a.e. on the set

$$Œ_0 = \left\{k \in \triangle : \lim_{n\to\infty} D_n(k) = \sum_{i=1}^{\infty} \bar{F}_{n,c}^{(2)}(k) < \infty\right\}. \tag{5.38}$$

On the complementary set

$$\bar{Œ}_0 = Œ_1 = \left\{k \in \triangle : \sum_{i=1}^{\infty} \bar{F}_{n,c}^{(2)}(k) < \infty\right\} \tag{5.39}$$

the sequence diverges

$$\limsup_{n\to\infty} S_{n,c}(k) = +\infty \qquad \text{a.e. on } Œ_1$$
$$\lim\ \inf S_{n,c}(k) = -\infty \qquad \text{a.e. on } Œ_1$$

**Remark**. Instead of the condition (75) it is enough to assume (see below) that uniformly over $n$

$$||F_n(\cdot,\cdot)||_\infty \leq A_0\ ,\ ||\frac{\partial F_n}{\partial \varphi}|| \leq A_1\ ,\ |F_n(k,\varphi) - F_n(k',\varphi)| \leq \tag{5.40}$$
$$\leq A_2|k-k'|^\delta\ ,\ A_0, \delta > 0$$

and

$$\lim_{n\to\infty} \frac{S_{n,c}(k)}{D_n^{1/2}(k)\ln^{1/2+\varepsilon} D_n} = 0. \qquad \text{a.e. on } Œ_1.$$

The proof is absolutely the same. This observation will be important in what follows (compare with Theorem 3)

**Proof.** Let $\mathcal{F}_n, n = 1, 2, ...$ be $\sigma$-algebras on $\triangle$ generated by partition $\sqcap_n$ with the set of endpoints

$$\sqcap_n = \left\{ \frac{2\pi i}{L_j} = k_{j,i} \in \triangle \ , \ j \leq n \right\}.$$

Of course,

$$|\sqcap_n| \leq \sum_{j=1}^{n} \frac{|\triangle| L_j}{2\pi} + 2 \leq C(\triangle) \sum_{j=1}^{n} L_j \leq C_1(\triangle) L_n$$

and $\{\mathcal{F}_1 \subset \mathcal{F}_2 \subset ...\} = \Phi$ is the filtration in the Borel's $\sigma$-algebra on $\triangle$.

Now we will transform the functions

$$G_n(k) = F_{n,c}(\tau_{n-1}(k) + L_n k, k) \tag{5.41}$$

to the piecewise sequence, adapted to filtration $\Phi$. We will make it in two steps. First, let us put

$$G_n(k) = F_{n,c}(\tau_{n-1}(k_{nj}) + L_n k, \ k_{nj}), x \in \triangle_{nj} \tag{5.42}$$

and

$$\triangle_{nj} = [k_{nj}, k_{n(j+1)}] = \left[ \frac{2\pi j}{L_n} \ , \ \frac{2\pi (j+1)}{L_n} \right]$$

(freezing of the slow variables!).

For the intervals $\triangle_n^{\pm}$, containing the edges $\alpha, \beta$ of the interval $\triangle$ function $\widetilde{G}_n(k)$ will again be equal to the value of $G_n(k)$ at the corresponding left edges.

Functions $F_{n,c}(\varphi, k)$ satisfy the smoothness conditions (5.30), (5.31) (with two times greater const: $2A_0, 2A_1$) so that

$$||G_n(\cdot) - \widetilde{G}_n(\cdot)||_\infty \leq \frac{2\pi}{L_n} || \frac{\partial F_n}{\partial k} ||_\infty + \frac{2\pi}{L_n} || \frac{\partial F_n}{\partial \varphi} ||_\infty || \frac{\partial \tau_{n-1}}{\partial k} ||_\infty \leq$$
$$\leq C(\triangle) \left( \frac{A_1 + A_1 A_2 L_{n-1}}{L_n} \right)$$

and $\sum_n ||G_n - \widetilde{G}_n||_\infty < \infty$. The series $\sum G_n, \sum \widetilde{G}_n$ have the same sets of convergence and divergence and the same rate of divergence.

On the second step of approximation we will put

$$\widetilde{\Gamma}_n(k) = E\left[\widetilde{G}_n^{(\cdot)}|\mathcal{F}_{n+1}\right] = \frac{1}{|\Delta_{n+1,i}|}\int_{\Delta_{n+1,i}} \widetilde{G}_n(k)dk \ , \ k \subset \Delta_{n+1,i} \quad (5.43)$$

and

$$\widetilde{a}_n(k) = E\left[\widetilde{G}_n(\cdot)|\mathcal{F}_n\right] = E\left[\widetilde{\Gamma}_n(k)|\mathcal{F}_n\right]$$

(as $\mathcal{F}_n \subset \mathcal{F}_{n+1}$).

Function $\widetilde{\Gamma}_n(k)$ are again "very close" to $\widetilde{G}_n(k)$. In fact,

$$||\widetilde{\Gamma}_n - \widetilde{G}_n||_\infty \le ||\frac{\partial F_n}{\partial \varphi}||_\infty \cdot \frac{2\pi L_n}{L_{n+1}} \le c \cdot \frac{L_n}{L_{n+1}}$$

i.e.,

$$\sum_n ||\widetilde{\Gamma}_n - \widetilde{G}_n||_\infty > \infty$$

and studies of convergence (divergence) for all three series $\sum G_n$, $\sum \widetilde{G}_n$, $\sum \Gamma_n$ are equivalent.

Let us introduce the sums

$$\Sigma_n(k) = \sum_{i=1}^{n} \widetilde{\Gamma}_i(k)$$

$$\Sigma_n'(k) = \sum_{i=1}^{n} \left(\widetilde{\Gamma}_i(k) - \widetilde{a}_i(k)\right).$$

Both pairs $(\Sigma_n, \mathcal{F}_{n+1})$, $(\Sigma_n', \mathcal{F}_{n+1})$ have an adaption property.

The second pair is a square integrable martingale, but the first one is closely related to the initial problem on the convergence of $\sum G_n(\cdot)$.

Let us prove that $\Sigma_n, \Sigma_n'$ converge at the same set (moduli set of the Lebesque measure 0). It is enough to check that sequence $\sum_{i=1}^{n} \widetilde{a}_i(k) = A_n(k)$ converges a.e. on $\Delta$.

The last sentence is the consequence of the following observations. Transition from the partition $\sqcap_{n-1}$ to $\sqcap_n$ requires to add points $\left\{k_{nj} = \frac{2\pi j}{L_n}\right\}$ to the set $\sqcap_{n-1} = \left\{\frac{2\pi i}{L_s}, s \le n-1\right\}$. Some of the intervals $\Delta_{nj} = \left(\frac{2\pi j}{L_n}, \frac{2\pi(j+1)}{L_n}\right)$ will not contain the points from $\sqcap_{n-1}$, we will call such intervals "good" ($\mathcal{G}_n$-intervals), other intervals (including, may be the intervals, containing endpoints $\alpha, \beta$ of $\triangle$) will be "bad" ($B_n$-intervals).

Then for each $\triangle_{nj} \in \mathcal{G}_n$

$$a_n(k) = \frac{1}{|\triangle_{nj}|} \int_{\triangle_{nj}} \widetilde{G}(k)dk = \frac{1}{|\triangle_{nj}|} \int_{\triangle_{nj}} \widetilde{G}(k)dk = 0$$

and

$$\varepsilon_n = \sum_{\triangle_{ni} \in B_n} |\triangle_{ni}| \leq \frac{2\pi}{L_n} |\Pi_{n-1}| \leq \frac{cL_{n-1}}{L_n},$$

i.e., $\sum\limits_n \varepsilon_n < \infty$.

The Borel-Cantelli argument shows the a.e. on $\triangle$ the series $\sum\limits_n \widetilde{a}_n(k)$ has only finite number of non-zero terms, i.e., $\sum \widetilde{a}_n < \infty$ a.e.

Now we reduced the initial problem of the convergence (divergence) of the series $\sum G_n$ to the problem of asymptotical behavior of the martingale $(\sum'_n, \mathcal{F}_{n+1}) = \left( \sum\limits_{i=1}^{n} \left( \widetilde{\Gamma}_i - \widetilde{a}_i \right), \mathcal{F}_{n+1} \right)$. According to the general theory everything depends on the quadratic characteristic

$$D_n(k) = \sum_{i=1}^{n} E\left[ (\widetilde{\Gamma}_i - \widetilde{a}_i)^2 | \mathcal{F}_i \right]. \tag{5.44}$$

But if $\triangle_{nj} \in \mathcal{G}_n$ and $k \in \triangle_{nj}$

$$E\left[ \left( \widetilde{\Gamma}_n - \widetilde{a}_n \right)^2 | \mathcal{F}_n \right] = E\left[ \widetilde{\Gamma^2}_n | F_n \right] = \frac{1}{|\triangle_{nj}|} \int_{\triangle_{nj}} \widetilde{\Gamma^2}_n dk.$$

Let us remember that

$$|\widetilde{\Gamma}_n - \widetilde{G}_n| \leq \frac{cL_n}{L_{n+1}}, \ |\widetilde{\Gamma^2}_n - \widetilde{G^2}_n| \leq CA_0 \cdot \frac{L_n}{L_{n+1}}$$

and

$$\frac{1}{|\triangle_{nj}|} \int_{\triangle_{nj}} \widetilde{G}^2_n(k)dk = \bar{F}^{(2)}_{nc}\left( \frac{2\pi j}{L_n} \right),$$

$$|F^{(2)}_{nc}(k) - F^{(2)}_{nc}\left( \frac{2\pi j}{L_n} \right)| \leq C\frac{A_0 A_1}{L_n}.$$

Collecting together all these estimations we will get

$$\widetilde{D}_n(k) = \sum_{i=1}^{n} \bar{F}_{ic}^{(2)}(k) + O\left(\sum_{i=1}^{n} \frac{L_i}{L_{i+1}}\right), k \in \Delta_{nj} \in g_n \tag{5.45}$$

Borel-Cantelli lemma together with super lacunarity of $\{L_n\}$ give

$$\widetilde{D}_n(k) = \sum_{i=1}^{n} \bar{F}_{ic}^{(2)}(k) + O\left(1\right) \tag{5.46}$$

and reference to the general martingale theory [27] (as in the Theorem 1) complete the proof of Theorem 2.

The next result is the direct consequence of the Theorem 2.

**Theorem 2′**. Let us consider the functional series

$$S(k) \sim \sum_{n=1}^{\infty} f_n(k) \cos L_n k + g_n(x) \sin L_n k \tag{5.47}$$

where

$$k \in \Delta \subset (0, \infty), \quad f_n, g_n \in C^1, \quad ||f_n||_\infty \leq 1, \quad ||g_n(\cdot)||_\infty \leq 1,$$

$$||f_n'(\cdot)||_\infty \leq L_{n-1}, \quad ||g_n'(\cdot)|| \leq L_{n-1}, \ \sum_n \frac{L_{n-1}}{L_n} < \infty.$$

$$If A_n(k) = \sum_{i=1}^{n} f_n^2(k) + g_n^2(k)$$

and $Œ_0 = \left\{K : \lim_{n\to\infty} A_n(k) < \infty\right\}$, $Œ_1 = \Delta \backslash \bar{Œ}_0$ then $S_n(x)$ converge a.e. on $Œ_0$ and has unbounded ($\pm\infty$) divergence a.e. on $Œ_1$ (as in the Theorems 1,2).

The further results on the pure a.c. spectrum will be based on Lemma 5 (Sec. 4). Technically they include the analytical lemma in the spirit of well-known Krilov-Bogoljubov method of averaging with respect to "fast variable". These lemmas reflect "independence" of the wave processes with the distant frequencies and conceptually are close to martingale theorems (see the first part of the section).

Let us begin with the simplest result of this type.

**Lemma 12**. Let $c(k), a(k),\ b(k),\ k \in \Delta \subset (0, \infty)$ be $C^1(\Delta)$ functions, $L$ and $\varepsilon$ be a large and a small parameter, respectively. Then under condition $\varepsilon||a||_\infty + \varepsilon||b||_\infty \leq 1$ and suitable const $C_i(\triangle) > 0$

$$\int\limits_{\triangle} \exp\left[c(k) + \varepsilon a(k)\cos kL + \varepsilon b(k)\sin kL\right] dk \tag{5.48}$$

$$= \int\limits_{\triangle} \exp\left[c(k) + \frac{\varepsilon^2}{4}\left(a^2(k) + b^2(k)\right)\right] dk \cdot \exp\left(R_0(\varepsilon, L)\right)$$

$$+R_1(\varepsilon, L)$$

and remainders $R_0, R_1$ admit the following estimations

$$\begin{aligned} R_0(\varepsilon, L) &\leq \frac{c_1}{L}\left(||c'||_\infty + \varepsilon(||a'||_\infty + ||b'||_\infty)\right) + \\ &\quad + c_2\, \varepsilon^4(||a||_\infty^4 + ||b||_\infty^4) \\ R_2(\varepsilon, L) &\leq \frac{c_3}{L}\exp(||c||_\infty). \end{aligned} \tag{5.49}$$

*Proof.* Let $\triangle = [\alpha, \beta]$ and $n_L^{\pm}(\triangle)$ be the minimal and maximal points of the lattice $Z_L = \left\{\frac{2\pi n}{L}\right\}$ inside $\triangle$.

Then

$$\int\limits_{\triangle} \exp[...]\, dk = \sum_{j=n_L^-}^{n_L^+ - 1} \int_{\frac{2\pi j}{L}}^{\frac{2\pi(j+1)}{L}} \exp[\ldots]\, dk + \int_\alpha^{\frac{2\pi n_L^-}{L}} \exp[\ldots]\, dk$$

$$+ \int_{\frac{2\pi n_L^+}{L}}^{\beta} \exp[\ldots]\, dk = \sum_{j=n_L^-}^{n_L^+ - 1} \int_{\frac{2\pi j}{L}}^{\frac{2\pi(j+1)}{L}} \exp[\ldots]\, dk + R_1(\varepsilon, \triangle).$$

It is clear that

$$|R_1| \leq \frac{4\pi}{L}\exp\left(||c||_\infty + \varepsilon||a||_\infty + \varepsilon||b||_\infty\right) \leq \frac{C_0}{L}\exp(||c||_\infty). \tag{5.50}$$

But

$$\int_{\frac{2\pi j}{L}}^{\frac{2\pi(j+1)}{L}} \exp\left[c(k) + \varepsilon(a(k)\cos kL + b(k)\sin kL)\right] dk$$

$$= \int_{\frac{2\pi j}{L}}^{\frac{2\pi(j+1)}{L}} \exp\left[c(\tfrac{2\pi j}{L}) + \varepsilon(a(\tfrac{2\pi j}{L})\cos kL + b(\tfrac{2\pi j}{L})\sin kL\right] dk \cdot$$

$$\cdot \exp R_2(\varepsilon, L) = I_j \cdot \exp R_2(\varepsilon, L),$$

where

$$|R_2(\varepsilon, L)| \leq \frac{C}{L}\left[||c'||_\infty + \varepsilon||a'||_\infty + \varepsilon||b'||_\infty\right]. \tag{5.51}$$

Integral

$$I_j = \int_0^{\frac{2\pi}{L}} \exp\left[\varepsilon a\left(\frac{2\pi j}{L}\right)\cos kL + \varepsilon b\left(\frac{2\pi j}{L}\right)\sin kL\right] dk$$

can be evaluated explicitly in terms of Bessel functions but it is simpler to use Taylor expansion:

$$\begin{aligned} I_j &= \frac{1}{2\pi L}\int_0^{2\pi} \exp\left(\varepsilon\sqrt{a^2(\cdot)+b^2(\cdot)}\cos\varphi\right) d\varphi \qquad (5.52)\\ &= \frac{1}{L}\left[1 + \frac{\varepsilon^2}{4\pi}\int_0^{2\pi}\cos^2\varphi\, d\varphi\left(a^2(\cdot)+b^2(\cdot)+O(\varepsilon^4(a^2(\cdot)+b^2(\cdot))^2)\right)\right]\\ &= \frac{1}{L}\exp\left[\frac{\varepsilon^2}{4\pi}(a^2+b^2)(\frac{2\pi j}{L}) + O(\varepsilon^4(||a||_\infty^4 + ||b||_\infty^4)\right]. \end{aligned}$$

Then

$$\sum_{j=n_L^-}^{n_L^+-1} I_j \exp\left(c(\frac{2\pi j}{L})\right) = \frac{1}{L}\sum_{n_L^-}^{n_L^+-1}\exp\left[c(\frac{2\pi j}{L}) + \frac{\varepsilon^2}{4}(a^2+b^2)(\frac{2\pi j}{L}) + \right.$$
$$\left. + O(\varepsilon^4||a||_\infty^4 + \varepsilon^4||b||_\infty^4)\right].$$

Now one can use the standard estimation of the difference between the integral in the integral sum (rectangular rule). It gives

$$\begin{aligned} &\frac{1}{L}\sum_{j=n_L^-}^{n_L^+-1}\exp\left[c(\frac{2\pi j}{L}) + \frac{\varepsilon^2(a^2+b^2)}{4}(\frac{2\pi j}{L})\right]\\ &= \left(\int_\alpha^\beta \exp\left[c(k) + \frac{\varepsilon^2(a^2+b^2)}{4}(k)\right] dk \cdot \exp R_3(\varepsilon, L)\right) + R_4(\varepsilon, L) \end{aligned}$$

where (compare with (5.50))

$$|R_4(\varepsilon, L)| \leq \frac{4\pi}{L}\exp\left(||c||_\infty + \frac{\varepsilon^2}{4}(||a||_\infty^2 + ||b||_\infty^2\right) \leq \frac{C_1}{L}\exp(||c||_\infty)$$

and

$$\begin{aligned} R_3(\varepsilon, L) &\leq \frac{C|\triangle|}{L}\left(||c'||_\infty + \frac{\varepsilon^2}{4}||(a^2+b^2)'||_\infty\right) \leq\\ &\leq \exp\left[(||c'||_\infty + \varepsilon||a'||_\infty + \varepsilon||b'||_\infty)\cdot\frac{C_2}{L}\right] \end{aligned}$$

(compare with (5.51)). Collecting together all inequalities we will get (5.49)).

We will also need the trivial generalization of the Lemma 12 to the case when exponent contains finite number of multiple frequencies: not only $Lk$, but $2Lk, 3Lk$ etc.

**Lemma 12′**. Assume that the functions $c(k), a_i(k), b_i(k), i = 1, 2, 3$ are from $C^1(\Delta)$, $L, \varepsilon$ are large and small parameters, $\varepsilon \cdot \sum_{i=1}^{3} ||a_i||_\infty + ||b_i||_\infty \leq c_0$.

Then for appropriate const $c_i$

$$\int_\Delta \exp\left[c(k) + \varepsilon \sum_{j=1}^{3} a_j(k) \cos jLk + b_j(k) \sin jLk\right] dk \tag{5.53}$$

$$= \exp R_0(\varepsilon, L) \cdot \int_\Delta \exp\left[c(k) + \frac{\varepsilon^2}{4} \sum_{i=1}^{3} (a_i^2 + b_i^2)(k)\right] dk + R_1(\varepsilon, L),$$

where

$$R_0(\varepsilon, L) \leq \frac{c_1}{L}\left[||c'||_\infty + \varepsilon(\sum_{i=1}^{3} ||a_i||_{C^1} + ||b_i||_{C'}\right] \tag{5.54}$$

$$R_1(\varepsilon, L) \leq \frac{c_2}{L} \exp(||c||_\infty).$$

The proof is practically the same.

The next important theorem is a simple consequence of the Lemma 12. Corresponding applications see in Sec. 7. Other applications (including number theory) are also possible.

**Theorem 3.** Assume that the coefficients $a_n(k), b_n(k), k \in \Delta$ in the functional "Fourier like" series

$$S(k) \sim \sum_{n=1}^{\infty} a_n(k) \cos kL_n + b_n(k) \sin kL_n \tag{5.55}$$

have the form (compare with Theorems 2, 2′).

$$\begin{aligned} a_n(k) &= \alpha_n F_n(k, \tau_{n-1}(k)) \\ b_n(k) &= \alpha_n \widetilde{F}_n(k, \tau_{n-1}(k) \end{aligned} \tag{5.56}$$

where $(k, \varphi) \in \Delta + S^{(1)}$ and for some positive const $S, c_i$

$$|F_n(k,\varphi) - F_n(k',\varphi)| \leq c_0|k - k'|^\delta \ , \ ||F_n(\cdot)||_\infty \leq C_1 \tag{5.57}$$

$$||\frac{\partial F_n}{\partial \varphi}||_\infty \leq c_2 \ , \ ||\frac{\partial \tau_{n-1}}{\partial k}||_\infty \leq c_3 L_{n-1}$$

and the same estimations for $\widetilde{F}_n(\cdot)$. Then under conditions

$$\sum_{n=1}^{\infty} \alpha_n^2 < \infty\,, \ \sum_{n=1}^{\infty} \frac{L_n}{L_{n+1}} < \infty$$

uniformly in $n$, $a \le D \le b \le \infty$

$$\int\limits_{\triangle} \exp\left(D\ S_n(k)\right) dk = \int\limits_{\triangle} \exp\left(D \sum_{i=1}^{n} (a_i(\cdot)\cos kL_i + b_i(\cdot)\sin kL_i\right) dk$$
$$\le \exp(C_4 D^2) + \widetilde{C}_4(D). \tag{5.58}$$

*Proof.* In the special case $\delta = 1$ it is direct corollary of lemma 12. In fact,

$$\int\limits_{\triangle} \exp\left(D\ S_n(k)\right) dk =$$

$$\int\limits_{\triangle} dk \exp\left[DS_{n-1}(k) + \left(a_n(k)\cos kL_n + b_n(k)\sin kL_n\right)\right]$$

But

$$\|D\ S_{n-1}(\cdot)\|_{\infty} \le CD \sum_{i=1}^{n-1} |\alpha_i| \le C \sqrt{\sum_{i=1}^{\infty} \alpha_i^2 \sqrt{n}} \le c_1 \sqrt{n}$$

$$\|D\ S'_{n-1}(\cdot)\|_{\infty} \le CD \sum_{i=1}^{n-1} \left(L_i|\alpha_i| + C_2|\alpha_i|\right) \le c_3 L_{n-1}.$$

Lemma 12 gives

$$\int\limits_{\triangle} \exp\left[D\ S_{n-1}(k) + a_n(\cdot)\cos kL_n + b_n(\cdot)\sin kL_n\right] dk$$

$$\le \int\limits_{\triangle} \exp\left[D\ S_{n-1}(k) + \frac{a_n^2(k) + b_n^2(k)}{4}\right] dk \cdot \exp\left(\frac{C_5 L_{n-1}}{L_n}|D| + \right.$$
$$\left. + \frac{C_6 D^4}{L_n}\right) + \frac{C_7 \exp(C_8 \sqrt{n} D)}{L_n} \le$$
$$\le \int\limits_{\triangle} \exp\left[D\ S_{n-1}(k)\right] dx \cdot \exp\left(D \frac{c_5\ L_{n-1}}{L_n} + C_8 \alpha_n^2 D^2 + \frac{C_5}{L_n} D^4\right)$$
$$+ \frac{C_7 \exp(c\sqrt{n} D)}{L_n}.$$

One can apply the same estimation to $\int_{\triangle} \exp(D\ S_{n-1})dk$ etc.

If functions $F_n(k,\varphi)$ have only Hölder continuity in the first argument (like in condition (5.56)) we have to return to the Lemma 12. The central elements there were the estimations of the type

$$|a_n(k) - a_n(\frac{2\pi i}{L_n})| \leq \frac{c}{L_n}\ ,\ k \in \left[\frac{2\pi i}{L_n}, \frac{2\pi(i+1)}{L_n}\right].$$

If $a_n = F_n(k,\ \tau_{k-1}(k))$ (like in Theorem 3) then

$$|a_n(k,\ \tau_{n-1}(k)) - a_n(\frac{2\pi i}{L_n}\ ,\ \tau_{n-1}(\frac{2\pi i}{L_n})| \leq \frac{c_1}{L_n^{\delta}} + \frac{c_2 \cdot L_{n-1}}{L_n}$$

(due to conditions (5.57)). Of course, $\sum_n \frac{1}{L_n^{\delta}} < \infty$ and after this remark all the proof of the Lemma 12 and Theorem 3 goes through word by word.

The next result is a simple paraphrase of the Lemma 12′.

**Theorem** 3′. Let

$$S_n(k) = \sum_{i=1}^{n} \sum_{j=1}^{3} (a_{i,j}(k) \cos jL_i k + b_{i,j}(k) \sin j_i k) \tag{5.59}$$

and for $i = 1, 2, ...,\ j = 1, 2, 3$

$$\begin{aligned} a_{i,j}(k) &= F_{i,j}(k,\ \tau_{i-1}(k)) \\ b_{i,j}(k) &= \widetilde{F}_{i,j}(k,\ \tau_{i-1}(k)) \end{aligned} \tag{5.60}$$

with estimation (5.57) for $F_{i,j}, \widetilde{F}_{i,j}$. Then

$$\int_{\triangle} \exp(D\ S_n(k))dk \leq \exp(C\ D^2) + C_1(D)$$

i.e., the integral is uniformly in $n$ bounded for bounded $D$.

The latter theorem is the most general. It is associated with the Theorem 2 in the same sense as Theorem 3 is similar to Theorem 2′.

**Theorem 4**. Let us consider the functional series

$$S(k) \sim \sum_{i=1}^{\infty} F_i(k, L_i k + \tau_{i-1}(k)) \tag{5.61}$$

with conditions (5.57) on the general term $F_n(\cdot)$ and extra conditions

$$\begin{aligned} ||F_n(\cdot)||_\infty &= \alpha_n \to 0,\ n \to \infty \\ \bar{F}_n(k) &= \frac{1}{2\pi}\int_0^{2\pi} F_n(k,\varphi)d\varphi \equiv 0. \end{aligned} \tag{5.62}$$

If also $\frac{1}{2\pi}\int_0^{2\pi} F_n^2(k,\varphi)d\varphi \le \beta_n^2,\ k \in \triangle$ then the convergence of the series $\sum_{n=1}^{\infty} \beta_n^2 < \infty$ implies uniform in $n$ boundedness of the integrals

$$I_n(D) = \int_\triangle \exp(D\, S_n(K))dk, \tag{5.63}$$

$$S_n(k) = \sum_{i=1}^{n} F_i(k, L_i k + \tau_{i-1}(k))\ ,\ k \in \triangle.$$

The proof is practically the same as in the case of Theorem 3. The central moment is the following calculation:

$$\begin{aligned} \frac{1}{2\pi}\int_0^{2\pi} \exp(D\, F_n(k,\varphi))d\varphi &= \frac{1}{2\pi}\int_0^{2\pi}\Big[1 + DF_n(\cdot) + \frac{D^2}{2}F_n^2(\cdot)(1+ \\ &+O(||F_n||_\infty D))\Big]\, d\varphi = 1 + \frac{D^2}{2}\frac{1}{2\pi}\int_0^{2\pi} F_n^2(k,\varphi)d\varphi\ (1+O(\alpha_n)) \\ &\le \exp\left(\frac{D^2}{2}\beta_n^2(1+O(\alpha_n)\right)\ \text{etc.} \end{aligned}$$

Other ingredients of the proof (partition of $\triangle$, approximation of $F_n(k,\ \tau_{n-1}(k))$ by the function $\widetilde{F}_n(\frac{2\pi i}{L_n},\ L_n k + \tau_{n-1}(\frac{2\pi i}{L_n})),\ k \in \left[\frac{2\pi i}{L_n}, \frac{2\pi(i+1)}{L_n}\right]$ and so forth are the same as in the proofs of Theorems 3, 3′.

## 6 Spectral theory of the one-dimensional Schrödinger operators with sparse potentials. Negative energies $\lambda < 0$.

Let us recall that the goal of our paper is the study of the spectral properties of Hamiltonian $H^{\theta_0}$ given on $L^2(R_+)$ by

$$H\psi = -\psi'' + q(x)\psi\ ,\ x \ge 0 \tag{6.1}$$

$$\cos\theta_0\ \psi(0) - \sin\theta_0\ \psi'(0) = 0$$

with the sparse potential

$$q(x) = \sum_{n=1}^{\infty} g_n(x - x_n) \tag{6.2}$$

and conditions

$$\sum_{n=1}^{\infty} \frac{x_n}{x_{n+1}} < \infty(\mathcal{SL}), \quad |g_n(x)| \leq \frac{c_0}{1 + |x|^{1+\delta}}, c_0, \delta > 0 \tag{6.3}$$

Operator $H^{\theta_0}$ is essentially self adjoint. Because potential $q(\cdot)$ is arbitrarily close to 0 on the increasing intervals (say $[x_{n-1} - \exp(A(n-1)$, $x_n - \exp(An)])$ $\sum_{ess}(H^{\theta_0}) \supseteq [0, \infty)$. Structure of the essential spectrum as a set depends on the signs of elementary bumps $g_n(\cdot)$.

Let $H_n = -\frac{d^2}{dx^2} + g_n(\cdot)$ be a Hamiltonian on $L^2(R)$ corresponding to the scattering on the $n$ th bump and $\lambda_{n,i} < 0$ be the discrete spectrum of $H^{\theta_0}$. Generally speaking, for $0 < \delta \leq 1$ this spectrum is infinite (for each $n$) with the accumulating point $\lambda_0 = 0$. The set of all negative eigenvalues

$$\Lambda = \{\lambda_{n,i}\ ,\ n \geq 1\}$$

and the set of limiting points

$$\widetilde{\Lambda} = \left\{\lambda :\ \lambda = \lim_{n' \to \infty} \lambda_{n'}, i(n')\right\} \tag{6.4}$$

are especially important. The following result gives a complete description of the negative part of the spectrum (compare with Theorem B).

**Theorem 5.** a) If the set $\widetilde{\Lambda}$ is empty or includes only one point $\lambda_0 = 0$ then

$$\sum\nolimits_{ess}(H) \cap (-\infty, 0) = \emptyset$$

and the negative spectrum (if any) of $H^{\theta_0}$ is discrete.

b) If $\widetilde{\Lambda} \cap (-\infty, 0) \neq \emptyset$ then negative essential spectrum is not trivial and

$$\sum\nolimits_{ess}(H) \cap (-\infty, 0] = \widetilde{\Lambda} \tag{6.5}$$

c) Let $\psi_\lambda(x)$ be the solution of the problem $H^{\theta_0}\psi = \lambda\psi\ ,\ \psi(0) = \sin\theta_0\ ,\ \psi'(0) = \cos\theta_0$ and $\rho_\lambda(x)$ be the corresponding amplitude (Sec 2). Then for a.e. $\lambda < 0$

$$\lim_{x' \to \infty} \frac{\ln\ \rho_\lambda(x)}{x} = \sqrt{|\lambda|} \tag{6.6}$$

i.e., Lyapunov exponent exists and strictly positive.

d) For a.e. $\theta_0 \in [0, \pi)$

$$\sum (H^{\theta_0}) \cap (-\infty, 0] = \sum\nolimits_p (H^{\theta_0}) \tag{6.7}$$

This spectrum includes dense p.p. spectrum in the Int $(\sum_{ess})$ = Int $(\widetilde{\Lambda})$ (if this interior is not empty) and discrete spectrum with possible accumulating points on the $\partial(\sum_{ess}) = \partial(\widetilde{\Lambda})$. This boundary contains the isolated points of $\widetilde{\Lambda}$ and point $\lambda_0 = 0$.

e) For topologically generic set $\mathbf{C}_0 \subset [0, \pi)$, $|\mathbf{C}_0| = 0$ the fact $\theta_0 \subset \mathbf{C}_0$ implies that

$$\sum\nolimits_{pp} (H^{\theta_0}) \cap (\sum\nolimits_{ess}) = \emptyset.$$

The proof of this theorem is relatively simple. L. Pasur notices that this proof works under essentially weaker conditions than (6.3). In fact, we need only $x_{n+1} - x_n \to \infty$ plus some power estimation for the tails of bumps $g_n(\cdot)$. (Conditions (6.3) are crucial for the analysis of the positive part of the spectrum.) We will not discuss this opportunity here.

Statement a) of the Theorem 5 can be proved by elementary application of the Sturm oscillation theorem. Assume that $\widetilde{\Lambda}$ is empty or it contains only one limiting point $\lambda_0 = 0$. Let us remark that $\lambda_{n,i} \geq -C_0$ (condition (93″)) and fix negative energy $\lambda_1 < 0$. One can remove (if necessary), few first bumps $g_m(\cdot)$, $m \leq M$ in such a way that a new set $\Lambda_M = \{\lambda_{n,i}, n > M\}$ will be in the interval $\left(\widetilde{\lambda}_1 = \lambda_1 + \varepsilon, 0\right)$ for fixed sufficiently small $\varepsilon$. Let us consider the evolution of the phase $\theta_{\widetilde{\lambda}_1}$ for one specific Hamiltonian $H_n$ (i.e., for $n$ th bump). Equation

$$H_n\psi = -\psi'' + g_n(x - x_n)\psi = \widetilde{\lambda}_1\psi$$

has two solutions $\psi^{(1,2)}$, one with exponential growth from (-$\infty$) to (+$\infty$), the second one ($\psi^{(2)}$) exponentially decaying in the same direction. Phase $\theta_{\widetilde{\lambda}_1}(x)$ of the first solution satisfies the standard equation

$$\theta'_{\widetilde{\lambda}_1}(x) = \cos^2\theta_{\widetilde{\lambda}_1} + \left(\widetilde{\lambda}_1 - g_n(x - x_n)\right)\sin^2\theta_{\widetilde{\lambda}_1}$$

and has asymptotics

$$\theta_{\widetilde{\lambda}_1}(x) \to \varphi_1 = \tan^{-1}\frac{1}{\sqrt{|\widetilde{\lambda}_1|}}, x \to \pm\infty.$$

Phase $\theta_{\widetilde{\lambda}_1}(\cdot)$ can not make a full turn (i.e., return to the point $\varphi_1$ on the circle $[0,\pi)$ through point 0) since such a turn indicates the existence of the eigenvalue $\widetilde{\lambda}_2 \le \widetilde{\lambda}_1 = \lambda_1 + \varepsilon < 0$ for the Hamiltonian $H_n$.

Compactness arguments show that for suitable $\delta > 0$ uniformly in $n \ge n_0(\delta)$ the phase $\theta_{\widetilde{\lambda}_1}(x)$, starting from the value $\theta_0 \in V_\delta = \left\{\theta : |\theta - \varphi_1| = |\theta - \tan^{-1}\frac{1}{\sqrt{|\widetilde{\lambda}_1|}}| \le \delta\right\}$ at the "moment" $x_n^- = x_n - \exp(A_n)$, $A > 0$ will return (without turns!) to the same neighborhood $V_\delta$ at the "moment" $x_n^+ = x_n + \exp(An)$.

Monodromy operator $M_{\widetilde{\lambda}_1}(\cdot)$ between points $\left(x_{n-1}^+, x_n^-\right)$ is strictly positive and (because $|q(x)| \le \exp(-\gamma n)$ , $\gamma > 0$ , $x \in [x_{n-1}^+, x_n^-]$)

$$M_{\widetilde{\lambda}_1}\left(x_{n-1}^+, x_n^-\right) = \begin{bmatrix} \operatorname{ch}\sqrt{|\widetilde{\lambda}_1|}\left(x_n^- - x_{n-1}^+\right) & \frac{\operatorname{sh}\sqrt{|\widetilde{\lambda}_1|}(x_n^- - x_{n-1}^+)}{\sqrt{|\widetilde{\lambda}_1|}} \\ \sqrt{|\widetilde{\lambda}_1|}\operatorname{sh}\sqrt{|\widetilde{\lambda}_1|}(x_n^- - x_{n-1}^+) & \operatorname{ch}\sqrt{|\widetilde{\lambda}_1|}(x_n^- - x_{n-1}^+) \end{bmatrix} \times$$

$$\times (I + O\left(\exp(-\gamma n)\right))$$

As result $M_{\widetilde{\lambda}_1}\left(x_{n-1}^+\, ,\, x_n^-\right)$ maps $V_\delta$ into $V_\delta$ without turns for $n \ge n_0(\delta, A)$).

It means that on the interval $\left[x_{n_0}^+, \infty\right)$phase $\theta_{\widetilde{\lambda}_1}(x)$ can not make more than one turn, i.e., $\widetilde{\lambda}_1 \notin \sum_{ess}$, i.e., $\sum_{ess} \cap (-\infty, 0) = \emptyset$.

Let us give another proof of the same statement a) of the Theorem 5. This proof is based on the ideology of the cluster expansions, and is important for the further analysis.

If $G_\lambda^{(n)}(x,y)$ is the kernal of the resolvent $(H - \lambda)^{-1}$, i.e., solution of the equation

$$H_n\psi = \lambda\psi - \delta_y(x) = -\psi'' + g_n(x - x_n)\psi \tag{6.8}$$

and $\lambda \ne \lambda_{n,i}$ , $\lambda < 0$ then (see [13])

$$|G_n(x,y)| \le C \, \exp(-\sqrt{|\lambda|}|x - y|) \tag{6.9}$$

where const $C$ depends only on $|\lambda|$, dist $(\lambda,\ \{\lambda_{n,i}\})$ and parameters in the estimation (93″).

It is also clear that

$$G_\lambda^{(n)}(x,y) = \widetilde{G}^{(n)}(x - x_n,\ y - x_n) \tag{6.10}$$

where $\widetilde{G}^{(n)}_\lambda(x,z)$

$$-\widetilde{G}''_\lambda + g_n(x)\,\widetilde{G}_\lambda = \lambda\widetilde{G}_\lambda - \delta_z(x). \tag{6.11}$$

Assume that $\lambda \in \triangle \subset (-\infty, 0)$, $\triangle$ does not contain the eigenvalues $\{\lambda_{n,i}\}$ and we already have removed finite number of "bad" bumps (see details below). We will try to construct the resolvent kernel $G_\lambda(x,y)$ for the initial Hamiltonian $H$ "gluing together" the resolvent kernels $G^{(n)}_\lambda(\cdot)$ and prove the analyticity in $\lambda$ and boundedness of this kernel in arbitrary closed subinterval $\widetilde{\triangle} \subset \triangle$.

Let us represent potential $q(\cdot)$ as a sum of compactly supported bumps. Put

$$q(x) = \sum_{n=1}^{\infty} q(x) I_{y_n} < x \le y_{n+1} = \sum_{n=1}^{\infty} V_n(x) \tag{6.12}$$

where $y_n = \frac{x_{n-1}+x_n}{2}$, $n \ge 2$, $y_1 = 0$.

We have to solve the equation

$$H\psi = -\psi'' + \left(\sum_{n=1}^{\infty} V_n(x)\right) \psi = \lambda\psi - \delta_y(x),\ \lambda \in \triangle \subset (-\infty, 0)$$
$$\psi(0)\cos\theta_0 - \psi'(0)\sin\theta_0 = 0. \tag{6.13}$$

If $G^{(0)}_\lambda(x,y)$ is the solution of the problem

$$-\psi'' = \lambda\psi - \delta_y(x), \cos\theta_0 \cdot \psi(0) - \sin\theta_0 \psi'(0) = 0$$

then

$$|G^{(0)}_\lambda(x,y)| \le c(\triangle)\exp(-\sqrt{|\lambda|}|x-y|),\ \lambda \in \triangle.$$

We are looking for the solution of the problem (6.13) in the form

$$G_\lambda(x,y) = G^{(0)}_\lambda(x,y) + F(x,y). \tag{6.14}$$

Of course,

$$-F''(x,\cdot) + \left(\sum_{n=1}^{\infty} V_n(x)\right) F_n = \lambda F_n - \sum_{n=1}^{\infty} f_n(x) \tag{6.15}$$

where

$$f_n(x) = G^{(0)}(x,\cdot)V_n(x)\ ,\ n = 1, 2, \ldots \tag{6.16}$$

Functions $f_n(x)$ are uniformly bounded, continuous and supported on $\Delta_n = [y_n, y_{n+1}]$. Let

$$F^{(1)}(x) = \sum_{n=1}^{\infty} \left(G_\lambda^{(n)} f_n\right)(x) = \sum_{n=1}^{\infty} \int_0^\infty G_\lambda^{(n)}(x,y)\, f_n(y)dy$$

then

$$F^{(2)}(x) = F(x) - F^{(1)}(x)$$

satisfies the equation

$$H\ F^{(2)}(x) = \lambda F^{(2)} - \sum_{n=1}^{\infty} f_n^{(1)}(x) \tag{6.17}$$

$$f_n^{(1)}(x) = \sum_{m \neq n, m \geq 1} \left(G_\lambda^{(m)} f_m(x)\right) I_{\Delta_n}. \tag{6.18}$$

The mapping $\mathcal{M}$

$$\left(f_n(x)\ ,\ n = 1,2,...\right) \overset{\mathcal{M}}{\rightarrow} \left(f_n^{(1)}(x),\ n = 1,2,...\right) \tag{6.19}$$

is the crucial element of the construction.

$$\text{Let } \|(f_n(\cdot)\ ,\ n = 1,2,...)\| = \sup_n \|f_n(\cdot)\|_\infty .$$

Then

$$\|\mathcal{M}(f_n(\cdot), n = 1,2,...)\| \leq \sup_n \sup_{x \in \Delta_n} \sum_{m \neq n} \left(\int_{\Delta_m} G_\lambda^{(m)}(x,y)\, f_m(y)dy\right)$$

$$\leq \sup_n \sup_{x \in \Delta_n} C(\lambda, \Delta) \sum_{m \neq n} \int_{\Delta_m} \exp\left(-\sqrt{|\lambda|}|x-y|\right) dy \cdot \|f_n(\cdot)\|$$

$$\leq C_1(\Delta, \lambda) \sup_n \exp\left(-\frac{\sqrt{|\lambda|}|x_n - x_{n-1}|}{2}\right).$$

and this mapping is the contraction with arbitrary small norm if we will eliminate sufficiently large number of initial bumps.

Then

$$G_\lambda(x,y) = G_\lambda^{(0)}(x,y) + G_\lambda f_0 + G_\lambda \mathcal{M} f_0 + G_\lambda \mathcal{M}^2 f_0 + ... \tag{6.20}$$

where $f_0 = (f_n(x),\ n = 1, 2...)$

$$(G_\lambda f_0)(x) = \sum_{n=1}^{\infty} G_\lambda^{(n)}(x,y)\ f_n(x)$$

and operator $\mathcal{M}$ is defined in (6.18). Series (6.20) converges in $\mathbf{C}([0,\infty))$ as well as in many different functional spaces ($L^p(0,\infty)$ , $p \geq 1$ etc.).

Part b) of the Theorem 5 is simple. If $\lambda < 0$ and $\lambda = \lim_{n' \to \infty} \lambda_{n'}, i(n')$ then the corresponding sequence of the eigenfunctions $\left\{\psi_{n',i(n')}(x)\right\}$ is the "almost eigenfunction" in the sense Weil's theory of the $\Sigma_{ess}$.

In fact,

$$|\psi_{n',i(n')}(x)| \leq C \exp\left((1-\delta)\sqrt{|\lambda|}\,(x - x_n)\right)$$

for suitable $C, \delta > 0$ and all $n \geq 1$, i.e.,

$$\text{a) } \psi_{n',i(n')} \Rightarrow 0 \text{ (weakly in } L^2(0,\infty))$$

Also $||\psi_{n',i(n')}|| = 1$ and

$$||H\psi_{n',i(n')} - \lambda\psi_{n',i(n')})||_2 \leq |\lambda - \lambda_{n',i(n')}| +$$
$$\sum_{n \neq n'} \left(\int_0^\infty g_n^2(x - x_n)\psi_{n',i(n')}^2(x)dx\right)^{\frac{1}{2}} \leq |\lambda - \lambda_{n',i(n')}| + C_1 \sum_{n \neq n'} \cdot$$
$$\left(\int_0^\infty g_n^2(x - x_n)\exp(-\gamma|x - x_{n'}|)\right)^{\frac{1}{2}} \to_{n' \to \infty} \to 0.$$

Inverse statement is also an elementary one (due to cluster expansion technique). If $\lambda < 0$ and $\lambda \notin \widetilde{\Lambda}$ (i.e., is not a limiting point for $\{\lambda_{n,i}\}$) then (eliminating if necessary finite number of bumps) we can assume that some interval $\Delta = (\lambda - \varepsilon,\ \lambda + \varepsilon)$ does not contain $\lambda_{n,i}$. Then the corresponding (perturbed) resolvent $\widetilde{G}_\lambda$ will be analytic in the inner part of $\Delta$ (see discussions above), i.e., $\lambda \notin \sum_{ess}(H)$.

Parts d), e) of the theorem are the consequences of the Kotani-Simon-Wolff theorem (Lemma 8 in Sec. 3, (see [20], [21]) and the general theory of instability of the $\sum_{pp}$ (see [22], [23]) if only we have the positiveness of the Lyapunov exponent, i.e., the statement c) of the Theorem 5.

Now we will give the sketch of the proof of the positiveness of the Lyapunov exponent for negative energies using different form of the cluster expansion (the so-called block expansion, see [8], [9]).

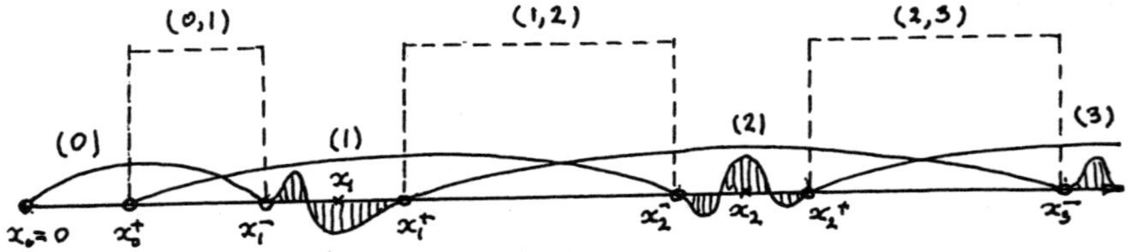


Figure 1:

Let $\lambda \in \Delta \subset (-\infty, 0)$ , $x_n^{\pm} = x_n \pm \exp(A_n)$ for some fixed $A > 0$. One can construct a special partition of [0,∞). This partition includes large (main) blocks around bumps and smaller (boundary) block between bumps. Notations are clear from the Fig 1.

Endpoints for the main blocks are

$$\begin{array}{c}(0) = [x_0, x_1^-] \\ (1) = [x_0^+, x_2^-] \\ (2) = [x_1^+, x_3^-] \\ \dots\dots\dots\dots\dots\dots ,\end{array}$$

corresponding points for the boundary blocks are

$$\begin{array}{c}(0,1) = [x_0^+, x_1^-] \\ (1,2) = [x_1^+, x_2^-] \\ \dots\dots\dots\dots\dots\dots .\end{array}$$

On the boundary blocks the potential $q(\cdot)$ is extremely small:

$$|q(x)| \leq \exp(-A_1 n) \ , \ x \in (n-1, n) = [x_{n-1}^+, x_n^-]$$

Analysis of the corresponding resolvent $R_\lambda^{(n-1,n)}(x,y)$ , $x, y \in (n-1, n)$ , $R_\lambda^{(\cdot)}(x_n^+, y) = R_\lambda^{(\cdot)}(x_n^-, y) = 0$ can be based either on perturbation arguments or on the integral equations:

$$\psi_\lambda^{(1)}(x) = \frac{\mathrm{sh}\sqrt{|\lambda|}(x - x_{n-1}^+)}{\sqrt{|\lambda|}} + \int_{x_{n-1}^+}^{x} \frac{\mathrm{sh}\sqrt{|\lambda|}(x-z)}{\sqrt{|\lambda|}} q(z)\psi_\lambda^{(1)}(z)dz$$

$$\psi_\lambda^{(2)}(x) = \frac{\mathrm{sh}\sqrt{|\lambda|}(x_n^- - x)}{\sqrt{|\lambda|}} + \int_{x}^{x_n^-} \frac{\mathrm{sh}\sqrt{|\lambda|}(z-x)}{\sqrt{|\lambda|}} q(z)\psi_\lambda^{(2)}(z)dz,$$

$$x \in (n-1, n) = [x_{n-1}^+, x_n^-] \ , \ \lambda \in \Delta \subset (-\infty, 0). \qquad (6.21)$$

It is easy to show

$$R_\lambda^{(n-1,n)}(x,y) \le C(\triangle)\exp(-\sqrt{|\lambda|}|x-y|) \tag{6.22}$$

and

$$K_\lambda^{(n-1,n)}(x_{n-1}^+, x_n^-) = \left.\frac{\partial^2 R_\lambda^{(n-1,n)}}{\partial x \partial y}\right| \begin{array}{l} x = x_{n-1}^+ \\ y = x_n^-. \end{array} \tag{6.23}$$
$$\le C_1(\triangle)\exp(-\sqrt{|\lambda|}(x_{n-1}^+ - x_n^-) = C_1(\triangle)\exp(-\sqrt{|\lambda|}L_n)$$

The latter function and estimation (6.23) are the crucial elements of the block expansion.

Resolvents $R_\lambda^{(n)}(x,y)$ of the main blocks have a more complicated structure. They are related to the scattering on the $n$ th bump, $n = 1, 2, \ldots.$ In the spirit of [8], [9] we will use only very rough information about $R_\lambda^{(n)}(\cdot)$.

Let us return to estimation of

$$\gamma(\lambda) = \lim_{x' \to \infty} \frac{||M_\lambda(0,x)||}{x}$$

and existence of the corresponding limit.

First of all,

$$||M_\lambda^{(n)}|| = ||M_\lambda(x_{n-1}^+, x_n^-)|| \le C\exp(\sqrt{|\lambda|}L_n). \tag{6.24}$$

Using trivial estimation

$$||M_\lambda(x_n^-, x_n^+)|| \le \exp(C(||q||_\infty)(x_n^+ - x_n^-|)$$

and submultiplicativity of the matrix norm from (6.24) and superlacunarity condition we can deduce that for $\lambda < 0$

$$\limsup_{x \to \infty} \frac{\ln ||M_\lambda(0,x)||}{x} \le \sqrt{|\lambda|}. \tag{6.25}$$

To prove the inverse estimation it is sufficient to check that for a.e. $\lambda \in \triangle \subset (-\infty, 0)$, arbitrary $\delta > 0$ and fixed $x_0$

$$|G_\lambda(x,x_0)| \le C(x_0,\lambda,\delta)\exp\left((-\sqrt{|\lambda|}+\delta)x\right) \tag{6.26}$$

(i.e., Green function for a.e. $\lambda < 0$ exists and decays roughly with the exponent -$\sqrt{|\lambda|}$).

If it is true complimentary solution $\psi_\lambda(x)$, $x > x_0$ will have amplitude increasing not slower then $\exp\left(\left(\sqrt{|\lambda|} - \delta\right) x\right)$ (because $W(\psi_\lambda(a)$, $G_\lambda(x_0, 0)) = 1$ and $|\frac{\partial G_\lambda}{\partial x}(x, x_0)| \leq c_1(x_0, \lambda, \delta) \exp(-(\sqrt{|\lambda|} + \delta)x)$ due to proposition (2.16)).

We can not guarantee estimation (6.26) for all negative $\lambda$ because of the presence of negative eigenvalues. Resolvent $G_\lambda(x, x_0)$ will be constructed in terms of $K_\lambda^{(n,n-1)}(\cdot)$ , $R_\lambda^{(n)}(\cdot)$. Corresponding expansion converges outside the set of the small denominators, generated by the poles of the resolvents $R_\lambda^{(n)}(\cdot)$.

Let us give the "physical" description of the block expansion, which converges a priory only for complex energies $\lambda$ (where resolvents are bounded and small). For an additional technical details see [8], [9].

"Quantum particle" starting from the point $x_0$ (let, for definiteness, $x \in (0)$, $x \notin (0,1)$) can reach distant point $x \ni (n)$ using different "paths" on the graph with vertices $(k)$, $n = 1, 2, \ldots$ and possible transitions $(k) \to (k \pm 1)$ through "boundary blocks" $(k, k-1)$, $(k, k+1)$. A typical path has a form

$$\widetilde{\gamma} : (0) \to (1) \to (2) \to (1) \to (2) \to (3) \to \ldots (n+1) \to (n)$$

Corresponding contribution to the resolvent $G_x^{(0)}(x_0, x)$ has a form:

$$\begin{aligned} G_\lambda^{(\widetilde{\gamma})}(x_0, x) &= R_\lambda^{(0)}(x_0, x_1^-)\, K_\lambda^{(0,1)}(x_1^-, x_1^+)\, R_\lambda^{(1)}\left(x_1^+, x_2^-\right) \cdot \\ &\cdot K_\lambda^{(0,1)}\left(x_2^-, x_2^+\right)\, R_\lambda^{(2)}\left(x_2^+, x_2^+\right)\, K_\lambda^{(2,1)}\left(x_2^+, x_2^-\right) R_\lambda^{(1)}\left(x_2^-, x_2^-\right) \cdot \\ &K_\lambda^{(1,2)}\left(x_2^-, x_2^+\right) R_\lambda^{(2)}\left(x_2^+, x_2^-\right) K_\lambda^{(2,3)}\left(x_3^-, x_3^+\right) R_\lambda^{(3)}\left(x_3^+, \cdot\right) \ldots \\ &\ldots K_\lambda^{(n+1,n)}\left(x_n^+, x_n^-\right) R_\lambda^{(n)}(x_n^-, x). \end{aligned} \tag{6.27}$$

(see details in [8], [9]).

At last,

$$G_\lambda^{(0)}(x_0, x) = \sum_{\widetilde{\gamma}:(0)\to(n)} G_\lambda^{(\widetilde{\gamma})}(x_0, x) \tag{6.28}$$

For the factors $K^{(n-1,n)}(x_n^-, x_n^+) = K^{(n,n-1)}(x_n^+, x_n^-)$ we will use excelent estimation (6.23) for the resolvent of the main blocks, we will use estimation, based on lemma 2.8 a) in [9], which in fact goes to A. Kolmogorov.

**Proposition.** For all $n \geq 1$ and for possible variants of the arguments in $R_\lambda^{(n)}(\cdot)$ :

$$R_\lambda^{(n)}\left(x_n^+, x_n^+\right) \ , \ R_\lambda^{(n)}\left(x_n^+, x_{n+1}^-\right) \ , \ R_\lambda^{(n)}\left(x_{n+1}^-, x_n^+\right), \tag{6.29}$$

$R_\lambda^{(n)}\left(x_{n+1}^-, x_{n+1}^-\right)$ one has $mes\left\{\lambda \in \Delta : |R_\lambda^{(n)}(\cdot)| > n^2\right\} \leq \frac{c_0}{n^2}$.

Borel-Cantelly argument gives that for a.e. $\lambda < 0$ all $n > n_0(\lambda)$ and all possible variants of the argument

$$|R_\lambda^{(n)}(\cdot,\cdot)| \leq n^2 \tag{6.30}$$

Resolvent kernels $R^{(n)}$ can be large but not too much.

Substitution of the estimations (6.30), (6.23) gives the upper estimation (6.26) for $|G_\lambda(x_0, x)|$. Main contribution gives "direct path" $(0) \to (1) \to (n)$. We completed the proof of the Theorem 5.

A few examples will illustrate the Theorem 5.

**Example 1**. Let

$$q(x) = \sum_{n=1}^{\infty} -\frac{2k_n^2}{\operatorname{ch}^2(k_n(x - x_n))} \tag{6.31}$$

Here $g(x) = -\frac{2k^2}{\operatorname{ch}^2(kx)}$ is the classical 1-solution. Corresponding wave $u(t,x) = g(x - 4k^2 t)$ is the solution of the $KdV$ equation $u_t + 6u_x u + u_{xxx} = 0$. Hamiltonian $H_n = -\psi'' + g_n(x)\psi$ has one negative eigenvalue $\lambda_{n,1} = -k_n^2$, see [12].

Assume that the set $\{k_n^2\}$ is dense on some interval, say $\Delta = [1, 2]$. Then

$$\sum\nolimits_{ess}(H^{\theta_0}) = [-1, -2] \cup [0, \infty).$$

For some $\theta_0 \subset \mathbf{C}_0$ (due to [22], [23]) spectrum on $[-1, -2]$ will be $\sum_{sc}$, for typical $\theta_0$ it is $\sum_{pp}$.

**Example 2**. Let $g(x) = -\frac{1}{1+|x|^{1+\delta}}$ and

$$q(x) = \sum_{n=1}^{\infty} k_n^2 \, g(k_n(x - x_n) \tag{6.32}$$

where $k_n$ are dense on some interval [1, A], $A >> 1$. Set $\{\lambda_{n,i}\}$ has a form

$$\lambda_{n,i} = k_n^2 \lambda_i \ , \ i = 1, 2, \ldots \ \text{(set is infinite)}$$

where $\lambda_i$ are the negative eigenvalues for test Hamiltonian on $L^2(R^1)$

$$H = -\psi'' - \frac{\psi}{1+|x|^{1+\delta}}.$$

Here $\lambda_i \sim \frac{c}{i^\gamma}$ , $\gamma > 2$ and for suitable $A >> 1$

$$\sum\nolimits_{ess}(H) \cap (-\infty, 0) = \left[-A^2\lambda_1, 0\right].$$

**Example 3**. Let

$$q(x) = \sum_{n=1}^{\infty} g(x - x_n) \tag{6.33}$$

for fixed $g(\cdot)$ from our class with nontrivial set $\{\lambda_i\ ,\ \lambda_i < 0\}$ . Then

$$\sum\nolimits_{ess}(H) = \{\lambda_i\}$$

i.e., discrete set.

As a result

$$\sum(H^{\theta_0}) \cap (-\infty, 0) = \sum\nolimits_d(H^{\theta_0})$$

for arbitrary $\theta_0 \in [0, \pi]$ . (Compare with Examples 2, 3.)

Further analysis will show that for positive $\lambda$ the spectrum of the Hamiltonian from Example 1 is a.c. (because $B_n(k) \equiv 0$ , solutions are reflectionless potentials). In the case of Examples 2, 3 everything depends on convergence or divergence of the series $\sum\limits_n |B_n^2(k)|$. It will be the subject of Sec. 6.

## 6.1 *Spectral theory of the one-dimensional Schrödinger operators with sparse potentials. Positive energies $\lambda = k^2 > 0$.*

As it has already been explained in the introduction to the Sec. 5 we will use for positive energies $k-$ phase-amplitude representations of the solutions:

$$\begin{aligned} t_k'(x) &= k - \frac{q(x)\sin^2 t_k(x)}{k} \\ r_k'(x) &= \frac{q(x)\sin 2t_k(x)}{2k} r_k(x) \end{aligned} \tag{6.34}$$

with the proper initial data (2.12). Due to cutoff Lemma 11 from the very beginning we can assume that

$$q(x) = \sum_{n=1}^{\infty} g_n(x - x_n), \quad \text{Supp } g_n(x - x_n) = \left[x_n^-, x_n^+\right] = \qquad (6.35)$$
$$= [x_n - l_n, x_n + l_n] \ , \ l_n = \exp(An), \quad A \text{ is large enough}$$

Then (see (5.18))

$$\ln r_k \ (x_n^+) = \sum_{j=1}^{n} \frac{1}{2k} \int_{x_n^-}^{x_n^+} q(z) \ \sin 2 \ t_k(z) dz =$$
$$= \sum_{j=1}^{n} F_i \ (k, \ t_k \ (x_i^-)) = \sum_{j=1}^{n} F_j \ (k, \ L_j k + t_k \ (x_{j-1}^+))$$

and function $F_j(k, \varphi)$ , $k \in \triangle \subset (0, \infty)$ , $\varphi \in S' = [0, 2\pi)$ has one of the following three equivalent representations

a) $F_j(k, \varphi) = \ln r_k(x_j^+)$ where

$(t_k(x), r_k(x))$ is the solution of the system (6.34) on $(x_j^-, x_j^+)$ with initial conditions $t_k(x_j^-) = \varphi$, $r_k(x_j^-) = 1$.

b) $F_j(k, \varphi) = \ln |M_k(x_j^-, x_j^+) \overrightarrow{e}_\varphi|$

c)

$$F_j(k, \varphi) = \frac{1}{2} \ln \left(1 + |A_j(k)|^2 + |B_j(k)|^2 + 2|A_j| \cdot \right. \qquad (6.36)$$
$$\left. \cdot |B_j| \cos(2kl + 2\varphi + \theta)\right) \ , \ \cos\theta = \frac{a_2^{(j)} b_2^{(j)} - a_1^{(j)} b_1^{(j)}}{|A_j||B_j|}$$

and

$$A_j(k) = a_1^{(j)}(k) + i b_1^{(j)}(k) \ ; \ B_j(k) = a_2^{(j)}(k) + i \ b_2^{(j)}(k)$$

are the transmission and reflection coefficients for the scattering problem with the potential

$$g_j(x - x_i) = g_j(x) I_{[x_j^-, x_j^+]}$$

(see formulas (3.9), (3.9))

To use the general results of the Section 5 (martingale and homogenization theorems) we have to complete our study of the analytical properties for

$$f_n(k) = F_n(k, \ L_n k + \tau_{n-1}(k)) \ , \ \tau_{n-1}(k) = t_k(x_{n-1}^+), \ k \in \triangle$$

We already know (Lemma 3, Section 2, formulas (3.15)) that

$$|F_n(k,\varphi) - F_n(k',\varphi| \,| \leq c_1|k-k'|^{\delta} \tag{6.37}$$
$$|\frac{\partial F_n}{\partial \varphi}| \leq c_2$$

uniformly in $n$. Constants $c_1, c_2 > 0$ depend only on $\triangle$ and $q(\cdot)$, constant $\delta$ in the Hölder condition obeys to the initial assumption $|g_n(x)| \leq \frac{c_0}{1+|x|^{1+\delta}}$.

**Lemma 13**. For suitable $c_3 = c_3(\triangle,\ q(\cdot))$

$$|\frac{\partial \tau_{n-1}(k)}{\partial k}| \leq c_3\ L_{n-1} \tag{6.38}$$

uniformly in $n$, $\theta_0 \in [0,\pi]$ and also

$$||M_k(0,x)|| \leq \exp\left(c_4(k,q)n(x)\right), \tag{6.39}$$

where $n(x) = \min(n : x_n > x)$

*Proof.* Inequality (6.39) is trivial. We have

$$||M_k(0,x)||^2 = |M_k(0,x)\overrightarrow{e}_1|^2 + |M_k(0,x)\overrightarrow{e}_2|^2,\ \ \overrightarrow{e}_1 = \begin{pmatrix}1\\0\end{pmatrix}, \overrightarrow{e}_2 = \begin{pmatrix}0\\1\end{pmatrix}.$$

But

$$|M_k(0,x)\overrightarrow{e}|^2 = r_k^2 = \exp\left(\frac{1}{k}\int_0^x q(x)\sin 2t_k(x)\right)dx \leq$$
$$\leq \exp\left(c(k)\sum_{i=1}^{n(x)}\int_{x_i^-}^{x_i^+}|q|dx\right) \leq \exp\left(c_4(k,q(\cdot))\, n(x)\right)$$

and so forth.

Inequality (6.38) is slightly more complicated. Let us put $\tau_k(x) = \frac{\partial t_k(x)}{\partial k}$. From the basic equation (6.34) it follows that

$$\tau_k' = 1 + \frac{q\sin^2 t_k}{k^2} - \frac{q\sin 2t_k}{k}\tau_k. \tag{6.40}$$

Linear ODE (6.40) has the obvious solution

$$\tau_k(x) = \tau_k(0)\exp\left(-\frac{1}{k}\int_0^x q(z)\sin 2t_k(z)dz\right) +$$
$$+\int_0^x \left[1 + \frac{q\sin^2 t_k(z)}{k^2}\right]\exp\left[-\frac{1}{k}\int_z^x q(\nu)\sin 2t_k(\nu)d\nu\right]dz$$

Under our conditions (including superlacunarity!) we have

$$|\tau_k(x_n^+)| \leq c_0 \exp\left(c_1(\triangle, q)n\right) + c_2\left(\triangle, q(\cdot)\right) \cdot \int_0^{x_n^+} dz \cdot$$
$$\cdot \exp\left[-\frac{2}{k}\int_z^{x_n^+} q(v)\sin 2t_k(v)dv\right] \leq c_0 \exp(c_1 n) +$$
$$+c_2 \int_0^{x_n^+} \exp\left(c_1\, n(x) - n(z)\right) dz \leq c_0 \exp(c_1 n) +$$
$$+c_2\left[c_3 L_n + c_3 L_{n-1}e^{c_1} + c_3 L_{n-2}e^{2c_1} + \ldots + L_1 e^{nc_1}\right]$$
$$\leq c_0 \exp(c_1 n) + c_4 L_n\left(1 + \frac{L_{n-1}}{L_n}c_5 + \frac{L_{n-2}}{L_n}c_5^2 + \ldots + \frac{L_1}{L_n}c_5^n\right)$$
$$\leq c_6\, L_n.$$

Now everything is ready to collect final results. The following simple theorem is the consequence of Lemma 13 and Lemma 10 (in the absence of $\sum_{pp}$).

**Theorem 6.** For the sparse potentials under conditions (1.5)

$$\sum\nolimits_{pp}(H^{\theta_0}) \cap (0,\infty) = \emptyset. \tag{6.41}$$

In fact, $\|M_k(0,x)\|^{-2} \geq \exp\left(-2c_4\, n(x)\right)$ and

$$\int_0^\infty \|M_k(0,x)\|^{-2}\, dx \geq \int_0^\infty \exp\left(-cn(x)\right) dx \geq$$
$$\geq \sum_n L_n \exp(-cn) = \infty.$$

The series $\sum\limits_n F_n(k,\ L_n k + \tau_{n-1}(k))$ satisfies all conditions of Theorems 2, 3 (Section 5). Now we only need estimations of a few integrals, which express $\bar{F}_n(k) = \frac{1}{2\pi}\int_0^{2\pi} F_n(k,\varphi)$, $\bar{F^2_{n,c}} = \frac{1}{2\pi}\int_0^{2\pi}(F_n(k,\varphi) - \bar{F}_n(k)^2)d\varphi$ in terms of reflection coefficients $B_n(k)$.

**Lemma 14**. Let $A > B \geq 0$ be const, $A^2 = 1 + B^2$, $\frac{B}{A} \leq \rho_0 < 1$. Then

a) $I_1 = \frac{1}{2\pi}\int_0^{2\pi} \ln\left(A^2 + B^2 + 2AB\cos\varphi\right) d\varphi = \ln(1 + B^2)$, (6.42)

$$c_1^-(\rho_0)\, B^2 \leq I_1 \leq B^2\,,\ c_1^- > 0. \tag{6.43}$$

If $|B| \le \varepsilon$ then

$$I_1 = I_1(B) = B^2 \left(1 + O(\varepsilon^2)\right). \tag{6.44}$$

$$\text{b) } I_2 = \frac{1}{2\pi}\int_0^{2\pi} \ln^2\left(\frac{A^2+B^2+2AB\cos\varphi}{1+B^2}\right) d\varphi \le c_1^+ \left(\frac{B}{A}\right)^2. \tag{6.45}$$

If $|B| \le \varepsilon$ then

$$I_2 = I_2(B) = 2B^2 \left(1 + O(\varepsilon^2)\right) \tag{6.46}$$

$$\text{c) } I_3 = \frac{1}{2\pi}\int_0^{2\pi} \frac{d\varphi}{(A^2+B^2+2AB\cos\varphi)^{1+\delta}} \le c_2^+(\rho_0) \left(\frac{B}{A}\right)^2. \tag{6.47}$$

*Proof.* Part a) is a simple corollary of the known formula

$$\frac{1}{2\pi}\int_0^{2\pi} \ln\left(a + b\cos\varphi\right) d\varphi = \ln\left(\frac{a+\sqrt{a^2-b^2}}{2}\right)$$

In the part b) let us remark first that

$$I_2 = \frac{1}{2\pi}\int_0^{2\pi} (f-\bar{f})^2 d\varphi, \quad f(\varphi) = \ln\left(A^2 + B^2 + 2AB\cos\varphi\right).$$

Because $A^2 = 1 + B^2$

$$I_2\left(\frac{B}{A}\right) = \frac{1}{2\pi}\int_0^{2\pi} \ln^2\left(1 + \left(\frac{B}{A}\right)^2 + 2\left(\frac{B}{A}\right)\cos\varphi\right) d\varphi.$$

For small ratio $\left(\frac{B}{A}\right)$ one can use the Taylor expansion to get (6.46). For $|B| \le B_0$, i.e., $\frac{B}{A} \le \frac{B_0}{\sqrt{1+B_0^2}} - \rho_0 < 1$ we may apply either continuity property of $\frac{I_2(B)}{B^2}$ or, for a better estimation, the series

$$\ln^2(1-x) = \sum_{n=1}^{\infty} a_n x^n, \; a_n = \sum_{m=1}^{n-1} \frac{1}{m(n-m)} \sim \frac{2\ln n}{n}$$

Last part c) has similar proof.

Let us formulate one of the central results of our paper.

**Theorem 7.** Let $H^{\theta_0}$ be our basic Hamiltonian (formula (1′) from the introduction), sparse potential $q = \sum_{n=1}^{\infty} g_n(x - x_n)$ satisfy the conditions

$$|g_n(z)| \le \frac{c_0}{1+|z|^{1+\delta}}\,, \quad \sum_{n=1}^{\infty} \frac{x_n}{x_{n+1}} < \infty \ \text{( SL)}.$$

a) On the set

$$Œ_0 = \left\{ k = \sqrt{\lambda} > 0 : \sum_{n=1}^{\infty} |B_n^2(k)| < \infty \right\}$$

for each $\theta_0 \in (0,\pi)$ the spectral measure $\mu^{\theta_0}(dk)$ has dense a.c. component, Supp $\mu_{ac}(d\lambda) = \widetilde{Œ}_0 = \{$essential closure of $Œ_0\}$.

b) The complement of $Œ_0$, i.e.,

$$Œ_1 = \bar{Œ}_0 = \left\{ k : \sum_n |B_n^2(k)| = \infty \right\}$$

contains only $\mu_{sc}$, i.e., $\mu_{ac}(Œ_1) = 0$.

c) Complement of $\bar{Œ}_0$, $(0,\infty)\backslash Œ_0$ belongs to $\sum_{sc}$. Each open interval $\triangle_i$ in contains dense s.c. spectral measure.

*Proof.* This result is the direct corollary of the general martingale Theorem 2, Lemma 3 (presenting $F_n(k,\varphi)$ in terms of reflection coefficient $B_n(k)$), Lemma 14 and our tests for $\sum_{sc}, \sum_{ac}$. Namely, we use Lemma 6 (on $\sum_{ac}$), Lemma 7 (on $\sum_{\text{sing}}$) and the absence of $\sum_{pp}$ (Theorem 6).

We have

$$r_k(x_n^+) = \sum_{i=1}^{n} F_i(k,\ L_i k + \tau_{i-1}(k)) = \sum_{i=1}^{n} \bar{F}_i(k) + \\ + \sum_{i=1}^{n} (F_i - \bar{F}_i).$$

Due to the Lemmas 3 and 14 $\sum_{i=1}^{n} \bar{F}_i(k) \le c|B_i^2(k)|$, i.e., $\sum_{i=1}^{\infty} \bar{F}_i(k)$ converges a.e. on $Œ_0$. The second series converges a.e. on $Œ_0$ because of the Theorem 2. In fact,

$$\sum_{n=1}^{\infty} \bar{F_{n,c}^2} \le c \sum_{n=1}^{\infty} |B_n^2(k)| < \infty.$$

It means that sequence $r_k(x_n^+)$ is bounded a.e. on $Œ_0$. But

$$r_k(x_{n-1}^+) = r_k(x_n^-)$$

and between edges $x_n^-, x_n^+$ of the $n^{th}$ bump the possible variation of the $r_k(x)$ is uniformly bounded. In other words the functions $r_k(x)$ for a.e. $k \in Œ_0$ are uniformly in $x$ bounded.

Now Hilbert-Pearson theory (i.e., Lemma 6) gives statements a).

On the complement $Œ_1 = \left\{k : \sum_{n=1}^{\infty} B_n^2(k) = \infty\right\}$

$$\begin{aligned} r_k(x_n^+) &= \sum_{i=1}^{n} \bar{F}_i(k) + \sum_{i=1}^{n}(F_i - \bar{F}_i) \geq \\ &\geq c\sum_{i=1}^{n} |B_i^2(k)| + O\left(\left(\sum_{i=1}^{n} |B_i^2(k)|\right)^{\frac{1}{2}} \left(\ln \sum_{i=1}^{n} |B_i^2(k)|\right)^{\frac{1}{2}+\varepsilon}\right) \end{aligned}$$

i.e., $r_k(x_n^+) \to +\infty$ a.e. on $Œ_1$ etc. It proves b).

Last statement is trivial. Each connected component $\triangle_i \subset (0,\infty)\backslash(\bar{Œ}_0)$ belongs to the support of $\mu^{\theta_0}(d\lambda)$. But due to Theorem 6 the spectral measure on $\triangle_i$ may be only s.c.

Generally speaking, Theorem 7 does not give the complete description of $\sum_{sc}$, as $\sum_{ac}$ can contain the dense $\mu_{\text{sing}}(d\lambda)$. Description of $\sum_{sc}$ is the problem of the $\mu_{sc}$ on $\sum_{ac}$. We only have partial results in this direction and no good examples (see discussion ahead). In some cases we can eliminate the $\mu_{sc}$ on $\sum_{ac}$.

**Theorem 8** a) Assume that for the closed interval $\triangle \subset (0, \infty)$

$$|B_n^2(k)| \leq \beta_n^2\,,\ k \in \triangle \text{ and } \sum_n \beta_n^2 < \infty$$

(dominated convergence). Then $\mu_{sc}(\triangle) = 0$.

b) If the series $\sum_n |B_n^2(k)|$ , $k \in \triangle$ converges in the Banach space $H\ddot{o}l_{\delta'}(\triangle)$, $\delta' \leq \delta$ (recall that for $f(k) \in H\ddot{o}l_\delta(\triangle)$

$$||f(\cdot)|| = ||f(\cdot)||_\infty + \sup_{k,k'\in\triangle} \frac{(f(k) - f(k'))}{|k-k'|^\delta}. \tag{6.48}$$

Then $\mu_{sc}(\triangle) = 0$.

The proof is based on the homogenization Theorem 4. In fact,

$$\int_{\triangle} \exp\left(D \ln r_k(x_n^+)\right) dk = \int_{\triangle} \exp\left(D\ S_n(k)\right) dk = \tag{6.49}$$

$$= \int_{\triangle} \exp\left[\left(D\left(\sum_{i=1}^{n} \bar{F}_i(k)\right) + D\left(\sum_{i=1}^{n}\left(F_i - \bar{F}_i\right)\right)\right)\right] dk$$

$$= \int_{\triangle} \exp\left[D\ S_{n-1}(k) + D\ \bar{F}_n(k) + \frac{D^2}{2}\bar{F^2_{n,c}}(k) + \right.$$

$$\left. + O\left(\frac{L_{n-1}}{L_n}\right) + O\left(\frac{1}{L_n^\delta}\right)\right] dk + O\left(\frac{\exp(c\sqrt{n})}{L_n}\right).$$

The second and the third terms in the exponent can be estimated from above by $C(D)\beta_n^2$. Continuation of the process, together with the converges of the series $\sum\limits_n \frac{L_n}{L_{n+1}}, \sum \frac{1}{L_n^\sigma}, \sum\limits_n \beta_n^2$, gives (due to Lemma 5) the statement a) of the theorem.

Slightly different approach is also available. If $|B_n^2(k)| \in H\ddot{o}l_{\delta'}(\triangle)$ $\delta' \leq \delta$ then the same is true for $\bar{F}_n(k)$, $\bar{F^2_{n,c}}(k)$. More, the converges of $\sum\limits_n |B_n^2(k)|$ in $H\ddot{o}l_{\delta'}(\triangle)$ implies the convergence in the same spaces for the series $\sum\limits_n \bar{F}_n(k)$ , $\sum\limits_n \bar{F^2_{n,c}}(k)$.

On the second step of the homogenization process we will approximate by piece-wise const functions (with the step $\frac{2\pi}{L_{n-1}}$) not $S_{n-1}(k)$ as earlier but $S_{n-1}(k) + D\ \bar{F}_n(k) + \frac{D^2 \bar{F^2_{n,c}}(k)}{2}$ with the error in the exponent of the order

$$O\left(\frac{L_{n-2}}{L_{n-1}}\right) + O\left(\frac{1}{L_{n-1}^\delta}\right) + O\left(\frac{||\bar{F}_n(\cdot)||_{H\ddot{o}l_{\delta'}} + ||\bar{F^2_{n,c}}(\cdot)||_{H\ddot{o}l_\delta}}{L_{n-1}^{\delta'}}\right)$$

etc. It proves part b).

**Example 4**. Let $g_n(\cdot) \equiv g$ and $B(k)$ be the reflection coefficient for the scattering problem

$$H\psi = -\psi'' + g(x)\psi = k^2\psi$$

and $Œ_0 = \{k:\ B(k) = 0\}$, $Œ_1 = \{k : |B(k)| > 0\}$.

On the open set $Œ_1$

$$\mu(H) = \mu_{\text{sing}}(H)$$

a) If $|Œ_0| = 0$ then $\mu(H) = \mu_{sc}(H)$ on $[0, \infty]$

b) If $|Œ_0| > 0$ and $Œ_0$ is nowhere dense (i.e., Int $Œ_0 = \emptyset$) then

$$\sum\nolimits_{sc}(H) = [0, \infty], \ \sum\nolimits_{ac}(H) = Œ_0$$

(in this case we can use not Theorem 7 but trivial observation that after cutoff $\sum_n |B_n(k)| \le c < \infty$ on $Œ_0$, i.e., momodromy operators $M_k(0, x)$ are uniformly bounded $k \in \Delta, \ x \ge 0$).

c) If $|Œ_0| > 0$ and Int $Œ_0 \ne \emptyset$ then

$$\sum\nolimits_{ac} = Œ_0, \ \sum\nolimits_{sc} = Œ_1^0(\text{ closure of } Œ_1).$$

Let us stress that in the cases b), c)

$$|\sum\nolimits_{sc} \cap \sum\nolimits_{ac}| > 0.$$

**Example 5.** If $q(x) = \sum -\frac{2k_n^2}{\text{ch}^2(k_n(x-x_n))}$ then $\sum_{sc}(H) = 0$, on $[0, \infty), \sum_{ac}(H) = [0, \infty]$.

Together with Example 1 (Sec. 6) it gives the complete structure of the spectrum for the potentials of the "solution-like" type. Conditions on the sequence of the centers $\{x_n\}$ in this case can be essentially weakened. Results, presented in the Examples 1, 5, in fact, have place if only $x_n - x_{n-1} \ge cn^\alpha, \alpha > 0$.

The central unsolved problem in the context of the Theorems 7 and 8 is the following one. Assume that for some set A, $|A| > 0$ we have

$$A \subset \sum\nolimits_{ac} = \widetilde{Œ}_0 \ , \ A \subset Œ_1$$

simultaneously. Is it true (or not) that $A \subset \sum_{sc}$? It follows from the formula

$$\frac{1}{\pi} \int_0^\pi \mu^{\theta_0}(d\lambda) d\theta_0 = d\lambda$$

that for some set $\sqcap \subset [0, \pi], \ |\sqcap| > 0$ and all $\theta_0 \subset \sqcap$ one has $\mu_{sc}(A) > 0$.

Assume for a second that $\sum_{sc}$ is stable with respect to variation of the boundary phase $\theta_0$. Then, of course, $A \subset \sum_{sc}$ and more, $\sum_{sc} = Œ_1^c$. What are conditions on the potential $q(\cdot)$ or sequence of $|B_n^2(\cdot)|$ which guarantee such stability? How is this stability related to fractal properties of $\mu_{sc}$ in terms, for instance, of the Hausdorff dimension of $\mu_{sc}$ etc.? We have no answers to these questions as well as good counter-examples. There are some partial results (with A. Gordon) in this direction for the random sparse potentials. It will be the subject of another publication.

At the end of the section we will formulate some extensions of the Theorems 7, 8. The key elements of these theorems are the technical Theorems 2 and 4 (Section 5). Applications of the martingale and homogenization ideas are based on the approximation of the functions

$$f_n(k) = F_n(k,\ L_n k + \tau_{n-1}(k))$$

by the piece-wise const functions on the intervals of the length $\frac{2\pi}{L_n}$. If the function $F_n(k, \varphi)$ has the properties (compare with (5.30), (5.31):

$$||F_n(\cdot,\cdot)|| \le A_0\ ,\ ||F_n(k,\varphi) - F_n(k',\varphi)|| \le c_0|k-k'|^\delta \qquad (6.50)$$

$$||\frac{\partial F_n}{\partial\varphi}|| \le A_1\ ,\ |\frac{\partial\, \tau_{n-1}}{\partial k}| \le c_1\, L_{n-1}$$

(like in the text) then the error of such approximation, i.e.,

$$|F_n(k,\ L_n k + \tau_{n-1}(k)) - F_n\left(\frac{2\pi i}{L_n},\ L_n k + \tau_{n-1}\left(\frac{2\pi i}{L_n}\right)\right)| \le$$

$$\frac{c_2}{L_n^\delta} + \frac{c_3\, L_{n-1}}{L_n} \text{ for } k \in \left[\frac{2\pi i}{L_n}, \frac{2\pi(i+1)}{L_n}\right]$$

Convergence of the series $\sum\limits_n \frac{1}{L_n^\delta}\ ,\ \sum\limits_n \frac{L_{n-1}}{L_n}$ guarantee the equiconvergence of the functional series

$$S_1 \sim \sum_n F_n\left(k,\ L_n k + \tau_{n-1}(k)\right),\ \tilde{S}_1 \sim \sum_n G_n(k)$$

where $G_n(k) = F_n\left(\frac{2\pi i}{L_n}\ ,\ L_n k + \tau_{n-1}\left(\frac{2\pi i}{L_n}\right)\right),\ k \in \left[\frac{2\pi i}{L_n}, \frac{2\pi(i+1)}{L_n}\right]$ etc. (see Theorems 2, 4).

Hölder condition for $F_n(k,\varphi)$ with respect to the first argument looks too strong. In fact, we can use the weaker condition:

Assume, that for $k, k' \in \triangle$ uniformly in $n$ and $\varphi \in [0, 2\pi]$

$$|F_n(k,\varphi) - F_n(k',\varphi)| \le C_0(\triangle)\omega\,(k-k') \qquad (6.51)$$

and $\omega(h) \downarrow 0$ if $h \downarrow 0$. Then Theorems 2, 4 and, as result, the Theorems 7, 8 are valid under condition

$$\sum_n \omega\left(\frac{1}{L_n}\right) < \infty\ ,\ \sum_n \frac{L_{n-1}}{L_n} < \infty. \qquad (6.52)$$

For example, if

$$|g_n(z)| \leq \frac{c_0}{1 + |z| \ln_+^{2+\delta}(1+z)} \tag{6.53}$$

then $\omega_\triangle(h) \leq \frac{c_1(\triangle)}{\ln^{1+\delta}(\frac{1}{h})}$ and superlacunarity condition implies the convergence of $\sum\limits_n \omega\left(\frac{1}{L_n}\right)$.

If the family of the bumps $\{g_n(\cdot)\}$ is uniformly bounded and uniformly integrable, i.e., for arbitrary $\varepsilon > 0$ one can find $L = L(\varepsilon) > 0$ such that for all $n \geq 1$

$$\int\limits_{|z| \geq L} |g(z)| dz \leq \varepsilon \tag{6.54}$$

then on each compact interval $\triangle$ the family of the corresponding reflection coefficients $\{B_n(k)\}$ has uniform continuity property: for appropriate $\omega_\triangle(h)$, $\omega(h) \downarrow 0$, $h \downarrow 0$

$$|B_n(k) - B_n(k')| \leq \omega_\triangle(k - k') \ , \ n \geq 1. \tag{6.55}$$

**Theorem 9**.Under conditions (6.54), (6.55) and

$$\sum_n \omega_\triangle \left(\frac{1}{L_n}\right) < \infty, \ \sum_n \frac{L_{n-1}}{L_n} < \infty \tag{6.56}$$

all the conclusions of the Theorems 7, 8 holds if the sequence of $\{x_n\}$ is increasing fast enough.

The proof includes the construction of the proper cutoff: $g_n(\cdot) \to g_n(x - x_n) I_{[x - l_n, x + l_n]}$, the definition of $L_n$ and application of the Theorems 2, 4 in the more general form (based on the estimation (6.50) and relations (6.52) or (6.56)).

For instance, if we have uniform majorant (6.53), one can put $l_n = \exp(An), A >> 1$ and results of the Theorems 7, 8 are valid under only superlacunarity condition $\sum\limits_n \frac{L_{n-1}}{L_n} < \infty$ (due to the remark after formula (6.53)).

**Example 6**. Construction of the potentials $q(\cdot)$, such that corresponding Hamiltonian has all three spectral component: $\mu_{ac}, \mu_{sc}, \mu_{pp}$ can be based on the study of the bumps $g(\cdot)$, such that $B_g(k) \equiv 0, \ k \geq k_0 > 0$. Such bumps exist according to the well-known results in the inverse scattering problem (see [13]). Unfortunately, explicit analytical

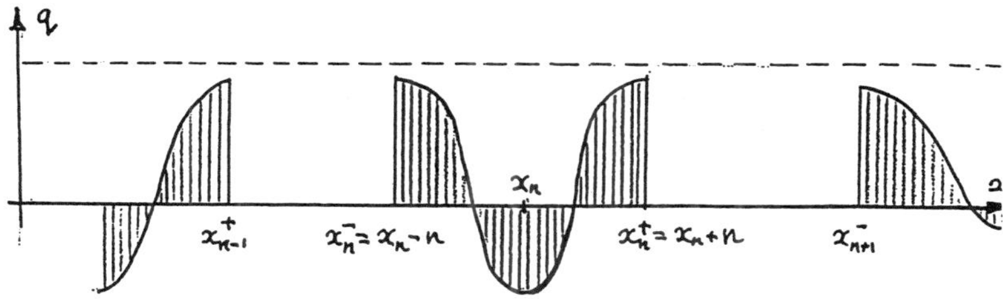


Figure 2:

formulas for such potentials (transparent for the waves with frequencies $k \geq k_0$!) are not known. We can propose the following palliative.

Let

$$q(x) = \left[ \sum_{n=1}^{\infty} - \frac{2k_n^2}{\operatorname{ch}^2(k_n(x - x_n))} + 1 \right] \cdot \sum_{i=1}^{\infty} I_{[x_n - x, x_n + n]}(x) \tag{6.57}$$

The potential looks like Fig 2.

Assume that the set $\{k_n^2\}$ is dense on [1,2]. We have a very simple spectral structure.

a) If $k > 1$, the monodromy operators between bumps are orthogonal matrices, on each bump they are orthogonal modulo exponentially small errors (solutions are reflectionless and exponentially decaying). It gives $\sum(H) \cap [1, \infty] = \sum_{ac}(H) = [1, \infty)$.

b) If $0 < k < 1$, the situation is different. Between bumps the monodromy operators are (as before) orthogonal, but on the bumps

$$|B_n^2(k)| \geq \exp(\gamma n), \gamma > 0.$$

It gives s.c. spectrum on [0,1].

c) At last, for $k < 0$ each bump has (for $n \geq n_0$) only one negative eigenvalue, which is due to a standard perturbation argument

$$\widetilde{\lambda}_n = \lambda_n = -k_n^2 + 1 + O(\exp(-\delta_n)), \delta > 0$$

The spectrum $H = H^{\theta_0}$ is p.p. for a.e. $\theta_0 \in [0, \pi]$ and the essential spectrum (as a set) has a form

$$\sum\nolimits_{ess}(H) = [-2, -1] \oplus 1 = [-1, 0].$$

In the last part of this section we will describe briefly the second extension of the Theorems 7, 8. Namely, our goal is to eliminate the condition of the uniform boundedness and integrability (1.6) which plays an important role in our analysis. Assume that

$$q(x) = \sum_{n=1}^{\infty} g_n(x - x_n) \tag{6.58}$$

$$\int_{R'} |g_n(z)|dz = \alpha_n < \infty.$$

**Theorem 10**. If the sequence of the bumps centers increases fast enough then all the statements of the Theorems 7, 8 are valid. For instance

a) $\text{Œ}_0 = \left\{k : \sum_n |B_n^2(k)| < \infty\right\}$ contains dense a.c. spectrum

b) $\text{Œ}_1 = \bar{\text{Œ}}_0 = \left\{k : \sum_n |B_n^2(k)| = \infty\right\}$ contains only singular continuous spectrum etc.

The proof of this theorem can be based on the pure "soft" arguments. Nevertheless, we will make an attempt of the quantitative study.

For each $n = 1, 2, \ldots$ let us introduce the cutoff levels $l_n$ by the relation

$$\int_{|z|\geq l_n} |g_n(z)|dz = \exp\left[-c_0 \sum_{i=1}^{n} \max(\alpha_n, 1)\right]. \tag{6.59}$$

If points $\{x_n\}$ have the property that the intervals $I_n = [x_n - l_n, x_n + l_n]$ are non intersected and $c_0$ is large enough (one can put $c_0 = 3$), then we can use Lemma 11 to compare

$$q(x) = \sum_n g_n(x - x_n),\ \tilde{q}(x) = \sum_n g_n(x - x_n) I_{|x - x_n| \leq l_n}. \tag{6.60}$$

According to this lemma for each $k \in \triangle$ ($\triangle \subset (0, \infty)$ is a fixed bounded interval) for $x \to \infty$ the ratio $\frac{||M_k(0,x)||}{||\widetilde{M}_k(0,x)||}$ is uniformly (in $k$) bounded.

Let us put (as earlier)

$$L_n = x_n^- - x_{n-1}^+ = x_n - l_n - x_{n-1} - l_{n-1} = x_n - x_{n-1} - (l_n + l_{n-1})$$

and introduce (for the cutoff potential $\tilde{q}$) the following monodromy operators (Fig 3)

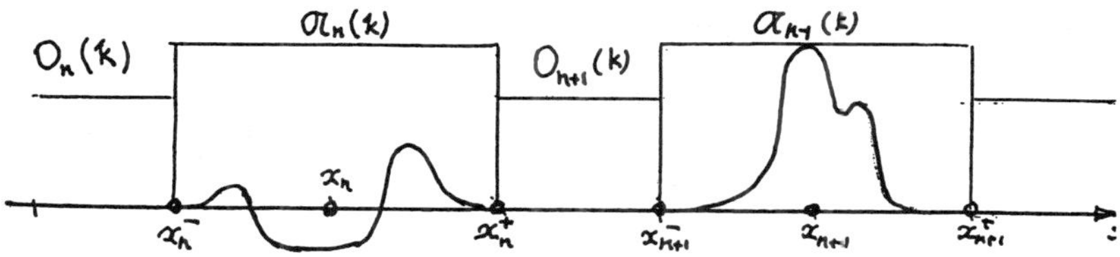


Figure 3:

Here

$$O_n(k) = \begin{vmatrix} \cos L_n k & \sin L_n k \\ -\sin L_n k & \cos L_n k \end{vmatrix}$$

and (in the basis $\exp(\pm ikx)$)

$$\text{Œ}_n(k) = \begin{vmatrix} A_n(k) & B_n(k) \\ \bar{B}_n(k) & \bar{A}_n(k) \end{vmatrix}.$$

Let

$$F_n(k,\varphi) = \ln ||\text{Œ}_n(k)\overrightarrow{e_\varphi}||, \ \overrightarrow{e_\varphi} = \begin{pmatrix} \cos\varphi \\ \sin\varphi \end{pmatrix}$$
$$= \frac{1}{2} \ln \left(\text{Œ}_n(k)\overrightarrow{e_\varphi}, \ \text{Œ}_n(k)\overrightarrow{e_\varphi}\right).$$

Let us remark that

$$\left|\frac{\partial F_n}{\partial \varphi}\right| = \left|\frac{\left(\text{Œ}_n \overrightarrow{e_\varphi}, \text{Œ}_n \overrightarrow{\widetilde{e_\varphi}}\right)}{\left(\text{Œ}_n \overrightarrow{e_\varphi}, \text{Œ}_n \overrightarrow{\widetilde{e_\varphi}}\right)}\right| \leq \|\text{Œ}_n\|^2 \tag{6.61}$$

Here $\widetilde{e_\varphi} = (e_\varphi)' = \begin{pmatrix} -\sin\varphi \\ \cos\varphi \end{pmatrix}$ and we used the trivial facts:

$$\left|\left(\overrightarrow{x}, \overrightarrow{y}\right)\right| \leq \left|\overrightarrow{x}\right| \left|\overrightarrow{y}\right| , \ \left|\left(\overrightarrow{x}, \overrightarrow{x}\right)\right| = \left|\overrightarrow{x}\right|^2$$
$$\max_\varphi \left(\frac{1}{\left|\text{Œ}\overrightarrow{e_\varphi}\right|}\right) \leq \left\|\text{Œ}^{-1}\right\| = \|\text{Œ}\|.$$

Similar calculations give

$$\left|\frac{\partial F_n}{\partial k}\right| = \left|\frac{\left(Œ_n \overrightarrow{e_\varphi},\ Œ'_n \overrightarrow{e_\varphi}\right)}{\left(Œ_n e_\varphi,\ Œ_n \overrightarrow{e_\varphi}\right)}\right| \le \|Œ_n\|\, \|Œ'_n\|\,. \tag{6.62}$$

If $\overrightarrow{\psi}_k(x_n^+) = \left(\psi_k(x_n^+),\ \frac{\psi'_k(x_n^+)}{k}\right)$ then

$$\begin{aligned}\overrightarrow{\psi_k}(x_n^+) &= Œ_n O_n Œ_{n-1} O_{n-1} ... Œ_1 O_1 \overrightarrow{e_\varphi} = \\ &= r_k(x_n^+)\ \overrightarrow{t_k}(x_n^+)\ ,\ \overrightarrow{t_k}(x_n^+) = \begin{pmatrix}\cos\ t_k(x_n^+)\\ \sin\ t_k(x_n^+)\end{pmatrix}.\end{aligned}$$

Here $r_k(x_n^+)$, $t_k(x_n^+)$ are the amplitude and the phase of the solution $\psi_k(x)$, $x = x_n^+$ , $\ln\ r_k(x_n^+) = \sum_{i=1}^{n} F_i(k,\ L_i k + \tau_{i-1}(k))$ (we use the old notations!).

Of course, (for $\|Œ\| = \|Œ\|_0$ i.e., operator norm)

$$\begin{aligned} r_k(x_n) &\le \|Œ_1\| ... \|Œ_n\| \ \text{ and} \\ \frac{\partial}{\partial K}\overrightarrow{\psi_k}\left(x_n^+\right) &= \Big[Œ'_n O_n Œ_{n-1} ... O_1 \overrightarrow{e_\varphi} + \\ &\quad + Œ_n O'_n Œ_{n-1} ... O_1 e_\varphi + ... Œ_n O_n Œ_{n-1} ... O'_1 \overrightarrow{e_\varphi}\Big] \\ &= \frac{\partial\, r_k}{\partial\, k}\overrightarrow{t_k} + r_k \left|\frac{\partial\, \overrightarrow{t_k}}{\partial\, k}\right| \overrightarrow{\tilde{t}_k}\ ,\ \left|\tilde{t}_k\right| = 1,\ \left(\overrightarrow{\tilde{t}_k}\,,\ \overrightarrow{t_k}\right) = 0. \end{aligned}$$

That is (after simple calculations)

$$\begin{aligned}\left|\frac{\partial\, \overrightarrow{t_k}}{\partial\, k}\right| = \left|\frac{\partial\, t_k\ (x_n^+)}{\partial\, k}\right| &\le \|Œ_n\|\, \|Œ'_n\| + L_n\, \|Œ_n\|^2 + \\ &+ \|Œ_n\|^2\, \|Œ_{n-1}\|\, \left\|Œ'_{n-1}\right\| + L_{n-1}\, \|Œ_n\|^2\, \|Œ_{n-1}\|^2 + ... \\ &+ \|Œ_n\|^2\, \left\|Œ^2_{n-1}\right\| ... \|Œ_1\|\, \|Œ'_1\| + L_1\, \left\|Œ^2_n\right\| ... \|Œ_1\|^2\,.\end{aligned} \tag{6.63}$$

Using the inequality

$$\begin{aligned} &\left|\tfrac{\partial\, F_n}{\partial\, k}\right| + \left|\tfrac{\partial\, F_n}{\partial\, \varphi}\right| \left|\tfrac{\partial\, \tau_{n-1}}{\partial\, k}\right| \le \\ \|Œ_n\|\, \|Œ'_n\| + \|Œ_n\|^2 &\Big[\left\|Œ_{n-1}\right\|\, \left\|Œ'_{n-1}\right\| + L_{n-1}\, \left\|Œ^2_{n-1}\right\|^2 + \\ &... \|Œ_{n-1}\|^2 ... \|Œ_1\|\, \|Œ'_1\| + L_1 \|Œ_{n-1}\|^2 ... \|Œ_1\|^2\Big] \end{aligned}$$

one can estimate (as in the Theorem 2) the difference (for $k \in \left[\frac{2\pi i}{L_n}, \frac{2\pi(i+1)}{L_n}\right]$)

$$\left|F_n(k,\ L_n k + \tau_{n-1}(k) - F_n(\frac{2\pi i}{L_n},\ L_n k + \tau_{n-1}\left(\frac{2\pi i}{L_n}\right)\right| \quad (6.64)$$
$$\leq \frac{2\pi}{L_n}\left(\left\|\frac{\partial F_n}{\partial k}\right\|_\infty + \left\|\frac{\partial F_n}{\partial \varphi}\right\|\left\|\frac{\partial \tau_{n-1}}{\partial k}\right\|_\infty\right).$$

Assume that $\|Œ_n'\| \leq L_n \|Œ_n\|$ (in many cases $\|Œ_n'\| \leq C_0 \|Œ_n\|$) and

$$L_n = \|Œ_n\|^2 \dots \left\|Œ_1^2\right\| \cdot \beta_n \quad (6.65)$$
$$\sum_n \frac{\beta_{n-1}}{\beta_n} < \infty.$$

Then all the estimations from the Theorem 2 work. Let us take into account two additional observations (see Lemma 14)

$$\bar{F_n}(k) = \frac{1}{2\pi}\int_0^{2\pi} F_n(k,\varphi)d\varphi = \ln\left(1 + B_n^2(k)\right)$$
$$\bar{F_{n,c}^2}(k) = \frac{1}{2\pi}\int_0^{2\pi} \ln^2\left(1 + \left|\frac{B_n}{A_n}\right|^2 + 2\left|\frac{B_n}{A_n}\right|\cos\varphi\right)d\varphi \leq c_1\left|\frac{B_n}{A_n}\right|^2.$$

It gives the following result (which we formulate only for cutoff potential).

**Theorem 11**. Let

$$q(x) = \sum g_n(x - x_n)I_{|x-x_n|} \leq l_n$$

$Œ\ _n(k)$ be the monodromy operators for the problems

$$-\psi'' + g_n(x - x_n)I_{|x-x_n|\leq l_n} = k^2\psi$$

(between points $x_n^- = x_n - l_n,\ x_n^+ = x_n + l_n$); $A_n(k), B_n(k)$be transmission and reflection coefficients for the corresponding scattering problem.

If

$$L_n = \|Œ_1\|^2 \dots \|Œ_n\|^2 \beta_n\ ,\ \sum_n \frac{\beta_{n-1}}{\beta_n} < \infty \quad (6.66)$$

$$\|Œ_n'\| \leq L_n \|Œ_n\|$$

then on the set

$$Œ_0 = \left\{ k : \sum_n \ln(1 + \left|B_n^2\right|) < \infty \right\} \tag{6.67}$$

there exists dense a.c. spectrum. On the set

$$Œ_1 = \left\{ k : \sum_n \ln \left(1 + |B_n^2|\right) = \infty \right\} \tag{6.68}$$

the spectrum is s.c.

An analogue of the Theorem 8 takes place also. The following two examples illustrate possible effects for the unbounded bumps.

**Example 7.** (compare with [10]). System of increasing barriers. Let

$$g_n(x) = n\, I_{|x|\leq a}\,,\ q(x) = \sum_n g_n(x - x_n) \tag{6.69}$$

Here

$$Œ_n(k) = \left| \begin{array}{cc} \mathrm{ch} 2a\sqrt{n-k^2} & \frac{k}{\sqrt{n-k^2}}\, \mathrm{sh} 2a\sqrt{n-k^2} \\ \frac{\sqrt{n-k^2}}{k} \mathrm{sh} 2a\sqrt{n-k^2} & \mathrm{ch} 2a\sqrt{n-k^2} \end{array} \right|$$

and

$$\|Œ_n(k)\| \leq c\,(\triangle) \exp\left(2a\sqrt{n}\right)\sqrt{n}$$
$$\|Œ_n'(k)\| \leq c\,(\triangle) \exp\left(2a\sqrt{n}\right)\sqrt{n}.$$

Then

$$\left\|Œ_1^2\right\| \ldots \left\|Œ_n^2\right\| \leq \exp\left(\frac{8}{3} a\, n^{\frac{3}{2}} + n \ln n + O(n)\right).$$

Let us put $\beta_n = \exp\left(An \ln n\right)$. Theorem 11 gives that under condition

$$L_n = x_n^- - x_{n-1}^+ = x_n - x_{n-1} - 2a = \exp\left(\frac{8}{3} a\, n^{\frac{3}{2}} + Bn \ln n\right),\ B > 2$$

the spectrum of the operator $H^{\theta_0}$ is singular continuous for positive energies. Negative part of the spectrum is either empty or (for some $\theta_0$) contains no more than one eigenvalue.

Direct calculations (based on Lemma 10) and Simon-Spencer theorem [7] give similar (but a slightly better, namely $B > 1$!) result.

If $L_n \leq \exp\left(\frac{8}{3} a\, n^{\frac{3}{2}} + Bn \ln n\right)$, $B < -1$, then method of the cluster expansion in the spirit of [8]-[10] gives dense p.p. spectrum for a.e. $\theta_0 \in [0, \pi]$.

**Example 8**. Sequence of negative potential wells. Let

$$g_n(x) = -n\, I_{|x|\le a} \qquad (6.70)$$
$$q(x) = \sum_{n=1}^{\infty} g_n(x - x_n)$$

In this case

$$Œ_n(k) = \begin{vmatrix} \cos 2a\sqrt{n+k^2} & \frac{k \sin 2a\sqrt{n+k^2}}{\sqrt{n+k^2}} \\ -\frac{\sqrt{n+k^2}}{k} \sin 2a\sqrt{n+k^2} & \cos 2a\sqrt{n+k^2} \end{vmatrix}$$

and

$$\|Œ_n(k)\|\,, \|Œ'_n(K)\| \le C(\triangle)\sqrt{n}.$$

I.e.,

$$\|Œ_1\|^2 \ldots \|Œ_n\|^2 = \exp\left(n \ln n + O(n)\right).$$

Condition

$$L_n \ge \exp(B\, n \ln n)\,,\ B > 2 \qquad (6.71)$$

is sufficient for the pure singular spectrum for $\lambda = k^2 > 0$. For $\lambda < 0$ and a.e. $\theta_0$ the spectrum is p.p.

Probably, in the case $B < 1$ in (6.71) the spectrum will be dense p.p. even for positive energies. Most interesting question for both Examples 7 and 8 is the possible coexistence of the s.c. and p.p. spectra near the "border line" between pure s.c. spectrum and p.p. spectrum. In the case of Example 7 this border line corresponds to

$$L_n = \exp\left(\frac{8}{3} a\, n^{\frac{3}{2}} + B\, n \ln n\right), -1 \le B \le 1.$$

There are some indications (and even partial results) that such coexistence is possible.

## 7 Perturbation theory and decaying sparse potentials

The last section of the paper is related to the direct generalization of the Pearson's results [3]. It will emphasize the special role of the square-integrability condition for potentials in the theory of absolutely continuous spectra. The theory presented in this section has nontrivial intersection with the recent preprints [4]-[5] of the B. Simon's group.

However there is an essential difference not only in the methods (absence of the martingale theory in [4]-[5]) but also in assumptions and results. In [4]-the bumps have a form $g_n(\cdot) = a_n \varphi(x - x_n)$ with compactly supported $\varphi(\cdot)$ like in original D. Pearson's paper [3]. Fourier transform $\widehat{\varphi}(k)$ is analytic and constructions similar to Example 10 (see below) are not possible. On the other hand, instead of the superlacunarity condition $\sum_i x_i/x_{i+1} < \infty$ they use a weaker condition $x_i/x_{i+1} \leq \gamma < 1$.

Let

$$q(x) = \sum_{n=1}^{\infty} a_n\, g_n(x - x_n) \tag{7.1}$$
$$a_n \to 0\,,\; n \to \infty$$

and the system of the functions $\{g_n(\cdot)\}$ satisfy our initial conditions (1.6), (1.7):

$$(SL) : \sum_n \frac{x_n}{x_{n+1}} < \infty\,,\; |g_n(z)| \leq \frac{c_0}{1 + |z|^{1+\delta}}. \tag{7.2}$$

Spectral properties of the operator $H^{\theta_0}$ given by (1.1) can be studied in the context of the general theorems of the Sec. 5, 6, but the decaying of the potential simplifies the situation.

First of all, $\Sigma_{ess}\left(H^{\theta_0}\right) \cap (-\infty, 0) = \emptyset$, negative discrete spectrum has, generally speaking, the accumulating point $\lambda_0 = 0$. Second, reflection coefficient $B_n(k)$ can be expanded into convergent series over the degrees of the amplitude factor $a_n$ :

$$B_n(k) = a_n\, d_1(k) + a_n^2\, d_2(k) + a_n^3\, d_3(k) + \ldots \tag{7.3}$$

Explicit formulas for the perturbation series (7.3) can be found from the integral equation

$$A(k) = 1 - \frac{\alpha(k)}{2ki}\,,\; \alpha(k) = \lim_{x \to +\infty} \alpha_k(x) = \alpha_k(+\infty) \tag{7.4}$$
$$B(k) = \frac{\beta(k)}{2ki}\,,\; \beta(k) = \lim_{x \to +\infty} \beta_k(x) = \beta_k(+\infty)$$

and

$$\overrightarrow{\alpha_k}(x) = \begin{pmatrix} \alpha_k(x) \\ \beta_k(x) \end{pmatrix} = \begin{pmatrix} \sigma \int_{-\infty}^{x} g(z)dz \\ \sigma \int_{-\infty}^{x} g(z)e^{-2ikz}dz \end{pmatrix} + \frac{\sigma}{2ki} \int_{-\infty}^{x} g(z) \cdot \tag{7.5}$$

$$\cdot \begin{pmatrix} -1 & e^{2ikz} \\ -e^{-2ikz} & 1 \end{pmatrix} \begin{pmatrix} \alpha_k(z) \\ \beta_k(z) \end{pmatrix} dz = \sigma \overrightarrow{h_k}(x) + \frac{\sigma}{2ki} \int\limits_{-\infty}^{x} g\, A_k(z)\, \overrightarrow{\alpha_k}(z) dz.$$

Formulas (7.4), (7.5) are equivalent to the relations (3.19), (3.20), (3.21) from Sec. 2. We omitted index $n$ and used the "neutral" parameter $\sigma$ instead of the small parameter $a_n$.

Since $\sigma << 1$ and $\int\limits_{R^1} |g| dx \leq c_1$ equation (7.5) can be solved by sequential approximations. Norm of the linear operator in the right part has order $\sigma$, i.e., these approximations give the expansion with respect to degrees of $\sigma$.

The first approximation (first order perturbation theory) gives

$$\beta_K^{(0)}(\infty) = \sigma \int\limits_{-\infty}^{+\infty} g(z) e^{-2kiz} dz = \sigma\, \widehat{g}(-2k)$$

i.e.,

$$|B_n^2(k)| = \frac{a_n^2 |\widehat{g}_n(2k)|^2}{4k^2} + O(|a_n|^3). \tag{7.6}$$

This formula was the central point of the Pearson's analysis [1].

The second order perturbation provides the simple expression

$$\overrightarrow{\alpha}_k^{(2)}(\infty) = \sigma \begin{pmatrix} \int\limits_{-\infty}^{+\infty} g(z) dz \\ \int\limits_{-\infty}^{+\infty} g(z) e^{-2ikz} \end{pmatrix} + \frac{\sigma^2}{2ki} \int\limits_{-\infty}^{+\infty} dz\, g(z) \begin{pmatrix} -1 & e^{2ikz} \\ -e^{-2ikz} & 1 \end{pmatrix} \cdot$$

$$\cdot \int\limits_{-\infty}^{z} dv \begin{pmatrix} g(v) \\ g(v) e^{-2ikv} \end{pmatrix}.$$

It gives

$$\beta_k(\infty) = \sigma \widehat{g}(-2k) + \frac{\sigma^2}{2ki} \int\limits_{-\infty}^{+\infty} dz\, g(z) \left[ \int\limits_{-\infty}^{z} [g(v) e^{-2ikv} dv - \right.$$

$$\left. -e^{-2ikz} \int\limits_{-\infty}^{z} g(v) dv \right] + O(\sigma^3) =$$

$$= \sigma\, \widehat{g}\,(-2k) + \frac{\sigma^2}{2ik} \left[ \int\limits_{-\infty}^{+\infty} g(v) e^{-2ikv} \int\limits_{v}^{\infty} g(z) dz - \int\limits_{-\infty}^{+\infty} g(z) e^{-2ikz} \int\limits_{-\infty}^{z} g(z) dv \right].$$

Substitution

$$g(v) = \frac{1}{2\pi} \int e^{i\lambda v} \widehat{g}(\lambda) d\lambda$$

gives after calculation the final formula

$$\beta_k(\infty) = \sigma \widehat{g}(-2k) + \frac{\sigma^2}{4k\pi i} \int_{-\infty}^{+\infty} \widehat{g}(\lambda) \left[ \frac{\widehat{g}(\lambda - 2k) - \widehat{g}(-2k)}{i\lambda} \right] d\lambda$$
$$+ O(\sigma^3). \tag{7.7}$$

The second term looks like the convolution of $\widehat{g}(\lambda)$ with itself.

Formula (7.7), together with relation $B(k) = \frac{\beta_k(\infty)}{2ki}$, gives two terms in the expansion of $|B_n(k)|^2$. To find three terms, i.e.,

$$|B_n(k)|^2 = a_n^2 \cdot D_1(k) + a_n^3 D_2(k) + a_n^4 D_3(k) + O(a_n^5)$$

we have to calculate the third approximation in the integral equation (7.5). We will not make these calculations but note, that in the special case

$$\widehat{g}(2k) \equiv 0, \ k \in \triangle$$

we have

$$|B_n(k)|^2 = c_0 \, a_n^4 \cdot \frac{1}{k^4} \Big| \int_{-\infty}^{+\infty} \widehat{g}(\lambda) \left[ \frac{\widehat{g}(\lambda - 2k) - \widehat{g}(-2k)}{\lambda} \right] d\lambda \Big|^2. \tag{7.8}$$

Now we can apply the results of our asymptotical analysis to the spectral properties of the operator $H^{\theta_0}$ (for $\lambda = k^2 > 0$). Due to Theorems 7, 8 we have

**Theorem 12.** For the Schrödinger operator $H^{\theta_0}$ with potential (7.1) and conditions (7.1), (7.2) for positive energies

a) If $\sum\limits_n |a_n^3| < \infty$, then

$$\sum\nolimits_{ac} \left( H^{\theta_0} \right) = \widetilde{Œ}_0, \text{ where } Œ_0 = \left\{ k : \sum_n a_n^2 |\widehat{g}_n(2k)|^2 < \infty \right\},$$

$\widetilde{Œ}_0$ is the essential closure of $Œ_0$

b) If for some interval $\triangle \subset (0, \infty)$

$$\sum_n a_n^2 |\widehat{g}_n(2k)|^2 = \infty \ \text{ a.e. on } \triangle$$

then $\mu_{ac}(\triangle) = 0$ , $\triangle \subset \sum_{sc}$.

c) If for some interval $\triangle$ the series $\sum_n a_n^2 \, |\widehat{g}_n(2k)|^2$ converges in the $\text{Höl}_{\delta'}(\triangle)$ norm, or $a_n^2|\widehat{g}_n(2k)|^2 \le \beta_n^2$ , $\sum_n \beta_n^2 < \infty$ (dominated convergence) then $\triangle$ contains only $\mu_{ac}$ (with bounded and strictly positive continuous spectral density).

d) In particular, if $\sum a_n^2 < \infty$ then

$$\sum\nolimits_{ac}(H^{\theta_0}) = [0,\infty) \ , \ \mu_{sc}\left([0,\infty)\right) = 0.$$

e) If $\widehat{g}_n(2k) \equiv 0, \ k \in \triangle$ , $n = 1, 2, \ldots$ and $\sum a_n^4 < \infty$ then $\triangle \subset \sum_{ac}$.

Let us discuss a few special cases.

**Example 9**. (Compare with [3]). Let $g_n(x) \equiv g(x)$ under conditions (SL) and (7.2).

a) If $\sum a_n^2 < \infty$ then $\sum(H) = \sum_{ac}(H) = [0,\infty), \sum_{sc} = \emptyset$.

b) If $\sum a_n^2 = \infty$ and $|\widehat{g}(2k)| > 0$ a.e. on $[0,\infty)$ then $\sum(H) = \sum_{sc}(H) = [0,\infty)$ , $\sum_{ac} = \emptyset$.

c) If $\sum a_n^2 = \infty$ and $Œ_1 = \{k : |\widehat{g}(2k)| > 0\}$ then $Œ_1 \subset \sum_{sc}$.

d) If $\sum a_n^2 = \infty$, $\sum_n |a_n^4| < \infty$ and $Œ_0 = \{k : |\widehat{g}(2k)| = 0\}$ and $|Œ_0| > 0$ then

$$Œ_0 = \sum\nolimits_{ac} \, , \ Œ_1 = \sum\nolimits_{sc} .$$

In the special case when $Œ_0$ is nowhere dense set of the positive Lebesque measure,

$$Œ_0 = \sum\nolimits_{ac} \, , \ [0,\infty) = \sum\nolimits_{sc} .$$

The statements a), b) contain result of the Theorem A (see introduction and [13]), because of the compactly supported bump $g(\cdot)$ corresponding Fourier transform $\widehat{g}(k)$ is an analytic function.

The next example is a specification of the general Example 9.

**Example 10**. Let $q(x) = \sum_n \ a_n \ g\left(x - x_n\right), g\left(x\right) = \frac{2\sin^2 \frac{x}{2}}{\pi \, x^2}, \{x_k\}$ be a superlacunar sequence and $a_n \to 0$. One has $\widehat{g}(2k) = (1-k)_+$. According to the results from Example 9, we have the following spectral picture:

a) If $\sum a_n^2 < \infty$ then $\sum(H) = \sum_{ac}(H) = [0,\infty)$

b) If $\sum a_n^2 = \infty, \sum a_n^4 < \infty$

$\sum_{sc} = [0,1]$ , $\sum_{ac} = [1,\infty)$.

c) If $\sum_n a_n^4 = \infty$ , $\sum_n |a_n|^5 < \infty$

$\sum_{sc} = [0,2]$ , $\sum_{ac} = [2,\infty)$

Let us stress that $\sum_{sc} = [0,1] \cup [1,2]$ and these two intervals contain the singular measures of the different Hausdorf dimensions.

Probably in c) instead of condition $\sum |a_n|^5 < \infty$ one can use weaker condition $\sum a_n^6 < \infty$ and one can continue the process of extension of $\sum_{sc}$ for more and more slowly decaying amplitudes $a_n$.

**Conclusion**. There are a few natural open problems, related with our study. We will formulate two of them, which seem especially important.

a) All the problems about mixed spectral type on the same interval are open. For instance: let under basic conditions (1.6), (1.7)

$$\sum_n |B_n^2(k)| < \infty \quad \text{a.e. on } \triangle \subset (0,\infty)$$

Construct the example when $\mu_{sc}$ is dense on $\triangle$. Such examples (in the spirit of constructions by Simon and Naboko) probably exist. The second question of the same type (see Theorem 11 and Examples 7, 8): construct a potential with increasing bumps such that on some interval $\triangle$ and for a.e. $\theta_0$ both components of $\mu_{\text{sing}}$ (i.e., $\mu_{pp}$ and $\mu_{sc}$) are dense.

b) What is the role of superlacunarity condition? We have some partial results when instead of (SL) one can use the standard lacunarity:

**Theorem**. If $q(x) = \sum_{n=1}^{\infty} a_n g_n(x - x_n)$, $\sum_n a_n^2 \ln^2 n < \infty$ , $|g_n(x)| \leq \frac{c_0}{1+|x|^{2+\delta}}$ , $\frac{x_{n+1}}{x_n} \geq \gamma > 1$ then $\sum (H) \cap [0,\infty) = \sum_{ac}(H) = [0,\infty)$

It is important to construct examples, which show that one cannot preserve the results of the Theorems 7, 8 or even 12 if $\frac{x_{n+1}}{x_n} \to 1$. For instance, it is probably true that for a given sequence $\varepsilon_n \to 0$ (arbitrary slowly) one can construct a sequence $\{x_n\}$ , family of the bumps under uniform integrability condition (1.6) and system of amplitudes $a_n \to 0$ such that

$$|\frac{x_{n+1}}{x_n} - 1| \geq \varepsilon_n$$

$$\sum_n a_n^2 = \infty$$

but $\sum(H) = \sum_{ac}(H) = [0,\infty]$.

## Acknowledgments

The part of this work was done at UNCC during the Spring of 96. The final version was prepared at Paris 6 (Summer 96). I gave a few talks on the subject of this paper at the seminars of UNCC and minicourse at Paris 7. I would like to thank A. Boutet de Monvel and her colleagues from Paris 7 (V. Georgescu, L. Pastur and others) for the hospitality and interesting discussions. I discussed these and similar results with B. Simon and his students A. Kiselev and C. Remling during the International conference at Atlanta (Spring 97). This conference stimulated further progress in the spectral theory of 1-D Schrödinger operators. I acknowledge the support and useful discussions and recommendations from B. Simon and his group as well as my friends from UNCC: R. Anderson, A. Figotin, A. Gordon and B. Vainberg. The special gratitude is addressed to J. Taylor for her patience in preparation of the manuscript.

This work is partially supported by NSF-Grant #93/0710.


## References


1. S. Kozlov and S. Molchanov, On conditions for the applicability of CLT to the walks on inhomogeneous lattices , *Sov. Math.Dokl.*, **278** (3), 531–534 (1984).
2. S. Kozlov, The method of averaging and walks in inhomogenous environments, *Russ. Math. Surveys*, **138** (A), 289–309 (1986).
3. D. Pearson, Singular continuous measures in scattering theory, *Comm. Math. Ph.*, **60**, 13–36 (1978).
4. B. Simon, Results on the spectral theory of 1-D Schrödinger operators, Preprint and talk on ICDEMP (Atlanta, Caltech, Spring 97).
5. A. Kiselev, Y. Last and B. Simon, Modified Prüfer transformations and the spectral analysis of 1-D Schrödinger operators, Preprint, Caltech–97.
6. M. Read and B. Simon, *Methods of Modern Mathematical Physics*, VI, New York, Acad. Press, 1978.
7. W. Kirsh, S. Molchanov and L. Pastur, One-dimensoinal Schrödinger operator with unbounded potential: pure point spectrum, *Funct. Anal. and Appl.*, **3**, 24–36 (1990).

8. S. Molchanov, Lectures on random media, (1994), In Lecture Notes in Math, #1581, 242-
9. A Gordon, A deterministic potential with a purely point spectrum, *Math. Notes*, **48** (5–6), 1197–1203 (1990).
10. S. Molvhanov, One-dimensional Schrödinger operator with sparse bumps Preprint, 1997, Paris VI, 111, 1–62.
11. S. Novikov, S. Manakov, L. Pitaevski and V. Zakharov, (1985), Theory of Solutions: The inverse problem method, Nauka, Moskov.
12. V. Marchenko, Sturm-Lionville operators and applications, (1986), Birkhäuser, Boston.
13. B. Levitan and I. Sargsyan, Introduction to spectral theory (1976), AMS, Providence, RI (Transl. Math. Monographs, 39).
14. F. Atkinson, Discrete and Continuous Boundary problems, (1964) Acad. Press, New-York and London.
15. F. Shnoll, On the behaviour of the eigenfunctions, *Dokl. Acad. Nank SSR*, **94**, 389–392 (1954).
16. I. Glazman, Direct methods of qualitative spectral analysis of singular differential operators, (1965), Israeli Progr. Scient. Transl., Jerusalem.
17. B. Simon, Bounded eigenfunctions and absolutely continuous spectra for one-dimensional Schrödinger operators, Preprint (1995) , Caltech, to appear in Proc. AMS.
18. D. Gilbert, and D. Pearson, On subordinary and analysis of the spectrum of one-dimensional Schrödinger operators, *J. Math. Anal. Appl.*, **128**, 30–56 (1987).
19. B. Simon and T. Wolff, Singular continuous spectrum under rank-one perturbations and localization for random Hamiltonians, , **125**, 113–125 (1989).
20. S. Kotani, Lyapunov exponents and spectra for one-dimensional operators, *Contemporary Math.*, **50**, 277–286 (1986).
21. A. Gordon, Pure point spectrum under 1-parameter perturbations and instability of Anderson localization, *Comm. Math. Phys.* , **164**, 489–505 (1994).
22. R. del Rio, N. Makarov and B. Simon, Operators with singular continuous spectrum, II, *Comm. Math. Phys.*, **164**, 59–67 (1994).
23. A. Gordon, V. Maslov and S. Molchanov, Unbounded eigenfunc-

tions, *Russian J. Math. Phys.*

24. I. Goldsheidt, Asymptotic properties of the product of random matrices depending on parameter in muticomponent systems, *Advance in Probab*, **8**, 239–289 (1980).
25. Zigmundt, Trigonometrical series, vol. I, II, (1968), Cambridge University Press.
26. J. Neveu, Bases mathematiques du calcul des probabilites (1964), Masson et Cie, Paris.

# REMARKS ON AN ESTIMATE OF SERGUEI KOZLOV

U. MOSCO
*Dipartimento di Matematica, Universita di Roma La Sapienza, I-00185, Roma*

## 1 Introduction

In his paper "Harmonization and Homogenization on Fractals" [3] Serguei Kozlov states a very precise estimate for functions of finite energy on the Sierpinski gasket. The estimate asserts that functions of finite energy on the gasket are Hölder continuous, with the Hölder exponent

$$\alpha = \frac{\log \frac{5}{3}}{\log 4}. \tag{1.1}$$

This property is reminiscent of the classic Morrey's imbedding of the Sobolev space $H^1$ on the real line. However, as we shall see, none of the known relevant fractal dimensionalities seems to legitimate $\alpha$ as a Morrey's exponent.

Our aim in this note is to explain the strange exponent in Kozlov's estimate. On the way, we shall get a better insight into the dynamic behavior of the gasket. We observe that the example of the Sierpinski gasket is of special relevance in fractal theories. It provides indeed the basic mathematical model of so called *"finitely ramified fractals"* of physics.

Its dynamic properties, i.e. *the properties involving its energy form and the related "Laplacian" and "Brownian motion"*, exhibit all major peculiarities of the dynamic behavior of this class of fractals. For example, Weyl's type spectral asymptotics and the space–time scaling of the "Brownian" process are seen both to be governed by "non-Euclidean" exponents, the so-called *spectral* and *walk dimensions*, see for example [2].

A further subtle discrepancy with the classic case — also related to the energy behavior — is the one connected with the "strange" Morrey's type exponent observed, as mentioned before, by Kozlov. Kozlov's estimate is indeed particularly intriguing, because, as we shall explain, it involves not only non-Euclidean "dimensional" exponents, but it also suggests that alternative non-Euclidean *metrics* of dynamic nature have to be considered inside a fractal.

The dynamic strange behavior of the Sierpinski gasket has its counterpart in the peculiar properties enjoined by the "harmonic functions" of the fractal. This is the aspect of fractal theory more directly investigated in Kozlov' paper. Serguei's approach to harmonic fractal theory is beautiful and somehow unique — in its geometric flavor — and we like to refer explicitly to his clear presentation of this subject.

In addressing Morrey's imbedding question, we shall reproduce below Kozlov's explicit construction of the gasket — as a subset of the Euclidean plane — and of the quadratic energy form that describes its dynamics. We shall also adopt his notation, only with minor changes.

## 2 Kozlov's estimate

The (unit) Sierpinski gasket in the plane is the closure $V = \bar{V}_*$ of the so-called pre-gasket

$$V_* = \bigcup_{n=0}^{\infty} V_n.$$

The finite subsets $V_0, V_1, \ldots$ are obtained recursively from the vertices $V_0 = \{a_1, a_2, a_3\}$ of an equilateral triangle — say, $a_1 = (0,0)$, $a_2 = (1,0)$, $a_3 = \left(\frac{1}{3}, \frac{\sqrt{3}}{2}\right)$ — by setting

$$V_{n+1} = \bigcup_{i=1}^{3} \frac{1}{2}\left(a_i + V_n\right).$$

Here we use the notation $x+A = \{x+y : \ y \in A\}, \quad cA = \{cy : \ y \in A\}$.

For every function $f$: $V_* \to R$, the renormalized energy $\varepsilon_n(f), n = 0, 1, \ldots$, is defined by

$$\varepsilon_n\left(f\right) = \left(\frac{5}{3}\right)^n \sum_{\substack{x,y \in V_n \\ |x-y| = 2^{-n}}} \left(f\left(x\right) - f\left(y\right)\right)^2.$$

Kozlov's estimate in [3] is

$$\sup_{x,y \in V_*} \frac{\left|f\left(x\right) - f\left(y\right)\right|}{\left|x-y\right|^{\alpha}} \leq 6\sqrt{\sup_{n \geq 0} \varepsilon_n\left(f\right)}, \tag{2.1}$$

where $\alpha$ has the value (1.1).

This estimate has important consequences, as nicely explained by Kozlov himself in the same paper.

Due to the special choice of the renormalization factor, the sequence $\varepsilon_n(f)$ is non-decreasing in $n$. Therefore, for given $f: V_* \to R$, we can define the limiting energy of $f$ as the — possibly infinite — limit

$$\varepsilon_n(f) = \lim_{n\to\infty} \varepsilon_n(f)$$

and we can then introduce, in Kozlov's notation, the space

$$H^1 = \{f : V_* \to R : \varepsilon_n(f) < \infty\} \ .$$

By (2.1), each $f \in H^1$ has a continuation to $V$, which belongs to a standard Hölder space $C^\alpha$. The space of all these extensions is still denoted by $H^1$, hence $H^1 \subset C^\alpha$ and

$$\sup_{x,y\in V} \frac{|f(x) - f(y)|}{|x-y|^\alpha} \leq 6\sqrt{\varepsilon_n(f)} \tag{2.2}$$

for every $f \in H^1$.

## 3 The strange exponent

The notation $H^1$ used by Kozlov clearly alludes to the standard notation for the classic Sobolev space $H^1 \equiv W^{1,2}$. In fact, the energy $\varepsilon(f)$ is the fractal analogue of the classic Dirichlet integral $\int |\text{grad } f|^2 dx$. So we may raise the question whether the imbedding $H^1 \subset C^\alpha$ on the gasket is of the same nature as Morrey's imbedding $H^1 \subset C^{\frac{1}{2}}$ on the real line ($1/2 = 1 - d/2$ for $d = 1$).

On purely arithmetic grounds, this analogy postulates a relation of the kind

$$\alpha = 1 - \frac{"d"}{2}$$

between Kozlov's Hölder exponent $\alpha$ in (1.1) and some relevant fractal dimension "$d$".

On the gasket, generally speaking, there are three distinct important dimensions which could be in the role of "$d$". They are the embedding Euclidean dimension $d$, the fractal dimension $d_f$ and the so-called spectral dimension $d_s$, $d > d_f > d_s$. In our case, $d = 2$, $d_f = \log 3/\log 2$, $d_s = \log 9/\log 5$. A simple check shows that "$d$" $\neq d, d_f, d_s$.

This seems to discredit our tentative of explaining (2.2) in terms of Morrey's imbeddings.

## 4 The change of metric

Despite this negative evidence, we decide to persist. A closer look at the three candidates $d, d_f, d_s$ and at the values they take on higher dimensional Sierpinski gaskets

$$d \geq 3, \quad d_f = \frac{\log(d+1)}{\log 2}, \quad d_s = 2\frac{\log(d+1)}{\log(d+3)}$$

shows that only $d_s$, if taken in place of "$d$" leads to a positive value of $\alpha$. Therefore, we summarize that the good "$d$" might indeed be $d_s$ and — going back to our case $d = 2$ — we introduce the virtual "Morrey exponent"

$$\beta := 1 - \frac{d_s}{2} = \frac{\log \frac{5}{3}}{\log 5}. \tag{4.1}$$

The exponent $\beta$ is related to Kozlov's exponent $\alpha$ by

$$\alpha = \beta\delta,$$

where

$$\delta = \frac{\log 5}{\log 4}. \tag{4.2}$$

We carry on our naive arithmetics a little further and we observe that, if, in Kozlov's estimate (2.2), we make the change of metric

$$d(x,y) := |x-y|^{\delta}, \tag{4.3}$$

then we get

$$\sup_{x,y\in V} \frac{|f(x) - f(y)|}{d(x,y)^{\beta}} \leq 6\sqrt{\varepsilon_n(f)}. \tag{4.4}$$

In view of (4.1), this inequality is what we would expect — *in the metric* $d(x,y)$ — from a $d_s$-dimensional "Morrey imbedding". Is that just arithmetic nonsense, or do we have here the evidence that (4.4) is a genuine *fractal* Morrey's inequality?

## 5 The physics of $V$

In order to better understand the nature of new metric $d(x,y)$, we must better understand the exponent (4.2). A simple check shows that

$$\delta \equiv \frac{\alpha}{\beta} = \frac{d_f}{d_s}. \tag{5.1}$$

For classic Euclidean media we have $d_s = d_f = d$, hence $\delta = 1$ and $d(x,y) \equiv |x-y|$. Therefore, we can trace back the origin of $d(x,y)$ to the fractal discrepancy between the exponent $d_f$, that governs the *statics* of the structure, and the exponent $d_s$, that governs its *dynamics*.

Before proceeding, let us briefly explain this elementary fractal "physics" ([2],[5]).

The fractal dimension $d_f$ is the Hausdorff dimension of $V$ and the $d_f$-dimensional Hausdorff measure — restricted to $V$ and normalized to have unit total mass — is the *invariant measure* $\mu$ of $V$, that is, $\mu$ is self-similar on $V$:

$$\mu(\cdot) = \frac{1}{3}\sum_{i=1}^{3} \mu(\psi_i(\cdot)), \tag{5.2}$$

where by $\psi_1, \psi_2, \psi_3$ we have denoted the basic self-similarities of $V$, $\psi_i(x) \equiv a_i + \frac{1}{2}(x - a_i)$. The measure $\mu$ describes the homogeneous distribution of the fractal volume in $V$, hence also of the mass. The exponent $d_f$ governs the decay of the mass $\mu(l)$ of a (Euclidean) spherical neighborhood $V \cap B_l^{eucl}$ in $V$, $B_l^{eucl} \equiv B_l^{eucl}(x) = \{y : |x-y| < l\}, x \in V, l > 0$. In fact, we have

$$\frac{\mu(l)}{\mu(L)} \approx \left(\frac{l}{L}\right)^{d_f} \tag{5.3}$$

for $0 < l \leq L$. This can be proved by relying on the self-similarity of $\mu$ and on a finite overlapping property of $V$ (each neighborhood $V \cap B_l^{eucl}, \quad l = 2^{-2n}$, can be covered for every $n$ by a fixed number of $M$ of self-similar copies $\psi_{i_1} \circ \psi_{i_2} \circ \cdots \circ \psi_{i_n}(V)$ of $V$ of size $2^{-n}$).

The spectral dimension $d_s$, on the other hand, is the exponent that governs Weyl's distribution function $N(\lambda)$ of the spectrum of the "Laplace" operator on the gasket,

$$N(\lambda) \approx \lambda^{\frac{d_s}{2}} \tag{5.4}$$

for $\lambda > 0$. Kozlov's paper contains a clear definition of the fractal Laplacian and also a nice derivation of the Weyl asymptotic formula. Invariance — this time of the *energy* — is at the root of the dynamic behavior (5.4). The invariance of the energy is expressed by the self-similarity property of the form $\varepsilon$:

$$\varepsilon(f) = \frac{5}{3}\sum_{i=1}^{3} \varepsilon(f \circ \psi_i) \tag{5.5}$$

for every $f \in H^1$. This identity can be obtained easily going back to the construction of $\varepsilon$ as the limit of the renormalized energies, $\varepsilon_n(f) = \lim_{n\to\infty} \varepsilon_n(f)$ (recall that 5/3 is the renormalization factor).

## 6 The invariance of $d(x,y)$

The preceding remarks show that the metric $d(x.y)$ depends on the dynamics, hence on the energy, of $V$. In fact, it can be easily checked that the scaling exponent $\delta$ — occurring in the definition (4.3) of $d(x,y)$ — is the *unique* value $\delta > 0$ for which the identity

$$d^2(x,y) = \frac{5}{3}\sum_{i=1}^{3} d^2(\psi_i(x), \psi_i(y)) \tag{6.1}$$

holds for all $x, y \in V$. The *new metric* $d(x,y)$ *inherits the self-similar invariance of* $\varepsilon$.

The function $d(x,y)$ is a *quasi-distance*, because the triangular inequality holds up to a multiplicative constant, $2^{\delta-1}$. With the quasi-distance $d(x,y)$, the fractal $V$ is a *quasi-metric* space. The (quasi-)balls of $d(x,y)$ in $V$ are the sets

$$B_r \equiv B_r(x) = \{y \in V : d(x,y) < r\}, \quad x \in V, \quad r > 0.$$

Since $B_r = V \cup B_l^{eucl}$ for $l = r^{\frac{1}{\delta}}$, by just recalling (5.3), we can estimate the value of the balls $B_r$ as

$$\frac{\mu(B_r)}{\mu(B_R)} \approx \left(\frac{r}{R}\right)^{\nu} \tag{6.2}$$

for every $0 < r \leq R$, with

$$\nu = \frac{d_f}{\delta} = \frac{\log 9}{\log 5}. \tag{6.3}$$

## 7 The homogeneous dimension

In abstract harmonic analysis, a triple like $V$, $d(x,y)$ and $\mu$, is called a *space of homogeneous type* and the exponent $\nu$ in (6.2) is the *homogeneous dimension* of the space. The homogeneity property enjoyed by the quasi-metric spaces is the following: the maximal number of point $B_r$ with mutual distance $\approx \varepsilon r$ grows up $\approx \varepsilon^{-\nu}$ as $\varepsilon \to 0$.

We now observe that the exponent (6.3) is equal to $d_s$. The spectral dimension is thus explained — more "geometrically" — as the homogeneous dimension of $V$,

$$d_s = \nu. \tag{7.1}$$

The equality (7.1) — first observed in [1] — is an interesting by-product of our analysis of Kozlov's inequality. In fact, the geometric interpretation of $d_s$ is one of the difficult problems in fractal theory, because it demands to explain the dynamics with the statics. The identity (7.1) shows that the key for this interpretation is the "effective" metric $d\,(x,y)$.

The connection just established, at a basic structural level, between fractal theory and harmonic analysis — together with the self-similar invariance of mass, energy and metric — turns out to be very useful. From our present point of view, the realization of $V$ as a space of homogeneous type under the metric $d\,(x,y)$ provides the correct setting — as we shall see below — for the interpretation of Kozlov's inequality as a true Morrey's imbedding.

## 8 Scaled Poincaré inequalities

The following global Poincaré inequality

$$\int_V |f - f_V|^2 d\mu \leq c\varepsilon\,(f) \tag{8.1}$$

for all $f \in H^1$, where $f_V = \int_V f d\mu$, is a much weaker estimate than Kozlov's estimate (2.1), and can be proved again by direct calculations on the pre-fractals $V_n$.

More work is needed — based on the invariance of $\mu, \varepsilon, d\,(x,y)$ and the finite overlapping property of $V$ — in order to prove the following *scaled Poincaré inequalities* on the balls of $d\,(x,y)$ for functions $f \in H^1$:

$$\int_{B_r} |f - f_{B_r}|^2 d\mu \leq cr^2 \int_{B_{\sqrt{5}r}} d\,|\text{grad}\, f|^2, \tag{8.2}$$

$f_{B_r} = \frac{1}{\mu(B_r)} \int_{B_r} f d\mu$. By $|\text{grad}\, f|^2$ we denote here the *local energy* of $f \in H^1$, defined as a (positive) *measure* in $V$ by the identity

$$\int_V \varphi d\,|grad\; f|^2 \equiv \varepsilon\,(\varphi f, f) - \frac{1}{2}\varepsilon\left(\varphi, f^2\right)$$

for every $\varphi \in H^1$.

The scaled Poincaré inequalities express the relation of the metric $d(x,y)$ with the energy $\varepsilon$ at a deeper level than their common invariance (5.5), (6.1), because they relate both energy and metric with the (invariant) measure $\mu$, hence, they relate dynamics with statics.

For example, based on the inequalities (8.2), a general theory of *Morrey-Sobolev imbeddings* can be developed for general homogeneous spaces [1]. This theory gives the Morrey imbedding $H^1 \subset C^\beta$, with $\beta = 1 - \nu/2$, whenever the homogeneous dimension $\nu$ of the space is — as in our case — strictly less than 2.

Let us explain with some more detail how the estimate (4.4) can be obtained — up to a multiplicative constant — from the inequalities (8.2).

The estimates (6.2) and (8.2) together give the following

$$\sup_B \left[\frac{1}{\mu(B)} \int_B |f - f_B|^2 d\mu\right]^{\frac{1}{2}} \leq C\mu(B)^\gamma, \tag{8.3}$$

for every ball $B \equiv B_r$ in $V$, where $\gamma = 1/\nu - 1/2$ and $C \equiv C(f) = c_1\sqrt{\varepsilon(f)}$. We now use a general result of the theory of homogeneous spaces, due to Macias-Segovia [4], which says that (8.3) implies that $f$ — up to $\mu$ — a.e. equivalence — satisfies the estimate

$$|f(x) - f(y)| \leq c_2 C(f)\, \mu(B)^\gamma \tag{8.4}$$

in $V$ for any ball $B$ containing $x$ and $y$. We choose $B \equiv B_r(x)$, with $r = d(x,y) + \eta, \eta > 0$. Then $x, y \in B$. Moreover, again by (6.2), $\mu(B) \leq c_3 r^\nu = c_3 (d(x,y) + \eta)^\nu$. By replacing this $B$ in (8.4), we get $|f(x) - f(y)| \leq c_4 C(f)(d(x,y) + \eta)^{\nu\gamma}$, therefore, by the arbitrariness of $\eta$, $|f(x) - f(y)| \leq c_4 C(f)\, d(x,y)|^{\nu\gamma}$. This gives the inequality (4.4), with the required exponent $\beta = 1 - \nu/2$.

We have seen, finally, that Kozlov's imbedding $H^1 \subset C^\alpha$ is indeed a Morrey's imbedding — $H^1 \subset C^\beta$ — in the intrinsic homogeneous metric $d(x,y)$ of $V$.

Kozlov's paper [3] contains further significant contributions to fractal theory; in particular, it contains very interesting *homogenization properties* of the Sierppinski gasket, that bring new deep insight into the asymptotic energy properties of fractal structure. The research of

Serguei on fractals will certainly long stimulate and strongly influence the research in this field, proving once again the impact of his work on the frontier of mathematics.

## References

1. M. Biroli and U. Mosco, Sobolev and isoperimetric inequalities for Dirichlet forms on homogeneous spaces, *Rend. Mat. Accad. Lincei*, **9** (6), 37–44 (1995).
2. A. Bunde and S. Havlin, (Eds.), Fractals and Disordered Systems, Springer-Verlag, Berlin Heidelberg, 1991.
3. S. Kozlov, Harmonization and Homogenization on Fractals, *Commun. Math. Phys.*, **153**, 339–357 (1993).
4. R.A. Macias and C. Segovia, Lipschitz functions on spaces of homogeneous type, *Adv. in Math.*, **33**, 257–270 (1979).
5. U. Mosco, Lagrangian metrics on fractals, *AMS Proc. Symp. Appl. Math.*, **54**, 301–323 (1998).

# CENTRAL LIMIT THEOREM AND SPECTRAL ASYMPTOTICS FOR NONLINEAR STOCHASTIC PARTIAL DIFFERENTIAL EQUATION WITH WEAK NONLINEARITY

A.L. PIATNITSKI
*P.N. Lebedev Physical Institute, Leninski prospect, 53, Moscow, 117924, Russia*

We prove the central limit theorem for one nonlinear stochastic partial differential equation with weak nonlinearity and study the spectral asymptotics of the infinitesimal generator of the corresponding Markov process.

## 1 Introduction

In 1994 S.M. Kozlov in the preprint [1] posed two interrelated problems: the one about the limiting behavior of normalized integrals of solutions to one nonlinear stochastic partial differential equation (SPDE) with weak nonlinearity (see equation (2.1) below), and the other one about the spectral asymptotics for the corresponding infinitesimal generator. He formulated the conjecture that the central limit theorem (CLT) holds for the solution of SPDE mentioned above, obtained the conditional result, found the covariance operator of the limit distribution and proposed a nonrigorous method of calculating the spectral asymptotics. He hypothized, in particular, the exponential growth of the counting function. It turns out, however, that this method does not yield the correct asymptotics.

In the present paper we propose another approach to the spectral asymptotics problem and, also, justify some results which has not been proved in [1]. In particular, we prove the relative compactness of the operator that corresponds to the nonlinear term.

Recent years are marked by the growing interest of mathematicians and physicists in SPDEs and in stochastic equations on Hilbert and Banach spaces. Many modern investigations in hydrodynamics, material sciences, and other applied fields, rely on the equations of this type. Also, many asymptotic methods used in random media description result in infinite dimensional stochastic problems.

Earliest mathematical works on the subject were mainly focused on the existence and uniqueness results for the simplest SPDEs and on the existence of an invariant measure. Further investigations were devoted to more complex equations and such qualitative properties as Markovity, irreducibility, ergodicity, the Feller property, and many others.

We mention here the books [2], [3], [4], [5], [6] that comprise main aspects of the theory and where further bibliography can be found.

One of the natural and important questions is the long-term behavior of solutions, in particular, the limits of applicability of the CLT. As shown in [1], the CLT holds for SPDEs with weak nonlinearity. On the other hand, there is a conjecture that, in contrast to the finite-dimensional case, the CLT fails to hold for SPDEs with strong nonlinearity.

## 2 Setting of the problem

Consider the SPDE

$$\begin{gathered}\frac{\partial}{\partial t}u(x,t) = \frac{\partial^2}{\partial x^2}u(x,t) + \alpha \sin u(x,t) + \dot{w}_t, \quad x \in (0,\pi); \\ u(0,t) = u(\pi,t) = 0, \qquad u|_{t=0} = u_0(x);\end{gathered} \tag{2.1}$$

here $w_t$ is functional Brownian motion in $L^2(0,\pi)$. It is well known (see, for example, [7], [8]) that a weak solution of (2.1) does exist for all $t > 0$, is unique, and belongs to the functional space $V = L^2(0,T;H^1(0,\pi)) \cap C(0,T;L^2_w(0,\pi))$, where the symbol $w$ indicates that the corresponding space is endowed with the weak topology.

Consider, also, the linear equation

$$\begin{gathered}\frac{\partial}{\partial t}u(x,t) = \frac{\partial^2}{\partial x^2}u(x,t) + \dot{w}_t, \quad x \in (0,\pi) \\ u(0,t) = u(\pi,t) = 0, \qquad u|_{t=0} = u_0(x);\end{gathered} \tag{2.2}$$

which, in fact, is an infinite-dimensional version of Ornstein–Ulenbeck equation. The invariant measure $\mu_0$ of the latter equation is unique (see [7]) and can be found explicitly: it is the Gaussian measure in $L^2(0,\pi)$ with a zero mean value and with a correlation operator $A = \left(\frac{d^2}{dx^2}\right)^{-1}$, which is diagonal in the basis $\left\{\sqrt{\frac{2}{\pi}}\sin kx\right\}\Big|_{k=1}^{\infty}$ with eigenvalues $k^{-2}$

$$(Au, u) = \sum_{k=1}^{\infty} k^{-2} u_k^2;$$

here $u_k$ are the corresponding Fourier coefficients of $u(x)$. Define the functional space $\mathcal{L}^2(L^2(0,\pi), \mu_0)$ of square–integrable functions over $L^2(0,\pi)$ with respect to the measure $\mu_0$ with the norm

$$\left\|F(\cdot)\right\|_{\mathcal{L}^2} = \int_{L^2(0,\pi)} F^2(u)\mu_0(du).$$

The solution of (2.2) defines a Markov process in $L^2(0,\pi)$, and the corresponding infinitesimal generator has the form

$$(\mathcal{A}_0 F(\cdot), G(\cdot))_{\mu_0} = \frac{1}{2}\int_{L^2(0,\pi)} \nabla_u F(u)\cdot\nabla_u G(u)\mu_0(du),$$

where we denote $\nabla_u F(u) = \dfrac{\delta F}{\delta u}(u)$. In the coordinates $u_1, u_2, \ldots, u_k, \ldots$, the operator $\mathcal{A}_0$ can be formally rewritten as follows:

$$\mathcal{A}_0 F(u) = \frac{1}{2}\sum_{k=1}^{\infty}\frac{\partial^2 F}{\partial u_k^2}(u) + \nabla F(u)\cdot\frac{d^2}{dx^2}u.$$

Now, let us define a new measure $\mu(du)$ absolutely continuous with respect to $\mu_0$, by the following relation

$$\frac{d\mu}{d\mu_0}(u) = \exp\left\{-\alpha\int_0^{\pi}\cos u(x)dx\right\}.$$

Taking into account the fact that the function $\alpha\sin u(x)$ is a variational derivative of $-\alpha\int_0^{\pi}\cos u(x)dx$, one can verify that $\mu(du)$ is an invariant measure of equation (2.1), and that the solution $u(x,t)$ of (2.1) is a Markov process (see [7] for more detail) which infinitesimal generator is defined as follows

$$(\mathcal{A}F(\cdot), G(\cdot))_{\mu} = \frac{1}{2}\int_{L^2(0,\pi)} \nabla_u F(u)\cdot\nabla_u G(u)\mu(du).$$

The purpose of this paper is to investigate the main properties of the operator $\mathcal{A}$.

## 3 Spectral asymptotics for $\mathcal{A}_0$

In the proper basis $\{e_j\} = \{\sqrt{\frac{2}{\pi}}\sin jx\}$ the operator $\mathcal{A}_0$ can be decomposed into direct sum of one–dimensional operators

$$\mathcal{A}_0 = \sum_{k=1}^{\infty} A_k,$$

here $A_k = \frac{1}{2}\frac{d^2}{du_k^2} + k^2 u_k \frac{d}{du_k}$ acts in the space $L^2(R, \exp(-k^2u_k^2)dx)$. The spectrum of $A_k$ is well-known: $\lambda_n^k = nk^2$, $n = 0, 1, 2, \ldots$; $\varphi_n^k(u_k) = H_n(ku_k)$ where $h_n(x)$ are Chebysheff-Hermite polynomials.

Now we can form the eigenbasis of $\mathcal{A}_0$ in $\mathcal{L}^2(L^2(0,\pi), \mu_0)$ as follows (see [9], Ch.VIII)

$$\nu_{k_1,\ldots,k_s}^{n_1,\ldots,n_s}(u) = \varphi_{n_1}^{k_1}(u_{k_1})\varphi_{n_2}^{k_2}(u_{k_2})\ldots\varphi_{n_s}^{k_s}(u_{k_s});$$

here $s \geq 0$, $n_j$ and $k_j$ are arbitrary natural numbers, $k_j \neq k_i$ for $j \neq i$. The corresponding eigenvalue $\Lambda_J = \Lambda_{k_1,\ldots,k_s}^{n_1,\ldots,n_s}$ is equal to $\sum\limits_{j=1}^{s} k_j^2 n_j$. For $s = 0$ we set $\nu_0(u) = 1$.

The above considerations show that the counting function $N_{\mathcal{A}_0}(\Lambda)$ coincides with the number of integer nonnegative solutions $(n_1, n_2, \ldots)$ of the following inequality

$$\sum_{k=1}^{\infty} n_k k^2 \leq \Lambda.$$

**Theorem 3.1** *The following limiting relation holds:*

$$\lim_{\Lambda\to\infty} \frac{\ln N_{\mathcal{A}_0}(\Lambda)}{\Lambda^{1/3}} = c_0,$$

*$c_0$ being the maximum in the variational problem*

$$\begin{gathered} c_0 = \max_{p\in\mathbf{B}} \int_0^{\infty} [(p(x)+1)\ln(p(x)+1) - p(x)\ln p(x)]dx, \\ \mathbf{B} = \left\{ p(x) \geq 0 \Big| \int_0^{\infty} x^2 p(x)dx = 1 \right\} \end{gathered} \tag{3.1}$$

**Proof.** First of all let us note that $\Theta(p) = ((p+1)\ln(p+1) - p\ln p)$ is a monotone convex function, $\Theta(0) = 0$, and $\Theta(p) \sim \ln p$ as $p \to \infty$. Thus, by the standard arguments, the problem (3.1) is well-posed and, using of the Lagrange multipliers method, we obtain

$$p(x) = \frac{1}{\exp(\lambda x^2) - 1}, \qquad \int_0^\infty \frac{x^2 dx}{\exp(\lambda x^2) - 1} = 1$$

and

$$c_0 = \int_0^\infty \left( \frac{\exp(\lambda x^2)}{\exp(\lambda x^2) - 1} \ln \frac{\exp(\lambda x^2)}{\exp(\lambda x^2) - 1} - \frac{1}{\exp(\lambda x^2) - 1} \ln \frac{1}{\exp(\lambda x^2) - 1} \right) dx$$

In order to estimate $N_{\mathcal{A}_0}(\Lambda)$ from below we fix an interval $(\delta, \delta^{-1})$ and divide this interval into $L = \Big((\delta^{-1} - \delta)/\varepsilon\Big)$ equal parts. In what follows we assume the relation $\varepsilon \ll \delta \ll 1$ holds. Denote $(n_1, n_2, \ldots)$ by $\bar{n}$. We say that $\mathrm{supp}(\bar{n}) \in (j_1, j_2)$ if $n_i = 0$ for all $i \notin (j_1, j_2)$. Clearly, for any continuous $p(x)$ such that $\int_\delta^{\delta^{-1}} x^2 p(x) dx \le 1$ we have

$$\#\Big\{\bar{n} \,\Big|\, \sum_{i=1}^\infty n_i i^2 \le \Lambda\Big\} \ge \prod_{k=0}^L \#\Big\{\bar{n} \,\Big|\, \mathrm{supp}(\bar{n}) \in \Lambda^{1/3} I_k, \\ \sum n_i i^2 \le \Lambda\varepsilon \min_{x \in I_k} x^2 p(x)\Big\}, \tag{3.2}$$

where the notation $I_k = (k\varepsilon + \delta, (k+1)\varepsilon + \delta)$ has been used and the symbol # stands for the number of elements in the set. Let us estimate each factor in the right hand side of (3.2)

$$\#\Big\{\bar{n} \,|\, \mathrm{supp}(\bar{n}) \in \Lambda^{1/3} I_k, \sum_i n_i i^2 \le \varepsilon\Lambda q_k\Big\} >$$

$$> \#\Big\{\bar{n} \,|\, \mathrm{supp}(\bar{n}) \in \Lambda^{1/3} I_k, \sum_i n_i \le \frac{\varepsilon\Lambda q_k}{\Lambda^{2/3}((k+1)\varepsilon + \delta)^2}\Big\} =$$

$$= \left(\Lambda^{1/3} \frac{\varepsilon q_k}{((k+1)\varepsilon + \delta)^2} + \Lambda^{1/3}\varepsilon\right)! \Big/ \left(\Lambda^{1/3} \frac{\varepsilon q_k}{((k+1)\varepsilon + \delta)^2}\right)! \left(\Lambda^{1/3}\varepsilon\right)!;$$

here we have also used the standard combinatorial formula. By the Stirling formula

$$\ln\left\{\left(\Lambda^{1/3}\Big(\frac{\varepsilon q_k}{((k+1)\varepsilon + \delta)^2} + \varepsilon\Big)\right)! \Big/ \left(\Lambda^{1/3}\Big(\frac{\varepsilon q_k}{((k+1)\varepsilon + \delta)^2}\Big)\right)! (\varepsilon\Lambda^{1/3})!\right\} =$$

$$= \ln\Big[\Big(\Lambda^{1/3}\big(\frac{\varepsilon q_k}{((k+1)\varepsilon+\delta)^2}+\varepsilon\big)\Big)^{\Lambda^{1/3}\left(\frac{\varepsilon q_k}{((k+1)\varepsilon+\delta)^2}+\varepsilon\right)}\times$$

$$\Big(\Lambda^{1/3}\big(\frac{\varepsilon q_k}{((k+1)\varepsilon+\delta)^2}\big)\Big)^{-\Lambda^{1/3}\left(\frac{\varepsilon q_k}{((k+1)\varepsilon+\delta)^2}\right)}(\varepsilon\Lambda^{1/3})^{-\varepsilon\Lambda^{1/3}}\Big](1+o(1)) =$$

$$= \Lambda^{1/3}\varepsilon\Big[\Big(\frac{q_k}{((k+1)\varepsilon+\delta)^2}+1\Big)\ln\Big(\frac{q_k}{((k+1)\varepsilon+\delta)^2}+1\Big)-$$

$$\frac{q_k}{((k+1)\varepsilon+\delta)^2}\ln\frac{q_k}{((k+1)\varepsilon+\delta)^2}\Big](1+o(1))$$

as $\Lambda\to\infty$. It remains to put $q_k=\min\limits_{I_k} x^2p(x)$ and substitute the last relation into (3.2):

$$\ln\#\Big\{\bar n\mid\sum_{i=1}^{\infty}n_ii^2\le\Lambda\Big\}\ge\varepsilon\Lambda^{1/3}\sum_{k=0}^{L}\Big[\Big(\frac{q_k}{((k+1)\varepsilon+\delta)^2}+1\Big)\times$$

$$\ln\Big(\frac{q_k}{((k+1)\varepsilon+\delta)^2}+1\Big)-\frac{q_k}{((k+1)\varepsilon+\delta)^2}\ln\frac{q_k}{((k+1)\varepsilon+\delta)^2}\Big]\ge$$

$$\ge\Lambda^{1/3}\Big(\int_{\delta}^{\delta^{-1}}(p(x)+1)\ln(p(x)+1)-p(x)\ln p(x))dx-\delta_1(\varepsilon)\Big)\ge$$

$$\ge\Lambda^{1/3}\Big(\int_0^{\infty}(p(x)+1)\ln(p(x)+1)-p(x)\ln p(x))dx-\delta_2(\varepsilon,\delta)\Big),$$

where $\delta_1\to0$ as $\varepsilon\to0$ and $\delta_2\to0$ as $\varepsilon$ and $\delta\to0$. Thus,

$$\liminf_{\Lambda\to\infty}\frac{\ln\#\Big\{\bar n\mid\sum\limits_{i=1}^{\infty}n_ii^2\le\Lambda\Big\}}{\Lambda^{1/3}}\ge c_0. \tag{3.3}$$

To establish the upper bound let us consider an arbitrary sequence $\{q_k\}$ such that $\sum\limits_{k=1}^{L}\varepsilon q_k\le1$, and estimate $\#\{\mathrm{supp}(\bar n)\in I_k\mid\sum\limits_i n_ii^2\le\varepsilon q_k\Lambda\}$. In the same way as above one can prove that

$$\ln\#\{\mathrm{supp}(\bar n)\in I_k\mid\sum_i n_ii^2\le\varepsilon q_k\Lambda\}\le$$

$$\le\varepsilon\Lambda^{1/3}\Big[\Big(\frac{q_k}{(k\varepsilon+\delta)^2}+1\Big)\ln\Big(\frac{q_k}{(k\varepsilon+\delta)^2}+1\Big)- \tag{3.4}$$

$$-\frac{q_k}{(k\varepsilon+\delta)^2}\ln\frac{q_k}{(k\varepsilon+\delta)^2}\Big](1+o(1))$$

as $\Lambda \to \infty$. Similarly,

$$\ln \#\{\operatorname{supp}(\bar{n}) \in (\frac{\delta\Lambda^{1/3}}{2^{j+1}}, \frac{\delta\Lambda^{1/3}}{2^j}) \mid \sum_i n_i i^2 \le \Lambda\} \le$$

$$\le \ln\Big[\Big(\frac{2^{2j+2}\Lambda^{1/3}}{\delta^2} + \frac{\delta\Lambda^{1/3}}{2^j}\Big)!\Big/\Big(\frac{2^{2j+2}\Lambda^{1/3}}{\delta^2}\Big)!\Big/\Big(\frac{\delta\Lambda^{1/3}}{2^j}\Big)!\Big] \le$$

$$\le \Lambda^{1/3}\Big(\frac{2^{2j+2}}{\delta^2}+\frac{\delta}{2^j}\Big)\ln\Big(\frac{2^{2j+2}}{\delta^2}+\frac{\delta}{2^j}\Big)-\Lambda^{1/3}\frac{2^{2j+2}}{\delta^2}\ln\frac{2^{2j+2}}{\delta^2}-\Lambda^{1/3}\frac{\delta}{2^j}\ln\frac{\delta}{2^j} =$$

$$= \Lambda^{1/3}\frac{2^{2j+2}}{\delta^2}\ln\Big(1+\frac{\delta^3}{2^{3j+2}}\Big)+\Lambda^{1/3}\frac{\delta}{2^j}\ln\Big(1+\frac{2^{3j+2}}{\delta^3}\Big) \le$$

$$\le \Lambda^{1/3}\frac{\delta}{2^{j-1}}((3j+2)\ln 2 - 3\ln\delta)$$

and summing the leftmost and the rightmost parts of the latter inequality over $j$, we find

$$\begin{gathered}\ln \#\{\operatorname{supp}(\bar{n}) \in (1, \delta\Lambda^{1/3}) \mid \textstyle\sum_i n_i i^2 \le \Lambda\} \le \\ \textstyle\sum_{j=1}^{\infty} \Lambda^{1/3}\frac{\delta}{2^{j-1}}((3j+2)\ln 2 - 3\ln\delta) \le \Lambda^{1/3}(4\delta\ln 2 - 3\delta\ln\delta)\end{gathered} \tag{3.5}$$

Similar estimates can be obtained for the interval $\operatorname{supp}(\bar{n}) \in (\Lambda^{1/3}/\delta, \Lambda^{1/2})$:

$$\begin{gathered}\ln \#\{\operatorname{supp}(\bar{n}) \in (\Lambda^{1/3}/\delta, \Lambda^{1/2}) \mid \sum_i n_i i^2 \le \Lambda\} \le \\ \sum_{j=1}^{\infty} \ln\Big[\Big(\frac{2^j}{\delta}\Lambda^{1/3}+\frac{\delta^2}{2^{2j-2}}\Lambda^{1/3}\Big)!\Big/\Big(\frac{2^j}{\delta}\Lambda^{1/3}\Big)!\Big/\Big(\frac{\delta^2}{2^{2j-2}}\Lambda^{1/3}\Big)!\Big] \le \\ \le \Lambda^{1/3}(8\delta^2\ln 2 - 12\delta^2\ln\delta)\end{gathered} \tag{3.6}$$

In view of (3.4)-(3.6) the number of solutions to the problem

$$\Big\{\bar{n} \mid \sum_{i=1}^{\infty} i^2 n_i \le \Lambda, \sum_{i=\Lambda^{1/3}(\delta+\varepsilon k)}^{\Lambda^{1/3}(\delta+\varepsilon(k+1))} i^2 n_i \le \varepsilon\Lambda q_k\Big\}$$

does not exceed

$$\exp\Big[\Lambda^{1/3}\Big\{(4\delta\ln 2 - 3\delta\ln\delta) + (8\delta^2\ln 2 - 12\delta^2\ln\delta) +$$

$$+\varepsilon\sum_{k=1}^{L}\Big(\frac{q_k}{(\delta+\varepsilon k)^2}+1\Big)\ln\Big(\frac{q_k}{(\delta+\varepsilon k)^2}+1\Big)$$
$$-\frac{q_k}{(\delta+\varepsilon k)^2}\ln\frac{q_k}{(\delta+\varepsilon k)^2}\}(1+o(1))\Big]$$

as $\Lambda\to\infty$. Denoting $\dfrac{q_k}{(\delta+k\varepsilon)^2}$ by $p_k$ and recalling the definition of $p(x)$ we derive

$$\ln\#\Big\{\bar{n}\mid\sum_{i=1}^{\infty}i^2n_i\leq\Lambda,\sum_{i=\Lambda^{1/3}(\delta+\varepsilon k)}^{\Lambda^{1/3}(\delta+\varepsilon(k+1))}i^2n_i\leq\varepsilon\Lambda q_k\Big\}\leq$$
$$\Lambda^{1/3}\Big(\int_0^{\infty}((p(x)+1)\ln(p(x)+1)-p(x)\ln p(x))dx\Big)(1+\delta_1(\delta,\varepsilon)) \tag{3.7}$$

where $\delta_1(\delta,\varepsilon)\to0$ as $\varepsilon$ and $\delta\to0$ and $\Lambda\to\infty$. It remains to estimate the number of integer vectors $\{\varepsilon\Lambda q_k\}$, $0\leq k\leq L$, such that $\sum\limits_{k=1}^{L}\varepsilon q_k\leq1$:

$$\#\Big\{\{\varepsilon\Lambda q_k\}\mid\sum_{k=1}^{L}\varepsilon q_k\leq1\Big\}\leq\big(\frac{\delta^{-1}}{\varepsilon}+\Lambda\big)!/\big(\frac{\delta^{-1}}{\varepsilon}\big)!/\Lambda!\leq$$

$$\leq\exp\Big\{\big(\frac{\delta^{-1}}{\varepsilon}+\Lambda\big)\ln\big(\frac{\delta^{-1}}{\varepsilon}+\Lambda\big)-\frac{\delta^{-1}}{\varepsilon}\ln\frac{\delta^{-1}}{\varepsilon}-\Lambda\ln\Lambda\Big\}\leq$$
$$\leq\exp\Big\{\frac{\ln\Lambda}{\delta\varepsilon}+\frac{1}{\Lambda(\delta\varepsilon)^2}+\frac{1}{\delta\varepsilon}\Big\}.$$

Finally, (3.7) and the last inequality together imply

$$\#\{\bar{n}\mid\sum_i n_ii^2\leq\Lambda\}\leq\sum_{\{q_k\}}\#\{\bar{n}\mid\sum_i n_ii^2\leq\Lambda,$$

$$\sum_{j=\Lambda^{1/3}(\delta+k\varepsilon)}^{\Lambda^{1/3}(\delta+(k+1)\varepsilon)}n_ii^2\leq\Lambda\varepsilon q_k\}\leq$$

$$\leq\exp\Big[\Lambda^{1/3}\Big(\int_0^{\infty}((p(x)+1)\ln(p(x)+1)-p(x)\ln p(x))dx+\delta_1(\delta,\varepsilon)+$$

$$+\frac{\ln\Lambda}{\Lambda^{1/3}\delta\varepsilon}+\frac{1}{\lambda^{4/3}(\delta\varepsilon)^2}\Big)\Big].$$

Passing to the limit in $\Lambda$ and choosing $\delta$ and $\varepsilon$ sufficiently small, we have

$$\limsup_{\Lambda\to\infty} \Lambda^{-1/3} \ln \#\{\bar{n} \mid \sum_{i=1}^{\infty} n_i i^2\} \leq c_0.$$

The theorem is proved.

## 4 Compactness results

In this section we prove that the operator $\mathcal{A}$ is a relatively compact perturbation of operator the $\mathcal{A}_0$. This allows us to study the spectral properties of $\mathcal{A}$.

We start with proving the following simple statement.

**Proposition 1** *There exists a constant $c > 0$ such that*

$$c(\mathcal{A}_0 F(\cdot), F(\cdot))_{\mu_0} \leq (\mathcal{A}F(\cdot), F(\cdot))_{\mu} \leq c^{-1}(\mathcal{A}_0 F(\cdot), F(\cdot))_{\mu_0}.$$

**Proof.**

This assertion is a direct consequence of the following evident relation

$$c \leq \frac{d\mu}{d\mu_0}(u) \leq c_{-1}.$$

By the Courant minimax principle, we derive from the last proposition that

$$N_{\mathcal{A}_0}(c\Lambda) \leq N_{\mathcal{A}}(\Lambda) \leq N_{\mathcal{A}_0}(c^{-1}\Lambda).$$

In particular, the resolvent of $\mathcal{A}$ is a compact operator. In fact, the operator $(\mathcal{A} - \mathcal{A}_0)$ is relatively compact with respect to $\mathcal{A}_0$.

**Lemma 4.1** *The operator $(\mathcal{A} - \mathcal{A}_0)(\mathcal{A}_0 + I)^{-1}$ is compact in $\mathcal{L}^2(L^2(0, \pi), \mu_0)$.*

**Proof.**

Let $\mathcal{H}$ be a dense subspace of $\mathcal{L}^2(L^2(0, \pi), \mu_0)$ endowed with the inner product

$$(F, G)_{\mathcal{H}} = (F, G)_{\mathcal{L}^2} + \int_{\mathcal{L}^2} (\nabla_u F, \nabla_u G)\mu_0(du) = ((\mathcal{A}_0 + I)F, G)_{\mathcal{L}^2}.$$

$\mathcal{H}$ is embedded compactly in $\mathcal{L}^2$. Denote by $\mathcal{H}^{-1}$ the adjoint space of $\mathcal{H}$. Then $(\mathcal{A}_0 + I)$ is a bounded operator from $\mathcal{H}$ to $\mathcal{H}^{-1}$. For any $F \in \mathcal{H}$ by the Schwartz inequality

$$\| \sin u \cdot \nabla_u F(u)\|^2_{\mathcal{L}^2} = \int_{\mathcal{L}^2} (\sin u \cdot \nabla_u F(u))^2 \mu_0(du) \leq$$

$$\leq \int_{\mathcal{L}^2} \| \sin u \|^2_{L^2(0,\pi)} |\nabla_u F(u)\|^2_{L^2(0,\pi)} \mu_0(du) \leq \|F\|^2_{\mathcal{H}}$$

Thus, $(\mathcal{A}-\mathcal{A}_0)$ is a bounded operator from $\mathcal{H}$ to $\mathcal{L}^2$. Clearly, $(\mathcal{A}_0+I)^{-1}$ is a compact operator from $\mathcal{L}^2$ to $\mathcal{H}$. Consequently, $(\mathcal{A}-\mathcal{A}_0)(\mathcal{A}_0+I)^{-1}$ is compact in $\mathcal{L}^2$. The lemma is proved.

Combining the last lemma with ([10], Theorem 8.2), we derive
**Corollary 4.1** *The following limiting relation takes place*

$$\lim_{\Lambda\to\infty} \frac{\ln N_{\mathcal{A}}(\Lambda)}{\Lambda^{1/3}} = c_0.$$

## 5 Central Limit Theorem

The main result of the section is the following.
**Theorem 5.1** *Let the initial distribution $u_0(\cdot)$ be absolutely continuous with respect to $\mu(du)$ (or $\mu_0(du)$) and let its density $U_0(u)$ belong to the space $\mathcal{L}^2(L^2(0,\pi), \mu(du))$. Then the family of normalized integrals $\frac{1}{\sqrt{t}} \int_0^t u(s,x)ds$ converges in distribution in the space $L^2(0,\pi)$ as $t \to \infty$ to a Gaussian measure with a zero mean value and the covariance operator*

$$(B\varphi,\varphi) = C \int_{L^2(0,\pi)} |\nabla_u n_\varphi(u)|^2 \mu(du) = C(\mathcal{A}n_\varphi, n_\varphi)_{\mathcal{L}^2(L^2,\mu)} =$$

$$= C\Big(\mathcal{A}^{-1}(u,\varphi)_{L^2(0,\pi)}, (u,\varphi)_{L^2(0,\pi)}\Big)_{\mathcal{L}^2(L^2,\mu)};$$

*here $n_\varphi$ is a solution of $\mathcal{A}n_\varphi = (u,\varphi)$ and $C = \int_{L^2(0,\pi)} U_0(u)\mu(du)$.*

**Proof.**

First of all let us note that $\int_{L^2(0,\pi)} (u,\varphi)_{L^2}\mu(du) = 0$ for any $\varphi \in L^2$. Hence, a solution $n_\varphi(u)$ indeed exists. We fix a solution by the following condition $\int_{L^2(0,\pi)} n_\varphi(u)\mu(du) = 0$. Using Ito's formula one can verify the identity

$$\int_0^t (u(s,\cdot),\varphi(\cdot))ds + n_\varphi(u(t,\cdot)) - n_\varphi(u_0(\cdot)) = \int_0^t \nabla_u n_\varphi(u(s,\cdot))dw_s$$

which implies, in particular, that the process in the left hand side is a martingale. Obviously, the limiting distribution of

$$\left(\frac{1}{\sqrt{t}}\left[\int_0^t (u(s,x),\varphi(x))ds + n_\varphi(u(t,\cdot)) - n_\varphi(u_0(\cdot))\right]\right)$$

in $V$ coincides with the limiting distribution of $\frac{1}{\sqrt{t}}\left[\int_0^t (u(s,x),\varphi(x))ds\right]$. Indeed, due to the properties of $\mathcal{A}$, the function $\int_{L^2(0,\pi)} n_\varphi^2(u(t,\cdot))\mu(du)$ decreases monotonically in $t$. The limiting behavior of

$$\frac{1}{\sqrt{t}}\int_0^t \nabla_u n_\varphi(u(s,\cdot))dw_s$$

can, in turn, be studied by the methods developed in [11], [12], and the theorem follows.

## References


1. S.M. Kozlov, Vibration d'une poutre soumise à des forces aléatoires, 1994, Preprint 94-19, LATP, Université de Provence, Marseille.
2. K. Ito, Foundations of Stochastic Differential Equations in Infinite Dimensional Spaces, CBMS Notes, V.47, SIAM Baton Ronge, 1984.
3. A.V. Fursikov and M.I. Vishik, Mathematical Problems of Stochastic Hydromechanics, Kluwer, 1988.
4. G. Da Prato and J. Zabczyk, Stochastic Equations in Infinite Dimensions, Cambridge Univ. Press, 1992.
5. G. Kallianpur, and J. Xiong, Stochastic Differential Equations in Infinite Dimensional Spaces, Lecture Notes – Monograph Series, V.26, 1995.
6. B. Rozovskii, Stochastic Evolution Equations, Linear Theory and Applications to Nonlinear Filtering, Kluwer, 1990.
7. S.M. Kozlov, Some questions on stochastic equations with partial derivatives, 1978, Moscow Univ. Publ. House, Transactions of Petrovskii seminar, N4, pp. 147–172.

8. S.M. Kozlov, Equivalence of measures in linear Ito partial differential equations, *Moscow Univ. Mathem. Bulletin*, 1977, Vol. 32, No. 4, pp. 36-40.
9. N. Dunford, and J.T. Schwartz, Linear Operators, Part I, 1963, John Wiley and Sons, New York.
10. A.S. Markus, Introduction to the spectral theory of polynimial operator pencils, AMS, Providence, 1988.
11. M. Viot, Solutions faibles d'equations aux derivees partielles stochastiques non lineaires, 1976, Thesis. Paris.
12. E. Pardoux, Equations aux derivees partielles stochastiques non lineaires monotone, 1975, Thesis. Paris.

# Series on Advances in Mathematics for Applied Sciences

## Aims and Scope

This Series reports on new developments in mathematical research relating to methods, qualitative and numerical analysis, mathematical modeling in the applied and the technological sciences. Contributions related to constitutive theories, fluid dynamics, kinetic and transport theories, solid mechanics, system theory and mathematical methods for the applications are welcomed.

This Series includes books, lecture notes, proceedings, collections of research papers. Monograph collections on specialized topics of current interest are particularly encouraged. Both the proceedings and monograph collections will generally be edited by a Guest editor.

High quality, novelty of the content and potential for the applications to modern problems in applied science will be the guidelines for the selection of the content of this series.

## Instructions for Authors

Submission of proposals should be addressed to the editors-in-charge or to any member of the editorial board. In the latter, the authors should also notify the proposal to one of the editors-in-charge. Acceptance of books and lecture notes will generally be based on the description of the general content and scope of the book or lecture notes as well as on sample of the parts judged to be more significantly by the authors.

Acceptance of proceedings will be based on relevance of the topics and of the lecturers contributing to the volume.

Acceptance of monograph collections will be based on relevance of the subject and of the authors contributing to the volume.

Authors are urged, in order to avoid re-typing, not to begin the final preparation of the text until they received the publisher's guidelines. They will receive from World Scientific the instructions for preparing camera-ready manuscript.

## SERIES ON ADVANCES IN MATHEMATICS FOR APPLIED SCIENCES

Vol. 17 The Fokker-Planck Equation for Stochastic Dynamical Systems and Its Explicit Steady State Solution
*by C. Soize*

Vol. 18 Calculus of Variation, Homogenization and Continuum Mechanics
*eds. G. Bouchitté et al.*

Vol. 19 A Concise Guide to Semigroups and Evolution Equations
*by A. Belleni-Morante*

Vol. 20 Global Controllability and Stabilization of Nonlinear Systems
*by S. Nikitin*

Vol. 21 High Accuracy Non-Centered Compact Difference Schemes for Fluid Dynamics Applications
*by A. I. Tolstykh*

Vol. 22 Advances in Kinetic Theory and Computing: Selected Papers
*ed. B. Perthame*

Vol. 23 Waves and Stability in Continuous Media
*eds. S. Rionero and T. Ruggeri*

Vol. 24 Impulsive Differential Equations with a Small Parameter
*by D. Bainov and V. Covachev*

Vol. 25 Mathematical Models and Methods of Localized Interaction Theory
*by A. I. Bunimovich and A. V. Dubinskii*

Vol. 26 Recent Advances in Elasticity, Viscoelasticity and Inelasticity
*ed. K. R. Rajagopal*

Vol. 27 Nonstandard Methods for Stochastic Fluid Mechanics
*by M. Capinski and N. J. Cutland*

Vol. 28 Impulsive Differential Equations: Asymptotic Properties of the Solutions
*by D. Bainov and P. Simeonov*

Vol. 29 The Method of Maximum Entropy
*by H. Gzyl*

Vol. 30 Lectures on Probability and Second Order Random Fields
*by D. B. Hernández*

Vol. 31 Parallel and Distributed Signal and Image Integration Problems
*eds. R. N. Madan et al.*

Vol. 32 On the Way to Understanding The Time Phenomenon: The Constructions of Time in Natural Science – Part 1. Interdisciplinary Time Studies
*ed. A. P. Levich*

## SERIES ON ADVANCES IN MATHEMATICS FOR APPLIED SCIENCES

Vol. 33 Lecture Notes on the Mathematical Theory of the Boltzmann Equation
*ed. N. Bellomo*

Vol. 34 Singularly Perturbed Evolution Equations with Applications to Kinetic Theory
*by J. R. Mika and J. Banasiak*

Vol. 35 Mechanics of Mixtures
*by K. R. Rajagopal and L. Tao*

Vol. 36 Dynamical Mechanical Systems Under Random Impulses
*by R. Iwankiewicz*

Vol. 37 Oscillations in Planar Dynamic Systems
*by R. E. Mickens*

Vol. 38 Mathematical Problems in Elasticity
*ed. R. Russo*

Vol. 39 On the Way to Understanding the Time Phenomenon: The Constructions of Time in Natural Science — Part 2. The "Active" Properties of Time According to N. A. Kozyrev
*ed. A. P. Levich*

Vol. 40 Advanced Mathematical Tools in Metrology II
*eds. P. Ciarlini et al.*

Vol. 41 Mathematical Methods in Electromagnetism
Linear Theory and Applications
*by M. Cessenat*

Vol. 42 Constrained Control Problems of Discrete Processes
*by V. N. Phat*

Vol. 43 Motor Vehicle Dynamics: Modeling and Simulation
*by G. Genta*

Vol. 44 Microscopic Theory of Condensation in Gases and Plasma
*by A. L. Itkin and E. G. Kolesnichenko*

Vol. 45 Advanced Mathematical Tools in Metrology III
*eds. P. Cianlini et al.*

Vol. 46 Mathematical Topics in Neutron Transport Theory — New Aspects
*by M. Mokhtar-Kharroubi*

Vol. 47 Theory of the Navier-Stokes Equations
*eds. J. G. Heywood et al.*

Vol. 48 Advances in Nonlinear Partial Differential Equations and Stochastics
*eds. S. Kawashima and T. Yanagisawa*

## SERIES ON ADVANCES IN MATHEMATICS FOR APPLIED SCIENCES

Vol. 49 Propagation and Reflection of Shock Waves
*by F. V. Shugaev and L. S. Shtemenko*

Vol. 50 Homogenization
*eds. V. Berdichevsky, V. Jikov and G. Papanicolaou*